Numerical Methods

Using MATLAB®

Numerical Methods
Using MATLAB®
Fourth Edition

George Lindfield
Aston University, School of Engineering and Applied Science,
Birmingham, England, United Kingdom

John Penny
Aston University, School of Engineering and Applied Science,
Birmingham, England, United Kingdom

Academic Press is an imprint of Elsevier
125 London Wall, London EC2Y 5AS, United Kingdom
525 B Street, Suite 1650, San Diego, CA 92101, United States
50 Hampshire Street, 5th Floor, Cambridge, MA 02139, United States
The Boulevard, Langford Lane, Kidlington, Oxford OX5 1GB, United Kingdom

Library of Congress Cataloging-in-Publication Data
A catalog record for this book is available from the Library of Congress

British Library Cataloguing-in-Publication Data
A catalogue record for this book is available from the British Library

ISBN: 978-0-12-812256-3

For information on all Academic Press publications
visit our website at https://www.elsevier.com/books-and-journals

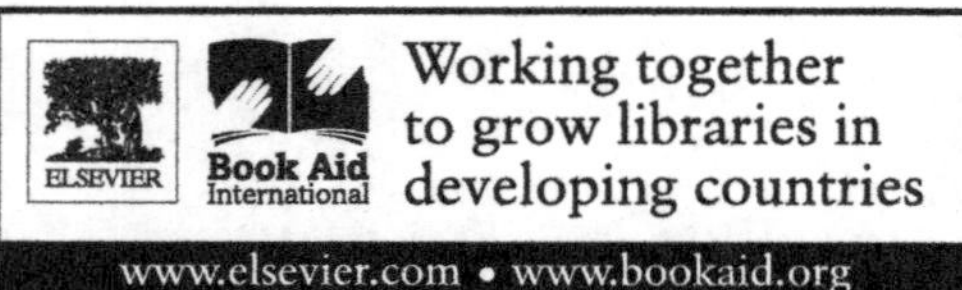

Publisher: Katey Birtcher
Acquisition Editor: Steve Merken
Editorial Project Manager: Peter Jardim / Michael Lutz
Production Project Manager: Sruthi Satheesh
Designer: Matthew Limbert

Typeset by VTeX

This book is for my wife Zena. With tolerance and patience, she has supported and encouraged me for many years and for our grown up children Katy and Helen.

George Lindfield

This book is for my wife Wendy, for her patience, support and encouragement, and for our grown up children, Debra, Mark and Joanne. Also to our cat Jeremy who provided me with company whilst I worked on this book.

John Penny

Contents

List of Figures

About the Authors

George Lindfield is a former lecturer in the Department of Computer Science at Aston University and is now retired. He taught courses in computer science and in optimization at bachelor- and master's-level. He has coauthored books on numerical methods and published many papers in various fields including optimization. He is a member of the Institute of Mathematics, a Chartered Mathematician, and a Fellow of the Royal Astronomical Society.

John Penny is an Emeritus Professor in the School of Engineering and Applied Science at Aston University, Birmingham. England. He is a former head of the Mechanical Engineering Department. He taught bachelor- and master's-level students in structural and rotor dynamics and related topics such as numerical analysis, instrumentation, and digital signal processing. His research interests were in topics in dynamics such as damage detection in static and rotating structures. He has published over 40 peer reviewed papers. He is a Fellow of the Institute of Mathematics and its Applications and is a coauthor of four text books.

Preface

Our primary aim in this text is unchanged from previous editions; it is to introduce the reader to a wide range of numerical algorithms, explain their fundamental principles and illustrate their application. The algorithms are implemented in the software package MATLAB which is constantly being enhanced and provides a powerful tool to help with these studies.

Many important theoretical results are discussed but it is not intended to provide a detailed and rigorous theoretical development in every area. Rather, we wish to show how numerical procedures can be applied to solve problems from many fields of application, and that the numerical procedures give the expected theoretical performance when used to solve specific problems.

When used with care MATLAB provides a natural and succinct way of describing numerical algorithms and a powerful means of experimenting with them. However, no tool, irrespective of its power, should be used carelessly or uncritically.

This text allows the reader to study numerical methods by encouraging systematic experimentation with some of the many fascinating problems of numerical analysis. Although MATLAB provides many useful functions this text also introduces the reader to numerous useful and important algorithms and develops MATLAB functions to implement them. The reader is encouraged to use these functions to produce results in numerical and graphical form. MATLAB provides powerful and varied graphics facilities to give a clearer understanding of the nature of the results produced by the numerical procedures.

Particular examples are given throughout the text to illustrate how numerical methods are used to study problems which include applications in the biosciences, chaos, neural networks, engineering, and science.

It should be noted that the introduction to MATLAB is relatively brief and is meant as an aid to the reader. It can in no way be expected to replace the standard MATLAB manual or text books devoted to MATLAB software. We provide a broad introduction to the topics, develop algorithms in the form of MATLAB functions and encourage the reader to experiment with these functions which have been kept as simple as possible for reasons of clarity. These functions can be improved and we urge readers to develop the ones that are of particular interest to them.

In addition to a general introduction to MATLAB, the text covers the solution of linear equations and eigenvalue problems; methods for solving non-linear equations; numerical integration and differentiation; the solution of initial value and boundary value problems; curve fitting including splines, least squares, and Fourier analysis, topics in optimization such as interior point methods, non-linear programming, and heuristic algorithms and, finally, we show how symbolic computing can be integrated with numeric algorithms. Specifically in this 4th edition, descriptions and examples of some functions recently added to MATLAB such as implicit functions and the Live Editor are given in Chapter 1. Chapter 4 now includes a section on adaptive integration. Chapter 5 now includes a brief introduction to Simulink; a toolbox which provides a visual interface to help the user simulating the process of solving differential equations. The old Chapter 7 has been split into two chapters and we have added the Kalman filter and principal component analysis, and the Hilbert, Walsh, and wavelet transforms. The old Chapter 8 has had the emphasis on the genetic algorithm reduced and replaced by the more modern and efficient differential evolution algorithm.

The text contains many worked examples, practice problems (some of which are new to this edition) and solutions and we hope we have provided an interesting range of problems.

For readers of this book, additional materials, including all .m file scripts and functions listed in the text, are available on the book's companion site: https://www.elsevier.com/books-and-journals/book-companion/9780128122563. For instructors using this book as a text for their courses, a solutions manual is available by registering at the textbook site: www.textbooks.elsevier.com.

The text is suitable for undergraduate and postgraduate students and for those working in industry and education. We hope readers will share our enthusiasm for this area of study. For those who do not currently have access to MATLAB, this text still provides a general introduction to a wide range of numerical algorithms and many useful and interesting examples and problems.

We would like to thank the many readers from all over the world for their helpful comments which have enhanced this edition and we would be pleased to hear from readers who note errors or have suggestions for improvements.

George Lindfield
John Penny
Aston University, Birmingham
March 2018

Acknowledgment

We thank Peter Jardim for his encouragement and support, Joe Hayton, the Publishing Director and the production team members.

CHAPTER

AN INTRODUCTION TO MATLAB® 1

Abstract

MATLAB® is a software package produced by The MathWorks, Inc. (http://www.mathworks.com) and is available on systems ranging from personal computers to super-computers and including parallel computing. In this chapter we aim to provide a useful introduction to MATLAB, giving sufficient background for the numerical methods we consider. The reader is referred to the MATLAB manual for a full description of the package.

1.1 THE SOFTWARE PACKAGE MATLAB

MATLAB is probably the world's most successful commercial numerical analysis software and the name MATLAB is derived from the phrase "matrix laboratory". It has evolved from some software written by Cleve Moler in the late 1970s to allow his students to access matrix routines in the LINPACK and EISPACK packages without the need to write FORTRAN programs. This first version of MATLAB had only 80 functions, primitive graphics and "matrix" was the only data type. Its use spread to other universities and, after it was reprogrammed in C, MATLAB was launched as a commercial product in 1984. MATLAB provides an interactive development tool for scientific and engineering problems and more generally for those areas where significant numeric computations have to be performed. The package can be used to evaluate single statements directly or a list of statements called a script can be prepared. Once named and saved, a script can be executed as an entity. The package was originally based on software produced by the LINPACK and EISPACK projects but in 2000, MATLAB was rewritten to use the newer BLAS and LAPACK libraries for fast matrix operations and linear algebra, respectively. MATLAB provides the user with:

1. Easy manipulation of matrix structures.
2. A vast number of powerful built-in routines which are constantly growing and developing.
3. Powerful two- and three-dimensional graphing facilities.
4. A scripting system which allows users to develop and modify the software for their own needs.
5. Collections of functions, called toolboxes, which may be added to the facilities of the core MATLAB. These are designed for specific applications: for example neural networks, optimization, digital signal processing, and higher-order spectral analysis.

It is not difficult to use MATLAB, although to use it with maximum efficiency for complex tasks requires experience. Generally MATLAB works with rectangular or square arrays of data, the elements of which may be real or complex. A scalar quantity is thus an array containing a single element. This is an elegant and powerful notion but it can present the user with an initial conceptual difficulty. A user schooled in many traditional computer languages is familiar with a pseudo-statement of the form A = B + C and can immediately interpret it as an instruction that A is assigned the sum of the

Numerical Methods. https://doi.org/10.1016/B978-0-12-812256-3.00010-5

values stored in B and C. In MATLAB the variables B and C may represent arrays so that *each element* of the array A will become the sum of the values of corresponding elements of B and C; that is the addition will follow the laws of matrix algebra.

There are several languages or software packages that have some similarities to MATLAB. These packages include:

Mathematica and Maple. These packages are known for their ability to carry out complicated symbolic mathematical manipulation but they are also able to undertake high precision numerical computation. In contrast MATLAB is known for its powerful numerical computational and matrix manipulation facilities. However, MATLAB also provides an optional symbolic toolbox. This is discussed in Chapter 10.

Other Matlab-style languages. Languages such as Scilab,[1] Octave,[2] and Freemat[3] are somewhat similar to MATLAB in that they implement a wide range of numerical methods, and, in some cases, use similar syntax to MATLAB.
It should noted that the languages do not necessarily have a range of toolboxes like MATLAB.

Julia. Julia[4] is a new high-level, high-performance dynamic programming language. The developers of Julia wanted, amongst other attributes, the speed of C, the general programming easy of Python, and the powerful linear algebra functions and familiar mathematical notation of MATLAB.

General purpose languages. General purpose languages such as Python and C. These languages don't have any significant numerical analysis capability in themselves but can load libraries of routines. For example Python+Numpy, Python+Scipy, C+GSL.

The current MATLAB release, version 9.4 (R2018a), is available on a wide variety of platforms. Generally MathWorks releases an upgraded version of MATLAB every six months.

When MATLAB is invoked it opens a command window. Graphics, editing, and help windows may also be opened if required. Users can design their MATLAB working environment as they see fit. MATLAB scripts and function are generally platform independent and they can be readily ported from one system to another. To install and start MATLAB, readers should consult the manual appropriate to their particular working environment.

The scripts and functions given in this book have been tested under MATLAB release, version 9.3.0.713579 (R2017b). However, most of them will work directly using earlier versions of MATLAB but some may require modification.

The remainder of this chapter is devoted to introducing some of the statements and syntax of MATLAB. The intention is to give the reader a sound but brief introduction to the power of MATLAB. Some details of structure and syntax are omitted and must be obtained from the MATLAB manual. A detailed description of MATLAB is given by Higham and Higham (2017). Other sources of information are the MathWorks website and Wikipedia. Wikipedia should be used with some care.

Before we begin a detailed discussion of the features of MATLAB, the meaning some terminology needs clarification. Consider the terms MATLAB statements, commands, functions, and keywords. If

---

[1] http://www.scilab.org.
[2] http://www.gnu.org/software/octave.
[3] http://freemat.sourceforge.net.
[4] https://julialang.org/.

we take a very simple MATLAB expression, like `y = sqrt(x)` then, if this is used in the command window for immediate execution, it is a command for MATLAB to determine the square root of the variable `x` and assign it to `y`. If it is used in a script, and is not for immediate execution, then it is usually called a statement. The expression `sqrt` is a MATLAB function, but it can also be called a keyword. The vast majority of MATLAB keywords *are* functions, but a few are not: for example `all`, `long`, and `pi`. The last of these is a reserved keyword to denote the mathematical constant π. Thus, the use of the four word discussed are often interchangeable.

1.2 MATRICES IN MATLAB

A two-dimensional array is effectively a table of data, not restricted to numeric data. If arrays are stacked in the third dimension, then they are three-dimensional arrays. Matrices are two-dimensional arrays that contain only numeric data or mathematical expressions where the variables of the expression have already been assigned numeric values. Thus, 23.2 and x^2 are allowed, *peter* is allowed if it is a numeric constant but not if it is a person's name. Thus a two dimension array of numeric data can legitimately be called an array or a matrix. Matrices can be operated on, using the laws of matrix algebra. Thus if **A** is a matrix, then 3**A** and $\mathbf{A}^{-1}$ have a meaning, whereas, if **A** is an alpha-numeric array these statements have no meaning. MATLAB supports matrix algebra, but also allows array operations. For example, an array of data might be a financial statement, and therefore, it might be necessary to sum the 3rd through 5th rows and place the result in the 6th row. This is a legitimate array operation that MATLAB supports.

The matrix is fundamental to MATLAB and we have provided a broad and simple introduction to matrices in Appendix A. In MATLAB the names used for matrices must start with a letter and may be followed by any combination of letters or digits. The letters may be upper or lower case. Note that throughout this text a distinctive font is used to denote MATLAB statements and output, for example `disp`.

In MATLAB the arithmetic operations of addition, subtraction, multiplication, and division can be performed in the usual way on scalar quantities, but they can also be used directly with matrices or arrays of data. To use these arithmetic operators on matrices, the matrices must first be created. There are several ways of doing this in MATLAB and the simplest method, which is suitable for small matrices, is as follows. We assign an array of values to `A` by opening the **command** window and then typing

```
>> A = [1 3 5;1 0 1;5 0 9]
```

after the prompt >>. Notice that the elements of the matrix are placed in square brackets, each row element separated by at least one space or comma. A semicolon (;) indicates the end of a row and the beginning of another. When the return key is pressed the matrix will be displayed thus:

```
A =

     1     3     5
     1     0     1
     5     0     9
```

All statements are executed by pressing the return or enter key. Thus, for example, by typing `B = [1 3 51;2 6 12;10 7 28]` after the >> prompt, and pressing the return key, we assign values to `B`. To add the matrices in the **command** window and assign the result to `C` we type `C = A+B` and similarly if we type `C = A-B` the matrices are subtracted. In both cases the results are displayed row by row in the **command** window. Note that terminating a MATLAB statement with a semicolon suppresses any output.

For simple problems we can use the **command** window. By simple we mean MATLAB statements of limited complexity – even MATLAB statements of limited complexity can provide some powerful numerical computation. However, if we require the execution of an ordered sequence of MATLAB statements (commands) then it is sensible for these statements to be typed in the MATLAB **editor** window to create a script which must be saved under a suitable name for future use as required. There will be no execution or output until the name of this script is typed into the **command** window and the script executed by pressing return.

A matrix which has only one row or column is called a vector. A row vector consists of one row of elements and a column vector consists of one column of elements. Conventionally in mathematics, engineering, and science an emboldened upper case letter is usually used to represent a matrix, for example **A**. An emboldened lower case letter usually represents a *column* vector, that is **x**. The transpose operator converts a row to a column and vice versa so that we can represent a row vector as a column vector transposed. Using the superscript T in mathematics to indicate a transpose, we can write a row vector as $\mathbf{x}^\mathsf{T}$. In MATLAB it is often convenient to ignore the convention that the initial form of a vector is a column; the user can define the initial form of a vector as a row or a column.

The implementation of vector and matrix multiplication in MATLAB is straightforward. Beginning with vector multiplication, we assume that row vectors having the same number of elements have been assigned to `d` and `p`. To multiply them together we write `x = d*p'`. Note that the symbol ' transposes the row `p` into a column so that the multiplication is valid. The result, `x`, is a scalar. Many practitioners use .' to indicate a transpose. The reason for this is discussed in Section 1.4.

Assuming the two matrices `A` and `B` have been assigned, for matrix multiplication the user simply types `C = A*B`. This computes `A` post-multiplied by `B`, assigns the result to `C` and displays it, providing the multiplication is valid. Otherwise MATLAB gives an appropriate error indication. The conditions for matrix multiplication to be valid are given in Appendix A. Notice that the symbol * must be used for multiplication because in MATLAB multiplication is not implied.

A very useful MATLAB function is `whos` (and the similar function, `who`).

These functions tell us the current content of the work space. For example, provided `A`, `B`, and `C` described above have not been cleared from the memory, then

```
>> whos
  Name      Size                     Bytes  Class
  A         3x3                         72  double array
  B         3x3                         72  double array
  C         3x3                         72  double array
```

This tells us that `A`, `B`, and `C` are all 3×3 matrices. They are stored as double precision arrays. A double precision number requires 8 bytes to store it, so each array of 9 elements requires 72 bytes; a grand total is 27 elements using 216 bytes. Consider now the following operations:

```
>> clear A
>> B = [ ];
>> C = zeros(4,4);
>> whos
  Name      Size                Bytes  Class
  B         0x0                     0  double array
  C         4x4                   128  double array
```

Here we see that we have cleared (i.e., deleted) `A` from the memory, assigned an empty matrix to `B` and a 4 × 4 array of zeros to `C`.

Note that the size of matrices can also be determined using the `size` and `length` functions thus:

```
>> A = zeros(4,8);
>> B = ones(7,3);
>> [p q] = size(A)

p =
    4


q =
    8

>> length(A)

ans =
    8

>> L = length(B)

L =
    7
```

It can be seen that `size` gives the size of the matrix whereas `length` gives the number of elements in the largest dimension.

1.3 MANIPULATING THE ELEMENTS OF A MATRIX

In MATLAB, matrix elements can be manipulated individually or in blocks. For example,

```
>> X(1,3) = C(4,5)+V(9,1)
>> A(1) = B(1)+D(1)
>> C(i,j+1) = D(i,j+1)+E(i,j)
```

are valid statements relating elements of matrices. Rows and columns can be manipulated as complete entities. Thus `A(:,3)`, `B(5,:)` refer respectively to the third column of `A` and fifth row of `B`. If `B` has 10 rows and 10 columns, i.e. it is a 10×10 matrix, then `B(:,4:9)` refers to columns 4 through 9 of the matrix. The : by itself indicates all the rows, and hence all elements of columns 4 through 9. Note that in MATLAB, by default, the *lowest matrix index starts at* 1. This can be a source of difficulty when implementing some algorithms.

The following examples illustrate some of the ways subscripts can be used in MATLAB. First we assign a matrix

```
>> A = [2 3 4 5 6;-4 -5 -6 -7 -8;3 5 7 9 1; ...
          4 6 8 10 12;-2 -3 -4 -5 -6]

A =
     2     3     4     5     6
    -4    -5    -6    -7    -8
     3     5     7     9     1
     4     6     8    10    12
    -2    -3    -4    -5    -6
```

Note the use of ... (an ellipsis) to indicate that the MATLAB statement continues on the next line. Executing the following statements

```
>> v = [1 3 5];
>> b = A(v,2)
```

gives

```
b =
     3
     5
    -3
```

Thus `b` is composed of the elements of the first, third, and fifth rows in the second column of `A`. Executing

```
>> C = A(v,:)
```

gives

```
C =
     2     3     4     5     6
     3     5     7     9     1
    -2    -3    -4    -5    -6
```

Thus `C` is composed of the first, third, and fifth rows of `A`. Executing

```
>> D = zeros(3);
>> D(:,1) = A(v,2)
```

gives

```
D =
     3     0     0
     5     0     0
    -3     0     0
```

Here `D` is a 3×3 matrix of zeros with column 1 replaced by the first, third, and fifth elements of column 2 of `A`.

Executing

```
>> E = A(1:2,4:5)
```

gives

```
E =
     5     6
    -7    -8
```

Note that if we index an existing square or rectangular array with a single index, then the elements of the array are identified as follows. Index 1 gives the top left element of the array, and the index is incremented down the columns in sequence, from left to right. For example, with reference to the preceding array `C`

```
C1 = C;
C1(1:4:15) = 10

C1 =
    10     3     4     5    10
     3    10     7     9     1
    -2    -3    10    -5    -6
```

Note that in this example the index is incremented by 4.

When manipulating very large matrices it is easy to become unsure of the size of the matrix. Thus, if we want to find the value of the element in the penultimate row and last column of `A` defined previously we could write

```
>> size(A)

ans =
     5     5

>> A(4,5)

ans =
    12
```

but it is easier to use end thus:

```
>> A(end-1,end)

ans =
    12
```

The reshape function may be used to manipulate a matrix. As the name implies, the function reshapes a given matrix into a new matrix of any specified size provided it has an identical number of elements. For example a 3 × 4 matrix can be reshaped into a 6 × 2 matrix but a 3 × 3 matrix cannot be reshaped into a 5 × 2 matrix. It is important to note that this function takes each column of the original matrix in turn until the new required column size is achieved and then repeats the process for the next column. For example, consider the matrix P.

```
>> P = C(:,1:4)

P =
     2     3     4     5
     3     5     7     9
    -2    -3    -4    -5

>> reshape(P,6,2)

ans =
     2     4
     3     7
    -2    -4
     3     5
     5     9
    -3    -5

>> s = reshape(P,1,12);
>> s(1:10)

ans =
     2     3    -2     3     5    -3     4     7    -4     5
```

1.4 TRANSPOSING MATRICES

A simple operation that may be performed on a matrix is transposition which interchanges rows and columns. Transposition of a vector is briefly discussed in Section 1.2. In MATLAB transposition is denoted by the symbol '. For example, consider the matrix A, where

```
>> A = [1 2 3;4 5 6;7 8 9]
```

```
A =

     1     2     3
     4     5     6
     7     8     9
```

To assign the transpose of A to B we write

```
>> B = A'

B =

     1     4     7
     2     5     8
     3     6     9
```

Had we used .' to obtain the transpose we would have obtained the same result. However, if A is complex then the MATLAB operator ' gives the complex conjugate transpose. For example

```
>> A = [1+2i 3+5i;4+2i 3+4i]

A =

   1.0000 + 2.0000i   3.0000 + 5.0000i
   4.0000 + 2.0000i   3.0000 + 4.0000i

>> B = A'

B =

   1.0000 - 2.0000i   4.0000 - 2.0000i
   3.0000 - 5.0000i   3.0000 - 4.0000i
```

To provide the transpose without conjugation we execute

```
>> C = A.'

C =

   1.0000 + 2.0000i   4.0000 + 2.0000i
   3.0000 + 5.0000i   3.0000 + 4.0000i
```

1.5 SPECIAL MATRICES

Certain matrices occur frequently in matrix manipulations and MATLAB ensures that these are generated easily. Some of the most common are ones(m,n), zeros(m,n), rand(m,n), randn(m,n), and randi(p,m,n). These MATLAB functions generate $m \times n$ matrices composed of ones, zeros, uniformly distributed random numbers, normally distributed random numbers and uniformly distributed random integers, respectively. In the case of randi(p,m,n), p is the maximum integer. If only a single scalar parameter is given, then these statements generate a square matrix of the size given by the parameter. The MATLAB function `eye(n)` generates the $n \times n$ unit matrix. The function `eye(m,n)` generates a matrix of m rows and n columns with a diagonal of ones; thus:

```
>> A = eye(3,4), B = eye(4,3)

A =
     1     0     0     0
     0     1     0     0
     0     0     1     0

B =
     1     0     0
     0     1     0
     0     0     1
     0     0     0
```

If we wish to generate a random matrix `C` of the same size as an already existing matrix `A`, then the statement `C = rand(size(A))` can be used. Similarly `D = zeros(size(A))` and `E = ones(size(A))` generates a matrix `D` of zeros and a matrix `E` of ones, both of which are the same size as matrix `A`.

Some special matrices with more complex features are introduced in Chapter 2.

1.6 GENERATING MATRICES AND VECTORS WITH SPECIFIED ELEMENT VALUES

Here we confine ourselves to some relatively simple examples thus:

`x = -8:1:8` (or `x = -8:8`) sets `x` to a vector having elements $-8, -7, \ldots, 7, 8$.
`y = -2:.2:2` sets `y` to a vector having elements $-2, -1.8, -1.6, \ldots, 1.8, 2$.
`z = [1:3 4:2:8 10:0.5:11]` sets `z` to a vector having the elements

[1 2 3 4 6 8 10 10.5 11]

The MATLAB function `linspace` also generates a vector. However, in this function the user defines the beginning and end values of the vector and the number of elements in the vector. For example

```
>> w = linspace(-2,2,5)

w =
    -2    -1     0     1     2
```

This is simple and could just as well have been created by `w = -2:1:2` or even `w = -2:2`. However, consider

```
>> w = linspace(0.2598,0.3024,5)

w =
    0.2598    0.2704    0.2811    0.2918    0.3024
```

Generating this sequence of values by other means would be more difficult. If we require logarithmic spacing then we can use

```
>> w = logspace(1,2,5)

w =
   10.0000   17.7828   31.6228   56.2341  100.0000
```

Note that the values produced are between 10^1 and 10^2, not 1 and 2. Again, generating these values by any other means would require some thought! The user of `logspace` should be warned that if the second parameter is `pi` the values run to π, not 10^{π}. Consider the following

```
>> w = logspace(1,pi,5)

w =
   10.0000    7.4866    5.6050    4.1963    3.1416
```

More complicated matrices can be generated by combining other matrices. For example, consider the two statements

```
>> C = [2.3 4.9; 0.9 3.1];
>> D = [C ones(size(C)); eye(size(C)) zeros(size(C))]
```

These two statements generate a new matrix `D` the size of which is double the row and column size of the original `C`; thus

```
D =
    2.3000    4.9000    1.0000    1.0000
    0.9000    3.1000    1.0000    1.0000
    1.0000         0         0         0
         0    1.0000         0         0
```

The MATLAB function `repmat` replicates a given matrix a required number of times. For example, assuming the matrix `C` is defined in the preceding statement, then

```
>> E = repmat(C,2,3)
```

replicates `C` as a block to give a matrix with twice as many rows and three times as many columns. Thus we have a matrix `E` of 4 rows and 6 columns:

```
E =
    2.3000    4.9000    2.3000    4.9000    2.3000    4.9000
    0.9000    3.1000    0.9000    3.1000    0.9000    3.1000
    2.3000    4.9000    2.3000    4.9000    2.3000    4.9000
    0.9000    3.1000    0.9000    3.1000    0.9000    3.1000
```

The MATLAB function `diag` allows us to generate a diagonal matrix from a specified vector of diagonal elements. Thus

```
>> H = diag([2 3 4])
```

generates

```
H =
     2     0     0
     0     3     0
     0     0     4
```

There is a second used of the function `diag` which is to obtain the elements on the leading diagonal of a given matrix. Consider

```
>> P = rand(3,4)

P =
    0.3825    0.9379    0.2935    0.8548
    0.4658    0.8146    0.2502    0.3160
    0.1030    0.0296    0.5830    0.6325
```

then

```
>> diag(P)

ans =
    0.3825
    0.8146
    0.5830
```

A more complicated form of diagonal matrix is the block diagonal matrix. This type of matrix can be generated using the MATLAB function `blkdiag`. We set matrices `A1` and `A2` as follows:

```
>> A1 = [1 2 5;3 4 6;3 4 5];
>> A2 = [1.2 3.5,8;0.6 0.9,56];
```

Then,

```
>> blkdiag(A1,A2,78)

ans =
  1.0000    2.0000    5.0000         0         0         0         0
  3.0000    4.0000    6.0000         0         0         0         0
  3.0000    4.0000    5.0000         0         0         0         0
       0         0         0    1.2000    3.5000    8.0000         0
       0         0         0    0.6000    0.9000   56.0000         0
       0         0         0         0         0         0   78.0000
```

The preceding functions can be very useful in allowing the user to create matrices with complicated structures, without detailed programming.

1.7 MATRIX ALGEBRA IN MATLAB

The matrix is fundamental to MATLAB and we have provided a broad and simple introduction to matrices in Appendix A.

MATLAB allows matrix equations to be simply expressed and evaluated. For example, to illustrate matrix addition, subtraction, multiplication, and scalar multiplication, consider the evaluation of the matrix equation

$$\mathbf{Z} = \mathbf{A}\mathbf{A}^{\mathsf{T}} + s\mathbf{P} - \mathbf{Q}$$

where $s = 0.5$ and

$$\mathbf{A} = \begin{bmatrix} 2 & 3 & 4 & 5 \\ 2 & 4 & 6 & 8 \end{bmatrix} \quad \mathbf{P} = \begin{bmatrix} 1 & 3 \\ 2 & -9 \end{bmatrix} \quad \mathbf{Q} = \begin{bmatrix} -7 & 3 \\ 5 & 1 \end{bmatrix}$$

Assigning `A`, `P`, `Q`, and `s`, and evaluating this equation in MATLAB we have

```
>> A = [2 3 4 5;2 4 6 8];
>> P = [1 3;2 -9];
>> Q = [-7 3;5 1];
>> s = 0.5;
>> Z = A*A'+s*P-Q

Z =
   61.5000   78.5000
   76.0000  114.5000
```

This result can be readily checked by hand!

MATLAB allows a single scalar value to be added to or subtracted from every element of a matrix. This is called explicit expansion. To illustrate this we first generate a fourth-order Riemann matrix using the MATLAB function `gallery`. This function gives the user access to a range of special matrices with useful properties. See Chapter 2 for further discussion. In the following piece of MATLAB code we use it to generate a 4 × 4 Riemann matrix, and then subtract 0.5 from every element of the matrix.

```
>> R = gallery('riemann',4)

R =

     1    -1     1    -1
    -1     2    -1    -1
    -1    -1     3    -1
    -1    -1    -1     4

>> A = R-0.5

A =

    0.5000   -1.5000    0.5000   -1.5000
   -1.5000    1.5000   -1.5000   -1.5000
   -1.5000   -1.5000    2.5000   -1.5000
   -1.5000   -1.5000   -1.5000    3.5000
```

In the 2016b release of MATLAB this feature has been extended to allow a row or column vector to be added to a matrix. For example

```
>> B = 2*ones(4)-[1 2 3 4]

B =
     1     0    -1    -2
     1     0    -1    -2
     1     0    -1    -2
     1     0    -1    -2

>> C = 2*ones(4)+[2 4 6 8]'

C =
     4     4     4     4
     6     6     6     6
     8     8     8     8
    10    10    10    10
```

Note that `2*ones(4)` produces a 4×4 matrix where each element is 2. In computing `B` the results show that 1 is subtracted from each element of the first column, the 2 from each element in the second column, the 3 from each element of the third column and so on. Similarly. in computing `C` the results show that 2 is added to each element of the first row, the 4 to each element in the second row, the 6 to each element of the third row and so on.

1.8 MATRIX FUNCTIONS

Some arithmetic operations are simple to evaluate for single scalar values but involve a great deal of computation for matrices. For large matrices such operations may take a significant amount of time. An example of this is where a matrix is raised to a power. We can write this in MATLAB as `A^p` where `p` is a scalar value and `A` is a square matrix. This produces the power of the matrix for any value of `p`. For the case where the power equals 0.5 it is better to use `sqrtm(A)` which gives the principal square root of the matrix `A`, (see Appendix A, Section A.13). Similarly, for the case where the power equals -1 it is better to use `inv(A)`. Another special operation directly available in MATLAB is `expm(A)` which gives the exponential of the matrix `A`. The MATLAB function `logm(A)` provides the principal logarithm to the base e of `A`. If `B=logm(A)` then the principal logarithm `B` is the unique logarithm for which every eigenvalue has an imaginary part lying strictly between $-\pi$ and π.

For example

```
>> A = [61 45;60 76]

A =
    61    45
    60    76
```

```
>> B = sqrtm(A)

B =
     7.0000    3.0000
     4.0000    8.0000

>> B^2

ans =
    61.0000   45.0000
    60.0000   76.0000
```

1.9 USING THE MATLAB OPERATOR FOR MATRIX DIVISION

As an example of the power of MATLAB we consider the solution of a system of linear equations. It is easy to solve the problem $ax = b$ where a and b are simple scalar constants and x is the unknown. Given a and b then $x = b/a$. However, consider the corresponding matrix equation

$$\mathbf{Ax} = \mathbf{b} \tag{1.1}$$

where **A** is a square matrix and **x** and **b** are column vectors. We wish to find **x**. Computationally this is a much more difficult problem and in MATLAB it is solved by executing the statement

```
x = A\b
```

This statement uses the important MATLAB division operator `\` and solves the linear equation system (1.1).

Solving linear equation systems is an important problem and the computational efficiency and other aspects of this type of problem are discussed in considerable detail in Chapter 2.

1.10 ELEMENT-BY-ELEMENT OPERATIONS

Element-by-element operations differ from the standard matrix operations but they can be very useful. They are achieved by using a period or dot (`.`) to precede the operator. If `X` and `Y` are matrices (or vectors), then `X.^ Y` raises each *element* of `X` to the power of the corresponding element of `Y`. Similarly `X.*Y` and `Y.\X` multiply or divide each element of `X` by the corresponding element in `Y` respectively. The form `X./Y` gives the same result as `Y.\X`. For these operations to be executed the matrices and vectors used must be the same size. Note that a period is *not* used in the operations `+` and `-` because ordinary matrix addition and subtraction *are* element-by-element operations. Examples of element-by-element operations are given as follows:

```
>> A = [1 2;3 4]

A =
     1     2
     3     4

>> B = [5 6;7 8]

B =
     5     6
     7     8
```

First we use normal matrix multiplication thus:

```
>> A*B

ans =
    19    22
    43    50
```

However, using the dot operator (.) we have

```
>> A.*B

ans =
     5    12
    21    32
```

which is element-by-element multiplication. Now consider the statement

```
>> A.^B

ans =
           1          64
        2187       65536
```

In the above, each element of A is raised to the corresponding power in B.

Element-by-element operations have many applications. An important use is in plotting graphs (see Section 1.14). For example

```
>> x = -1:0.1:1;
>> y = x.*cos(x);
>> y1 = x.^3.*(x.^2+3*x+sin(x));
```

Notice here that using the vector x of many values, allows a vector of corresponding values for y and y1 to be computed simultaneously from single statements. Element-by-element operations are in effect operations on scalar quantities performed simultaneously.

1.11 SCALAR OPERATIONS AND FUNCTIONS

In MATLAB we can define and manipulate scalar quantities, as in most other computer languages, but no distinction is made in the naming of matrices and scalars. Thus A could represent a scalar or matrix quantity. The process of assignment makes the distinction. For example

```
>> x = 2;
>> y = x^2+3*x-7

y =

     3

>> x = [1 2;3 4]

x =

     1     2
     3     4

>> y = x.^2+3*x-7

y =

    -3     3
    11    21
```

Note that in the preceding examples, when vectors are used the dot must be placed before the operator. This is not required for scalar operations, but does not cause errors if used.

In the case where we multiply a square matrix by itself, for example, in the form x^2 we get the full matrix multiplication as shown below, rather than element-by-element multiplication as given by x.^2.

```
>> y = x^2+3*x-7

y =

     3     9
    17    27
```

A very large number of mathematical functions are directly built into MATLAB. They act on scalar quantities, arrays or vectors on an element-by-element basis. They may be called by using the function name together with the parameters that define the function. These functions may return one or more values. A small selection of MATLAB functions is given in the following table which lists the function name, the function use and an example function call. Note that all function names must be in lower case letters.

All MATLAB functions are not listed in Table 1.1, but MATLAB provides a complete range of trigonometric and inverse trigonometric functions, hyperbolic and inverse hyperbolic functions and

Table 1.1 Selected MATLAB mathematical functions

Function	Function gives	Example
`sqrt(x)`	square root of x	`y = sqrt(x+2.5);`
`abs(x)`	if x is real, is the positive value of x	
	if x is complex, is the scalar measure of x	`d = abs(x)*y;`
`real(x)`	real part of x when x is complex	`d = real(x)*y;`
`imag(x)`	imaginary part of x when x is complex	`d = imag(x)*y;`
`conj(x)`	the complex conjugate of x	`x = conj(y);`
`sin(x)`	sine of x in radians	`t = x+sin(x);`
`asin(x)`	inverse sine of x returned in radians	`t = x+sin(x);`
`sind(x)`	sine of x in degrees	`t = x+sind(x);`
`log(x)`	log to base e of x	`z = log(1+x);`
`log10(x)`	log to base 10 of x	`z = log10(1-2*x);`
`cosh(x)`	hyperbolic cosine of x	`u = cosh(pi*x);`
`exp(x)`	exponential of x, i.e., e^x	`p = .7*exp(x);`
`gamma(x)`	gamma function of x	`f = gamma(y);`
`bessel(n,x)`	nth-order Bessel function of x	`f = bessel(2,y);`

logarithmic functions. The following examples illustrate the use of some of the functions listed before:

```
>> x = [-4 3];
>> abs(x)

ans =
     4     3

>> x = 3+4i;
>> abs(x)

ans =
     5

>> imag(x)

ans =
     4

>> y = sin(pi/4)

y =
    0.7071
```

and

```
>> x = linspace(0,pi,5)

x =
         0    0.7854    1.5708    2.3562    3.1416

>> sin(x)

ans =
         0    0.7071    1.0000    0.7071    0.0000
```

and

```
>> x = [0 pi/2;pi 3*pi/2]

x =
         0    1.5708
    3.1416    4.7124

>> y = sin(x)

y =
         0    1.0000
    0.0000   -1.0000
```

Some functions perform special calculations for important and general mathematical processes. These functions often require more than one input parameter and may provide several outputs. For example, `bessel(n,x)` gives the nth-order Bessel function of x. The statement `y = fzero('fun',x0)` determines the root of the function `fun` near to `x0` where `fun` is a function defined by the user that provides the equation for which we are finding the root. For examples of the use of `fzero`, see Section 3.1. The statement `[Y,I] = sort(X)` is an example of a function that can return two output values. `Y` is the sorted matrix and `I` is a matrix containing the indices of the sort.

In addition to a large number of mathematical functions, MATLAB provides several utility functions that may be used for examining the operation of scripts. These are:

`pause` causes the execution of the script to pause until the user presses a key. Note that the cursor is turned into the symbol P, warning the script is in pause mode. This is often used when the script is operating with `echo on`.

`echo on` displays each line of script in the **command** window before execution. This is useful for demonstrations. To turn it off, use the statement `echo off`.

`who` lists the variables in the current workspace.

`whos` lists all the variables in the current workspace, together with information about their size and class, and so on.

MATLAB also provides functions related to time:

`clock` returns the current date and time in the form: <year month day hour min sec>.
`etime(t2,t1)` calculates elapsed time between `t1` and `t2`. Note that `t1` and `t2` are output from the clock function. When timing the duration of an event `tic ... toc` should be used.
`tic ... toc` times an event. For example, finding the time taken to execute a segment of script. The statement `tic` starts the timing and `toc` gives the elapsed time since the last `tic`.
`cputime` returns the total time in seconds since MATLAB was launched.
`timeit` times the operation of a function. Suppose we carry out a 8192 point Fourier transform using the MATLAB function `fft` (described in Chapter 8) then we run `fft_time = timeit(@()fft(8192))`.

The script `e4s101.m` uses the timing functions described previously to estimate the time taken to solve a 1000 × 1000 system of linear equations:

```
% e4s101.m  Solves a 5000 x 5000 linear equation system
A = rand(5000); b = rand(5000,1);
T_before = clock;
tic
t0 = cputime;
y = A\b;
tend = toc;
t1 = cputime-t0;
t2 = timeit(@() A\b);
disp('     tic-toc     cputime       timeit')
fprintf('%10.2f %10.2f %10.2f \n\n', tend,t1,t2)
```

Running script `e4s101.m` on a particular computer gave the following results:

```
    tic-toc     cputime       timeit
     2.52         5.09        2.60
```

The output shows that the three alternative methods of timing give essentially the same value. When measuring computing times the displayed times vary from run to run and the shorter the run time, the greater the percentage variation.

1.12 STRING VARIABLES

We have found that MATLAB makes no distinction in naming matrices and scalar quantities. This is also true of string variables or strings. For example, `A = [1 2; 3 4]`, `A = 17.23`, or `A = 'help'` are each valid statements and assign an array, a scalar or a text string respectively to `A`.

Characters and strings of characters can be assigned to variables directly in MATLAB by placing the string in quotes and then assigning it to a variable name. Strings can then be manipulated by specific MATLAB string functions which we list in this section. Some examples showing the manipulation of strings using standard MATLAB assignment are given below.

```
>> s1 = 'Matlab ', s2 = 'is ', s3 = 'useful'

s1 =
Matlab

s2 =
is

s3 =
useful
```

Strings in MATLAB are represented as vectors of the equivalent ASCII code numbers; it is only the way that we assign and access them that makes them strings. For example, the string `'is '` is actually saved as the vector [105 115 32]. Hence, we can see that the ASCII codes for the letters `i`, `s`, and a space are 105, 115, and 32 respectively. This vector structure has important implications when we manipulate strings. For example, we can concatenate strings, because of their vector nature, by using the square brackets as follows

```
>> sc = [s1 s2 s3]

sc =
Matlab is useful
```

Note the spaces are recognized. To identify any item in the string array we can write:

```
>> sc(2)

ans =
a
```

To identify a subset of the elements of this string we can write:

```
>> sc(3:10)

ans =
tlab is
```

we can display a string vertically, by transposing the string vector thus:

```
>> sc(1:3)'

ans =
M
a
t
```

We can also reverse the order of a substring and assign it to another string as follows:

```
>>a = sc(6:-1:1)

a =
baltaM
```

We can define string arrays as well. For example, using the string `sc` as defined previously:

```
>> sd = 'Numerical method'
>> s = [sc; sd]
Matlab is useful
Numerical method
```

To obtain the 12th column of this string we use

```
>> s(:,12)

ans =
s
e
```

Note that the string lengths must be the same in order to form a rectangular array of ASCII code numbers. In this case the array is 2×16. We now show how MATLAB string functions can be used to manipulate strings. To replace one string by another we use `strrep` as follows:

```
>> strrep(sc,'useful','super')

ans =
Matlab is super
```

Notice that this statement causes `useful` in `sc` to be replaced by `super`.

We can determine if a particular character or string is present in another string by using `findstr`. For example

```
>> findstr(sd,'e')

ans =
     4    12
```

This tells us that the 4th and 12th characters in the string are 'e'. We can also use this function to find the location of a substring of this string as follows

```
>> findstr(sd, 'meth')

ans =
    11
```

The string `'meth'` begins at the 11th character in the string. If the substring or character is not in the original string, we have the result illustrated by the example below:

```
>> findstr(sd,'E')

ans =
     [ ]
```

We can convert a string to its ASCII code equivalent by either using the function `double` or by invoking any arithmetic operation. Thus, operating on the existing string `sd` we have

```
>> p = double(sd(1:9))

p =
    78   117   109   101   114   105    99    97   108

>> q = 1*sd(1:9)

q =
    78   117   109   101   114   105    99    97   108
```

Note that in the case where we are multiplying the string by 1, MATLAB treats the string as a vector of ASCII equivalent numbers and multiplies it by 1. Recalling that `sd(1:9) = 'Numerical '` we can deduce that the ASCII code for N is 78 and for u it is 117, etc.

We convert a vector of ASCII code to a string using the MATLAB `char` function. For example

```
>> char(q)

ans =
Numerical
```

To increase each ASCII code number by 3, and then to convert to the character equivalent we have

```
>> char(q+3)

ans =
Qxphulfdo

>> char((q+3)/2)

ans =
(<84:6327

>> double(ans)

ans =
    40    60    56    52    58    54    51    50    55
```

`char(q)` has converted the ASCII string back to characters. Here we have shown that it is possible to do arithmetic on the ASCII code numbers and, if we wish, convert back to characters. If after manipulation the ASCII code values are non-integer, they are rounded down.

It is important to appreciate that the string '123' and the number 123 are not the same. Thus

```
>> a = 123

a =
   123

>> s1 = '123'

s1 =
123
```

Using `whos` shows the class of the variables `a` and `s1` as follows:

```
>> whos
  Name        Size                         Bytes  Class
  a           1x1                              8  double array
  s1          1x3                              6  char array
```

A total of 4 elements using 14 bytes. Thus, a character requires 2 bytes, a double precision number requires 8 bytes. We can convert strings to their numeric equivalent using the functions `str2num`, `str2double` as follows:

```
>> x=str2num('123.56')

x =
  123.5600
```

Appropriate strings can be converted to complex numbers but the user should take care, as we illustrate below:

```
>> x = str2num('1+2j')

x =
1.0 + 2.0000i
```

but

```
>> x = str2num('1+2 j')

x =
   3.0000                  0 + 1.0000i
```

Note that `str2double` can be used to convert to complex numbers and is more tolerant of spaces.

```
>> x = str2double('1+2 j')

x =
1.0 + 2.0000i
```

There are many MATLAB functions which are available to manipulate strings; see the appropriate MATLAB manual. Here we illustrate the use of some functions.

`bin2dec('111001')` or `bin2dec('111 001')` returns 57.

`dec2bin(57)` returns the string '111001'.

`int2str([3.9 6.2])` returns the string '4 6'.

`num2str([3.9 6.2])` returns the string '3.9 6.2'.

`str2num('3.9 6.2')` returns 3.9000 6.2000.

`strcat('how ','why ','when')` returns the string 'howwhywhen'.

`strcmp('whitehouse','whitepaint')` returns 0 because strings are not identical.

`strncmp('whitehouse','whitepaint',5)` returns 1 because first the 5 characters of strings are identical.

`date` returns the current date, in the form `24-Aug-2011`.

A useful and common application of the function `num2str` is in the `disp` and `title` functions see Sections 1.13 and 1.14 respectively.

1.13 INPUT AND OUTPUT IN MATLAB

To output the names and values of variables, the semicolon can be omitted from assignment statements. However, this does not produce clear scripts or well-organized and tidy output. It is often better practice to use the function `disp` since this leads to clearer scripts. The `disp` function allows the display of text and values on the screen. To output the contents of the matrix `A` on the screen we write `disp(A)`. Text output must be placed in single quotes, for example,

```
>> disp('This will display this test')
This will display this test
```

Combinations of strings can be printed using square brackets `[ ]`, and numerical values can be placed in text strings if they are converted to strings using the `num2str` function. For example,

```
>> x = 2.678;
>> disp(['Value of iterate is ', num2str(x), ' at this stage'])
```

will place on the screen

```
Value of iterate is 2.678 at this stage
```

The more flexible `fprintf` function allows formatted output to the screen or to a file. It takes the form

```
fprintf('filename','format_string',list);
```

Here `list` is a list of variable names separated by commas. The filename parameter is optional; if not present, output is to the screen rather than to the filename. The format string formats the output. The basic elements that may be used in the format string are

`%P.Qe` for exponential notation
`%P.Qf` fixed point
`%P.Qg` becomes `%P.Qe` or `%P.Qf` whichever is shorter
`\n` gives a new line

Note that `P` and `Q` in the preceding are integers. The integer string characters, including a period (`.`), must follow the `%` symbol and precede the letter `e`, `f`, or `g`. The integer before the period (`P`) sets the field width; the integer after the period (`Q`) sets the number of decimal places after the decimal point. For example, `%8.4f` and `%10.3f` give field width 8 with four decimal places and 10 with three decimal places, respectively. Note that one space is allocated to the decimal point. For example,

```
>> x = 1007.461; y = 2.1278; k = 17;
>> fprintf('\n x = %8.2f y = %8.6f k = %2.0f \n',x,y,k)
```

outputs

```
 x =  1007.46 y = 2.127800 k = 17
```

whereas

```
>> p = sprintf('\n x = %8.2f y = %8.6f k = %2.0f \n',x,y,k)
```

gives

```
p =

 x =  1007.46 y = 2.127800 k = 17
```

Note that `p` is a string vector, and can be manipulated if required.

The degree to which the MATLAB user will want to improve the style of MATLAB output will depend on circumstances. Is the output generated for other persons to read, perhaps requiring a clearly structured output, or is it just for the user alone and therefore requiring only a simple output? Will the output be filed away for future use, or is it a quick result that is rapidly discarded? In this text we have given examples of very simple output and sometimes quite elaborate output.

We now consider the input of text and data via the keyboard. An interactive way of obtaining input is to use the function `input`. One form of this function is

```
>> variable = input('Enter data: ');
Enter data: 67.3
```

The `input` function displays the text as a prompt and then waits for a numeric entry from the keyboard, `67.3` in this example. This is assigned to `variable` when the return key is pressed. Scalar values or arrays can be entered in this way. The alternative form of the `input` function allows string input thus:

```
>> variable = input('Enter text: ','s');
Enter text: Male
```

This assigns the string `Male` to `variable`.

For large amounts of data, perhaps saved in a previous MATLAB session, the function `load` allows the loading of files from disk using

```
load filename
```

The filename normally ends in `.mat` or `.dat`. A file of sunspot data already exists in the MATLAB package and can be loaded into memory using the command

```
>> load sunspot.dat
```

In the following example, we save the values of `x`, `y`, and `z` in file `test001`, clear the workspace and the reload `x`, `y`, and `z` into the workspace, thus

```
>> x = 1:5; y = sin(x); z = cos(x);
>> whos
  Name      Size        Bytes  Class
  x         1x5            40  double array
  y         1x5            40  double array
  z         1x5            40  double array
>> save test001
>> clear all, whos      Nothing listed
>> load test001
>> whos
  Name      Size        Bytes  Class
  x         1x5            40  double array
  y         1x5            40  double array
  z         1x5            40  double array
>> x = 1:5; y = sin(x); z = cos(x);
```

Here we only save `x`, `y` in file `test002` and then we clear the workspace and reload `x`, `y` thus:

```
>> save test002 x y
>> clear all,  whos  Nothing listed
>> load test002 x y, whos
  Name      Size            Bytes  Class
  x         1x5                40  double array
  y         1x5                40  double array
```

Note that the statement `load test002` has the same effect as `load test002 x y`. Finally we clear the workspace and reload `x` into the workspace thus:

```
>> clear all, whos   Nothing listed
>> load test002 x, whos
  Name      Size            Bytes  Class
  x         1x5                40  double array
```

Files composed of Comma Separated Values (CSV) are commonly used to exchange large amounts of tabular data between software applications. The data is stored in plain text and the fields are separated by commas. The files are easily editable using common spread sheet applications (e.g. MS Excel). If data has been generated elsewhere and saved as a CSV file it can be imported into MATLAB using csvread. We use csvwrite to generate a CSV file from MATLAB. In the following MATLAB statements we save the vector p, we clear the workspace and then reload p, but now call it the vector g:

```
>> p = 1:6;
>> whos
  Name       Size           Bytes  Class
  p          1x6               48  double array

>> csvwrite('test003',p)
>> clear
>> g = csvread('test003')
g =
     1     2     3     4     5
```

As a further example of the load and save commands, consider the following example:

```
% e4s110
x = [1 0 2 3 1]; y = [-1 0 4 1 1];
A = [x;x.^2; x.^3;x.^4;x.^5];
B = [x;2*x; 3*x; 4*x;5*x];
save test004 A B
clear all
whos
disp('nothing listed')
load test004
C = A*B;
disp('Results of matrix multiplication')
disp(C)
```

Running script e4s110 gives

```
>> e4s110
nothing listed
Results of matrix multiplication
          24           0          48          72          24
          54           0         108         162          54
         138           0         276         414         138
         378           0         756        1134         378
        1074           0        2148        3222        1074
```

1.14 MATLAB GRAPHICS

MATLAB provides a wide range of graphics facilities which may be called from within a script or used simply in command mode for direct execution. We begin by considering the plot function. This function takes several forms. For example,

Table 1.2 Symbols and characters used in plotting

Line	Symbol	Point	Symbol	Color	Character
Solid	-	point	.	yellow	y
Dashed	- -	plus	+	red	r
Dotted	:	star	*	green	g
Dashdot	-.	circle	o	blue	b
		x mark	×	black	k

`plot(x,y)` plots the vector `x` against `y`. If `x` and `y` are matrices the first column of `x` is plotted against the first column of `y`. This is then repeated for each pair of columns of `x` and `y`.
`plot(x1,y1,'type1',x2,y2,'type2')` plots the vector `x1` against `y1` using the line or point type given by `type1`, and the vector `x2` against `y2` using the line or point type given by `type2`.

The `type` is selected by using the required symbol from Table 1.2. This symbol may be preceded by a character indicating a color.

Semilog and log–log graphs can be obtained by replacing `plot` by `semilogx`, `semilogy`, or `loglog` functions and various other replacements for `plot` are available to give special plots. Titles, axis labels and other features can be added to a given graph using the functions `xlabel`, `ylabel`, `title`, `grid`, and `text`. These functions have the following forms:

`title('title')` displays the title which is enclosed between quotes, at the top of the graph.
`xlabel('x_axis_name')` displays the name which is enclosed between quotes for the x-axis.
`ylabel('y_axis_name')` displays the name which is enclosed between quotes for the y-axis.
`grid` superimposes a grid on the graph.
`text(x,y,'text-at-x,y')` displays text at position (x, y) in the graphics window where x and y are measured in the units of the current plotting axes. There may be one point or many at which text is placed depending on whether or not `x` and `y` are vectors.
`gtext('text')` allows the placement of text using the mouse by positioning it where the text is required and then pressing the mouse button.
`ginput` allows information to be taken from a graphics window. The function takes two main forms. The simplest is

```
[x,y] = ginput
```

This inputs an unlimited number of points into the vectors `x` and `y` by positioning the mouse cross-hairs at the points required and then pressing the mouse button. To exit ginput the return key must be pressed. If a specific number of points `n` are required, then we write

```
[x,y] = ginput(n)
```

In addition, the function axis allows the user to set the limits of the axes for a particular plot. This takes the form `axis(p)` where `p` is a four-element row vector specifying the lower and upper limits of the axes in the x and y directions. The axis statement must be placed *after* the plot statement to which it refers. Similarly the functions `xlabel`, `ylabel`, `title`, `grid`, `text`, `gtext`, and `axis` must *follow* the plot to which they refer.

Script `e4s102.m` gives the plot which is output as Fig. 1.1. The function `hold` is used to ensure that the two graphs are superimposed.

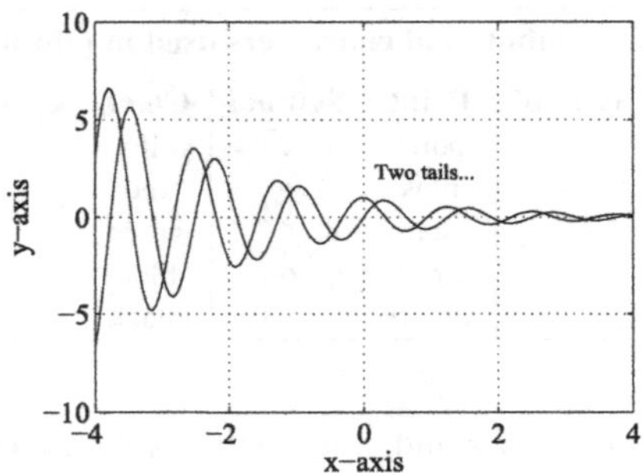

FIGURE 1.1

Superimposed graphs obtained using `plot(x,y)` and `hold` statements.

```
% e4s102.m
x = -4:0.05:4;
y = exp(-0.5*x).*sin(5*x);
figure(1), plot(x,y)
xlabel('x-axis'), ylabel('y-axis')
hold on
y = exp(-0.5*x).*cos(5*x);
plot(x,y), grid
gtext('Two tails...')
hold off
```

Script `e4s102.m` illustrates how few MATLAB statements are required to generate a graph.

The function `fplot` allows the user to plot a previously defined function between given limits. The important difference between `fplot` and `plot` is that `fplot` chooses the plotting points in the given range adaptively depending on the rate of change of the function at that point. Thus, more points are chosen when the function is changing more rapidly. This is illustrated by executing the MATLAB script `e4s103.m`:

```
% e4s103.m
y = @(x) sin(x.^3);
x = 2:.04:4;
figure(1)
plot(x,y(x),'o-')
xlabel('x'), ylabel('y')
figure(2)
fplot(y,[2 4])
xlabel('x'), ylabel('y')
```

Note `figure(1)` and `figure(2)` direct the graphic output to separate windows. The interpretation of the anonymous function `@(x) sin(x.^3)` is explained in Section 1.19.

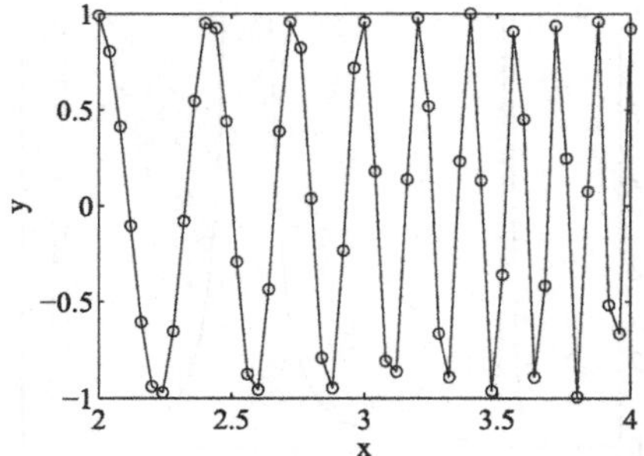

FIGURE 1.2

Plot of $y = \sin(x^3)$ using 51 equispaced plotting points.

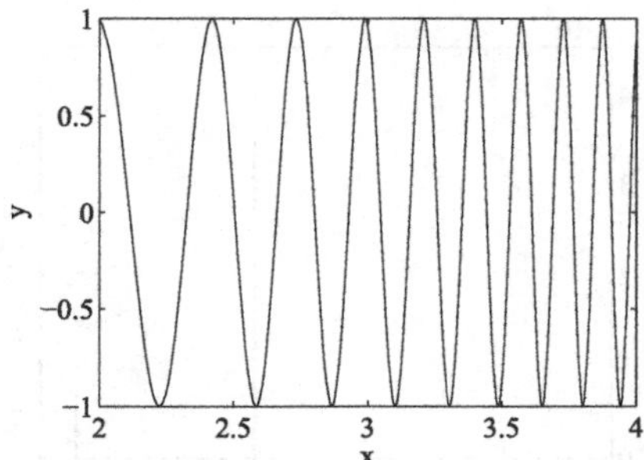

FIGURE 1.3

Plot of $y = \sin(x^3)$ using the function `fplot` to choose plotting points adaptively.

Running script `e4s103.m` produces Fig. 1.2 and Fig. 1.3. In the plot example, we have deliberately chosen an inadequate number of plotting points and this is reflected in the quality of Fig. 1.2. The function `fplot` produces a smoother and more accurate curve. Note that `fplot` only allows a function or functions to be plotted against an independent variable. Parametric plots cannot be created by `fplot`.

The MATLAB function `ezplot` is similar to `fplot` in the sense that we only have to specify the function, but has the disadvantage that the step size is fixed. However, `ezplot` does allow parametric plots and three-dimensional plots. For example

```
>> ezplot(@(t) (cos(3*t)), @(t) (sin(1.6*t)), [0 50])
```

is a parametric plot but the plot is rather coarse.

We have seen how `fplot` helps in plotting difficult functions. Other functions which help to clarify when the plot of a function is unclear or unpredictable are `ylim` and `xlim`. The function `ylim` allows the user to easily limit the range of the y-axis in the plot and `xlim` does the same for the x-axis. Their use is illustrated by the following example. Fig. 1.4 (without the uses of `xlim` and `ylim`) is unsatisfactory since it gives little understanding about how the function behaves except at the specific points $x = -2.5$, $x = 1$, and $x = 3.5$.

```
>> x = -4:0.0011:4;
>> y =1./(((x+2.5).^2).*((x-3.5).^2))+1./((x-1).^2);
>> plot(x,y)
>> ylim([0,10])
```

Fig. 1.5 shows how the MATLAB statement `ylim([0,10])` restricts the y-axis to maximum value of 10. This gives a clear picture of the behavior of the graph. There are a number of special features available in MATLAB for the presentation and manipulation of graphs and some of these are now be discussed. The `subplot` function takes the form `subplot(p,q,r)` where `p` and `q` split the figure window into a $p \times q$ grid of cells and places the plot in the rth cell of the grid, numbered consecutively along

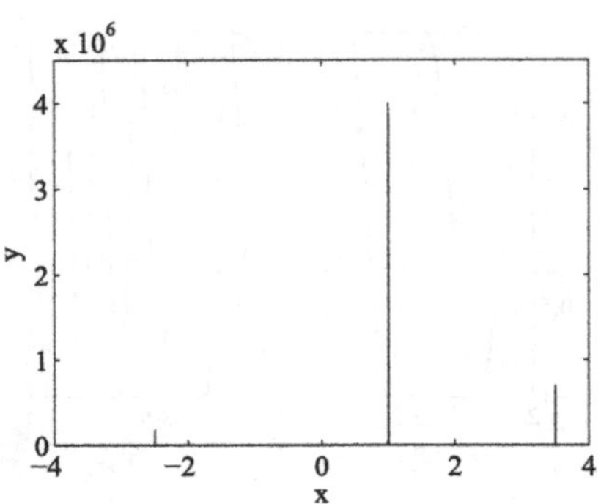

FIGURE 1.4

Function plotted over the range from −4 to 4. It has a maximum value of 4×10^6.

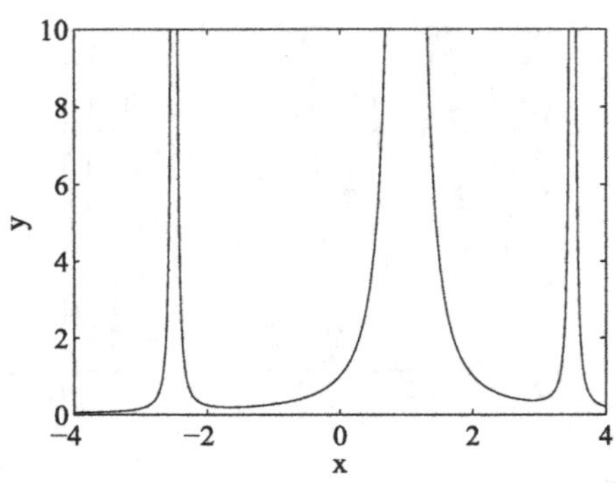

FIGURE 1.5

The same function as plotted in Fig. 1.4 but with a limit on the range of the y-axis.

the rows. This is illustrated by running script `e4s104.m` which generates six different plots, one in each of the six cells. These plots are given in Fig. 1.6.

```
% e4s104.m
x = 0.1:.1:5;
subplot(2,3,1), plot(x,x)
title('plot of x'), xlabel('x'), ylabel('y')
subplot(2,3,2), plot(x,x.^2)
title('plot of x^2'), xlabel('x'), ylabel('y')
subplot(2,3,3), plot(x,x.^3)
title('plot of x^3'), xlabel('x'), ylabel('y')
subplot(2,3,4), plot(x,cos(x))
title('plot of cos(x)'), xlabel('x'), ylabel('y')
subplot(2,3,5), plot(x,cos(2*x))
title('plot of cos(2x)'), xlabel('x'), ylabel('y')
subplot(2,3,6), plot(x,cos(3*x))
title('plot of cos(3x)'), xlabel('x'), ylabel('y')
```

The current plot can be held on screen by using the function `hold` and subsequent plots are drawn over it. The function `hold on` switches the hold facility on while `hold off` switches it off. The figure window can be cleared using the function `clf`.

MATLAB provides many other plot functions and styles. To illustrate two of these, the `polar` and `compass` plots, we display the roots of $x^5 - 1 = 0$ which have been determined using the MATLAB function `roots`. This function is described in detail in Section 3.11. Having determined the five roots of this equation we plot them using both `polar` and `compass`. The function `polar` requires the absolute values and phase angles of the roots, whereas as the function `compass` plots the real parts of the roots against their imaginary parts.

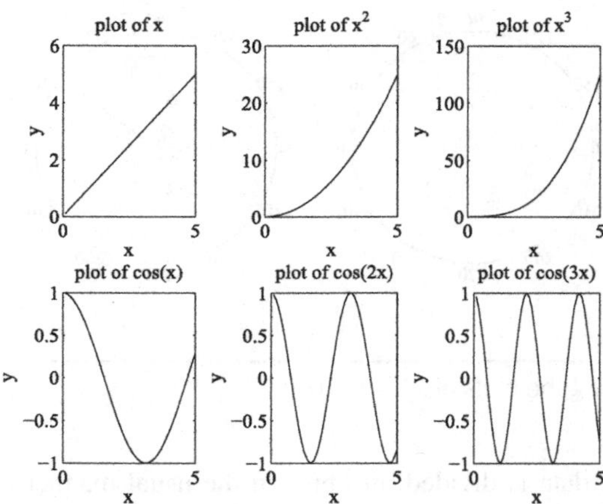

FIGURE 1.6

An example of the use of the `subplot` function.

```
>> p=roots([1 0 0 0 0 1])

p =
  -1.0000
  -0.3090 + 0.9511i
  -0.3090 - 0.9511i
   0.8090 + 0.5878i
   0.8090 - 0.5878i

>> pm = abs(p.')

pm =
    1.0000    1.0000    1.0000    1.0000    1.0000

>> pa = angle(p.')

pa =
    3.1416    1.8850   -1.8850    0.6283   -0.6283

>> subplot(1,2,1), polar(pa,pm,'ok')
>> subplot(1,2,2), compass(real(p),imag(p),'k')
```

Fig. 1.7 shows these subplots.

An interesting development introduced in the 2016b release of MATLAB are polar scatter and polar histogram plots that allow data to be plotted in polar coordinates. For the polar scatter function the points are plotted on a circle, each point being denoted by a circle, or some other selected symbol.

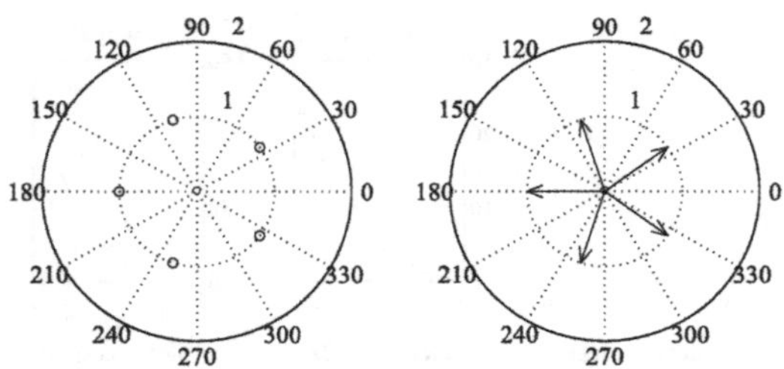

FIGURE 1.7

`polar` and `compass` plots showing the roots of $x^5 - 1 = 0$.

For the polar histogram the data is divided into bins in the usual manner of histogram plots but the bins are plotted as appropriate size segments of a circle. These processes are implemented using the MATLAB functions `polarscatter` and `polarhistogram`. The simplest form of `polarscatter` is given by:

```
polarscatter(th,r)
```

Here the `th` and `r` parameters are vectors of the same size providing the polar coordinates, so a call would take the form:

```
>> th = -2*pi:0.1:2*pi; r= 2*cos(th.*th);
>> polarscatter(th,r,'black')
```

This produces the diagram shown on the left side of Fig. 1.8. Additional parameters may be included in the function parameter list to increase marker size, provide color and fill the markers, and change marker symbol. Remarkably, different marker sizes can be set by using a vector for the marker sizes.

The following commands will produce the same figure but with much larger black, filled circles.

```
>> th2 = -2*pi:pi/4:2*pi; r2= 2*th2;
>> polarscatter(th2,r2,150,'black','filled')
```

Note the extra parameter values used are, `150,'black','filled'`. These provide the marker size, 150, the color, black, and require that the plotting symbol is filled. These commands produce the diagram on the right side of Fig. 1.8.

We now consider the polar histogram plot this divides a data set into a specific number of groups called bins. The data values of equal value are placed in the separate bins so in general the bins will contain a different numbers of values according to the frequency with which the data values occur. The polar histogram function plots a histogram of the data in polar coordinates. As an example we give the following commands which are executed in the command window:

```
>> x = 2*pi*rand(1,100);
>> polarhistogram(x,20)
```

which produces the polar histogram shown on the left side of Fig. 1.9.

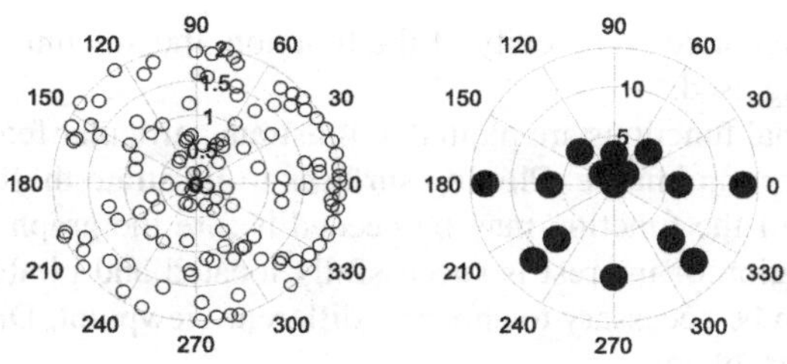

FIGURE 1.8

Polar scatter plots. Left diagram with default size circle markers. Right diagram with larger filled black circles.

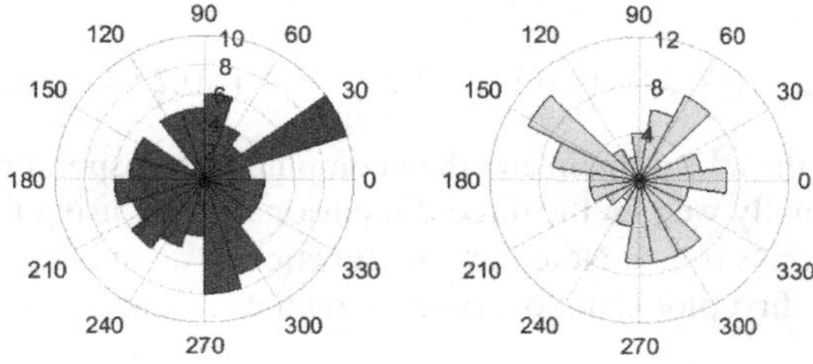

FIGURE 1.9

Polar scatter histogram.

The polar histogram function can include additional parameters. A new color can be set using `FaceColor` together with an indicator of color, for example 'red' or 'black', which plots the histogram blocks in red or black; transparency can be altered using `FaceAlpha` and a numerical value between 0 and 1. The higher the numerical value associated with 'FaceAlpha' the less transparent the image. As an example we give the commands:

```
>> x = 2*pi*randn(1,100);
>> polarhistogram(x,20,'Facecolor', 'black', 'FaceAlpha', 0.1)
```

These commands produce the output shown on the right side of Fig. 1.9. In this plot the bin segments are clearly transparent because of the low value, 0.1 used for the parameter `FaceAlpha` in the command.

1.15 THREE-DIMENSIONAL GRAPHICS

It is often convenient to draw a three-dimensional graph of a function or set of data to gain a deeper insight into the nature of the function or data. MATLAB provides powerful and extensive facilities to allow the user to draw a wide range of three-dimensional graphs. Here we only briefly introduce a small selection of these functions. These are the functions `meshgrid`, `mesh`, `surfl`, `contour`, and `contour3`. It should be noted that the more complex graphs of this type may take a significant time to draw on the screen, depending on the algebraic complexity of the function, the amount of detail required, and the power of the computer being used.

Usually three-dimensional functions are plotted to illustrate particular features of the function such as regions where maxima or minima lie. Plotting surfaces to illustrate these features can be difficult and some careful analysis of the function may be needed before the graph is drawn successfully. In addition, even when the region of interest is successfully located and plotted, the feature of interest may be hidden and it is then be necessary to choose a different viewpoint. Discontinuities may also be present and cause plotting problems.

For the function $z = f(x, y)$ the MATLAB function `meshgrid` is used to *generate a complete set of points in the x–y plane for the three-dimensional plotting functions.* We can then compute the values of z and these are finally plotted by using one of the functions `mesh`, `surf`, `surfl`, or `surfc`. For example, to plot the function

$$z = (-20x^2 + x)/2 + (-15y^2 + 5y)/2 \quad \text{for} \quad x = -4:0.2:4 \quad \text{and} \quad y = -4:0.2:4$$

we first set up the values of the x–y domain and then compute z corresponding to these x and y values using the given function. Finally we plot the three-dimensional graph using the function `surfl`. This is achieved by using the script `e4s105.m`. Note how the function `figure` is used to direct the output to a graphics window so that the first plot is not overwritten by the second.

```
% e4s105.m
[x,y] = meshgrid(-4.0:0.2:4.0,-4.0:0.2:4.0);
z = 0.5*(-20*x.^2+x)+0.5*(-15*y.^2+5*y);
figure(1)
surfl(x,y,z); axis([-4 4 -4 4 -400 0])
xlabel('x-axis'), ylabel('y-axis'), zlabel('z-axis')
figure(2)
contour3(x,y,z,15); axis([-4 4 -4 4 -400 0])
xlabel('x-axis'), ylabel('y-axis'), zlabel('z-axis')
figure(3)
contourf(x,y,z,10)
xlabel('x-axis'), ylabel('y-axis')
```

Running script `e4s105.m` generates the plots shown in Fig. 1.10, Fig. 1.11, and Fig. 1.12. The first plot is created using `surfl` and shows the function as a surface; the second is created by `contour3` and is a three-dimensional contour plot of the surface; and the third, created using `contourf`, provides a two-dimensional filled contour plot.

When plotting surfaces a very useful function is `view`. This function allows the surface or mesh to be viewed from different positions. The function has the form `view(az,el)` where `az` is the azimuth and `el` is the elevation of the viewpoint required. Azimuth may be interpreted as the viewpoint rotation about the z-axis and elevation as the rotation of the viewpoint about the x–y plane. A positive value of the elevation gives a view from above the object and a negative value a view from below. Similarly a positive value of azimuth gives a counterclockwise rotation of the viewpoint about the z-axis while a negative value gives a clockwise rotation. If the view function is not used, the default values are $-37.5°$ for the azimuth and $30°$ for the elevation.

There are many other three-dimensional plotting facilities but they are not described here.

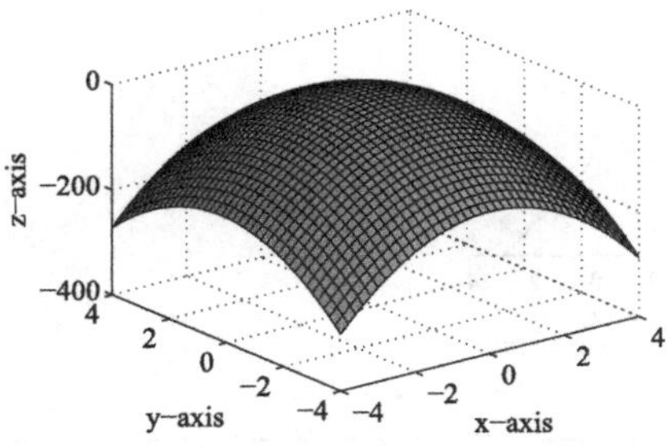

FIGURE 1.10

Three-dimensional surface using default view.

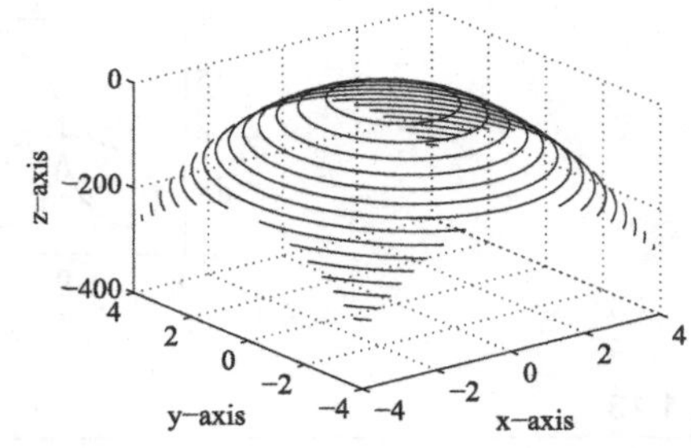

FIGURE 1.11

Three-dimensional contour plot.

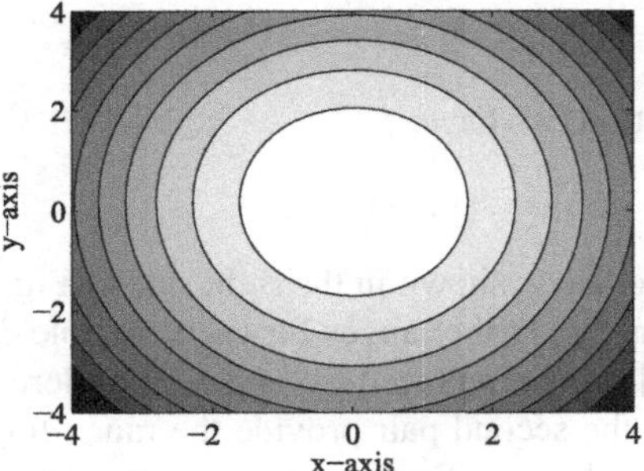

FIGURE 1.12

Filled contour plot.

1.16 IMPLICIT GRAPHICS

Since the 2016b release, MATLAB has the facility to plot implicit functions. An explicit function in two dimensions takes the form $y = f(x)$ whereas an implicit function takes the form $f(x, y) = 0$. Plotting a two-dimensional implicit function is achieved by using the MATLAB function `fimplicit`. To plot an implicit function of the form $f(x, y, z) = 0$, that is in three dimensions, the MATLAB function `fimplicit3` is used.

An example of a two-dimensional implicit function is the *quadrafolium*, given by the implicit function:

$$(x^2 + y^2)^3 - 4x^2y^2 = 0$$

This function can be plotted for x and y the range -1 to 1 by using the MATLAB statement:

```
>> fimplicit(@ (x,y) (x.^2+y.^2).^3-4*x.^2 .*y.^2,[-1  1])
```

The result of executing this command is shown in the left side of Fig. 1.13.

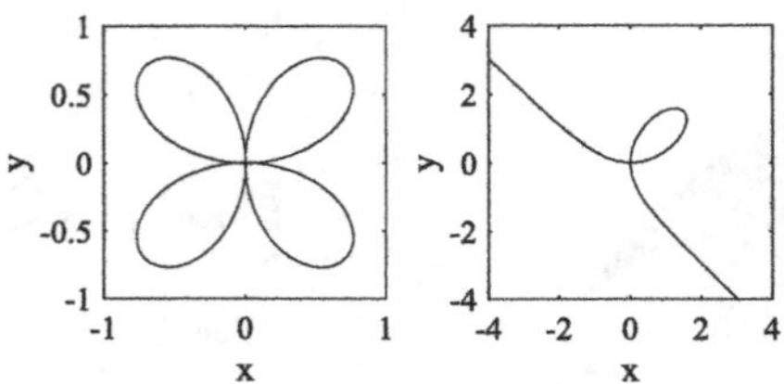

FIGURE 1.13

Implicit quadrafolium and folium of Descartes.

Another function, called the *folium of Descartes*, is defined by:

$$x^3 + y^3 - 3y = 0$$

Using MATLAB to plot this function, we have:

```
>> fimplicit(@ (x,y) x.^3+y.^3-3*x.*y,[-4  4])
```

The result of executing this command is shown in the right side of Fig. 1.13.

Note that for the function `fimplicit` the ranges for each variable can be different. In this case the single range for example [−4 4] can be replaced by [−5 5 0 6]. Here the first pair of values provide the range for the x variable and the second pair provide the range for the y variable. In addition the line type, color, and thickness can be specified. For example including the parameter `-g`, will ensure a continuous green line is use for the graph and using the parameter pair `LineWidth, 4` will ensure a thicker line is used.

We can use the MATLAB function `fimplicit3` to plot three-dimensional implicit functions. The following three-dimensional implicit function represents the general form of a torus:

$$\left(\sqrt{x^2 + y^2} - a\right)^2 + z^2 - b^2 = 0$$

where the distance form the center of the torus to the center of the torus ring is a and the radius of the torus tube is b.

As an example, suppose we plot one torus inside another. The distance from the center of the torus to the center of the torus tubes are $a = 1$ and $a = 3$ and the radius of both torus tube is $b = 0.4$. This is achieved by using the script `e4s106.m`:

```
% e4s106.m
fimplicit3(@ (x,y,z) (sqrt(x.^2 +y.^2)-3).^2 ...
+z.^2-0.4^2,[-2*pi  2*pi -2*pi 2*pi, -0.4,0.4],...
'EdgeColor', 'none', 'FaceAlpha',0.9);
hold on
fimplicit3(@ (x,y,z) (sqrt(x.^2 +y.^2)-1).^2 ...
+z.^2-0.4^2,[-2*pi  2*pi -2*pi 2*pi, -0.4,0.4],...
```

```
'EdgeColor', 'none', 'FaceAlpha',0.9);
view([-5.036 -75.565])
```

In this example, we use the MATLAB function `fimplicit3`, together with a number of parameters. These are: the implicit function we wish to plot, the range of each of the variables `x`, `y`, and `z`, the parameter `EdgeColor`, which gives the color of any edge lines on the surface, and the parameter `FaceAlpha`, which gives the transparency of the surface. In this case we have used `EdgeColor` as zero, so that no edges appear and `FaceAlpha` as 0.9 to provide limited transparency. The variables x and y are defined in the range -2π to 2π and z is between -0.4 and 0.4. Notice a view statement is also included which a direction of view favored by the user. This may be introduced by selecting the figure in its window and then it may be rotated, panned, and zoomed or viewed in the standard window. Once an acceptable view is obtained the code may be updated to provide this specific view.

We do not show the figure generated by this code. If the reader runs the code a pair of tori will be plotted.

1.17 MANIPULATING GRAPHICS – HANDLE GRAPHICS

Handle graphics allows the user to choose for a particular plot, font type, line thickness, symbol type and size, axes form, and many other features for a particular figure. It introduces more complexity into MATLAB but has considerable benefits. Here we give a very brief introduction to some of the main features. There are two key functions, `get` and `set`. The `get` function allows the user to obtain detailed information about a particular graphics function such as: `plot`, `title`, `xlabel`, `ylabel`, and others. The function `set` allows the user to modify the standard setting for the particular graphics element such as `xlabel` or `plot`. In addition, `gca` can be used with `set` to retrieve the handles of the axes of the current figure and with `get` to manipulate the properties of the axes of that figure.

To illustrate the details involved in a simple graphics statement, consider the following statements, where handles `h` and `h1` have been introduced for the plot and title functions:

```
>> x = -4:.1:4;
>> y = cos(x);
>> h = plot(x,y);
>> h1 = title('cos graph')
```

To obtain information about the detailed structure of the `plot` and `title` functions we use `get` and the appropriate handle as follows. Note that only a selection of the properties produced by `get` are shown.

```
>> get(h)
                Color: [0 0 1]
            EraseMode: 'normal'
            LineStyle: '-'
            LineWidth: 0.5000
               Marker: 'none'
           MarkerSize: 6
      MarkerEdgeColor: 'auto'
      .........................[etc]
```

However for the `title` function we have:

```
>> get(h1)
FontName = Helvetica
FontSize = [10]
FontUnits = points
  HorizontalAlignment = center
LineStyle = -
LineWidth = [0.5]
Margin = [2]
Position = [-0.00921659 1.03801 1.00011]
Rotation = [0]
String = cos graph
..........................[etc]
```

Notice there are different properties for `plot` and `title`. Script `e4s106.m` illustrates the use of handle graphics:

```
% e4s107.m
% Example for handle graphics
x = -5:0.1:5;
subplot(1,3,1)
e1 = plot(x,sin(x)); title('sin x')
subplot(1,3,2)
e2 = plot(x,sin(2*round(x))); title('sin round x')
subplot(1,3,3)
e3 = plot(x,sin(sin(5*x))); title('sin sin 5x')
```

Running script `e4s107.m` gives Fig. 1.14.

We now modify the script `e4s106.m` using a sequence of `set` statements to give script `e4s108.m`.

```
% e4s108.m
% Example for handle graphics
x = -5:0.1:5;
s1 = subplot(1,3,1);
e1 = plot(x,sin(x)); t1 = title('sin(x)');
s2 = subplot(1,3,2);
e2 = plot(x,sin(2*round(x))); t2 = title('sin(round(x))');
s3 = subplot(1,3,3);
e3 = plot(x,sin(sin(5*x))); t3 = title('sin(sin(5x))');
% change dimensions of first subplot
set(s1,'Position',[0.1 0.1 0.2 0.5]);
%change thickness of line of first graph
set(e1,'LineWidth',6)
set(s1,'XTick',[-5 -2  0 2  5])
%Change all titles to italics
```

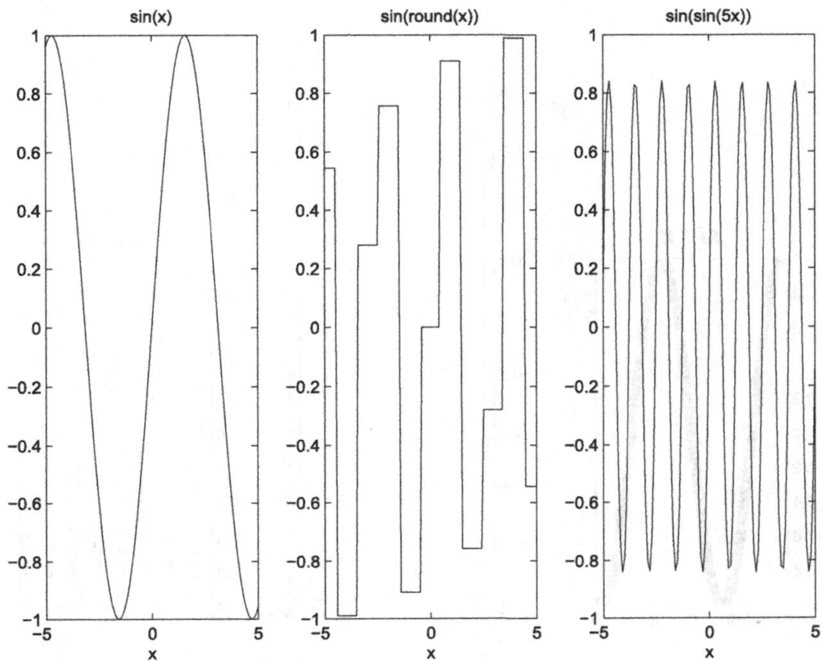

FIGURE 1.14

Plots illustrating aspects of handle graphics.

```
set(t1,'FontAngle','italic'), set(t1,'FontWeight','bold')
set(t1,'FontSize',16)
set(t2,'FontAngle','italic')
set(t3,'FontAngle','italic')
%change dimensions of last subplot
set(s3,'Position',[0.7 0.1 0.2 0.5]);
```

The `position` statement has the values

`[shift from left, shift from bottom, width, height].`

The size of the plotting area is taken as a unit square. Thus

```
set(s3,'Position',[0.7 0.1 0.2 0.5]);
```

shifts the figure 0.7 from the left and 0.1 from the bottom; its width is 0.2 and its height is 0.5. It may need some experimentation to get the required effect. Executing script `e4s107.m` gives Fig. 1.15. Notice the differing sizes of the boxes, thicker line in the first graph, bold title different ticks on x-axis and that all the titles are in italics, many other aspects could have been changed.

The following example shows how we can manipulate the various properties of the axes in Fig. 1.16 using `gca` with `set` which gets the current axes properties. The examples that follow show the use of `gca` in altering various properties of the axes:

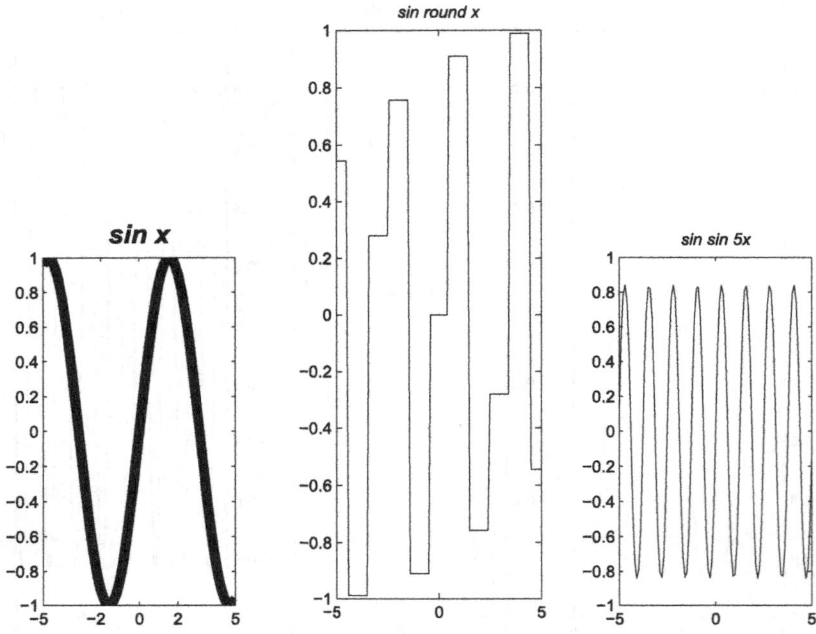

FIGURE 1.15

Plot of functions shown in Fig. 1.14 illustrating further handle graphs features.

```
>> x = -1:0.1:2; h = plot(x,cos(2*x));
```

These statements give left plot in Fig. 1.16.

```
>> get(gca,'FontWeight')

ans =
normal

>> set(gca,'FontWeight','bold')
>> set(gca,'FontSize',16)
>> set(gca,'XTick',[-1 0 1 2])
```

These additional statements provide the right plot of Fig. 1.16. Note that the differences produced larger bold font.

An alternative approach to manipulating font styles and other features is illustrated in script `e4s109.m`.

```
% e4s109.m
% Example of the use of special graphics parameters in MATLAB
% illustrates the use of superscripts, subscripts,
```

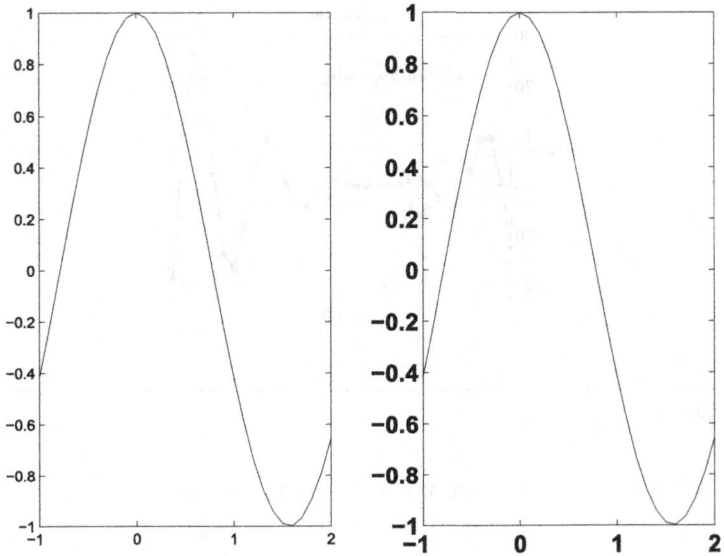

FIGURE 1.16

Plot of $\cos(2x)$. The axes of the right-hand plot are enhanced using handle graphics.

```
% Fontsize and special characters
x = -5:.3:5;
plot(x,(1+x).^2.*cos(3*x),...
'linewidth',1,'marker','hexagram','markersize',12)
title('(\omega_2+x)^2\alpha cos(\omega_1x)','fontsize',14)
xlabel('x-axis'), ylabel('y-axis','rotation',0)
gtext('graph for \alpha = 2,\omega_2 = 1, and \omega_1 = 3')
```

Executing script `e4s109.m` provides the graph shown in Fig. 1.17.

We now describe the features that were used in this script. We have used Greek characters from an extensive range of symbols that can be introduced using the backslash character "\". The following gives examples of how these characters may be introduced:

- `\alpha` gives α,
- `\beta` gives β,
- `\gamma` gives γ.

Any of the Greek symbols may be obtained by typing the backslash followed by the standard English name of the Greek letter. Titles and axis labels may include superscripts and subscripts by preceding the subscript character by "_" and the superscript by "∧". Font sizes may be specified by placing the additional parameter `'fontsize'` in the `xlabel`, `ylabel`, or `title` statements, followed by and separated by a comma from the actual font size required. For example,

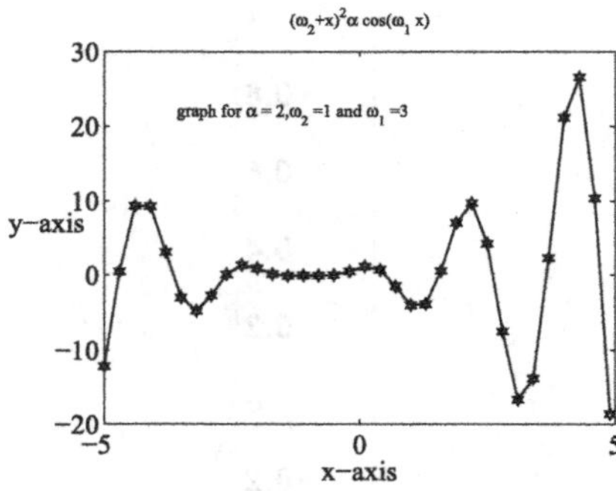

FIGURE 1.17

Plot of $(\omega_2 + x)^2 \alpha \cos(\omega_1 x)$.

```
title('(\omega_2+x)^2\alpha*cos(\omega_1*x)','fontsize',14)
```

gives

$$(\omega_2 + x)^2 \alpha * \cos(\omega_1 * x)$$

in 14 point font. In the `plot` function itself, additional markers for the graph points are available and may be indicated by using the additional parameter `'marker'` followed by the name of the marker, for example,

```
'marker','hexagram'
```

The size of the marker may also be specified using the additional parameter 'markersize' followed by the required marker size thus:

```
'markersize',12
```

The line thickness may also be adjusted using the parameter `'linewidth'`, for example,

```
'linewidth',1
```

Finally, the orientation of any label may be changed using `'rotation'`. For example,

```
'rotation',0
```

This additional parameter with the setting zero makes the label horizontal to the y-axis rather than the usual vertical orientation; the value of the parameter gives the angle in degrees.

A further more complex example involving reference to a partial differential equation in a MATLAB `text` statement is as follows:

```
gtext('Solution of \partial^2V/\partialx^2+\partial^2V/\partialy^2 = 0')
```

This leads to the text

$$\text{Solution of } \partial^2 V/\partial x^2 + \partial^2 V/\partial y^2 = 0$$

being placed in the current graphics window at a point selected using the cross-hairs cursor and clicking the mouse button.

In addition, features may be included in conjunction with the \ followed by a font name parameter which allows the specification of any available font. Examples are `\bf` that gives a bold style and `\it` that gives an italic style.

An important issue in placing figures in a manuscript is that they must have a consistent position and size and must be easy to read. The listed graphics scripts work satisfactorily but would not provide the quality required if directly imported into this manuscript. An example of this is shown in Fig. 1.14. To ensure that size and position of the figures generated by the MATLAB scripts are generally consistent and their fonts easily read, the following statements are added to all scripts producing graphical output except for Figs. 1.14, 1.15, and 1.16.

```
set(0,'defaultaxesfontsize',16)
set(0,'defaultaxesfontname','Times New Roman')
set(0,'defaulttextfontsize',12)
set(0,'defaulttextfontname','Times New Roman')
axes('position',[0.30 0.30 0.50 0.50])
```

These statements are examples of handle graphics. The first and second statements set the fonts used for the axes to 16 point Times New Roman. The third and fourth statements set the fonts used in the plot to 12 point Times New Roman and the fifth statement controls the size of the graph within the graphs window. Finally, we add a statement such as `print -depsc Fig101.eps` at the end of each script that generates a graph. This statement saves the plot as an extended postscript (eps) format file for inclusion in the manuscript.

This was used in the creation and placement of the MATLAB graphs in this book. These are not shown in the text listings because they are the same for each script.

1.18 SCRIPTING IN MATLAB

In some of the previous sections we have created some simple MATLAB scripts that have allowed a series of commands to be executed sequentially. However, many of the features usually found in programming languages are also provided in MATLAB to allow the user to create versatile scripts. The more important of these features are described in this section. It must be noted that scripting is done in the **edit** window using a text editor appropriate to the system, not in the **command** window which only allows the execution of statements one at a time or several statements provided that they are on the same line.

MATLAB *does not require the declaration of variable types*, but for the sake of clarity the role and nature of key variables may be indicated by using comments. Any text following the symbol `%` is considered a comment. In addition, there are certain variable names which have predefined special values for the convenience of the user. They can, however, be redefined if required. These are

`pi`	which equals π
`inf`	the result of dividing by zero
`eps`	which is set to the particular machine accuracy

`realmax`	largest positive floating-point number
`realmin`	smallest positive floating-point number
`NaN`	which is not a number produced on dividing zero by zero
`i,j`	both equal $\sqrt{-1}$.

Assignment statements in a MATLAB script take the form

```
variable = <expression>;
```

The expression is calculated and the value assigned to the variable on the left-hand side. If the semicolon is omitted from the end of these statements, the names of the variable(s) and the assigned value(s) are displayed on the screen. If an expression is not assigned explicitly to a variable then the value of the expression is calculated, assigned to the variable `ans` and displayed.

In previous sections it was stated that generally a variable in MATLAB is assumed to be a matrix of some kind; its name must start with a letter and may be followed by any combination of digits and letters; a maximum of 32 characters is recognized. It is good practice to use a meaningful variable name. The variable name must not include spaces or hyphens. However, the underscore character is a useful replacement for a space. For example, `test_run` is acceptable; `test run` and `test-run` are not. It is very important to avoid the use of existing MATLAB commands, function names, or even the word MATLAB itself! MATLAB does not prevent their use but using them can lead to problems and inconsistencies. An expression in MATLAB is a valid combination of variables, constants, operators, and functions. Brackets can be used to alter or clarify the precedence of operations. The precedence of operation for simple operators is first \, second *, third /, and finally + and - where

^	raises to a power
*	multiplies
/	divides
+	adds
-	subtracts

The effects of these operators in MATLAB have already been discussed.

Unless there are instructions to the contrary, a set of MATLAB statements in a script is executed in sequence. This is the case in script `e4s110.m`.

```
% e4s110.m
% Matrix calculations for two matrices A and B
A = [1 2 3;4 5 6;7 8 9];
B = [5 -6 -9;1 1 0;24 1 0];
% Addition. Result assigned to C
C = A+B; disp(C)
% Multiplication. Result assigned to D
D = A*B; disp(D)
% Division. Result assigned to E
E = A\B; disp(E)
```

To allow the repeated execution of one or more statements, a for loop is used. This takes the form

```
for <loop_variable> = <loop_expression>
    <statements>
end
```

The <loop_variable> is a suitably named variable and <loop_expression> is usually of the form n:m or m:i:n where n, i, and m are the initial, incremental, and final values of <loop_variable>. They may be constants, variables, or expressions; they can be negative or positive but clearly they should be chosen to take values consistent with the logic of the script. This structure should be used when the loop is to be repeated a predetermined number of times.

Examples:

```
for i = 1:n
    for j = 1:m
        C(i,j) = A(i,j)+cos((i+j)*pi/(n+m))*B(i,j);
    end
end
```

```
for k = n+2:-1:n/2
    a(k) = sin(pi*k);
    b(k) = cos(pi*k);
end
```

```
p = 1;
for a = [2 13 5 11 7 3]
    p = p*a;
end
p
```

```
p = 1;
prime_numbs = [2 13 5 11 7 3];
for a = prime_numbs
    p = p*a;
end
p
```

The first example illustrates the use of nested for loops, the second illustrates that upper and lower limits can be expressions and the step value can be negative. The third example shows that the loop does not have to use a uniform step and the fourth example, which gives an identical result to the third, illustrates that the <loop_expression> can be any previously defined vector.

When assigning values to a vector in a for loop, the reader should note that the vector generated is a *row* vector. For example,

```
for i = 1:4
    d(i) = i^3;
end
```

gives the row vector `d = 1 8 27 64` but in this case it is not displayed.

The `while` statement is used when the repetition is subject to a condition being satisfied which is dependent on values generated within the loop. This has the form

```
while <while_expression>
    <statements>
end
```

The `<while_expression>` is a *relational expression* of the form e1 o e2 where e1 and e2 are ordinary arithmetic expressions as described before and o is a relational operator defined as follows:

`==`	equals
`<=`	less than or equals
`>=`	greater than or equals
`~=`	not equals
`<`	less than
`>`	greater than

Relational expressions may be combined using the following *logical operators*:

`&`	the and operator
`|`	the or operator
`~`	the not operator
`&&`	the scalar and operator. If the first condition is false then the second is not evaluated.
`||`	the scalar or operator. If the first condition is true then the second is not evaluated.

Note that false is zero and true is non-zero. Relational operators have a higher order of precedence than logical operators.

Examples of while loops:

```
dif = 1;
x2 = 1;
while dif>0.0005
    x1 = x2-cos(x2)/(1+x2);
    dif = abs(x2-x1);
    x2 = x1;
end
```

```
x = [1 2 3];
y = [4 5 8];
while sum(x) ~= max(y)
    x = x.^2;
    y = y+x;
end
```

Note also that break stops the execution (and hence allows exit from) a while or for loop and that break cannot be used outside of a while or for loop. The statement return must be used in these circumstances.

A vital feature of all programming languages is the ability to change the sequence in which instructions are executed within the program. In MATLAB the if statement is used to achieve this and has the general form

```
if < if_expression1 >
    < statements >
elseif < if_expression2 >
< statements >
elseif < if_expression3 >
    < statements >
...
...
else
< statements >
end
```

Here < if_expression1 >, etc., are relational expressions of the form e1 o e2 where e1 and e2 are ordinary arithmetic expressions and o is a relational operator as described before. Relational expressions may be combined using logical operators.

Examples:

```
for k = 1:n
    for p = 1:m
        if k == p
            z(k,p) = 1;
      total = total+z(k,p);
        elseif k<p
            z(k,p) = -1;
            total = total+z(k,p);
        else
            z(k,p) = 0;
        end
    end
end

if (x~=0) & (x<y)
  b = sqrt(y-x)/x;
  disp(b)
end
```

The MATLAB function switch provides an alternative to the if structure and is particularly useful when many options must be considered. This has the form:

```
switch < condition >
  case ref1
    < statements >
  case ref2
    < statements >
  case ref3
    < statements >
  otherwise
    < statements >
end
```

The following fragment of code allows the user to choose a particular plot, dependent on the value of n. In the following script, the second plot has been chosen by setting `n = 2`.

```
 x = 1:.01:10; n = 2
switch n
case 1
plot(x,log(x));
case 2
plot(x,x.*log(x));
case 3
plot(x,x./(1+log(x)));
otherwise
disp('That was an invalid selection.')
end
```

As a further example of the `switch` function, the following fragment of code allows the user to convert an astronomical distance `x` given in AU (an astronomical unit), LY (a light year) or pc (a parsec) to km by setting the string variable `units` to `AU`, `LY`, or `pc` respectively.

```
x = 2;
units = 'LY'
switch units
    case {'AU' 'Astronomical  Units'}
        km = 149597871*x
    case  {'LY','lightyear'}
        km = 149597871*63241*x
    case  {'pc' 'parsec'}
        km = 149597871*63241*3.26156*x
    otherwise
        disp('That was an invalid selection.')
end
```

Note that if any statement in a MATLAB script is longer than one line then it must be continued by using an ellipsis (...) at the end of the line.

The `menu` function creates a menu window with buttons to allow the user to select options. For example:

```
frequency = 123;
units = menu('Select units for output data', 'rad/s','Hz', 'rev/min')
switch units
    case 1
        disp(frequency)
    case 2
        disp(frequency/(2*pi))
    case 3
    disp(frequency*60/(2*pi))
end
```

creates a small window (called **MENU**) with three buttons, labeled 'rad/s', 'Hz', and 'rev/min'. 'clicking' a particular button with the mouse provides a frequency converted to the chosen units.

1.19 USER-DEFINED FUNCTIONS IN MATLAB

MATLAB allows users to define their own functions but a specific form of definition must be followed. The first form of function is the function m-file and is described as follows:

```
function < output_params > = func_name(< input_params >)
< func body >
```

`< input_params >` is either a single variable or a set of variable names separated by commas and `< output_params >` is either a single variable or a list of variables separated by commas or spaces and placed in square brackets. The function body consists of the statements defining the user's function. These statements will utilize the values of the input arguments and *must include statements assigning values to the output parameters.* Once the function is defined *it must be saved as an M-file* under the same name as the given `< function_name >`. Then the function can then be used as required. It is good practice to put some comments describing the nature of the function immediately after the function heading. Writing `help` followed by the function name in the command window will access these comments.

To execute the function for specific parameters we write

```
< specific_out_params > = < func_name >(< assigned_input_params >)
```

where the `< assigned_input_params >` term is either a single parameter or a list of parameters separated by commas. The `< assigned_input_params >` must match the `<input_params >` in the function definition.

We now provide two examples of named functions.

Example 1.1. The Fourier series for a sawtooth wave is

$$y(t) = \frac{1}{2} - \sum_{n=1}^{\infty} \frac{1}{n\pi} \sin\left(\frac{2\pi nt}{T}\right)$$

where T is the period of the waveform. We can create a function to evaluate this for given values of t and T. Since we cannot sum to infinity, we will sum to m terms, where m is a relatively large value. Thus we can define the MATLAB function `sawblade` as follows. Note that this function has three input arguments and one output.

```
function y = sawblade(t,T,n_trms)
% Evaluates, at instant t, the Fourier approximation of a sawtooth wave of
% period T using the first n_trms terms in the infinte series.
y = 1/2;
for n = 1:n_trms
    y = y - (1/(n*pi))*sin(2*n*pi*t/T);
end
```

We can now use this function for a specific purpose. For example, if we wish to plot this waveform and a period of $T = 2$ over the range $t = 0$ to 4 using only 50 terms in the series, we have

```
>> c = 1;
>> for t = 0:0.01:4, y(c) = sawblade(t,2,50); c = c+1; end
>> plot([0:0.01:4],y)
```

Further valid function calls are

```
>> y = sawblade(0.2*period,period,terms)
```

where `period` and `terms` have previously assigned values, or

```
>> y = sawblade(2,5.7,60)
```

or using the function `feval`

```
>> y = feval('sawblade',2,5.7,60)
```

A more important application of `feval`, which is widely used in this text, is in the process of defining functions which themselves have functions as parameters. These function m-files can be evaluated internally in the body of the calling function by using `feval`.

Example 1.2. We now consider a further example which involves the generation of a matrix within a function. The essential features of the finite element method applied to the static and/or dynamic analysis of structures is to express the stiffness and inertia properties of a small section or element of the structure in matrix form. These *element* matrices are then assembled to obtain matrices that describe the overall stiffness and inertia for the whole structure. Knowing the forces acting on the structure, we can obtain the static or dynamic response of the structure. One such element is a uniform circular shaft. For this element the inertia matrix and the stiffness matrix, relating angular accelerations and displacements, respectively, to applied torques are given by

$$\boldsymbol{K} = \frac{GJ}{L}\begin{bmatrix} 1 & -1 \\ -1 & 1 \end{bmatrix}$$

and

$$M = \frac{\rho J L}{6}\begin{bmatrix} 2 & 1 \\ 1 & 2 \end{bmatrix}$$

where L is the length of the shaft, G and ρ are material properties and J is the polar second moment of area of the shaft; a geometric property related to the diameter of the shaft, d. If we intend to create a finite element package in MATLAB, which includes torsional elements then these matrices are required. The following function generates them from the shaft properties as follows:

```
function [K,M] = tors_el(L,d,rho,G)
J = pi*d^4/32;
K = (G*J/L)*[1 -1;-1 1];
M = (rho*J*L/6)*[2 1;1 2];
```

Note that this function has four input arguments and outputs two matrices.

MATLAB allows the inclusion of the definition of any functions, called by a function, at the end of the main function definition. This is convenient for the user since it allows all the elements of the function definition to be presented together, rather than saved separately. This is only useful if the nested function is only required by the main function. An example of this is shown in Section 3.11.1. Here the function `solveq` is not a generally useful function but it is required by the function `bairstow`. Therefore it is nested in `bairstow`. This arrangement has the advantage that `bairstow` is a complete entity, it does not require `solveq` to be stored and available. Conversely, it has the minor disadvantage that since it is not stored separately it cannot be used independently of `bairstow`.

In MATLAB release 2016b this feature was extended to allow scripts to call one or more functions. Thus, if a particular script includes five function calls, then each of these functions can be defined after the main calling script. This is a very useful development since it allows all the elements can be kept and saved together rather than saved separately.

To illustrate this feature we provide the script `e4s111.m`, which uses two functions called `fexlive` and `fitnesslive` to calculate a range of numerical values. The first group of statements is the script in which the two functions are called. This is followed by the definitions of the two functions.

```
%e4s111.m
% Calling script
x = rand(5,2);
for i = 1:5
    p(i) = x(i,2);
    fv(i) = fexlive(p(i));
end
for i = 1:5
    fitv(i) = fitnesslive(fv(i));
end
for i = 1:5
    pr(i) = fitv(i)/sum(fitv);
end
fv
```

```
fitv
disp('Probabilities')
pr
disp('sum of probabilities = ')
spr = sum(pr)
% The functions called in the above script are defined here
% ------------------------
function res1 = fexlive(x)
res1 = x.^2+cos(x);
end
% ----------------------------
function res2 = fitnesslive(val)
if val<=0
    res2 = abs(val);
else
    res2 = 1/val;
end
end
```

The script `e4s111.m`, including the functions, are saved together with an appropriate name and run. Output from running script `e4s111.m` is shown here.

```
fv =
    1.1223    1.1009    1.2160    1.2620    1.2980

fitv =
    0.8910    0.9083    0.8223    0.7924    0.7704

Probabilities

pr =
    0.2129    0.2171    0.1965    0.1894    0.1841

sum of probabilities =

spr =
     1
```

We note that this is the expected valid output and illustrates that the script and the two function definitions have been saved as a single entity.

MATLAB allows functions to be defined recursively. The technique provides a useful approach to the definition of a range of functions, although in many cases a recursive approach is less efficient than the standard iterative approach. Essentially in a recursive function definition the function definition calls within it the function itself. This is a natural approach for some mathematical functions which are defined in this way. For example the Fibonacci series is

$$F_{k+1} = F_k + F_{k-1}, \quad \text{for } k = 2, 3, 4, \ldots$$

where $F_2 = F_1 = 1$.

One of the simplest recursive function definitions is based on the factorial formula $f(n) = nf(n-1)$ where $f(0) = 1$. The following MATLAB function defines the factorial function recursively:

```
function y = facr(n)
% Factorial function implemented recurcively
if n<0
    disp('n must be positive')
    return
end
if n==0
    y=1;
    return
end
for i = 1:n
    % here the function calls itself
y = n*facr(n-1);
end
end
```

Using this function we obtain the results

```
>> facr(4)

ans =
    24

>> facr(-5)
n must be positive

>> facr(0)

ans =
    1
```

Thus we obtain the expected results for various cases of using the factorial function. A more interesting example is given for the evaluation of trigonometric integrals of the form:

$$I_n = \int_a^b \cos^n \theta \, d\theta$$

or, alternatively, the same expression but with the sine function. The expression is evaluated using simple integration by parts and takes the form:

$$I_n = \left[\sin\theta\cos^{(n-1)}\theta/n\right]_a^b + \frac{n-1}{n}\int_a^b \cos^{(n-2)}\theta\, d\theta$$

Taking limits between 0 and $\pi/2$, the first term vanishes and this equation may be rewritten in the form:

$$I_n = \left(\frac{n-1}{n}\right) I_{n-2}$$

The same formula applies for the sine integral. We now provide a MATLAB recursive function for this integral.

```
function I = cosnr(n)
% This function written recursively provides the integral of
% the nth degree cosine function in the range 0 to pi/2 if n >=2
if n==1
    I = 1;
    return
end
if n==2
    I = pi/4;
    return
end
% Here we call the function within the defintion of the function
I = (n-1)/n*cosnr(n-2);
end
```

We now calculate some values of the cosine integral for a range of values of the power n using the recursive function `cosnr` and compare the results with those obtained using a MATLAB numerical integration function `integral`. This is achieved using the script `e4s112.m`:

```
%e4s112.m
fprintf('%1s %12s %12s\n','n','I1','I2');
for n = 1:10
    I1 = cosnr(n); I2 = integral(@(x) cos(x).^n,0,pi/2);
    fprintf('%2.0f %12.4f %12.4f\n',n,I1,I2);
end
```

Running script `e4s112.m` produces the table of results for the integration of $\cos^n(x)$ to the powers of $n = 0$ to 10.

```
n          I1           I2
 1     1.0000       1.0000
 2     0.7854       0.7854
 3     0.6667       0.6667
```

```
  4        0.5890        0.5890
  5        0.5333        0.5333
  6        0.4909        0.4909
  7        0.4571        0.4571
  8        0.4295        0.4295
  9        0.4063        0.4063
 10        0.3866        0.3866
```

Note that there is complete agreement to 4 decimal places between the two sets of results for the integration.

An alternative and simpler form of the MATLAB user-defined function is the anonymous function. This function is not saved as an m-file; it is either entered into the workspace from the command window or from a script. For example, suppose we wish to define the function

$$\left(\frac{x}{2.4}\right)^3 - \frac{2x}{2.4} + \cos\left(\frac{\pi x}{2.4}\right)$$

This mathematical expression may be defined as an anonymous function in MATLAB as follows:

```
>> f = @(x) (x/2.4).^3-2*x/2.4+cos(pi*x/2.4);
```

An example calls of this function is `f([1 2])` which produces two values corresponding to $x = 1$ and $x = 2$. Another way of using this function is as an input parameter to another function. For example,

```
>> solution = fzero(f,2.9)
solution =
    3.4825
```

This gives the zero of `f` closest to 2.9. A further example of its use is

```
>> x = 0:0.1:5; plot(x,f(x))
```

Here we must call `f(x)` because the `plot` function needs all the values of the function over the range of `x`. Another form is

```
>> solution = fzero(@(x) (x/2.4).^3-2*x/2.4+cos(pi*x/2.4), 2.9)
```

Here we have used the anonymous function definition directly, rather than assigning it to a handle (`f` in this case) and then using the handle.

If an m-file function has an anonymous function as one of its input arguments, then this anonymous function can be evaluated directly without the use of the MATLAB function `feval`. If, however, the function in the parameter list requires a multi-statement definition, a function m-file must be used, and in this case `feval` must be used. In this text, to allow flexibility, when defining function m-files we have used `feval` so that the user can input a function as an m-file function or as an anonymous function.

For example, we define the m-file functions `sp_cubic` and `minandmax` thus:

```
function y = sp_cubic(x)
y = x.^3-2*x.^2-6;
```

```
function [minimum maximum] = minandmax(f,v)
% v is a vector with the start, increment and end value
y = feval(f,v); minimum = min(y); maximum = max(y);
```

Using this definition of `minandmax` means that `f` can be an anonymous or an m-file function. Thus, using the anonymous function's definition for `f` given before, we have

```
>> [lo hi] = minandmax(f,[-5:0.1:5]);
>> fprintf('lo = %8.4f hi = %8.4f\n',lo,hi)

lo = -181.0000 hi =  69.0000
```

Alternatively, using the other form of the function

```
>> [lo hi] = minandmax('sp_cubic',[-5:0.1:5]);
>> fprintf('lo = %8.4f hi = %8.4f\n',lo,hi)

lo = -181.0000 hi =  69.0000
```

give identical answers. However, suppose we define m-file function `minandmax` without the use of `feval` as follows

```
function [minimum maximum] = minandmax(f,v)
% v is a vector with the start, increment and end value
y = f(v); minimum = min(y); maximum = max(y);
```

Then, if `f` is an anonymous function, we obtain the preceding results, but if `f` is the m-file function `sp_cubic`, the function `minandmax` fails as shown below:

```
>> [lo hi] = minandmax('sp_cubic',[-5:0.1:5]);
>> fprintf('lo = %8.4f hi = %8.4f\n',lo,hi)
??? Subscript indices must either be real positive integers or logicals.

Error in ==> minandmax at 4
y = f(v);
```

1.20 DATA STRUCTURES IN MATLAB

Previous sections discussed the use of numerical and non-numerical data. We now introduce the cell array structure which allows a more complex data structure. The cell data structure is indicated by curly brackets, i.e., { }. As an example,

```
>> A = cell(4,1);
>> A = {'maths'; 'physics'; 'history'; 'IT'}
```

```
A =
    'maths'
    'physics'
    'history'
    'IT'
```

We may refer to the individual components

```
>> p = A(2)

p =
    'physics'

>> A(3:4)

ans =
    'history'
    'IT'
```

To access the contents of the cell we use curly brackets thus:

```
>> cont = A{3}

cont =
history
```

Note that `history` is no longer in quotes and thus we can reference individual characters as follows:

```
>> cont(4)

ans =
t
```

A cell array can include both numeric and string data and can also be generated using the `cell` function. For example, to generate a cell with 2 rows and 2 columns we have

```
>> F = cell(2,2)

F =
     [ ]     [ ]
     [ ]     [ ]
```

To assign a scalar, an array, or a character string to a cell we write

```
>> F{1,1} = 2;
>> F{1,2} = 'test';
>> F{2,1} = ones(3);
>> F
```

```
F =
    [         2]    'test'
    [3x3 double]       [ ]
```

An equivalent way of generating `F` is

```
>> F = {[2] 'test'; [ones(3)] [ ]}
```

Because we cannot see the detailed content of `F{2,1}` in the preceding, we use the function `celldisp` thus:

```
>> celldisp(F)

F{1,1} =
     2

F{2,1} =
     1     1     1
     1     1     1
     1     1     1

F{1,2} =
test

F{2,2} =
     [ ]
```

The cell array allows us to group data of different sizes and types together in the form of an array and access its elements by using subscripts.

The last form of data we consider is the structure, implemented in MATLAB using `struct`. This is similar to a cell array but individual cells are indexed by name. A structure combines a number of fields, each of which may be a different type. There is a general name for the field, for example, `'name'` or `'phone number'`. Each of these fields can have specific values such as `'George Brown'` or `'12719'`. To illustrate these points consider the following example which sets up a structure called `StudentRecords` containing three fields: `NameField`, `FeesField`, and `SubjectField`.

Note that we begin by setting up the information for three students as specific values held in the cell arrays: names, fees, and subjects.

```
>> names = {'A Best', 'D Good', 'S Green', 'J Jones'}

names =
    'A Best'    'D Good'    'S Green'    'J Jones'

>> fees = {333 450 200 800}
```

```
fees =
    [333]    [450]    [200]    [800]

>> subjects = {'cs','cs','maths','eng'}

subjects =
    'cs'    'cs'    'maths'    'eng'

>> StudentRecords = struct('NameField',names,'FeesField',fees,...
                           'SubjectField',subjects)

StudentRecords =
1x4 struct array with fields:
    NameField
    FeesField
    SubjectField
```

Now, having set up our structure, we can refer to each individual record using a subscript thus:

```
>> StudentRecords(1)

ans =
       NameField: 'A Best'
       FeesField: 333
    SubjectField: 'cs'
```

Further we can examine the contents of the components of each record thus:

```
>> StudentRecords(1).NameField

ans =
A Best

>> StudentRecords(2).SubjectField

ans =
cs
```

We can change or update the values of the components of the records thus:

```
>> StudentRecords(3).FeeField = 1000;
```

Now we check the contents of this student's `FeesField`

```
>> StudentRecords(3).FeeField

ans =
        1000
```

MATLAB provides functions that allow us to convert from one data structure to another and some of these are listed here.

```
cell2struct
struct2cell
num2cell
str2num
num2str
int2str
double
single
```

Most of these conversions are self-explanatory. For example, `num2str` converts a double precision number to an equivalent string. The function `double` converts to double precision and examples of its usage are given in Chapter 10.

The use of cells and structures is not usually essential in the development of numerical algorithms although they can be used to enhance an algorithm's ease of use. There is an example of the use of structures in Chapter 10.

1.21 EDITING MATLAB SCRIPTS

To help the user develop scripts MATLAB provides a comprehensive selection of debugging tools. These can be listed using the command `help debug`.

When a typing in a script into the MATLAB editor the user should note that the small colored square is displayed at the top right of the text window. This square is colored red if the script contains one or more fatal syntactic errors, orange warns of possible non-fatal problems, and green indicates no syntactic errors. Each error or warning is also indicated by an appropriately colored dash beneath the square. Touching these dashes will provide a description of the error or warning and the line in which it occurs.

Errors can be found by using `checkcode`. The `mlint` function can also be used but is now obsolete and is replaced by `checkcode`. The following script, `e4s113.m`, contains numerous errors and is provided to illustrate the use of `checkcode`:

```
% e4s113.m A script full of errors!!!
A = [1 2 3; 4 5 6
B = [2 3; 7 6 5]
c(1) = 1; c(1) = 2;
for k = 3:9
    c(k) = c(k-1)+c(k-2)
    if k = 3
        displ('k = 3, working well)
end
c
```

Running and checking script `e4s113.m` gives the following output:

```
>> e4s113
Error: File: e4s112.m Line: 3 Column: 3
The expression to the left of the equals sign is not a valid target
                                                   for an assignment.
```

Applying `checkcode` to script `e4s113.m` gives

```
>> checkcode e4s113
L 3 (C 3): Invalid syntax at '='. Possibly, a ), }, or ] is missing.
L 3 (C 16): Parse error at ']': usage might be invalid MATLAB syntax.
L 5 (C 1-3): Invalid use of a reserved word.
L 7 (C 5-6): IF might not be aligned with its matching END (line 9).
L 7 (C 10): Parse error at '=': usage might be invalid MATLAB syntax.
L 8 (C 15-35): A quoted string is unterminated.
L 11 (C 0): Program might end prematurely (or an earlier error
                                              confused Code Analyzer).
```

Note how both line (L), character position (C), and nature of the error are given so the errors are clearly identified. Of course, some errors cannot be detected at this stage. For example, the following script, `e4s113c.m`, is a partially corrected version of script `e4s113.m`.

```
% e4s113c.m A script less full of errors!!!
A = [1 2 3; 4 5 6];
B = [2 3; 7 6];
c(1) = 1; c(1) = 2;
for k = 3:9
    c(k) = c(k-1)+c(k-2)
    if k == 3
        disp('k = 3, working well')
    end
end
c
```

Running script `e4s113c.m` gives

```
>> e4s113c
Attempted to access c(2); index out of bounds because numel(c)=1.

Error in e4s112c (line 6)
    c(k) = c(k-1)+c(k-2)

>> checkcode e4s112c
L 6 (C 5): The variable 'c' appears to change size on every loop
         iteration (within a script). Consider pre-allocating for speed.
L 6 (C 10): Terminate statement with semicolon to suppress output
                                                      (within a script).
L 11 (C 1): Terminate statement with semicolon to suppress output
                                                      (within a script).
```

Now the script can be run as far as line 6 other possible errors now are detected.
In addition, a menu option "Debug" is provided in the MATLAB text editor.

1.22 SOME PITFALLS IN MATLAB

We now list five important points which if observed enable the MATLAB user to avoid some significant difficulties. This list is not exhaustive.

- It is important to take care when naming files and functions. File names and function names follow the rules for variable names; that is, they must start with a letter followed by a combination of letters or digits and names of existing functions must not be used.
- Do not use MATLAB function names or commands for variable names. For example, if we were so foolish as to assign a number to a variable which we will call `sin`, access to the sine function would be lost. For example,

```
>> sin = 4

sin =
     4

>> 3*sin

ans =
    12

>> sin(1)

ans =
     4

>> sin(2)
??? Index exceeds matrix dimensions.

>> sin(1.1)
??? Subscript indices must either be real positive integers or logicals.
```

- Matrix sizes are set by assignment so it is vital to ensure that matrix sizes are compatible. Often it is a good idea initially to assign a matrix to an appropriately sized matrix of zeros; this also makes code execution more efficient. For example, consider the following simple script:

```
for i = 1:2
    b(i) = i*i;
end
A = [4 5; 6 7];
A*b'
```

We assign two elements to b in the for loop and define A to be a 2×2 array, so we would expect this script to succeed. However, if b had in the same session been previously set to be a different size matrix, then this script would fail. To ensure that it works correctly we must either assign b to be a null matrix using `b = [ ]`, or make b a column vector of two elements by using `b = zeros(2,1)` or by using the clear statement to clear all variables from the system.

- Take care with dot products. For example, when creating a user-defined function where any of the input parameters may be vectors, dot products must be used. Note also that `2.^x` and `2. ^x` are different because the space is important. The first example gives the dot power while the second gives 2.0 to the power x, not the dot power. Similar care with spaces must be taken when using complex numbers. For example, `A = [1 2-4i]` assigns two elements: 1 and the complex number $2 - \imath 4$. In contrast `B = [1 2 -4i]` assigns three elements: 1, 2, and the imaginary number $-4\imath$.
- At the beginning of a script, it is often good practice to clear variables or set arrays equal to the empty matrix e.g., `A = [ ]`. This avoids incompatibility in matrix operations.

1.23 SPEEDING UP CALCULATIONS IN MATLAB

Calculations can be greatly speeded up by using vector operations rather than using a loop to repeat a calculation. To illustrate this consider the following simple examples.

Example 1.3. The script e4s114.m fills the vector **b** using a for loop.

```
% e4s114.m
% Fill b with square roots of 1 to 100000 using a for loop
tic;
for i = 1:200000
    b(i) = sqrt(i);
end
t = toc;
disp(['Time taken for loop method is ', num2str(t)]);
```

Example 1.4. The script e4s115.m fills the vector **b** using a vector operation.

```
% e4s115.m
% Fill b with square roots of 1 to 200000 using a vector
tic
a = 1:200000; b = sqrt(a);
t = toc;
disp(['Time taken for vector method is ',num2str(t)]);
```

If the reader runs scripts e4s114.m and e4s115.m and compares the time taken they will notice the vector method is generally faster than the loop method. There is a need to think very carefully about the way algorithms are implemented in MATLAB, particularly with regard to the use of vectors and arrays.

1.24 LIVE EDITOR

Here we give a brief description of the Live Editor which was first introduced into MATLAB in release 2016a.

The Live Editor allows the user to turn their MATLAB code into an interactive document and thereby provide a vivid platform for project and other presentations. Bare code can be enhanced to create a story. Text, equations, images, numerical and graphical output from the code, and even hyperlinks can all be added to the code in the Live Editor document. Thus, for example, a Live Editor document might begin with some text and possibly some images, describing the background to a problem. This might be followed by text and equations, providing detailed description of the proposed mathematical model for the problem. Having established the mathematical model the MATLAB code required to solve the problem is given, together with any numerical or graphical output from the code. This output is created as the code is executed. It is not a piece of text or a diagram inserted into the file, it is actually generated by running the code and it changes if the code is changed.

The Live Editor is extremely easy to use. Opening an existing Live Editor file automatically activates the Live Editor. Then the code, text, equations, and image options can be accessed from the Live Editor menu. The output from the MATLAB code is included in the Live Editor file and may be placed in the same column as the code or placed in the other column to suit the user's requirements.

1.25 SUMMARY

In this chapter we have provided some of the key features of MATLAB. Clearly a user new to MATLAB would need to support this description of the language with reference to the user manuals and lots of computer based experimentation. In the introduction to this chapter it is stated that the original version of MATLAB had only 80 functions. Currently, including the many toolboxes, it has over 8000 functions. Obviously users are, or will become, familiar only with the subset of these functions appropriate to the problems they wish to solve. This might, for example, be the core of MATLAB and two or three toolboxes.

1.26 PROBLEMS

1.1. (a) Start up MATLAB. In the **command** window type `x = -1:0.1:1` and then execute each of the following statements by typing them in and pressing return:

```
sqrt(x)              cos(x)
sin(x)               2./x
x.\ 3                plot(x, sin(x.^3))
plot(x, cos(x.^4))
```

Examine the effects of each statement carefully.

(b) Execute the following and explain the results:

```
x = [2 3 4 5]
y = -1:1:2
x.^y
x.*y
x./y
```

1.2. (a) Set up the matrix `A = [1 5 8;84 81 7;12 34 71]` in the **command** window and examine the contents of `A(1,1)`, `A(2,1)`, `A(1,2)`, `A(3,3)`, `A(1:2,:)`, `A(:,1)`, `A(3,:)`, and `A(:,2:3)`.
(b) What do the following MATLAB statements produce?

```
x = 1:1:10
z = rand(10)
y = [z;x]
c = rand(4)
e = [c eye(size(c)); eye(size(c)) ones(size(c))]
d = sqrt(c)
t1 = d*d
t2 = d.*d
```

1.3. Set up a 4 × 4 matrix. Given that the function `sum(x)` gives the sum of the elements of the vector `x`, use the function `sum` to find the sums of the first row and second column of the matrix.

1.4. Solve the following system of equations using the MATLAB function `inv` and also using the operators \ and / in the **command** window:

$$2x + y + 5z = 5$$
$$2x + 2y + 3z = 7$$
$$x + 3y + 3z = 6$$

Verify the solution is correct using matrix multiplication.

1.5. Write a simple script to input two square matrices **A** and **B**; then add, subtract, and multiply them. Comment the script and use `disp` to output suitable titles.

1.6. Write a MATLAB script to set up a 4 × 4 random matrix `A` and a four-element column vector `b`. Calculate `x = A\b` and display the result. Calculate `A*x` and compare it with `b`.

1.7. Write a simple script to plot the two functions $y_1 = x^2 \cos x$ and $y_2 = x^2 \sin x$ on the same graph. Use comments in your script and take $x = -2 : 0.1 : 2$.

1.8. Write a MATLAB script to produce graphs of the functions $y = \cos x$ and $y = \cos(x^3)$ in the range $x = -4 : 0.02 : 4$ using the same axes. Use the MATLAB functions `xlabel`, `ylabel`, and `title` to annotate your graphs clearly.

1.9. Draw the function $y = \exp(-x^2)\cos(20x)$ in the range $x = -2 : 0.1 : 2$. All axes should be labeled and a title included. Compare the results of using the functions `fplot` and `plot` to plot this function.

1.10. Write a MATLAB script to draw the functions $y = 3\sin(\pi x)$ and $y = \exp(-0.2x)$ on the same graph for $x = 0 : 0.02 : 4$. All axes should be labeled. Use `gtext` to label one of the several points of intersection of the graphs.

1.11. Use the functions `meshgrid` and `mesh` to obtain a three-dimensional plot of the function

$$z = 2xy/(x^2 + y^2) \quad \text{for } x = 1:0.1:3 \text{ and } y = 1:0.1:3$$

Redraw the surface using the function `surf`, `surfl`, and `contour`.

1.12. An iterative equation for solving the equation $x^2 - x - 1 = 0$ is given by

$$x_{r+1} = 1 + (1/x_r) \quad \text{for } r = 0, 1, 2, \ldots$$

Given x_0 is 2 write a MATLAB script to solve the equation. Sufficient accuracy is obtained when $|x_{r+1} - x_r| < 0.0005$. Include a check on the answer.

1.13. Given a 4×5 matrix **A**, write a script to find the sums of each of the columns using:
(a) the `for ... end` construction
(b) the function `sum`

1.14. Given a vector **x** with n elements, write a MATLAB script to form the products

$$p_k = x_1 x_2 \ldots x_{k-1} x_{k+1} \ldots x_n$$

for $k = 1, 2, \ldots, n$. That is, p_k contains the products of all the vector elements except the kth. Run your script with specific values of x and n.

1.15. The series for $\log_e(1 + x)$ is given by

$$\log_e(1 + x) = x - x^2/2 + x^3/3 - \ldots + (-1)^{k+1} x^k/k \ldots$$

Write a MATLAB script to input a value for x and sum the series while the value of the current term is greater than or equal to the variable *tol*. Use values of $x = 0.5$ and 0.82 and $tol =$ 0.005 and 0.0005. The result should be checked by using the MATLAB function `log`. The script should display the value of x and *tol* and the value of $\log_e(1 + x)$ obtained. Use `input` and `disp` functions to obtain clear output and prompts.

1.16. Write a MATLAB script to generate a matrix which has the values d along the main diagonal and the values c on the diagonals above and below the main diagonal and zero elsewhere. Your script should allow the user to input any values for c and d and work for any size of matrix n. The script should give clear prompts for input and display the results with a suitable heading.

1.17. Write a MATLAB function to solve the quadratic equation

$$ax^2 + bx + c = 0$$

The function will use three input parameters a, b, c and output the values of the two roots. You should take account of the three cases:
(a) no real roots
(b) real and different roots
(c) equal roots
Hint: Develop the function `rootquad` given in Section 1.19.

1.18. Adjust the function of Problem 1.17 to deal with the case when $a = 0$. That is, when the equation is non-quadratic. In this case include a third output parameter which will have the value 1 if the equation is quadratic and 0 otherwise.

1.19. Write a simple function to define $f(x) = x^2 - \cos(x) - x$ and plot the graph of the function in the range 0 to 2. Use this graph to find an initial approximation to the root and then apply the function `fzero` to find the root to tolerance 0.0005.

1.20. Write a script to generate the sequence of values given by

$$x_{r+1} = \begin{cases} x_r/2 & \text{if } x_r \text{ is even} \\ 3x_r + 1 & \text{if } x_r \text{ is odd} \end{cases} \quad \text{where } r = 0, 1, 2, \ldots$$

where x_0 is any positive integer. The sequence terminates when $x_r = 1$. Show after a sufficient number of steps that the sequence terminates for any value of x_0 you choose. It is interesting to plot the values of x_r against r.

1.21. Write a MATLAB script to plot the surface $z = f(x, y)$ over the ranges $x = -4 : 0.1 : 4$ and $y = -4 : 0.1 : 4$ where z is given by:

$$z = f(x, y) = (1 - x^2)e^{-p} - pe^{-p} - e^{-(x+1)^2 - y^2}$$

and $p = x^2 + y^2$. The script should provide `mesh`, `contour`, and `surf` plots and use the function `subplot` to layout the three plots one above the other.

1.22. The following three functions are presented in parametric form:

$$x = a(t - \sin(t)) \text{ and } y = a(1 - \cos(t))$$

$$x = 2at \text{ and } y = 2a/(1 + t^2)$$

$$x = a\cos(t) - b\cos(at/b) \text{ and } y = a\sin(t) - b\sin(at/b)$$

Write a MATLAB script to plot each of these functions one above the other using the MATLAB `subplot` function given $a = 2$ and $b = 3$; t is assigned the range of values $-10 : 0.1 : 10$.

1.23. The Riemann ζ function may be defined as the sum of an infinite series:

$$\zeta(s) = 1 + \frac{1}{2^s} + \frac{1}{3^s} + \frac{1}{4^s} + \ldots + \frac{1}{n^s} + \ldots$$

Write a MATLAB script `zetainf(s,acc)` to sum terms of this series until a term is less than `acc` and where `s` is an integer

1.24. Write a MATLAB function to sum the series:

$$S = 1 + 2^2/2! + 3^2/3! + \ldots + n^2/n!$$

to n terms. The function should take the form `sumfac(n)` where `n` is the number of terms used. You may use the MATLAB function `factorial` to evaluate the factorial terms. Write MATLAB statements using this function to sum the series to 5 and 10 terms.
Rewrite your script avoiding the need to use the function `factorial` by noting that the $k + 1$th term T_{k+1} is given by $T_k \times (k + 1)/k^2$.

1.25. Given the matrix `D = [1 -1; 3 2]`, give the values which will be assigned to `A`, `B`, `C`, and `E` by executing the following MATLAB statements:

(a) `A = D*(D*inv(D))`
(b) `B = D.*D`
(c) `C = [D,ones(2);eye(2),zeros(2)]`
(d) `E = D'*ones(2)*eye(2)`

1.26. The following matrices, called the Dirac matrices are defined by:

$$\mathbf{P}_1 = \begin{bmatrix} \mathbf{0} & \mathbf{I}_2 \\ \mathbf{I}_2 & \mathbf{0} \end{bmatrix}, \mathbf{P}_2 = \begin{bmatrix} \mathbf{0} & -\imath\mathbf{I}_2 \\ \imath\mathbf{I}_2 & \mathbf{0} \end{bmatrix}, \mathbf{P}_3 = \begin{bmatrix} \mathbf{I}_2 & \mathbf{0} \\ \mathbf{0} & -\mathbf{I}_2 \end{bmatrix}$$

where the $\mathbf{0}$ represents a 2×2 matrix of zeros and $\mathbf{I}_2$ represents a 2×2 unit matrix and $\imath = \sqrt{(-1)}$. A related set of matrices is given by:

$$\mathbf{Q}_k = \begin{bmatrix} \mathbf{0} & \mathbf{P}_k \\ -\mathbf{P}_k & \mathbf{0} \end{bmatrix} \text{ for } k = 1, 2, 3$$

Write MATLAB statements to generate the matrices $\mathbf{P}_1$, $\mathbf{P}_2$, $\mathbf{P}_3$ and the matrices $\mathbf{Q}_k$ for $k = 1,\ 2,\ 3$. Note that in $\mathbf{Q}_k$ the $\mathbf{0}$ represents a 4×4 matrix of zeros.

1.27. Plot the function

$$y = \frac{1}{(x+2.5)^2(x-3.5)^2}$$

for values of $x = -4 : 0.001 : 4$. Then use the MATLAB functions `xlim` and `ylim` in the form: `ylim([0,20])` and `xlim([-3,-2])` to illustrate how this allows considerable clarification of the nature of the function.

1.28. Write user-defined functions for the following functions:

(a) $$y = x^2 \cos(1 + x^2)$$

(b) $$y = \frac{1 + e^x}{\cos(x) + \sin(x)}$$

(c) $$z = \cos(x^2 + y^2)$$

Rewrite each of the previous functions as anonymous functions and illustrate of the use of these anonymous functions by using the function the MATLAB `subplot` to plot graphs of the functions (a) and (b) in the range for $x = 0$ to 2.

1.29. Consider the following MATLAB script which contains some errors. Use the MATLAB function `checkcode` to find these errors.

```
function sol = solvepoly(x0, acc)
%poly solver
d = 1+acc;
whil abs(d)>acc
    x1 = (2*x0^2-1))/x0^2;
    d = x1-x0;
    x0 = x1/x2
end
sol = x0;
```

1.30. The symmetric hyperbolic Fibonacci sine and cosine functions are defined as follows:

$$\text{sFs}(x) = \frac{\gamma^x - \gamma^{-x}}{\sqrt{5}} \quad \text{and } \text{cFs}(x) = \frac{\gamma^x + \gamma^{-x}}{\sqrt{5}}$$

where $\gamma = (1 + \sqrt{5})/2$. Also, the complex quasi-sine Fibonacci function is defined as

$$\text{cqsF}(x, n) = \frac{\gamma^x - \cos(n\pi x)\gamma^{-x}}{\sqrt{5}} + \imath \frac{\sin(n\pi x)\gamma^{-x}}{\sqrt{5}}$$

where γ is defined as before.
Write a MATLAB script that begins by defining these three functions as anonymous functions. Then, using these anonymous functions, carry out the following operations within the script:
(a) In a single figure, plot the graphs of sFs(x) and cFs(x) against x over the range -5 to 5.
(b) Plot the real and imaginary parts of the function cqsF(x,5) in 3D space. Plot the real part of the function in the y direction, the imaginary part in the z direction. Plot the function over the range -5 to 5. Use the MATLAB function `plot3`.
Stakhov and Rozin (2005, 2007) provide more information on these functions.

1.31. The Fibonacci series is defined by the equations:

$$F_{k+1} = F_k + F_{k-1}, \text{ for } k = 2, 3, 4, ..., n$$

where $F_2 = F_1 = 1$.
Write a MATLAB recursive function to evaluate the elements of this series to any number of steps n. Use this function to produce a table of the values of the Fibonacci series, F_k for $k = 1, 2, ..., 10$.

1.32. Use the MATLAB function `fimplicit` to provide a plot of the two-dimensional implicit function:

$$\cos(x^2 + y^2) + xy = 0$$

for values of x and y in the range $x = -4$ to 4; $y = -4$ to 4.

1.33. For the three-dimensional implicit function

$$\sin(x^2 + y^2 + z^2) - x^2y^2z^2 = 0$$

use the MATLAB function `fimplicit3` to plot this function for x and y in the range $x = -4$ to 4; $y = -4$ to 4.

1.34. For the three-dimensional implicit function

$$(x^2 + y^2 + z^2) - 6xyz = 0$$

use the MATLAB function `fimplicit3` to plot this function for x and y in the range $x = -4$ to 4; $y = -4$ to 4.

1.35. Use the MATLAB polar scatter function to plot the data given by:

$$r = \cos(\theta)\sin(\theta) \text{ for } \theta = -2\pi \text{ to } 2\pi$$

using the MATLAB `polarscatter` function. Replot this data again but introducing extra parameter values so that the face color is green and the size of the points is 200.

1.36. Draw a polar histogram figure for the data:

$$x_i = 2\pi r_i \quad i = 1, 2, 3, ..., 100$$

where r_i is a random number in the range 0 to 1. Using the MATLAB `polarhistogram` function you should use 15 bins and the necessary parameters to provide a face color red and transparency of 0.5.

1.37. Write a MATLAB function to calculate the values of $f(x)$ where

$$f(x) = x^2 \sin(x^2)$$

for any vector x. In addition write a function which takes the vector x and finds the element with the minimum value, x_{min}, and the element with the maximum value, x_{max}, and returns the value of the scalar r where:

$$r = (x_{max} - x_{min})/(x_{max} + x_{min})$$

Write a MATLAB script which will obtain the values of the function, $f(x)$, for $x = 0$ to 10 in increments 0.1 and plot a graph of this function. Then use the function you have written to calculate the value of r. Save the MATLAB script together with the functions using the name `functiontest` and run it.

1.38. Use the function `fimplicit3` to draw the implicit function

$$(x^2 + y^2 - 3x)^2 - 4x^2(2 - x) = 0$$

Use the ranges of values for x and y as 0 to 2, and -3 to 3 respectively. Label the x and y axes appropriately and give the graph a title.

1.39. Write a MATLAB script to determine the sum of the series

$$S = (1/e)[1 + 1/(3 \times 1!) + 1/(5 \times 2!) + 1/(7 \times 3!) + ...\infty]$$

to a prescribed accuracy. Take sufficient terms to obtain 6 decimal places of accuracy. The sum to infinity is equal to the value of the integral given in Problem 4.35. See Jolley (1961), Eq. (1004).

1.40. Determine the output from executing the MATLAB command `why`. You should execute this command more than once.
When you have become bored with this, try `logo`, `spy`, `penny`, and `image`.

CHAPTER

LINEAR EQUATIONS AND EIGENSYSTEMS

2

Abstract

When physical systems are modeled mathematically, they are sometimes described by linear equation systems or alternatively by eigensystems and in this chapter we examine how such equation systems are solved.

MATLAB is an ideal environment for studying linear algebra, including linear equation systems and eigenvalue problems, because MATLAB functions and operators can work directly on vectors and matrices. MATLAB is rich in functions and operators which facilitate the manipulation of matrices.

2.1 INTRODUCTION

We now discuss linear equation systems and defer discussion of eigensystems until Section 2.15. To illustrate how linear equation systems arise in the modeling of certain physical problems, we will consider how current flows are calculated in a simple electrical network. The necessary equations can be developed using one of several techniques; here we use the loop-current method together with Ohm's law and Kirchhoff's voltage law. A *loop current* is assumed to circulate around each loop in the network. Thus, in the network given in Fig. 2.1, the loop current I_1 circulates around the closed loop *abcd*. Thus the current $I_1 - I_2$ flows in the link connecting b to c. Ohm's law states that the voltage across an ideal resistor is proportional to the current flow through the resistor. For example, for the link connecting b to c

$$V_{bc} = R_2(I_1 - I_2)$$

where R_2 is the value of the resistor in the link connecting b to c. Kirchhoff's voltage law states that the algebraic sum of the voltages around a loop is zero. Applying these laws to the circuit *abcd* of Fig. 2.1 we have

$$V_{ab} + V_{bc} + V_{cd} = V$$

Substituting the product of current and resistance for voltage gives

$$R_1 I_1 + R_2(I_1 - I_2) + R_4 I_1 = V$$

Numerical Methods. https://doi.org/10.1016/B978-0-12-812256-3.00011-7

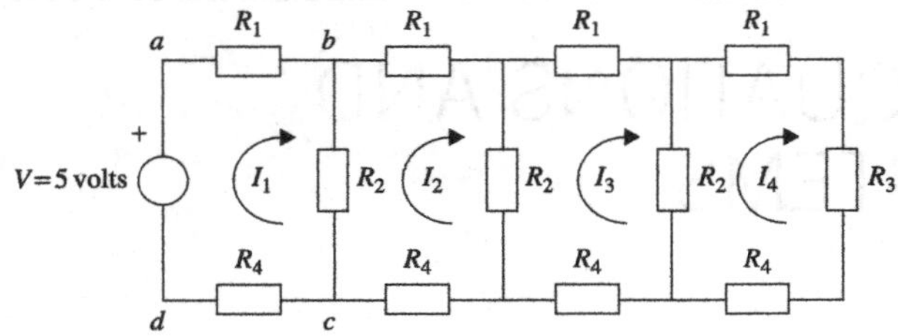

FIGURE 2.1

Electrical network.

We can repeat this process for each loop to obtain the following four equations:

$$
\begin{aligned}
(R_1 + R_2 + R_4)\, I_1 - R_2 I_2 &= V \\
(R_1 + 2R_2 + R_4)\, I_2 - R_2 I_1 - R_2 I_3 &= 0 \\
(R_1 + 2R_2 + R_4)\, I_3 - R_2 I_2 - R_2 I_4 &= 0 \\
(R_1 + R_2 + R_3 + R_4)\, I_4 - R_2 I_3 &= 0
\end{aligned}
\tag{2.1}
$$

Letting $R_1 = R_4 = 1\ \Omega$, $R_2 = 2\ \Omega$, $R_3 = 4\ \Omega$ and $V = 5$ volts, (2.1) becomes

$$
\begin{aligned}
4I_1 - 2I_2 &= 5 \\
-2I_1 + 6I_2 - 2I_3 &= 0 \\
-2I_2 + 6I_3 - 2I_4 &= 0 \\
-2I_3 + 8I_4 &= 0
\end{aligned}
$$

This is a system of linear equations in four variables, $I_1, \dots, I_4$. In matrix notation it becomes

$$
\begin{bmatrix} 4 & -2 & 0 & 0 \\ -2 & 6 & -2 & 0 \\ 0 & -2 & 6 & -2 \\ 0 & 0 & -2 & 8 \end{bmatrix}
\begin{bmatrix} I_1 \\ I_2 \\ I_3 \\ I_4 \end{bmatrix}
=
\begin{bmatrix} 5 \\ 0 \\ 0 \\ 0 \end{bmatrix}
\tag{2.2}
$$

This equation has the form $\mathbf{Ax} = \mathbf{b}$ where $\mathbf{A}$ is a square matrix of known coefficients, in this case relating to the values of the resistors in the circuit. The vector $\mathbf{b}$ is a vector of known coefficients, in this case the voltage applied to each current loop. The vector $\mathbf{x}$ is the vector of unknown currents. Although this set of equations can be solved by hand, the process is time consuming and error prone. Using MATLAB we simply enter matrix `A` and vector `b` and use the command `A\b` as follows:

```
>> A = [4 -2 0 0;-2 6 -2 0;0 -2 6 -2;0 0 -2 8];
>> b = [5 0 0 0].';
>> x = A\b
```

```
x =
    1.5426
    0.5851
    0.2128
    0.0532

>> A*x

ans =
    5.0000
   -0.0000
    0.0000
   -0.0000
```

The sequence of operations that are invoked by this apparently simple command is examined in Section 2.3.

In many electrical networks the ideal resistors of Fig. 2.1 are more accurately represented by electrical impedances. When a harmonic alternating current (AC) supply is connected to the network, electrical engineers represent the impedances by complex quantities. This is to account for the effect of capacitance and/or inductance. To illustrate this we will replace the 5 volt DC supply to the network of Fig. 2.1 by a 5 volt AC supply and replace the ideal resistors $R_1, \ldots, R_4$ by impedances $Z_1, \ldots, Z_4$. Thus, (2.1) becomes

$$\begin{aligned}(Z_1+Z_2+Z_4)I_1-Z_2I_2&=V\\(Z_1+2Z_2+Z_4)I_2-Z_2I_1-Z_2I_3&=0\\(Z_1+2Z_2+Z_4)I_3-Z_2I_2-Z_2I_4&=0\\(Z_1+Z_2+Z_3+Z_4)I_4-Z_2I_3&=0\end{aligned} \tag{2.3}$$

At the frequency of the 5 volt AC supply we will assume that $Z_1 = Z_4 = (1+0.5\jmath)\ \Omega$, $Z_2 = (2+0.5\jmath)\ \Omega$, $Z_3 = (4+1\jmath)\ \Omega$ where $\jmath = \sqrt{-1}$. Electrical engineers prefer to use $\jmath$ rather than $\imath$ for $\sqrt{-1}$. This avoids any possible confusion with i (or I) which is normally used to denote the current in a circuit. Thus, (2.3) becomes

$$\begin{aligned}(4+1.5\jmath)I_1-(2+0.5\jmath)I_2&=5\\-(2+0.5\jmath)I_1+(6+2.0\jmath)I_2-(2+0.5\jmath)I_3&=0\\-(2+0.5\jmath)I_2+(6+2.0\jmath)I_3-(2+0.5\jmath)I_4&=0\\-(2+0.5\jmath)I_3+(8+2.5\jmath)I_4&=0\end{aligned}.$$

This system of linear equations becomes, in matrix notation,

$$\begin{bmatrix}(4+1.5\jmath) & -(2+0.5\jmath) & 0 & 0\\-(2+0.5\jmath) & (6+2.0\jmath) & -(2+0.5\jmath) & 0\\0 & -(2+0.5\jmath) & (6+2.0\jmath) & -(2+0.5\jmath)\\0 & 0 & -(2+0.5\jmath) & (8+2.5\jmath)\end{bmatrix}\begin{bmatrix}I_1\\I_2\\I_3\\I_4\end{bmatrix}=\begin{bmatrix}5\\0\\0\\0\end{bmatrix} \tag{2.4}$$

Note that the coefficient matrix is now complex. This does not present any difficulty for MATLAB because the operation `A\b` works directly with both real and complex numbers. Thus:

```
>> p = 4+1.5i; q = -2-0.5i;
>> r = 6+2i; s = 8+2.5i;
>> A = [p q 0 0;q r q 0;0 q r q;0 0 q s];
>> b = [5 0 0 0].';
>> A\b

ans =
   1.3008 - 0.5560i
   0.4560 - 0.2504i
   0.1530 - 0.1026i
   0.0361 - 0.0274i
```

Note that strictly we have no need to re-enter the values in vector `b`, assuming that we have not cleared the memory, reassigned the vector `b` or quit MATLAB. The answer shows that currents flowing in the network are complex. This means that there is a phase difference between the applied harmonic voltage and the currents flowing.

We will now begin a more detailed examination of linear equation systems.

2.2 LINEAR EQUATION SYSTEMS

In general, a linear equation system can be written in matrix form as

$$\mathbf{Ax} = \mathbf{b} \tag{2.5}$$

where $\mathbf{A}$ is an $n \times n$ matrix of known coefficients, $\mathbf{b}$ is a column vector of n known coefficients and $\mathbf{x}$ is the column vector of n unknowns. We have already seen an example of this type of equation system in Section 2.1 where the matrix equation (2.2) is the matrix equivalent of the set of linear equations (2.1).

The equation system (2.5) is called homogeneous if $\mathbf{b} = \mathbf{0}$ and inhomogeneous if $\mathbf{b} \neq \mathbf{0}$. Before attempting to solve an equation system it is reasonable to ask if it has a solution and if so is it unique? A linear inhomogeneous equation system may be *consistent* and have one or an infinity of solutions or be *inconsistent* and have no solution. This is illustrated in Fig. 2.2 for a system of three equations in three variables x_1, x_2, and x_3. Each equation represents a plane surface in the x_1, x_2, x_3 space. In Fig. 2.2A the three planes have a common point of intersection. The coordinates of the point of intersection give the unique solution for the three equations. In Fig. 2.2B the three planes intersect in a line. Any point on the line of intersection represents a solution so there is no unique solution but an infinite number of solutions satisfying the three equations. In Fig. 2.2C two of the surfaces are parallel to each other and therefore they never intersect while in Fig. 2.2D the line of intersection of each pair of surfaces is different. In both of these cases there is no solution and the equations these surfaces represent are inconsistent.

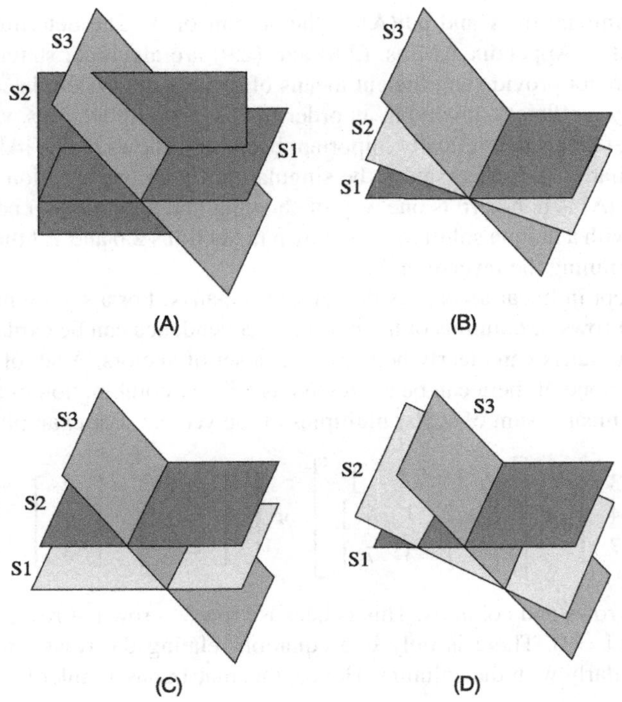

FIGURE 2.2

Three intersecting planes representing three equations in three variables. (A) Three plane surfaces intersecting in a point. (B) Three plane surfaces intersecting in a line. (C) Three plane surfaces, two of which do not intersect. (D) Three plane surfaces intersecting in three lines.

To obtain an algebraic solution to the inhomogeneous equation system (2.5) we multiply both sides of (2.5) by a matrix called the inverse of **A**, denoted by $\mathbf{A}^{-1}$, thus:

$$\mathbf{A}^{-1}\mathbf{A}\mathbf{x} = \mathbf{A}^{-1}\mathbf{b} \tag{2.6}$$

where $\mathbf{A}^{-1}$ is defined by

$$\mathbf{A}^{-1}\mathbf{A} = \mathbf{A}\mathbf{A}^{-1} = \mathbf{I} \tag{2.7}$$

and **I** is the identity matrix. Thus, we obtain

$$\mathbf{x} = \mathbf{A}^{-1}\mathbf{b} \tag{2.8}$$

The standard algebraic formula for the inverse of **A** is

$$\mathbf{A}^{-1} = \mathrm{adj}(\mathbf{A})/|\mathbf{A}| \tag{2.9}$$

where $|\mathbf{A}|$ is the determinant of $\mathbf{A}$ and $\text{adj}(\mathbf{A})$ is the adjoint of $\mathbf{A}$. The determinant and the adjoint of a matrix are defined in Appendix A. Eqs. (2.8) and (2.9) are algebraic statements allowing us to determine $\mathbf{x}$ but they do not provide an efficient means of solving the system because computing $\mathbf{A}^{-1}$ using (2.9) is extremely inefficient, involving an order $(n+1)!$ multiplications where n is the number of equations. However, (2.9) is theoretically important because it shows that if $|\mathbf{A}| = 0$ then $\mathbf{A}$ does not have an inverse. The matrix $\mathbf{A}$ is then said to be singular and a unique solution for $\mathbf{x}$ does not exist. Thus, establishing that $|\mathbf{A}|$ is non-zero is one way of showing that an inhomogeneous equation system is a consistent system with a unique solution. It is shown in Sections 2.6 and 2.7 that (2.5) can be solved without formally determining the inverse of $\mathbf{A}$.

An important concept in linear algebra is the rank of a matrix. For a square matrix, the rank is the number of independent rows or columns of the matrix. Independence can be explained as follows. The rows (or columns) of a matrix can clearly be viewed as a set of vectors. A set of vectors is said to be linearly independent if none of them can be expressed as a linear combination of any of the others. By linear combination we mean a sum of scalar multiples of the vectors. For example, the matrix

$$\begin{bmatrix} 1 & 2 & 3 \\ -2 & 1 & 4 \\ -1 & 3 & 7 \end{bmatrix} \text{ or } \begin{bmatrix} [\,1 \quad 2 \quad 3\,] \\ [\,-2 \quad 1 \quad 4\,] \\ [\,-1 \quad 3 \quad 7\,] \end{bmatrix} \text{ or } \left[\begin{bmatrix} 1 \\ -2 \\ -1 \end{bmatrix} \begin{bmatrix} 2 \\ 1 \\ 3 \end{bmatrix} \begin{bmatrix} 3 \\ 4 \\ 7 \end{bmatrix} \right]$$

has linearly *dependent* rows and columns. This is because row 3 − row 1 − row 2 = 0 and column 3 − 2(column 2) + column 1 = 0. There is only one equation relating the rows and thus there are two independent rows. Similarly with the columns. Hence, this matrix has a rank of 2. Now consider

$$\begin{bmatrix} 1 & 2 & 3 \\ 2 & 4 & 6 \\ 3 & 6 & 9 \end{bmatrix}$$

Here row 2 = 2(row 1) and row 3 = 3(row 1). There are two equations relating the rows and hence only one row is independent so that the matrix has a rank of 1. Note that the number of independent rows and columns in a square matrix is identical; that is, its row rank and column rank are equal. In general, matrices may be non-square and the rank of an $m \times n$ matrix $\mathbf{A}$ is written $\text{rank}(\mathbf{A})$. Matrix $\mathbf{A}$ is said to be of full rank if $\text{rank}(\mathbf{A}) = \min(m, n)$; otherwise $\text{rank}(\mathbf{A}) < \min(m, n)$ and $\mathbf{A}$ is said to be rank deficient.

MATLAB provides the function `rank` which works with both square and non-square matrices to find the rank of a matrix. In practice, MATLAB determines the rank of a matrix from its singular values; see Section 2.10. For example, consider the following MATLAB statements:

```
>> D = [1 2 3;3 4 7;4 -3 1;-2 5 3;1 -7 6]

D =

     1     2     3
     3     4     7
     4    -3     1
    -2     5     3
     1    -7     6
```

```
>> rank(D)

ans =
     3
```

Thus D is of full rank, since its rank is equal to the minimum number of rows or column of D.

A useful operation in linear algebra is the conversion of a matrix to its reduced row echelon form (RREF). The RREF is defined in Appendix A. In MATLAB we can use the rref function to compute the RREF of a matrix thus:

```
>> rref(D)

ans =
     1     0     0
     0     1     0
     0     0     1
     0     0     0
     0     0     0
```

It is a property of the RREF of a matrix that the number of rows with at least one non-zero element equals the rank of the matrix. In this example we see that there are three rows in the RREF of the matrix containing a non-zero element, confirming that the matrix rank is 3. The RREF also allows us to determine whether a system has a unique solution or not.

We have discussed a number of important concepts relating to the nature of linear equations and their solutions. We now summarize the equivalencies between these concepts. Let $\mathbf{A}$ be an $n \times n$ matrix. If $\mathbf{Ax} = \mathbf{b}$ is consistent and has a unique solution, then all of the following statements are true:

$\mathbf{Ax} = \mathbf{0}$ has only the trivial solution $\mathbf{x} = \mathbf{0}$.
$\mathbf{A}$ is non-singular and $\det(\mathbf{A}) \neq 0$.
The RREF of $\mathbf{A}$ is the identity matrix.
$\mathbf{A}$ has n linearly independent rows and columns.
$\mathbf{A}$ has full rank, that is $\text{rank}(\mathbf{A}) = n$.

In contrast, if $\mathbf{Ax} = \mathbf{b}$ is either inconsistent or consistent but with more than one solution, then all of the following statements are true:

$\mathbf{Ax} = \mathbf{0}$ has more than one solution.
$\mathbf{A}$ is singular and $\det(\mathbf{A}) = 0$.
The RREF of $\mathbf{A}$ contains at least one zero row.
$\mathbf{A}$ has linearly dependent rows and columns.
$\mathbf{A}$ is rank deficient, that is $\text{rank}(\mathbf{A}) < n$.

So far we have only considered the case where there are as many equations as unknowns. Now we consider the cases where there are fewer or more equations than the number of unknown variables.

If there are fewer equations than unknowns, then the system is said to be under-determined. The equation system does not have a unique solution; it is either consistent with an infinity of solutions, or inconsistent with no solution. These conditions are illustrated by Fig. 2.3. The diagram shows two plane surfaces in three-dimensional space, representing two equations in three variables. It is seen that

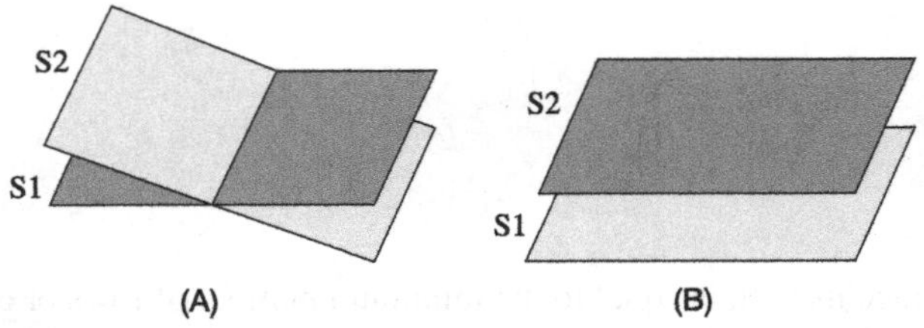

FIGURE 2.3

Planes representing an under-determined system of equations. (A) Two plane surfaces intersecting in a line. (B) Two plane surfaces which do not intersect.

the planes either intersect in a line so that the equations are consistent with an infinity of solutions represented by the line of intersection, or the surfaces do not intersect and the equations they represent are inconsistent.

Consider the following system of equations:

$$\begin{bmatrix} 1 & 2 & 3 & 4 \\ -4 & 2 & -3 & 7 \end{bmatrix} \begin{bmatrix} x_1 \\ x_2 \\ x_3 \\ x_4 \end{bmatrix} = \begin{bmatrix} 1 \\ 3 \end{bmatrix}$$

This under-determined system can be rearranged thus:

$$\begin{bmatrix} 1 & 2 \\ -4 & 2 \end{bmatrix} \begin{bmatrix} x_1 \\ x_2 \end{bmatrix} + \begin{bmatrix} 3 & 4 \\ -3 & 7 \end{bmatrix} \begin{bmatrix} x_3 \\ x_4 \end{bmatrix} = \begin{bmatrix} 1 \\ 3 \end{bmatrix}$$

or

$$\begin{bmatrix} 1 & 2 \\ -4 & 2 \end{bmatrix} \begin{bmatrix} x_1 \\ x_2 \end{bmatrix} = \begin{bmatrix} 1 \\ 3 \end{bmatrix} - \begin{bmatrix} 3 & 4 \\ -3 & 7 \end{bmatrix} \begin{bmatrix} x_3 \\ x_4 \end{bmatrix}$$

Thus, we have reduced this to a system of two equations in two unknowns, provided values are assumed for x_3 and x_4. Since x_3 and x_4 can take an infinity of values, the problem has an infinity of solutions.

If a system has more equations than unknowns, then the system is said to be over-determined. Fig. 2.4 shows four plane surfaces in three-dimensional space, representing four equations in three variables. Fig. 2.4A shows all four planes intersecting in a single point so that the system of equations is consistent with a unique solution. Fig. 2.4B shows all the planes intersecting in a line and this represents a consistent system with an infinity of solutions. Fig. 2.4D shows planes that represent an inconsistent system of equations with no solution. In Fig. 2.4C the planes do not intersect in a single point and so the system of equations is inconsistent. However, in this example the points of intersection of groups of three planes, (S1, S2, S3), (S1, S2, S4), (S1, S3, S4), and (S2, S3, S4), are close to each other and a mean point of intersection could be determined and used as an approximate solution. This example of marginal inconsistency often arises because the coefficients in the equations are determined experimentally; if the coefficients were known exactly, it is likely that the equations would be consistent with a unique solution. Rather than accepting that the system is inconsistent, we

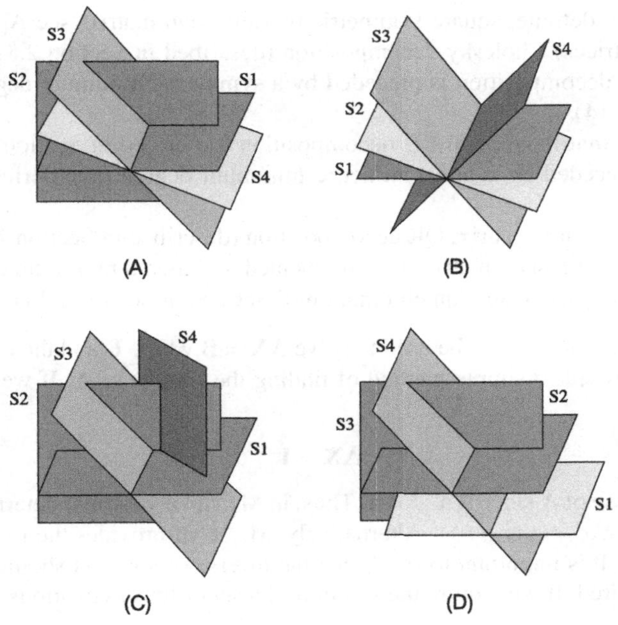

FIGURE 2.4

Planes representing an over-determined system of equations. (A) Four plane surfaces intersecting in a point. (B) Four plane surfaces intersecting in a line. (C) Four plane surfaces not intersecting at a single point. Point of intersection of (S1, S2, S3) and (S1, S2, S4) are visible. (D) Four plane surfaces representing inconsistent equations.

may ask which solution best satisfies the equations approximately. In Sections 2.11 and 2.12 we deal with the problem of over-determined and under-determined systems in more detail.

2.3 OPERATORS \ AND / FOR SOLVING Ax = b

The purpose of this section is to introduce the reader to the MATLAB operator `\`. A detailed discussion of the algorithms behind its operation will be given in later sections. This operator is a very powerful one which provides a unified approach to the solution of many categories of linear equation systems. The operators `/` and `\` perform matrix "division" and have identical effects. Thus, to solve $\mathbf{Ax} = \mathbf{b}$ we may write either `x=A\b` or `x'=b'/A'`. In the latter case the solution $\mathbf{x}$ is expressed as a row rather than a column vector. The algorithm chosen to solve a system of linear equations is dependent on the form of the matrix $\mathbf{A}$. These cases are outlined as follows:

- if $\mathbf{A}$ is a triangular matrix, the system is solved by back or forward substitution alone, described in Section 2.6.

- elseif **A** is a positive definite, square symmetric or Hermitian matrix (see Appendix A.7 for definitions of these matrices) Cholesky decomposition (described in Section 2.8) is applied. When **A** is sparse, Cholesky decomposition is preceded by a symmetric minimum degree preordering (described in Section 2.14).
- elseif **A** is a square matrix, general LU decomposition (described in Section 2.7) is applied. If **A** is sparse, this is preceded by a non-symmetric minimum degree preordering (described in Section 2.14).
- elseif **A** is a full non-square matrix, QR decomposition (described in Section 2.9) is applied.
- elseif **A** is a sparse non-square matrix, it is augmented and then a minimum degree preordering is applied, followed by sparse Gaussian elimination (described in Section 2.14).

The MATLAB \ operator can also be used to solve $\mathbf{AX} = \mathbf{B}$ where **B** and the unknown **X** are $m \times n$ matrices. This could provide a simple method of finding the inverse of **A**. If we make **B** the identity matrix **I** then we have

$$\mathbf{AX} = \mathbf{I}$$

and **X** must be the inverse of **A** since $\mathbf{AA}^{-1} = \mathbf{I}$. Thus, in MATLAB we could determine the inverse of **A** by using the statement `A\eye(size(A))`. Alternatively, MATLAB provides the function `inv(A)` to find the inverse of a matrix. It is important to stress that the inverse of a matrix should only be determined if it is specifically required. If we require the solution of a set of linear equations it is more efficient to use the operators \ or /.

We now give some example which show how the \ operator works, beginning with the solution of a system where the system matrix is triangular. The experiment in this case examines the time taken by the operator \ to solve a system when it is full and then when the same system is converted to triangular form by zeroing appropriate elements to produce a triangular matrix. The script `e4s201`, used for this experiment, is

```
% e4s201.m
disp('   n    full-time  full-time/n^3   tri-time  tri-time/n^2');
A = [ ]; b = [ ];
for n = 2000:500:6000
    A = 100*rand(n); b = [1:n].';
    tic, x = A\b; t1 = toc;
    t1n = 5e9*t1/n^3;
    for i = 1:n
        for j = i+1:n
            A(i,j) = 0;
        end
    end
    tic, x = A\b; t2 = toc;
    t2n = 1e9*t2/n^2;
    fprintf('%6.0f %9.4f %12.4f %12.4f %11.4f\n',n,t1,t1n,t2,t2n)
end
```

The results for a series of randomly generated $n \times n$ matrices are as follows:

```
  n    full-time  full-time/n^3   tri-time  tri-time/n^2
2000    0.3845       0.2403        0.0080     2.0041
2500    0.7833       0.2507        0.0089     1.4182
3000    0.9172       0.1698        0.0122     1.3560
3500    1.3808       0.1610        0.0178     1.4504
4000    2.0325       0.1588        0.0243     1.5215
4500    2.4460       0.1342        0.0277     1.3696
5000    3.5037       0.1401        0.0344     1.3741
5500    4.7222       0.1419        0.0394     1.3016
6000    5.9742       0.1383        0.0556     1.5444
```

Column 1 of this table gives n, the size of the square matrix. Columns 2 and 4 give the times taken to solve the full system and the triangular system respectively. Note that the triangular times are much smaller. Column 3 gives the time divided by n^3 and shows that the time take is closely proportional to n^3, whereas Column 5 is multiplied by 10^9 and divided by n^2. We see that the time required to solve a triangular system is approximately proportional to n^2.

We now perform experiments to examine the effects of using the operator \ with positive definite symmetric systems. This is a more complex problem than those previously discussed and the script `e4s202.m` implements this test. It is based on comparing the application of the \ operator to a positive definite system and a non-positive definite system of equations. We can create a positive define matrix by letting `A = M*M'` where `M` is any matrix, but in this case `M` will be a matrix of random numbers, `A` will then be a positive define matrix. To generate a non-positive definite system we add a random matrix to the positive define matrix and we compare the time required to solve the two forms of matrix. The script for `e4s202.m` is of the form

```
% e4s202.m
disp(' n       time-pos    time-pos/n^3  time-npos    time-b/n^3');
for n = 500:500:5000
    A = [ ]; M = 100*randn(n,n);
    A = M*M'; b = [1:n].';
    tic, x = A\b; t1 = toc*1000;
    t1d = t1/n^3;
    A = A+rand(size(A));
    tic, x = A\b; t2 = toc*1000;
    t2d = t2/n^3;
    fprintf('%4.0f %10.4f %14.4e %11.4f %13.4e\n',n,t1,t1d,t2,t2d)
end
```

The result of running this script is

```
 n       time-pos   time-pos/n^3  time-npos    time-b/n^3
 500      9.9106     7.9285e-08      10.6241    8.4993e-08
1000     46.7403     4.6740e-08      60.0436    6.0044e-08
1500    122.6701     3.6347e-08     164.9225    4.8866e-08
2000    191.2798     2.3910e-08     334.4048    4.1801e-08
2500    374.3233     2.3957e-08     736.9797    4.7167e-08
```

```
3000    619.7430     2.2953e-08     974.7610     3.6102e-08
3500    769.8043     1.7955e-08    1354.9658     3.1603e-08
4000   1040.6866     1.6261e-08    1976.1037     3.0877e-08
4500   1474.6506     1.6183e-08    2518.8017     2.7641e-08
5000   1934.6456     1.5477e-08    3437.0971     2.7497e-08
```

Column 1 of this table gives n, the size of the matrix. Column 2 gives the time taken multiplied by 1000 for the positive definite matrix and column 4 gives the time taken, again multiplied by 1000, for the non-positive definite matrix. These results show that the time taken to determine the solution for the system is somewhat faster for the positive definite system. This is because the operator \ checks to see if the matrix is positive definite and if so uses the more efficient Cholesky decomposition. Columns 3 and 5 give the times divided by the size of the matrix cubed to illustrate that the processing times are approximately proportional to n^3.

The next test we perform examines how the operator \ succeeds with the very badly conditioned Hilbert matrix. The test gives the time taken to solve the system and the accuracy of the solution given by the Euclidean norm of the residuals, that is norm($\mathbf{Ax} - \mathbf{b}$). For the definition of the norm see Appendix A, Section A.10. In addition the test compares these results for the \ operator with the results obtained using the inverse, that is $\mathbf{x} = \mathbf{A}^{-1}\mathbf{b}$. The script `e4s203.m` performs this test.

```
% e4s203.m
disp(' n  time-slash acc-slash   time-inv   acc-inv   condition');
for n = 4:2:20
    A = hilb(n); b = [1:n].';
    tic, x = A\b; t1 = toc; t1 = t1*10000;
    nm1 = norm(b-A*x);
    tic, x = inv(A)*b; t2 = toc; t2 = t2*10000;
    nm2 = norm(b-A*x);
    c = cond(A);
    fprintf('%2.0f %10.4f %10.2e %8.4f %11.2e %11.2e \n',n,t1,nm1,t2,nm2,c)
end
```

Script `e4s203.m` produces the following table of results:

```
 n  time-slash acc-slash  time-inv   acc-inv   condition
 4     0.8389   4.11e-14  0.6961    8.89e-13   1.55e+04
 6     0.4730   2.89e-12  0.6649    3.74e-10   1.50e+07
 8     0.4462   1.89e-10  0.7229    5.60e-07   1.53e+10
10     0.4641   9.81e-09  0.8479    2.97e-04   1.60e+13
12    10.2279   5.66e-07  9.4157    3.42e-01   1.62e+16
14     9.0498   1.33e-05  8.5991    5.33e+01   2.55e+17
16     8.7687   4.57e-05  8.6035    1.01e+03   4.89e+17
18     8.7776   2.25e-05  8.6660    1.92e+02   1.35e+18
20     9.5094   6.91e-05  8.9784    7.04e+01   2.11e+18
```

This output has been edited to remove warnings about the ill-conditioning of the matrix for $n >= 10$. Column 1 gives the size of the matrix. Columns 2 and 3 give the time taken multiplied by 10,000 and the accuracy when using the \ operator, respectively. Columns 4 and 5 give the same information when using the `inv` function. Column 6 gives the condition number of the system matrix. When the condition number is large, the matrix is nearly singular and the equations are ill-conditioned. Ill-conditioning is fully described in Section 2.4.

The results in the preceding table demonstrate convincingly the superiority of the \ operator over the `inv` function for solving a system of linear equations. It is considerably more accurate than using matrix inversion. However, it should be noted that the accuracy falls off as the matrix becomes increasingly ill-conditioned.

The MATLAB operator \ can also be used to solve under- and over-determined systems. In this case the operator \ uses a least squares approximation, discussed in detail in Section 2.12.

2.4 ACCURACY OF SOLUTIONS AND ILL-CONDITIONING

We now consider factors which affect the accuracy of the solution of $\mathbf{Ax} = \mathbf{b}$ and how any inaccuracies can be detected. A further discussion on the accuracy of the solution of this equation system is given in Appendix B, Section B.3. We begin with the following examples.

Example 2.1. Consider the following MATLAB statements:

```
>> A = [3.021 2.714 6.913;1.031 -4.273 1.121;5.084 -5.832 9.155]

A =
    3.0210    2.7140    6.9130
    1.0310   -4.2730    1.1210
    5.0840   -5.8320    9.1550

>> b = [12.648 -2.121 8.407].'

b =
   12.6480
   -2.1210
    8.4070

>> A\b

ans =
    1.0000
    1.0000
    1.0000
```

This result is correct and easily verified by substitution into the original equations.

Example 2.2. This example is obtained from Example 2.1 with `A(2,2)` changed from −4.2730 to −4.2750 thus:

```
>> A(2,2) = -4.2750

A =
    3.0210    2.7140    6.9130
    1.0310   -4.2750    1.1210
    5.0840   -5.8320    9.1550

>> A\b

ans =
   -1.7403
    0.6851
    2.3212
```

Here we have a solution which is very different from that of Example 2.1, even though the only change in the equation system is less than 0.1% in coefficient `a(2,2)`.

The two examples just shown have dramatically different solutions because the coefficient matrix **A** is ill-conditioned. Ill-conditioning can be interpreted graphically by representing each of the equation systems by three plane surfaces, in the manner shown in Fig. 2.2. In an ill-conditioned system at least two of the surfaces will be almost parallel so that the point of intersection of the surfaces will be very sensitive to small changes in slope, caused by small changes in coefficient values.

A system of equations is said to be ill-conditioned if a relatively small change in the elements of the coefficient matrix **A** causes a relatively large change in the solution. Conversely, a system of equations is said to be well-conditioned if a relatively small change in the elements of the coefficient matrix **A** causes a relatively small change in the solution. Clearly, we require a measure of the condition of a system of equations. We know that a system of equations without a solution – the very worst condition possible – has a coefficient matrix with a determinant of zero. It is therefore tempting to think that the size of the determinant of **A** can be used as a measure of condition. However, if $\mathbf{Ax} = \mathbf{b}$ and **A** is an $n \times n$ diagonal matrix with each element on the leading diagonal equal to s, then **A** is perfectly conditioned, regardless of the value of s. But the determinant of **A** in this case is s^n. Thus, the size of the determinant of **A** is not a suitable measure of condition because in this example it changes with s even though the condition of the system is constant.

Two of the functions MATLAB provides to estimate the condition of a matrix are `cond` and `rcond`. The function `cond` is a sophisticated function and is based on singular value decomposition, discussed in Section 2.10. For a perfect condition `cond` is unity but gives a large value for a matrix which is ill-conditioned. The function `rcond` is less reliable but usually faster. This function gives a value between zero and one. The smaller the value, the worse the conditioning. The reciprocal of `rcond` is usually of the same order of magnitude as `cond`. We now illustrate these points with two examples.

Example 2.3. Illustration of a perfectly conditioned system:

```
>> A = diag([20 20 20])
A =

    20     0     0
     0    20     0
     0     0    20

>> [det(A) rcond(A) cond(A)]

ans =
        8000            1            1
```

Example 2.4. Illustration of a badly conditioned system:

```
>> A = [1 2 3;4 5 6;7 8 9.000001];
>> format short e
>> [det(A) rcond(A) 1/rcond(A) cond(A)]

ans =
 -3.0000e-06   6.9444e-09   1.4400e+08   1.0109e+08
```

Note that the reciprocal of the `rcond` value is close to the value of `cond`. Using the MATLAB functions `cond` and `rcond` we now investigate the condition number of the Hilbert matrix (defined in Problem 2.1), using the script `e4s204.m`:

```
% e4s204.m Hilbert matrix test.
disp('    n             cond             rcond        log10(cond)')
for n = 4:2:20
    A = hilb(n);
    fprintf('%5.0f %16.4e',n,cond(A));
    fprintf('%16.4e %10.2f\n',rcond(A),log10(cond(A)));
end
```

Running script `e4s204.m` gives

```
    n          cond               rcond       log10(cond)
    n          cond               rcond       log10(cond)
    4       1.5514e+04       3.5242e-05          4.19
    6       1.4951e+07       3.4399e-08          7.17
    8       1.5258e+10       2.9522e-11         10.18
   10       1.6025e+13       2.8285e-14         13.20
   12       1.6212e+16       2.5334e-17         16.21
   14       2.5515e+17       6.9286e-19         17.41
   16       4.8875e+17       1.3275e-18         17.69
   18       1.3500e+18       2.5084e-19         18.13
   20       2.1065e+18       1.2761e-19         18.32
```

This shows that the Hilbert matrix is ill-conditioned even for relatively small values of n, the size of the matrix. The last column of the output of `e4s204.m` gives the value of $\log_{10}$ of the condition number of the appropriate Hilbert matrix. This gives a rule of thumb estimate of the number of significant figures lost in solving an equation system with this matrix or inverting the matrix.

The Hilbert matrix of order n was generated in the preceding script using the MATLAB function `hilb(n)`. Other important matrices with interesting structure and properties, such as the Hadamard matrix and the Wilkinson matrix can be obtained using, in these cases the MATLAB function `hadamard(n)` and `wilkinson(n)` where `n` is the required size of the matrix. In addition, many other interesting matrices can be accessed using the `gallery` function. In almost every case, we can choose the size of the matrix and in many cases we can also choose other parameters within the matrix. Example calls are

```
gallery('hanowa',6,4)
gallery('cauchy',6)
gallery('forsythe',6,8)
```

The next section begins the detailed examination of one of the algorithms used by the \ operator.

2.5 ELEMENTARY ROW OPERATIONS

We now examine the operations that can usefully be carried out on each equation of a system of equations. Such a system will have the form

$$
\begin{gathered}
a_{11}x_1 + a_{12}x_2 + \ldots + a_{1n}x_n = b_1 \\
a_{21}x_2 + a_{22}x_2 + \ldots + a_{2n}x_n = b_2 \\
\ldots\ldots\ldots\ldots\ldots \\
a_{n1}x_n + a_{n2}x_2 + \ldots + a_{nn}x_n = b_n
\end{gathered}
$$

or in matrix notation

$$\mathbf{Ax} = \mathbf{b}$$

where

$$
\mathbf{A} = \begin{bmatrix} a_{11} & a_{12}\ldots & a_{1n} \\ a_{21} & a_{22}\ldots & a_{2n} \\ \vdots & \vdots & \vdots \\ a_{n1} & a_{n2}\ldots & a_{nn} \end{bmatrix} \quad \mathbf{b} = \begin{bmatrix} b_1 \\ b_2 \\ \vdots \\ b_n \end{bmatrix} \quad \mathbf{x} = \begin{bmatrix} x_1 \\ x_2 \\ \vdots \\ x_n \end{bmatrix}
$$

A is called the coefficient matrix. Any operation performed on an equation must be applied to both its left and right sides. With this in mind it is helpful to combine the coefficient matrix **A** with the right side vector **b** thus:

$$\mathbf{A} = \begin{bmatrix} a_{11} & a_{12}\dots & a_{1n} & b_1 \\ a_{21} & a_{22}\dots & a_{2n} & b_2 \\ \vdots & \vdots & \vdots & \vdots \\ a_{n1} & a_{n2}\dots & a_{nn} & b_n \end{bmatrix}$$

This new matrix is called the augmented matrix and we will write it as [**A b**]. We have chosen to adopt this notation because it is consistent with MATLAB notation for combining **A** and **b**. Note that if **A** is an $n \times n$ matrix, then the augmented matrix is an $n \times (n+1)$ matrix. Each row of the augmented matrix holds all the coefficients of an equation and any operation must be applied to every element in the row. The three elementary row operations described in **1**, **2**, and **3** can be applied to individual equations in a system without altering the solution of the equation system. They are:

1. Interchange the position of any two rows (that is any two equations).
2. Multiply a row (that is an equation) by a non-zero scalar.
3. Replace a row by the sum of the row and a scalar multiple of another row.

These elementary row operations can be used to solve some important problems in linear algebra and we now discuss an application of them.

2.6 SOLUTION OF Ax = b BY GAUSSIAN ELIMINATION

Gaussian elimination is an efficient way to solve equation systems, particularly those with a non-symmetric coefficient matrix having a relatively small number of zero elements. The method depends entirely on using the three elementary row operations, described in Section 2.5. Essentially the procedure is to form the augmented matrix for the system and then reduce the coefficient matrix part to an upper triangular form. To illustrate the systematic use of the elementary row operations we consider the application of Gaussian elimination to solve the following equation system:

$$\begin{bmatrix} 3 & 6 & 9 \\ 2 & (4+p) & 2 \\ -3 & -4 & -11 \end{bmatrix} \begin{bmatrix} x_1 \\ x_2 \\ x_3 \end{bmatrix} = \begin{bmatrix} 3 \\ 4 \\ -5 \end{bmatrix} \tag{2.10}$$

where the value of p is known. Table 2.1 shows the sequence of operations, beginning at Stage 1 with the augmented matrix. In Stage 1 the element in the first column of the first row (enclosed in a box in the table) is designated the pivot. We wish to make the elements of column 1 in rows 2 and 3 zero. To achieve this, we divide row 1 by the pivot and then add or subtract a suitable multiple of the modified row 1 to or from rows 2 and 3. The result of this is shown in Stage 2 of the table. We then select the next pivot. This is the element in the second column of the new second row, which in Stage 2 is equal to p. If p is large, this does not present a problem, but if p is small, then numerical problems may arise because we will be dividing all the elements of the new row 2 by this small quantity p. If p is small then we are in effect multiplying the pivot by a large number. This not only multiplies the pivot element by this value but also amplifies any rounding error in the element by the same amount. If p

Table 2.1 Gaussian elimination used to transform an augmented matrix to upper triangular form

A1	[3]	6	9	3	**Stage 1**:
A2	2	$(4+p)$	2	4	Initial matrix
A3	−3	−4	−11	−5	
A1	3	6	9	3	**Stage 2**: Reduce
B2 = A2 − 2(A1)/3	0	p	−4	2	col 1 of row
B3 = A3 + 3(A1)/3	0	2	−2	−2	2 & 3 to zero
A1	3	6	9	3	**Stage 3**:
B3	0	[2]	−2	−2	Interchange
B2	0	p	−4	2	rows 2 & 3
A1	3	6	9	3	**Stage 4**: Reduce
B3	0	2	−2	−2	col 2 of row
C3 = B2 − p(B3)/2	0	0	[$(-4+p)$]	$(2+p)$	3 to zero

is zero, then we have an impossible situation because we cannot divide by zero. This difficulty is not related to ill-conditioning; indeed this particular equation system is quite well conditioned when p is zero. To circumvent these problems the usual procedure is to interchange the row in question with the row containing the element of largest modulus in the column *below the pivot*. In this way we provide a new and larger pivot. This procedure is called partial pivoting. If we assume in this case that $p < 2$, then we interchange rows 2 and 3 as shown in Stage 3 of the table to replace p by 2 as the pivot. From row 3 we now subtract row 2 divided by the pivot and multiplied by a coefficient in order to make the element of column 2, row 3, zero. Thus, in Stage 4 of the table it can be seen that the original coefficient matrix has been reduced to an upper triangular matrix. If, for example, $p = 0$, we obtain

$$3x_1 + 6x_2 + 9x_3 = 3 \tag{2.11}$$

$$2x_2 - 2x_3 = -2 \tag{2.12}$$

$$-4x_3 = 2 \tag{2.13}$$

We can obtain the values of the unknowns x_1, x_2, and x_3 by a process called back substitution. We solve the equations in reversed order. Thus, from (2.13), $x_3 = -0.5$. From (2.12), knowing x_3, we have $x_2 = -1.5$. Finally from (2.11), knowing x_2 and x_3, we have $x_1 = 5.5$.

It can be shown that the determinant of a matrix can be evaluated from the product of the elements on the main diagonal provided at Stage 3 in Table 2.1. This product must be multiplied by $(-1)^m$ where m is the number of row interchanges used. For example, in the preceding problem, with $p = 0$, one row interchange is used so that $m = 1$ and the determinant of the coefficient matrix is given by $3 \times 2 \times (-4) \times (-1)^1 = 24$.

A method for solving a linear equation system that is closely related to Gaussian elimination is Gauss–Jordan elimination. The method uses the same elementary row operations but differs from Gaussian elimination because elements both below and above the leading diagonal are reduced to zero. This means that back substitution is avoided. For example, solving system (2.10) with $p = 0$ leads to the following augmented matrix:

$$\begin{bmatrix} 3 & 0 & 0 & 16.5 \\ 0 & 2 & 0 & -3.0 \\ 0 & 0 & -4 & 2.0 \end{bmatrix}$$

Thus $x_1 = 16.5/3 = 5.5$, $x_2 = -3/2 = -1.5$, and $x_3 = 2/-4 = -0.5$.

Gaussian elimination requires order $n^3/3$ multiplications followed by back substitution requiring order n^2 multiplications. Gauss–Jordan elimination requires order $n^3/2$ multiplications. Thus, for large systems of equations (say $n > 10$), Gauss–Jordan elimination requires approximately 50% more operations than Gaussian elimination.

2.7 LU DECOMPOSITION

LU decomposition (or factorization) is a similar process to Gaussian elimination and is equivalent in terms of elementary row operations. The matrix **A** can be decomposed so that

$$\mathbf{A} = \mathbf{LU} \tag{2.14}$$

where **L** is a lower triangular matrix with a leading diagonal of ones and **U** is an upper triangular matrix. Matrix **A** may be real or complex. Compared with Gaussian elimination, LU decomposition has a particular advantage when the equation system we wish to solve, $\mathbf{Ax} = \mathbf{b}$, has more than one right side or when the right sides are not known in advance. This is because the factors **L** and **U** are obtained explicitly and they can be used for any right sides as they arise without recalculating **L** and **U**. Gaussian elimination does not determine **L** explicitly but rather forms $\mathbf{L}^{-1}\mathbf{b}$ so that all right sides must be known when the equation is solved.

The major steps required to solve an equation system by LU decomposition are as follows. Since $\mathbf{A} = \mathbf{LU}$, then $\mathbf{Ax} = \mathbf{b}$ becomes

$$\mathbf{LUx} = \mathbf{b}$$

where **b** is not restricted to a single column. Letting $\mathbf{y} = \mathbf{Ux}$ leads to

$$\mathbf{Ly} = \mathbf{b}$$

Because **L** is a lower triangular matrix this equation is solved efficiently by forward substitution. To find **x** we then solve

$$\mathbf{Ux} = \mathbf{y}$$

Because **U** is an upper triangular matrix, this equation can also be solved efficiently by back substitution.

We now illustrate the LU decomposition process by solving (2.10) with $p = 1$. We are not concerned with **b** and we do not form an augmented matrix. We proceed exactly as with Gaussian elimination, see Table 2.1, except that we keep a record of the elementary row operations performed at the ith stage in $\mathbf{T}^{(i)}$ and place the results of these operations in a matrix $\mathbf{U}^{(i)}$ rather than over-writing **A**.

We begin with the matrix

$$\mathbf{A} = \begin{bmatrix} 3 & 6 & 9 \\ 2 & 5 & 2 \\ -3 & -4 & -11 \end{bmatrix} \tag{2.15}$$

Following the same operations as used in Table 2.1, we will create a matrix $\mathbf{U}^{(1)}$ with zeros below the leading diagonal in the first column using the following elementary row operations:

$$\text{row 2 of } \mathbf{U}^{(1)} = \text{row 2 of } \mathbf{A} - 2(\text{row 1 of } \mathbf{A})/3 \tag{2.16}$$

and

$$\text{row 3 of } \mathbf{U}^{(1)} = \text{row 3 of } \mathbf{A} + 3(\text{row 1 of } \mathbf{A})/3 \tag{2.17}$$

Now $\mathbf{A}$ can be expressed as the product $\mathbf{T}^{(1)}\mathbf{U}^{(1)}$ as follows:

$$\begin{bmatrix} 3 & 6 & 9 \\ 2 & 5 & 2 \\ -3 & -4 & -11 \end{bmatrix} = \begin{bmatrix} 1 & 0 & 0 \\ 2/3 & 1 & 0 \\ -1 & 0 & 1 \end{bmatrix} \begin{bmatrix} 3 & 6 & 9 \\ 0 & 1 & -4 \\ 0 & 2 & -2 \end{bmatrix}$$

Note that row 1 of $\mathbf{A}$ and row 1 of $\mathbf{U}^{(1)}$ are identical. Thus row 1 of $\mathbf{T}^{(1)}$ has a unit entry in column 1 and zero elsewhere. The remaining rows of $\mathbf{T}^{(1)}$ are determined from (2.16) and (2.17). For example, row 2 of $\mathbf{T}^{(1)}$ is derived by rearranging (2.16); thus:

$$\text{row 2 of } \mathbf{A} = \text{row 2 of } \mathbf{U}^{(1)} - 2(\text{row 1 of } \mathbf{A})/3 \tag{2.18}$$

or

$$\text{row 2 of } \mathbf{A} = 2(\text{row 1 of } \mathbf{U}^{(1)})/3 + \text{row 2 of } \mathbf{U}^{(1)} \tag{2.19}$$

since row 1 of $\mathbf{U}^{(1)}$ is identical to row 1 of $\mathbf{A}$. Hence row 2 of $\mathbf{T}^{(1)}$ is [2/3 1 0].

We now move to the next stage of the decomposition process. In order to bring the largest element of column 2 in $\mathbf{U}^{(1)}$ onto the leading diagonal we must interchange rows 2 and 3. Thus $\mathbf{U}^{(1)}$ becomes the product $\mathbf{T}^{(2)}\mathbf{U}^{(2)}$ as follows:

$$\begin{bmatrix} 3 & 6 & 9 \\ 0 & 1 & -4 \\ 0 & 2 & -2 \end{bmatrix} = \begin{bmatrix} 1 & 0 & 0 \\ 0 & 0 & 1 \\ 0 & 1 & 0 \end{bmatrix} \begin{bmatrix} 3 & 6 & 9 \\ 0 & 2 & -2 \\ 0 & 1 & -4 \end{bmatrix}$$

Finally, to complete the process of obtaining an upper triangular matrix we make

$$\text{row 3 of } \mathbf{U} = \text{row 3 of } \mathbf{U}^{(2)} - (\text{row 2 of } \mathbf{U}^{(2)})/2$$

Hence, $\mathbf{U}^{(2)}$ becomes the product $\mathbf{T}^{(3)}\mathbf{U}$ as follows:

$$\begin{bmatrix} 3 & 6 & 9 \\ 0 & 2 & -2 \\ 0 & 1 & -4 \end{bmatrix} = \begin{bmatrix} 1 & 0 & 0 \\ 0 & 1 & 0 \\ 0 & 1/2 & 1 \end{bmatrix} \begin{bmatrix} 3 & 6 & 9 \\ 0 & 2 & -2 \\ 0 & 0 & -3 \end{bmatrix}$$

Thus $\mathbf{A} = \mathbf{T}^{(1)}\mathbf{T}^{(2)}\mathbf{T}^{(3)}\mathbf{U}$, implying that $\mathbf{L} = \mathbf{T}^{(1)}\mathbf{T}^{(2)}\mathbf{T}^{(3)}$ as follows:

$$\begin{bmatrix} 1 & 0 & 0 \\ 2/3 & 1 & 0 \\ -1 & 0 & 1 \end{bmatrix} \begin{bmatrix} 1 & 0 & 0 \\ 0 & 0 & 1 \\ 0 & 1 & 0 \end{bmatrix} \begin{bmatrix} 1 & 0 & 0 \\ 0 & 1 & 0 \\ 0 & 1/2 & 1 \end{bmatrix} = \begin{bmatrix} 1 & 0 & 0 \\ 2/3 & 1/2 & 1 \\ -1 & 1 & 0 \end{bmatrix}$$

Note that owing to the row interchanges **L** is not strictly a lower triangular matrix but it can be made so by interchanging rows.

MATLAB implements LU factorization by using the function `lu` and may produce a matrix that is not strictly a lower triangular matrix. However, a permutation matrix **P** may be produced, if required, such that **LU** = **PA** with **L** lower triangular.

We now show how the MATLAB function `lu` solves the example based on the matrix given in (2.15):

```
>> A = [3 6 9;2 5 2;-3 -4 -11]

A =

     3     6     9
     2     5     2
    -3    -4   -11
```

To obtain the **L** and **U** matrices, we must use that MATLAB facility of assigning two parameters simultaneously as follows:

```
>> [L1 U] = lu(A)

L1 =

    1.0000         0         0
    0.6667    0.5000    1.0000
   -1.0000    1.0000         0

U =

    3.0000    6.0000    9.0000
         0    2.0000   -2.0000
         0         0   -3.0000
```

Note that the `L1` matrix is not in lower triangular form, although its true form can easily be deduced by interchanging rows 2 and 3 to form a triangle. To obtain a true lower triangular matrix we must assign three parameters as follows:

```
>> [L U P] = lu(A)

L =

    1.0000         0         0
   -1.0000    1.0000         0
    0.6667    0.5000    1.0000
```

```
U =

    3.0000    6.0000    9.0000
         0    2.0000   -2.0000
         0         0   -3.0000

P =

     1     0     0
     0     0     1
     0     1     0
```

In the preceding output, `P` is the permutation matrix such that `L*U = P*A` or `P'*L*U = A`. Thus `P'*L` is equal to `L1`.

The MATLAB operator \ determines the solution of $\mathbf{Ax} = \mathbf{b}$ using LU factorization. As an example of an equation system with multiple right sides we solve $\mathbf{AX} = \mathbf{B}$ where

$$\mathbf{A} = \begin{bmatrix} 3 & 4 & -5 \\ 6 & -3 & 4 \\ 8 & 9 & -2 \end{bmatrix} \text{ and } \mathbf{B} = \begin{bmatrix} 1 & 3 \\ 9 & 5 \\ 9 & 4 \end{bmatrix}$$

Performing LU decomposition such that $\mathbf{LU} = \mathbf{A}$ gives

$$\mathbf{L} = \begin{bmatrix} 0.375 & -0.064 & 1 \\ 0.750 & 1 & 0 \\ 1 & 0 & 0 \end{bmatrix} \text{ and } \mathbf{U} = \begin{bmatrix} 8 & 9 & -2 \\ 0 & -9.75 & 5.5 \\ 0 & 0 & -3.897 \end{bmatrix}$$

Thus $\mathbf{LY} = \mathbf{B}$ is given by

$$\begin{bmatrix} 0.375 & -0.064 & 1 \\ 0.750 & 1 & 0 \\ 1 & 0 & 0 \end{bmatrix} \begin{bmatrix} y_{11} & y_{12} \\ y_{21} & y_{22} \\ y_{31} & y_{32} \end{bmatrix} = \begin{bmatrix} 1 & 3 \\ 9 & 5 \\ 9 & 4 \end{bmatrix}$$

We note that implicitly we have two systems of equations which when separated can be written:

$$\mathbf{L} \begin{bmatrix} y_{11} \\ y_{21} \\ y_{31} \end{bmatrix} = \begin{bmatrix} 1 \\ 9 \\ 9 \end{bmatrix} \text{ and } \mathbf{L} \begin{bmatrix} y_{12} \\ y_{22} \\ y_{32} \end{bmatrix} = \begin{bmatrix} 3 \\ 5 \\ 4 \end{bmatrix}$$

In this example, $\mathbf{L}$ is not strictly a lower triangular matrix owing to the reordering of the rows. However, the solution of this equation is still found by forward substitution. For example, $1y_{11} = b_{31} = 9$, so that $y_{11} = 9$. Then $0.75y_{11} + 1y_{21} = b_{21} = 9$. Hence $y_{21} = 2.25$, etc. The complete $\mathbf{Y}$ matrix is

$$\mathbf{Y} = \begin{bmatrix} 9.000 & 4.000 \\ 2.250 & 2.000 \\ -2.231 & 1.628 \end{bmatrix}$$

Finally solving **UX** = **Y** by back substitution gives

$$\mathbf{X} = \begin{bmatrix} 1.165 & 0.891 \\ 0.092 & -0.441 \\ 0.572 & -0.418 \end{bmatrix}$$

The MATLAB function `det` determines the determinant of a matrix using LU factorization as follows. Since **A** = **LU** then $|\mathbf{A}| = |\mathbf{L}|\,|\mathbf{U}|$. The elements of the leading diagonal of **L** are all ones so that $|\mathbf{L}| = 1$. Since **U** is upper triangular, its determinant is the product of the elements of its leading diagonal. Thus, taking account of row interchanges the appropriately signed product of the diagonal elements of **U** gives the determinant.

2.8 CHOLESKY DECOMPOSITION

Cholesky decomposition or factorization is a form of triangular decomposition that can only be applied to either a positive definite symmetric matrix or a positive definite Hermitian matrix. A symmetric matrix **A** is said to be positive definite if $\mathbf{x}^\mathsf{T}\mathbf{A}\mathbf{x} > 0$ for any non-zero **x**. Similarly, if **A** is Hermitian, then $\mathbf{x}^\mathsf{H}\mathbf{A}\mathbf{x} > 0$. A more useful definition of a positive definite matrix is one that has all eigenvalues greater than zero. The eigenvalue problem is discussed in Section 2.15. If **A** is symmetric or Hermitian, we can write

$$\mathbf{A} = \mathbf{P}^\mathsf{T}\mathbf{P} \quad (\text{or } \mathbf{A} = \mathbf{P}^\mathsf{H}\mathbf{P} \text{ when } \mathbf{A} \text{ is Hermitian}) \tag{2.20}$$

where **P** is an upper triangular matrix. The algorithm computes **P** row by row by equating coefficients of each side of (2.20). Thus $p_{11}, p_{12}, p_{13}, \ldots, p_{22}, p_{23}, \ldots$ are determined in sequence, ending with p_{nn}. Coefficients on the leading diagonal of **P** are computed from expressions that involve determining a square root. For example,

$$p_{22} = \sqrt{a_{22} - p_{12}^2}$$

A property of positive definite matrices is that the term under the square root is always positive and so the square root will be real. Furthermore, row interchanges are not required because the dominant coefficients will always be on the main diagonal. The whole process requires only about half as many multiplications as LU decomposition. Cholesky factorization is implemented for positive definite symmetric matrices in MATLAB by the function `chol`. For example, consider the Cholesky factorization of the following positive definite Hermitian matrix:

```
>> A = [2 -i 0;i 2 0;0 0 3]

A =

   2.0000 + 0.0000i   0.0000 - 1.0000i   0.0000 + 0.0000i
   0.0000 + 1.0000i   2.0000 + 0.0000i   0.0000 + 0.0000i
   0.0000 + 0.0000i   0.0000 + 0.0000i   3.0000 + 0.0000i

>> P = chol(A)
```

```
P =

   1.4142 + 0.0000i   0.0000 - 0.7071i   0.0000 + 0.0000i
   0.0000 + 0.0000i   1.2247 + 0.0000i   0.0000 + 0.0000i
   0.0000 + 0.0000i   0.0000 + 0.0000i   1.7321 + 0.0000i
```

When the operator \ detects a symmetric positive definite or Hermitian positive definite system matrix, it solves $\mathbf{Ax} = \mathbf{b}$ using the following sequence of operations. $\mathbf{A}$ is factorized into $\mathbf{P}^\mathsf{T}\mathbf{P}$, $\mathbf{y}$ is set to $\mathbf{Px}$; then $\mathbf{P}^\mathsf{T}\mathbf{y} = \mathbf{b}$. The algorithm solves for $\mathbf{y}$ by forward substitution since $\mathbf{P}^\mathsf{T}$ is a lower triangular matrix. Then $\mathbf{x}$ can be determined from $\mathbf{y}$ by backward substitution since $\mathbf{P}$ is an upper triangular matrix. We can illustrate the steps in this process by the following example:

$$\mathbf{A} = \begin{bmatrix} 2 & 3 & 4 \\ 3 & 6 & 7 \\ 4 & 7 & 10 \end{bmatrix} \text{ and } \mathbf{b} = \begin{bmatrix} 2 \\ 4 \\ 8 \end{bmatrix}$$

Then by Cholesky factorization

$$\mathbf{P} = \begin{bmatrix} 1.414 & 2.121 & 2.828 \\ 0 & 1.225 & 0.817 \\ 0 & 0 & 1.155 \end{bmatrix}$$

Now since $\mathbf{P}^\mathsf{T}\mathbf{y} = \mathbf{b}$ solving for $\mathbf{y}$ by forward substitution gives

$$\mathbf{y} = \begin{bmatrix} 1.414 \\ 0.817 \\ 2.887 \end{bmatrix}$$

Finally solving $\mathbf{Px} = \mathbf{y}$ by back substitution gives

$$\mathbf{x} = \begin{bmatrix} -2.5 \\ -1.0 \\ 2.5 \end{bmatrix}$$

We now compare the performance of the operator \ with the function `chol`. Clearly, their performance should be similar in the case of a positive definite matrix. To generate a symmetric positive definite matrix in script `e4s205.m`, we multiply a matrix by its transpose:

```
% e4s205.m
disp('  n        time-backslash  time-chol');
for n = 1000:100:2000
    A = [ ]; M = 100*randn(n,n);
    A = M*M'; b = [1:n].';
    tic, x = A\b; t1 = toc;
    tic, R = chol(A);
    v = R.'\b; x = R\b;
```

```
    t2 = toc;
    fprintf('%4.0f %14.4f %13.4f \n',n,t1,t2)
end
```

Running script e4s205.m gives

```
  n      time-backslash  time-chol
1000        0.0320        0.0232
1100        0.0388        0.0246
1200        0.0447        0.0271
1300        0.0603        0.0341
1400        0.0650        0.0475
1500        0.0723        0.0420
1600        0.0723        0.0521
1700        0.0965        0.0552
1800        0.1142        0.0679
1900        0.1312        0.0825
2000        0.1538        0.1107
```

The similarity in performance of the function chol and the operator \ is borne out in the preceding table. In this table, column 1 gives the size of the matrix and column 2 gives the time taken using the \ operator. Column 3 gives the time taken using Cholesky decomposition to solve the same problem.

Cholesky factorization *can* be applied to a symmetric matrix which is not positive definite but the process does not possess the numerical stability of the positive definite case. Furthermore, one or more rows in **P** may be purely imaginary. For example,

$$\text{If } \mathbf{A} = \begin{bmatrix} 1 & 2 & 3 \\ 2 & -5 & 9 \\ 3 & 9 & 4 \end{bmatrix} \text{ then } \mathbf{P} = \begin{bmatrix} 1 & 2 & 3 \\ 0 & 3\imath & -\imath \\ 0 & 0 & 2\imath \end{bmatrix}$$

This is not implemented in MATLAB.

2.9 QR DECOMPOSITION

We have seen how a square matrix can be decomposed or factorized into the product of a lower and an upper triangular matrix by the use of elementary row operations. An alternative decomposition of **A** is into an upper triangular matrix and an orthogonal matrix if **A** is real, or into an upper triangular matrix and a unitary matrix if **A** is complex. This is called QR decomposition. Thus

$$\mathbf{A} = \mathbf{QR}$$

where **R** is the upper triangular matrix and **Q** is the orthogonal, or the unitary matrix. If **Q** is orthogonal, $\mathbf{Q}^{-1} = \mathbf{Q}^{\mathsf{T}}$, and if **Q** is unitary, $\mathbf{Q}^{-1} = \mathbf{Q}^{\mathsf{H}}$. The preceding properties are very useful.

There are several procedures which provide QR decomposition; here we present Householder's method. To decompose a real matrix, Householder's method begins by defining a matrix $\mathbf{P}$ thus:

$$\mathbf{P} = \mathbf{I} - 2\mathbf{w}\mathbf{w}^{\mathsf{T}} \tag{2.21}$$

where $\mathbf{w}$ is a column vector and $\mathbf{P}$ is symmetrical matrix. Provided $\mathbf{w}^{\mathsf{T}}\mathbf{w} = 1$, $\mathbf{P}$ is also orthogonal. The orthogonality can easily be verified by expanding the product $\mathbf{P}^{\mathsf{T}}\mathbf{P} = \mathbf{P}\mathbf{P}$ as follows:

$$\begin{aligned}\mathbf{P}\mathbf{P} &= \left(\mathbf{I} - 2\mathbf{w}\mathbf{w}^{\mathsf{T}}\right)\left(\mathbf{I} - 2\mathbf{w}\mathbf{w}^{\mathsf{T}}\right) \\ &= \mathbf{I} - 4\mathbf{w}\mathbf{w}^{\mathsf{T}} + 4\mathbf{w}\mathbf{w}^{\mathsf{T}}\left(\mathbf{w}\mathbf{w}^{\mathsf{T}}\right) \\ &= \mathbf{I} - 4\mathbf{w}\mathbf{w}^{\mathsf{T}} + 4\mathbf{w}\left(\mathbf{w}^{\mathsf{T}}\mathbf{w}\right)\mathbf{w}^{\mathsf{T}} = \mathbf{I}\end{aligned}$$

since $\mathbf{w}^{\mathsf{T}}\mathbf{w} = 1$.

To decompose $\mathbf{A}$ into $\mathbf{QR}$, we begin by forming the vector $\mathbf{w}_1$ from the coefficients of the first column of $\mathbf{A}$ as follows:

$$\mathbf{w}_1^{\mathsf{T}} = \mu_1 [(a_{11} - s_1)\ a_{21}\ a_{31} \dots\ a_{n1}]$$

where

$$\mu_1 = \frac{1}{\sqrt{2s_1(s_1 - a_{11})}} \text{ and } s_1 = \pm\left(\sum_{j=1}^{n} a_{j1}^2\right)^{1/2}$$

By substituting for μ_1 and s_1 in $\mathbf{w}_1$ it can be verified that the necessary orthogonality condition, $\mathbf{w}_1^{\mathsf{T}}\mathbf{w}_1 = 1$, is satisfied. Substituting $\mathbf{w}_1$ into (2.21) we generate an orthogonal matrix $\mathbf{P}^{(1)}$.

The matrix $\mathbf{A}^{(1)}$ is now created from the product $\mathbf{P}^{(1)}\mathbf{A}$. It can easily be verified that all elements in the first column of $\mathbf{A}^{(1)}$ are zero except for the element on the leading diagonal which is equal to s_1. Thus,

$$\mathbf{A}^{(1)} = \mathbf{P}^{(1)}\mathbf{A} = \begin{bmatrix} s_1 & + & \cdots & + \\ 0 & + & \cdots & + \\ \vdots & \vdots & & \vdots \\ 0 & + & \cdots & + \\ 0 & + & \cdots & + \end{bmatrix}$$

In the matrix $\mathbf{A}^{(1)}$, + indicates a non-zero element.

We now begin the second stage of the orthogonalization process by forming $\mathbf{w}_2$ from the coefficients of the second column of $\mathbf{A}^{(1)}$ thus:

$$\mathbf{w}_2^{\mathsf{T}} = \mu_2 \left[0\ \left(a_{22}^{(1)} - s_2\right)\ a_{32}^{(1)}\ a_{42}^{(1)} \cdots a_{n2}^{(1)}\right]$$

where a_{ij} are the coefficients of **A** and

$$\mu_2 = \frac{1}{\sqrt{2s_2(s_2 - a_{22}^{(1)})}} \text{ and } s_2 = \pm\left(\sum_{j=2}^{n}(a_{j2}^{(1)})^2\right)^{1/2}$$

Then the orthogonal matrix $\mathbf{P}^{(2)}$ is generated from

$$\mathbf{P}^{(2)} = \mathbf{I} - 2\mathbf{w}_2\mathbf{w}_2^{\mathsf{T}}$$

The matrix $\mathbf{A}^{(2)}$ is then created from the product $\mathbf{P}^{(2)}\mathbf{A}^{(1)}$ as follows:

$$\mathbf{A}^{(2)} = \mathbf{P}^{(2)}\mathbf{A}^{(1)} = \mathbf{P}^{(2)}\mathbf{P}^{(1)}\mathbf{A} = \begin{bmatrix} s_1 & + & \cdots & + \\ 0 & s_2 & \cdots & + \\ \vdots & \vdots & & \vdots \\ 0 & 0 & \cdots & + \\ 0 & 0 & \cdots & + \end{bmatrix}$$

Note that $\mathbf{A}^{(2)}$ has zero elements in its first two columns except for the elements on and above the leading diagonal. We can continue this process $n-1$ times until we obtain an upper triangular matrix **R**. Thus,

$$\mathbf{R} = \mathbf{P}^{(n-1)} \ldots \mathbf{P}^{(2)}\mathbf{P}^{(1)}\mathbf{A} \tag{2.22}$$

Note that since $\mathbf{P}^{(i)}$ is orthogonal, the product $\mathbf{P}^{(n-1)}$ ... $\mathbf{P}^{(2)}\mathbf{P}^{(1)}$ is also orthogonal.

We wish to determine the orthogonal matrix **Q** such that $\mathbf{A} = \mathbf{QR}$. Thus $\mathbf{R} = \mathbf{Q}^{-1}\mathbf{A}$ or $\mathbf{R} = \mathbf{Q}^{\mathsf{T}}\mathbf{A}$. Hence, from (2.22),

$$\mathbf{Q}^{\mathsf{T}} = \mathbf{P}^{(n-1)} \ldots \mathbf{P}^{(2)}\mathbf{P}^{(1)}$$

Apart from the signs associated with the columns of **Q** and the rows of **R**, the decomposition is unique. These signs are dependent on whether the positive or negative square root is taken in determining s_1, s_2, etc. Complete decomposition of the matrix requires $2n^3/3$ multiplications and n square roots. To illustrate this procedure consider the decomposition of the matrix

$$\mathbf{A} = \begin{bmatrix} 4 & -2 & 7 \\ 6 & 2 & -3 \\ 3 & 4 & 4 \end{bmatrix}$$

Thus,

$$s_1 = \sqrt{4^2 + 6^2 + 3^2} = 7.8102$$

$$\mu_1 = \frac{1}{\sqrt{2 \times 7.8102 \times (7.8102 - 4)}} = 0.1296$$

$$\mathbf{w}_1^{\mathsf{T}} = 0.1296[(4 - 7.8102) \ 6 \ 3] = [-0.4939 \ 0.7777 \ 0.3889]$$

Using (2.21) we generate $\mathbf{P}^{(1)}$ and hence $\mathbf{A}^{(1)}$ thus:

$$\mathbf{P}^{(1)} = \begin{bmatrix} 0.5121 & 0.7682 & 0.3841 \\ 0.7682 & -0.2097 & -0.6049 \\ 0.3841 & -0.6049 & 0.6976 \end{bmatrix}$$

$$\mathbf{A}^{(1)} = \mathbf{P}^{(1)}\mathbf{A} = \begin{bmatrix} 7.8102 & 2.0486 & 2.8168 \\ 0 & -4.3753 & 3.5873 \\ 0 & 0.8123 & 7.2936 \end{bmatrix}$$

Note that we have reduced the elements of the first column of $\mathbf{A}^{(1)}$ below the leading diagonal to zero. We continue with the second stage thus:

$$s_2 = \sqrt{-4.3753^2 + 0.8123^2} = 4.4501$$

$$\mu_2 = \frac{1}{\sqrt{2 \times 4.4501 \times (4.4501 + 4.3753)}} = 0.1128$$

$$\mathbf{w}_2^{\mathrm{T}} = 0.1128\,[0\ \ (-4.3753 - 4.4501)\ \ 0.8123] = [0\ \ -0.9958\ \ 0.0917]$$

$$\mathbf{P}^{(2)} = \begin{bmatrix} 1 & 0 & 0 \\ 0 & -0.9832 & 0.1825 \\ 0 & 0.1825 & 0.9832 \end{bmatrix}$$

$$\mathbf{R} = \mathbf{A}^{(2)} = \mathbf{P}^{(2)}\mathbf{A}^{(1)} = \begin{bmatrix} 7.8102 & 2.0486 & 2.8168 \\ 0 & 4.4501 & -2.1956 \\ 0 & 0 & 7.8259 \end{bmatrix}$$

Note that we have now reduced the first two columns of $\mathbf{A}^{(2)}$ below the leading diagonal to zero. This completes the process to determine the upper triangular matrix $\mathbf{R}$. Finally, we determine the orthogonal matrix $\mathbf{Q}$ as follows:

$$\mathbf{Q} = \left(\mathbf{P}^{(2)}\mathbf{P}^{(1)}\right)^{\mathrm{T}} = \begin{bmatrix} 0.5121 & -0.6852 & 0.5179 \\ 0.7682 & 0.0958 & -0.6330 \\ 0.3841 & 0.7220 & 0.5754 \end{bmatrix}$$

It is not necessary for the reader to carry out the preceding calculations since MATLAB provides the function `qr` to carry out this decomposition. For example,

```
>> A = [4 -2 7;6 2 -3;3 4 4]

A =

     4    -2     7
     6     2    -3
     3     4     4

>> [Q R] = qr(A)
```

```
Q =

   -0.5121    0.6852    0.5179
   -0.7682   -0.0958   -0.6330
   -0.3841   -0.7220    0.5754

R =

   -7.8102   -2.0486   -2.8168
         0   -4.4501    2.1956
         0         0    7.8259
```

Note that the matrices **Q** and **R** in the MATLAB output are the negative of the hand calculations of **Q** and **R** above. This is not significant since their product is equal to **A**, and in the multiplication, the signs cancel.

One advantage of QR decomposition is that it can be applied to non-square matrices, decomposing an $m \times n$ matrix into an $m \times m$ orthogonal matrix and an $m \times n$ upper triangular matrix. Note that if $m > n$, the decomposition is not unique.

2.10 SINGULAR VALUE DECOMPOSITION

The singular value decomposition (SVD) of an $m \times n$ matrix **A** is given by

$$\mathbf{A} = \mathbf{USV}^{\mathsf{T}} \text{ (or } \mathbf{A} = \mathbf{USV}^{\mathsf{H}} \text{ if } \mathbf{A} \text{ is complex)} \tag{2.23}$$

where **U** is an orthogonal $m \times m$ matrix and **V** is an orthogonal $n \times n$ matrix. If **A** is complex then **U** and **V** are unitary matrices. In all cases **S** is a real diagonal $m \times n$ matrix. The elements of the leading diagonal of this matrix are called the singular values of **A**. Normally they are arranged in decreasing value so that $s_1 > s_2 > \ldots > s_n$. Thus,

$$\mathbf{S} = \begin{bmatrix} s_1 & 0 & \ldots & 0 \\ 0 & s_2 & \ldots & 0 \\ \vdots & \vdots & & \vdots \\ 0 & 0 & \ldots & s_n \\ 0 & 0 & \ldots & 0 \\ \vdots & \vdots & & \vdots \\ 0 & 0 & \ldots & 0 \end{bmatrix}$$

The singular values are the non-negative square roots of the eigenvalues of $\mathbf{A}^{\mathsf{T}}\mathbf{A}$. Because $\mathbf{A}^{\mathsf{T}}\mathbf{A}$ is symmetric or Hermitian these eigenvalues are real and non-negative so that the singular values are also real and non-negative. Algorithms for computing the SVD of a matrix are given by Golub and Van Loan (1989).

The SVD of a matrix has several important applications. In Section 2.4, we introduced the reduced row echelon form of a matrix and explained how the MATLAB function `rref` gave information from which the rank of a matrix can be deduced. However, rank can be more effectively determined from the SVD of a matrix since its rank is equal to the number of its non-zero singular values. Thus, for a 5×5 matrix of rank 3, s_4 and s_5 would be zero. In practice, rather than counting the non-zero singular values, MATLAB determines rank from the SVD by counting the number of singular values greater than some tolerance value. This is a more realistic approach to determining rank than counting any non-zero value, however small.

To illustrate how singular value decomposition helps us to examine the properties of a matrix we will use the MATLAB function `svd` to carry out a singular value decomposition and compare it with the function `rref`. Consider the following example in which a Vandermonde matrix is created using the MATLAB function `vander`. The Vandermonde matrix is known to be ill-conditioned. SVD allows us to examine the nature of this ill-conditioning. In particular a zero or a very small singular value indicates rank deficiency and this example shows that the singular values are becoming relatively close to this condition. In addition SVD allows us to compute the condition number of the matrix. In fact, the MATLAB function `cond` uses SVD to compute the condition number and this gives the same values as obtained by dividing the largest singular value by the smallest singular value. Additionally, the Euclidean norm of the matrix is supplied by the first singular value. Comparing the SVD with the RREF process in the following script, we see that the result of using the MATLAB functions `rref` and `rank` give the rank of this special Vandermonde matrix as 5 but tells us nothing else. There is no warning that the matrix is badly conditioned.

```
>> c = [1 1.01 1.02 1.03 1.04];
>> V = vander(c)

V =

    1.0000    1.0000    1.0000    1.0000    1.0000
    1.0406    1.0303    1.0201    1.0100    1.0000
    1.0824    1.0612    1.0404    1.0200    1.0000
    1.1255    1.0927    1.0609    1.0300    1.0000
    1.1699    1.1249    1.0816    1.0400    1.0000

>> format long
>> s = svd(V)

s =

   5.210367051037899
   0.101918335876689
   0.000699698839445
   0.000002352380295
   0.000000003294983

>> norm(V)
```

```
ans =
   5.210367051037898

>> cond(V)

ans =
     1.581303244929480e+09

>> s(1)/s(5)

ans =
     1.581303244929480e+09

>> rank(V)

ans =
     5

>> rref(V)

ans =
     1     0     0     0     0
     0     1     0     0     0
     0     0     1     0     0
     0     0     0     1     0
     0     0     0     0     1
```

The following example is very similar to the preceding one but the Vandermonde matrix has now been generated to be rank deficient. The smallest singular value, although not zero, is zero to machine precision and `rank` returns the value of 4.

```
>> c = [1 1.01 1.02 1.03 1.03];
>> V = vander(c)

V =
    1.0000    1.0000    1.0000    1.0000    1.0000
    1.0406    1.0303    1.0201    1.0100    1.0000
    1.0824    1.0612    1.0404    1.0200    1.0000
    1.1255    1.0927    1.0609    1.0300    1.0000
    1.1255    1.0927    1.0609    1.0300    1.0000

>> format long e
>> s = svd(V)
```

```
s =

       5.187797954424026e+00
       8.336322098941423e-02
       3.997349250041025e-04
       8.462129966394409e-07
       1.261104176071465e-23

>> format short
>> rank(V)

ans =
     4

>> rref(V)

ans =
    1.0000         0         0         0   -0.9424
         0    1.0000         0         0    3.8262
         0         0    1.0000         0   -5.8251
         0         0         0    1.0000    3.9414
         0         0         0         0         0

>> cond(V)

ans =
   4.1137e+23
```

The `rank` function does allow the user to vary the tolerance using a second parameter, that is `rank(A,tol)` where `tol` gives the tolerance for acceptance of the singular values of `A` to be taken as non-zero. However, tolerance should be used with care since the `rank` function counts the number of singular values greater than tolerance and this gives the rank of the matrix. If tolerance is very small, that is, smaller than the machine precision, the rank may be miscounted.

2.11 THE PSEUDO-INVERSE

Here we discuss the pseudo-inverse and in Section 2.12 we apply it to solve over- and under-determined systems.

If $\mathbf{A}$ is an $m \times n$ rectangular matrix, then the system

$$\mathbf{Ax} = \mathbf{b} \tag{2.24}$$

cannot be solved by inverting $\mathbf{A}$, since $\mathbf{A}$ is not a square matrix. Assuming an equation system with more equations than variables, that is $m > n$, then by pre-multiplying (2.24) by $\mathbf{A}^\top$ we can convert the system matrix to a square matrix as follows:

$$\mathbf{A}^{\mathsf{T}}\mathbf{A}\mathbf{x} = \mathbf{A}^{\mathsf{T}}\mathbf{b}$$

The product $\mathbf{A}^{\mathsf{T}}\mathbf{A}$ is square and, provided it is non-singular, it can be inverted to give the solution to (2.24) thus:

$$\mathbf{x} = \left(\mathbf{A}^{\mathsf{T}}\mathbf{A}\right)^{-1}\mathbf{A}^{\mathsf{T}}\mathbf{b} \tag{2.25}$$

Let

$$\mathbf{A}^{+} = \left(\mathbf{A}^{\mathsf{T}}\mathbf{A}\right)^{-1}\mathbf{A}^{\mathsf{T}} \tag{2.26}$$

The matrix $\mathbf{A}^{+}$ is called the Moore–Penrose pseudo-inverse of $\mathbf{A}$ or just the pseudo inverse. Thus, the solution of (2.24) is

$$\mathbf{x} = (\mathbf{A}^{+})\,\mathbf{b} \tag{2.27}$$

This definition of the pseudo-inverse, $\mathbf{A}^{+}$, requires $\mathbf{A}$ to have full rank. If $\mathbf{A}$ is full rank and $m > n$, then rank($\mathbf{A}$) $= n$. Now rank($\mathbf{A}^{\mathsf{T}}\mathbf{A}$) $=$ rank($\mathbf{A}$) and hence rank($\mathbf{A}^{\mathsf{T}}\mathbf{A}$) $= n$. Since $\mathbf{A}^{\mathsf{T}}\mathbf{A}$ is an $n \times n$ array, $\mathbf{A}^{\mathsf{T}}\mathbf{A}$ is automatically full rank and $\mathbf{A}^{+}$ is then a unique $m \times n$ array. If $\mathbf{A}$ is rank deficient, then $\mathbf{A}^{\mathsf{T}}\mathbf{A}$ is rank deficient and cannot be inverted.

If $\mathbf{A}$ is square and non-singular, then $\mathbf{A}^{+} = \mathbf{A}^{-1}$. If $\mathbf{A}$ is complex then

$$\mathbf{A}^{+} = \left(\mathbf{A}^{\mathsf{H}}\mathbf{A}\right)^{-1}\mathbf{A}^{\mathsf{H}} \tag{2.28}$$

where $\mathbf{A}^{\mathsf{H}}$ is the conjugate transpose, described in Appendix A, Section A.6. The product $\mathbf{A}^{\mathsf{T}}\mathbf{A}$ has a condition number which is the square of the condition number of $\mathbf{A}$. This has implications for the computations involved in $\mathbf{A}^{+}$.

It can be shown that the pseudo-inverse $\mathbf{A}^{+}$ has the following properties:

1. $\mathbf{A}(\mathbf{A}^{+})\mathbf{A} = \mathbf{A}$,
2. $(\mathbf{A}^{+})\mathbf{A}(\mathbf{A}^{+}) = \mathbf{A}^{+}$,
3. $(\mathbf{A}^{+})\mathbf{A}$ and $\mathbf{A}(\mathbf{A}^{+})$ are symmetrical matrices.

We must now consider the situation that occurs when $\mathbf{A}$ of (2.24) is an $m \times n$ array with $m < n$, that is an equation system with more variables than equations. If $\mathbf{A}$ is a full rank, then rank($\mathbf{A}$) $= m$. Now rank($\mathbf{A}^{\mathsf{T}}\mathbf{A}$) $=$ rank($\mathbf{A}$) and hence rank($\mathbf{A}^{\mathsf{T}}\mathbf{A}$) $= m$. Since $\mathbf{A}^{\mathsf{T}}\mathbf{A}$ is an $n \times n$ matrix, $\mathbf{A}^{\mathsf{T}}\mathbf{A}$ is rank deficient and cannot be inverted, even though $\mathbf{A}$ is of full rank. We can avoid this problem by recasting (2.24) as follows:

$$\mathbf{A}\mathbf{x} = \left(\mathbf{A}\mathbf{A}^{\mathsf{T}}\right)\left(\mathbf{A}\mathbf{A}^{\mathsf{T}}\right)^{-1}\mathbf{b}$$

and hence, multiplying by $\mathbf{A}^{-1}$

$$\mathbf{x} = \mathbf{A}^{\mathsf{T}}\left(\mathbf{A}\mathbf{A}^{\mathsf{T}}\right)^{-1}\mathbf{b}$$

Thus,

$$\mathbf{x} = (\mathbf{A}^{+})\,\mathbf{b}$$

where $\mathbf{A}^{+} = \mathbf{A}^{\mathsf{T}}(\mathbf{A}\mathbf{A}^{\mathsf{T}})^{-1}$ and is the pseudo-inverse. Note that $\mathbf{A}\mathbf{A}^{\mathsf{T}}$ is an $m \times m$ array with rank m and can thus be inverted.

Example 2.5. Consider the following matrix:

$$\mathbf{A} = \begin{bmatrix} 1 & 2 & 3 \\ 4 & 5 & 9 \\ 5 & 6 & 7 \\ -2 & 3 & 1 \end{bmatrix}$$

Computing the pseudo-inverse of **A** using a MATLAB implementation of (2.26) we have

```
>> A = [1 2 3;4 5 9;5 6 7;-2 3 1];
>> rank(A)

ans =
     3
```

We note that `A` is full rank. Thus,

```
>> A_plus = inv(A.'*A)*A.'

A_plus =
   -0.0747   -0.1467    0.2500   -0.2057
   -0.0378   -0.2039    0.2500    0.1983
    0.0858    0.2795   -0.2500   -0.0231
```

The MATLAB function `pinv` provides this result directly and with greater accuracy.

```
A*A_plus*A

ans =
    1.0000    2.0000    3.0000
    4.0000    5.0000    9.0000
    5.0000    6.0000    7.0000
   -2.0000    3.0000    1.0000

>> A*A_plus

ans =
    0.1070    0.2841    0.0000    0.1218
    0.2841    0.9096    0.0000   -0.0387
    0.0000    0.0000    1.0000   -0.0000
    0.1218   -0.0387   -0.0000    0.9834
```

```
>> A_plus*A

ans =
    1.0000    0.0000    0.0000
    0.0000    1.0000    0.0000
   -0.0000   -0.0000    1.0000
```

Note that these calculations verify that `A*A_plus*A` equals `A` and that both `A*A_plus` and `A_plus*A` are symmetrical.

It has been shown that if **A** is rank deficient, then (2.26) cannot be used to determine the pseudo-inverse of **A**. This doesn't mean that the pseudo-inverse does not exist; it always exists but we must use a different method to evaluate it. When **A** is rank deficient, or close to rank deficient, $\mathbf{A}^+$ is best calculated from the singular value decomposition (SVD) of **A**. If **A** is real, the SVD of **A** is $\mathbf{USV}^\mathsf{T}$ where **U** is an orthogonal $m \times m$ matrix and **V** is an orthogonal $n \times n$ matrix and **S** is an $n \times m$ matrix of singular values. Thus, the SVD of $\mathbf{A}^\mathsf{T}$ is $\mathbf{VS}^\mathsf{T}\mathbf{U}^\mathsf{T}$ so that

$$\mathbf{A}^\mathsf{T}\mathbf{A} = (\mathbf{VS}^\mathsf{T}\mathbf{U}^\mathsf{T})(\mathbf{USV}^\mathsf{T}) = \mathbf{VS}^\mathsf{T}\mathbf{SV}^\mathsf{T} \text{ since } \mathbf{U}^\mathsf{T}\mathbf{U} = \mathbf{I}$$

Hence,

$$\begin{aligned}\mathbf{A}^+ &= (\mathbf{VS}^\mathsf{T}\mathbf{SV}^\mathsf{T})^{-1}\mathbf{VS}^\mathsf{T}\mathbf{U}^\mathsf{T} = \mathbf{V}^{-\mathsf{T}}(\mathbf{S}^\mathsf{T}\mathbf{S})^{-1}\mathbf{V}^{-1}\mathbf{VS}^\mathsf{T}\mathbf{U}^\mathsf{T} \\ &= \mathbf{V}(\mathbf{S}^\mathsf{T}\mathbf{S})^{-1}\mathbf{S}^\mathsf{T}\mathbf{U}^\mathsf{T}\end{aligned} \tag{2.29}$$

We note that because of orthogonality $\mathbf{VV}^\mathsf{T} = \mathbf{I}$ and hence $\mathbf{VV}^\mathsf{T}\mathbf{V}^{-\mathsf{T}} = \mathbf{V}^{-\mathsf{T}}$.

Since **V** is an $m \times m$ matrix, **U** is an $n \times n$ matrix and **S** is an $n \times m$ matrix, then (2.29) is conformable, that is matrix multiplication is possible; see Appendix A, Section A.5. In this situation, however, $\mathbf{S}^\mathsf{T}\mathbf{S}$ cannot be inverted because of the very small or zero singular values. To deal with this problem we take only the r non-zero singular values of the matrix so that **S** is an $r \times r$ matrix where r is the rank of **A**. To make the multiplications of (2.29) conformable (that is, possible) we take the first r columns of **V** and the first r rows of $\mathbf{U}^\mathsf{T}$, that is the first r columns of **U**. This is illustrated in Example 2.6 where the pseudo-inverse of **A** is determined.

Example 2.6. Consider the following rank deficient matrix:

$$\mathbf{A} = \begin{bmatrix} 1 & 2 & 3 \\ 4 & 5 & 9 \\ 7 & 11 & 18 \\ -2 & 3 & 1 \\ 7 & 1 & 8 \end{bmatrix}$$

Using MATLAB, we have

```
>> A = [1 2 3;4 5 9;7 11 18;-2 3 1;7 1 8]

A =
     1     2     3
     4     5     9
     7    11    18
    -2     3     1
     7     1     8

>> rank(A)

ans =
     2
```

Since A has a rank of 2 and is thus rank deficient. Hence we cannot use (2.26) to determine its pseudo-inverse. We now find the SVD of A as follows

```
>> [U S V] = svd(A)

U =
   -0.1381    0.0839   -0.1198    0.4305   -0.8799
   -0.4115    0.0215    0.9092    0.0489   -0.0333
   -0.8258    0.2732   -0.3607   -0.3341    0.0413
   -0.0524    0.5650   -0.0599    0.7077    0.4165
   -0.3563   -0.7737   -0.1588    0.4469    0.2224

S =
   26.8394         0         0
         0    6.1358         0
         0         0    0.0000
         0         0         0
         0         0         0

V =
   -0.3709   -0.7274    0.5774
   -0.4445    0.6849    0.5774
   -0.8154   -0.0425   -0.5774
```

We now select the two significant singular values for use in the subsequent computation:

```
>> SS = S(1:2,1:2)

SS =
   26.8394         0
         0    6.1358
```

To make the multiplication conformable we use only the first two columns of U and V thus:

```
>> A_plus = V(:,1:2)*inv(SS.'*SS)*SS.'*U(:,1:2).'

A_plus =
  -0.0080    0.0031   -0.0210   -0.0663    0.0966
   0.0117    0.0092    0.0442    0.0639   -0.0805
   0.0036    0.0124    0.0232   -0.0023    0.0162
```

This result can be obtained directly using the pinv function which is based on the singular value decomposition of G.

```
>> A*A_plus

ans =
   0.0261    0.0586    0.1369    0.0546   -0.0157
   0.0586    0.1698    0.3457    0.0337    0.1300
   0.1369    0.3457    0.7565    0.1977    0.0829
   0.0546    0.0337    0.1977    0.3220   -0.4185
  -0.0157    0.1300    0.0829   -0.4185    0.7256

>> A_plus*A

ans =
   0.6667   -0.3333    0.3333
  -0.3333    0.6667    0.3333
   0.3333    0.3333    0.6667
```

Note that A*A_plus and A_plus*A are symmetric.

In Section 2.12, we will apply these methods to solve over- and under-determined systems and discuss the meaning of the solution.

2.12 OVER- AND UNDER-DETERMINED SYSTEMS

We will begin by examining examples of over-determined systems, that is, systems of equations in which there are more equations than unknown variables.

Although over-determined systems may have a unique solution, most often we are concerned with equation systems that are generated from experimental data which can lead to a relatively small degree of inconsistency between the equations. For example, consider the following over-determined system of linear equations:

$$\begin{aligned} x_1 + x_2 &= 1.98 \\ 2.05x_1 - x_2 &= 0.95 \\ 3.06x_1 + x_2 &= 3.98 \\ -1.02x_1 + 2x_2 &= 0.92 \\ 4.08x_1 - x_2 &= 2.90 \end{aligned} \tag{2.30}$$

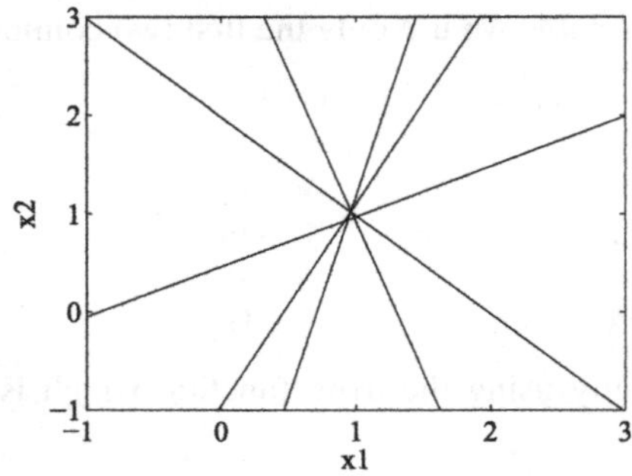

FIGURE 2.5

Plot of an inconsistent equation system (2.30).

Fig. 2.5 shows that (2.30) is such a system; the lines do not intersect in a point, although there is a point that *nearly* satisfies all the equations.

We would like to choose the best of all possible solutions of (2.30) in the region defined by the intersections. One criterion for doing this is that the chosen solution should minimize the sum of squares of the residual errors (or residuals) of the equations. For example, consider the equation system (2.30). Letting $r_1, ..., r_5$ be the residuals, then

$$\begin{aligned} x_1 + x_2 - 1.98 &= r_1 \\ 2.05x_1 - x_2 - 0.95 &= r_2 \\ 3.06x_1 + x_2 - 3.98 &= r_3 \\ -1.02x_1 + 2x_2 - 0.92 &= r_4 \\ 4.08x_1 - x_2 - 2.90 &= r_5 \end{aligned}$$

In this case the sum of the residuals squared is given by

$$S = \sum_{i=1}^{5} r_i^2 \tag{2.31}$$

We wish to minimize S and we can do this by making

$$\frac{\partial S}{\partial x_k} = 0, \quad k = 1, 2$$

Now

$$\frac{\partial S}{\partial x_k} = \sum_{i=1}^{5} 2r_i \frac{\partial r_i}{\partial x_k}, \quad k = 1, 2$$

and thus

$$\sum_{i=1}^{5} r_i \frac{\partial r_i}{\partial x_k} = 0, \quad k = 1, 2 \tag{2.32}$$

It can be shown that minimizing the sum of the squares of the residuals using (2.32) gives an identical solution to that given by the pseudo-inverse method for solving the equation system.

When solving a set of over-determined equations, determining the pseudo-inverse of the system matrix is only part of the process and normally we do not require this interim result. The MATLAB operator \ solves over-determined systems automatically. Thus, the operator may be used to solve any linear equation system.

In the following example, we compare the results obtained using the operator \ and using the pseudo-inverse for solving (2.30). The MATLAB script `e4s206.m` follows:

```
% e4s206.m
A = [1 1;2.05 -1;3.06 1;-1.02 2;4.08 -1];
b = [1.98;0.95;3.98;0.92;2.90];
x = pinv(A)*b
norm_pinv = norm(A*x-b)
x = A\b
norm_op = norm(A*x-b)
```

Running script `e4s206.m` gives the following numeric output:

```
x =
    0.9631
    0.9885

norm_pinv =
    0.1064

x =
    0.9631
    0.9885

norm_op =
    0.1064
```

Here, both the MATLAB operator \ and the function `pinv` have provided the same "best fit" solution for the inconsistent set of equations. Fig. 2.6 shows the region where these equations intersect in greater detail than Fig. 2.5. The symbol "+" indicates the MATLAB solution which can be seen to lie in this region. The norm of $\mathbf{Ax} - \mathbf{b}$ is the square root of the sum of the squares of the residuals and provides a measure of how well the equations are satisfied.

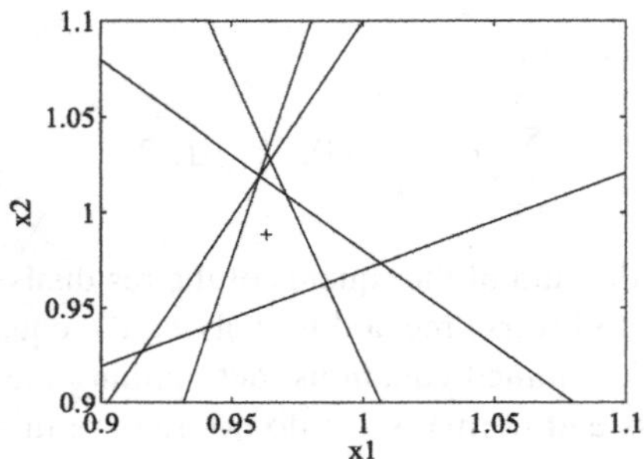

FIGURE 2.6

Plot of inconsistent equation system (2.30) showing the region of intersection of the equations, where + indicates "best" solution.

The MATLAB operator \ does not solve an over-determined system by using the pseudo-inverse, as given in (2.26). Instead, it solves (2.24) directly by QR decomposition. QR decomposition can be applied to both square and rectangular matrices providing the number of rows is greater than the number of columns. For example, applying the MATLAB function `qr` to solve the over-determined system (2.30) we have

```
>> A = [1 1;2.05 -1;3.06 1;-1.02 2;4.08 -1];
>> b = [1.98 0.95 3.98 0.92 2.90].';
>> [Q R] = qr(A)

Q =

   -0.1761    0.4123   -0.7157   -0.2339   -0.4818
   -0.3610   -0.2702    0.0998    0.6751   -0.5753
   -0.5388    0.5083    0.5991   -0.2780   -0.1230
    0.1796    0.6839   -0.0615    0.6363    0.3021
   -0.7184   -0.1756   -0.3394    0.0857    0.5749

R =

   -5.6792    0.7237
         0    2.7343
         0         0
         0         0
         0         0
```

In the equation $\mathbf{Ax} = \mathbf{b}$ we have replaced $\mathbf{A}$ by $\mathbf{QR}$ so that $\mathbf{QRx} = \mathbf{b}$. Let $\mathbf{Rx} = \mathbf{y}$. Thus, we have $\mathbf{y} = \mathbf{Q}^{-1}\mathbf{b} = \mathbf{Q}^{\mathsf{T}}\mathbf{b}$ since $\mathbf{Q}$ is orthogonal. Once $\mathbf{y}$ is determined we can efficiently determine $\mathbf{x}$ by back substitution since $\mathbf{R}$ is upper triangular. Thus, continuing the previous example,

```
>> y = Q.'*b

y =
   -4.7542
    2.7029
    0.0212
   -0.0942
   -0.0446
```

Using the second row of **R** and the second row of **y** we can determine x_2. From the first row of **R** and the first row of **y** we can determine x_1 since x_2 is known. Thus,

$$\begin{aligned} -5.6792x_1 + 0.7237x_2 &= -4.7542 \\ 2.7343x_2 &= 2.7029 \end{aligned}$$

gives $x_1 = 0.9631$ and $x_2 = 0.9885$, as before. The MATLAB operator \ implements this sequence of operations.

We now consider a case where the coefficient matrix of the over-determined system is rank deficient. The following example is rank deficient and represents a system of parallel lines.

$$\begin{aligned} x_1 + 2x_2 &= 1.00 \\ x_1 + 2x_2 &= 1.03 \\ x_1 + 2x_2 &= 0.97 \\ x_1 + 2x_2 &= 1.01 \end{aligned}$$

In MATLAB this becomes

```
>> A = [1 2;1 2;1 2;1 2]

A =
     1     2
     1     2
     1     2
     1     2

>> b = [1 1.03 0.97 1.01].'

b =
    1.0000
    1.0300
    0.9700
    1.0100

>> y = A\b
Warning: Rank deficient, rank = 1,  tol =   3.552714e-015.
```

```
y =
         0
    0.5012

>> norm(y)

ans =
    0.5012
```

The user is warned that this system is rank deficient. We have solved the system using the \ operator and now solve it using the `pinv` function as follows:

```
>> x = pinv(A)*b

x =
    0.2005
    0.4010

>> norm(x)

ans =
    0.4483
```

We see that when the `pinv` function and \ operator are applied to rank deficient systems the `pinv` function gives the solution with the smallest Euclidean norm; see Appendix A, Section A.10. Clearly there is no unique solution to this system since it represents a set of parallel lines.

We now turn to the problem of under-determined systems. Here there is insufficient information to determine a unique solution. For example, consider the equation system

$$\begin{aligned} x_1 + 2x_2 + 3x_3 + 4x_4 &= 1 \\ -5x_1 + 3x_2 + 2x_3 + 7x_4 &= 2 \end{aligned}$$

Expressing these equations in MATLAB we have

```
>> A = [1 2 3 4;-5 3 2 7];
>> b = [1 2].';
>> x1 = A\b

x1 =
   -0.0370
         0
         0
    0.2593

>> x2 = pinv(A)*b
```

```
x2 =
   -0.0780
    0.0787
    0.0729
    0.1755
```

We calculate the norms:

```
>> norm(x1)

ans =
    0.2619

>> norm(x2)

ans =
    0.2199
```

The first solution vector, `x1`, is a solution which satisfies the system; the second solution vector, `x2`, satisfies the system of equations but also gives the solution with the minimum norm.

The definition of the Euclidean or 2-norm of a vector is the square root of the sum of the squares of the elements of the vector; see Appendix A, Section A.10. The shortest distance between a point in space and the origin is given by Pythagoras's theorem as the square root of the sum of squares of the coordinates of the point. Thus, the Euclidean norm of a vector which is a point on a line, surface, or hypersurface may be interpreted geometrically as the distance between this point and the origin. The vector with the minimum norm must be the point on the line, surface, or hypersurface which is closest to the origin. It can be shown that the line joining this vector to the origin must be perpendicular to the line, surface, or hypersurface.

Giving the minimum norm solution has the advantage that, whereas there are an infinite number of solutions to an under-determined problem, there is only one minimum norm solution. This provides a standard result.

To complete the discussion of over- and under-determined systems we consider the use of the `lsqnonneg` function which solves the least squares problem with a non-negative solution. This solves the problem of finding a solution $\mathbf{x}$ to the linear equation system

$$\mathbf{Ax} = \mathbf{b}, \text{ subject to } \mathbf{x} \geq \mathbf{0}$$

where $\mathbf{A}$ and $\mathbf{b}$ must be real. This is equivalent to the problem of finding the vector $\mathbf{x}$ that minimizes norm($\mathbf{Ax} - \mathbf{b}$) subject to $\mathbf{x} \geq \mathbf{0}$.

We can call the MATLAB function `lsqnonneg` for a specific problem using the statement

```
x = lsqnonneg(A,b)
```

where A corresponds to **A** in our definition and b to **b**. The solution is given by x. Consider the example which follows. Solve

$$\begin{bmatrix} 1 & 1 & 1 & 1 & 0 \\ 1 & 2 & 3 & 0 & 1 \end{bmatrix} \begin{bmatrix} x_1 \\ x_2 \\ x_3 \\ x_4 \\ x_5 \end{bmatrix} = \begin{bmatrix} 7 \\ 12 \end{bmatrix}$$

subject to $x_i \geq 0$, where $i = 1, 2, ..., 5$. In MATLAB this becomes

```
>> A = [1 1 1 1 0;1 2 3 0 1];
>> b = [7 12].';
```

Solving this system gives

```
>> x = lsqnonneg(A,b)

x =
     0
     0
     4
     3
     0
```

We can also solve this using \ but this will not ensure non-negative values for x.

```
>> x2 = A\b

x2 =
         0
         0
    4.0000
    3.0000
         0
```

In this case, we do obtain a non-negative solution but this is fortuitous and cannot be guaranteed.

The following example illustrates how the lsqnonneg function forces a non-negative solution which best satisfies the equations:

$$\begin{bmatrix} 3.0501 & 4.8913 \\ 3.2311 & -3.2379 \\ 1.6068 & 7.4565 \\ 2.4860 & -0.9815 \end{bmatrix} \begin{bmatrix} x_1 \\ x_2 \end{bmatrix} = \begin{bmatrix} 2.5 \\ 2.5 \\ 0.5 \\ 2.5 \end{bmatrix} \tag{2.33}$$

```
>> A = [3.0501 4.8913;3.2311 -3.2379; 1.6068 7.4565;2.4860 -0.9815];
>> b = [2.5 2.5 0.5 2.5].';
```

We can compute the solution using \ or lsqnonneg function thus:

```
>> x1 = A\b

x1 =
    0.8307
   -0.0684

>> x2 = lsqnonneg(A,b)

x2 =
    0.7971
         0

>> norm(A*x1-b)

ans =
    0.7040

>> norm(A*x2-b)

ans =
    0.9428
```

Thus, the best fit is given by using the operator `\`, but if we require all components of the solution to be non-negative, then we must use the `lsqnonneg` function.

2.13 ITERATIVE METHODS

Except in special circumstances, it is unlikely that any function or script developed by the user will outperform a function or operator that is an integral part of MATLAB. We cannot expect to develop a function that will determine the solution of $\mathbf{Ax} = \mathbf{b}$ more efficiently than by using the MATLAB operation `A\b`. However, we describe iterative methods here for the sake of completeness. These methods are attractive if $\mathbf{A}$ is diagonally dominant.

Iterative methods of solution are developed as follows. We begin with a system of linear equations

$$\begin{array}{llll} a_{11}x_1+ & a_{12}x_2+ & \ldots & +a_{1n}x_n & = b_1 \\ a_{21}x_1+ & a_{22}x_2+ & \ldots & +a_{2n}x_n & = b_2 \\ \vdots & \vdots & & \vdots & \vdots \\ a_{n1}x_1+ & a_{n2}x_2+ & \ldots & +a_{nn}x_n & = b_n \end{array}$$

These can be rearranged to give

$$
\begin{aligned}
x_1 &= (b_1 - a_{12}x_2 - a_{13}x_3 - \ldots - a_{1n}x_n)/a_{11} \\
x_2 &= (b_2 - a_{21}x_1 - a_{23}x_3 - \ldots - a_{2n}x_n)/a_{22} \\
&\vdots \\
x_n &= \left(b_n - a_{n1}x_1 - a_{n2}x_2 - \ldots - a_{n,n-1}x_{n-1}\right)/a_{nn}
\end{aligned}
\tag{2.34}
$$

If we assume initial values for x_i, where $i = 1, \ldots, n$, and substitute these values into the right side of the preceding equations, we may determine new values for the x_i from (2.34). The iterative process is continued by substituting these values of x_i into the right side of the equations, etc. There are several variants of the process. For example, we can use old values of x_i in the right side of the equations to determine *all* the new values of x_i in the left side of the equation. This is called Jacobi or simultaneous iteration. Alternatively, we may use a new value of x_i in the right side of the equation as soon as it is determined, to obtain the other values of x_i in the right side. For example, once a new value of x_1 is determined from the first equation of (2.34), it is used in the second equation, together with the old $x_3, \ldots, x_n$ to determine x_2. This is called Gauss–Seidel or cyclic iteration.

The conditions for convergence for this type of iteration are

$$
|a_{ii}| >> \sum_{j=1,\ j\neq i}^{n} |a_{ij}| \text{ for } i = 1, 2, \ldots, n
$$

Thus, these iterative methods are only guaranteed to work when the coefficient matrix is diagonally dominant. An iterative method based on conjugate gradients for the solution of systems of linear equations is discussed in Chapter 9.

2.14 SPARSE MATRICES

Sparse matrices arise in many problems of science and engineering – for example, in linear programming and the analysis of structures. Indeed, most large matrices that arise in the analysis of physical systems are sparse and the recognition of this fact and its efficient exploitation makes the solution of linear systems with millions of coefficients feasible. The aim of this section is to give a brief description of the extensive sparse matrix facilities available in MATLAB and to give practical illustrations of their value through examples. For background information on how MATLAB implements the concept of sparsity, see Gilbert et al. (1992).

It is difficult to give a simple quantitative answer to the question: when is a matrix sparse? A matrix is sparse if it contains a high proportion of zero elements. However, this is significant only if the sparsity is of such an extent that we can utilize this feature to reduce the computation time and storage facilities required for operations used on such matrices. One major way in which time can be saved in dealing with sparse matrices is to avoid unnecessary operations on zero elements.

MATLAB *does not automatically treat a matrix as sparse* and the sparsity features of MATLAB are not introduced until invoked. Thus, the user determines whether a matrix is in the sparse class or the full class. If the user considers a matrix to be sparse and wants to use this fact to advantage, the matrix must first be converted to sparse form. This is achieved by using the function `sparse`. Thus `B = sparse(A)` converts the matrix `A` to sparse form and assigns it to `B` and MATLAB operations on B will take account of this sparsity. If we wish to return this matrix to full form, we simply use C = full(B). However, the sparse function can also be used directly to generate sparse matrices.

It is important to note that binary operators *, +, -, /, and \ produce sparse results if *both* operands are sparse. Thus, the property of sparsity may survive a long sequence of matrix operations. In addition such functions as chol(A) and lu(A) produce sparse results if the matrix A is sparse. However, in mixed cases, where one operand is sparse and the other is full, the result is generally a full matrix. Thus, the property of sparsity may be inadvertently lost. Notice in particular that eye(n) is not in the sparse class of matrices in MATLAB but a sparse identity matrix can be created using speye(n). Thus, the latter should be used in manipulations with sparse matrices.

We will now introduce some of the key MATLAB functions for dealing with sparse matrices, describe their use and, where appropriate, give examples of their application. The simplest MATLAB function which helps in dealing with sparsity is the function nnz(a) which provides the number of non-zero elements in a given matrix a, regardless of whether it is sparse or full. A function which enables us to examine whether a given matrix has been defined or has been propagated as sparse is the function issparse(a) which returns the value 1 if the matrix a is sparse or 0 if it is not sparse. The function spy(a) allows the user to view the structure of a given matrix a by displaying symbolically only its non-zero elements; see Fig. 2.7 later in the chapter for examples.

Before we can illustrate the action of these and other functions, it is useful to generate some sparse matrices. This is easily done using a different form of the sparse function. This time the function is supplied with the location of the non-zero entries in the matrix, the value of these entries, the size of the sparse matrix and the space allocated for the non-zero entries. This function call takes the form sparse(i, j, nzvals, m, n, nzmax). This generates an m × n matrix and allocates the non-zero values in the vector nzvals to the positions in the matrix given by the vectors i and j. The row position is given by i and the column position by j. Space is allocated for nzmax non-zeros. Since all but one parameter is optional, there are many forms of this function. We cannot give examples of all these forms but the following cases illustrate its use.

```
>> colpos = [1 2 1 2 5 3 4 3 4 5];
>> rowpos = [1 1 2 2 2 4 4 5 5 5];
>> value = [12 -4 7 3 -8 -13 11 2 7 -4];
>> A = sparse(rowpos,colpos,value,5,5)
```

These statements give the following output:

```
A =
   (1,1)        12
   (2,1)         7
   (1,2)        -4
   (2,2)         3
   (4,3)       -13
   (5,3)         2
   (4,4)        11
   (5,4)         7
   (2,5)        -8
   (5,5)        -4
```

We see that a 5 × 5 sparse matrix with 10 non-zero elements has been generated with the required coefficient values in the required positions. This sparse matrix can be converted to a full matrix as follows:

```
>> B = full(A)

B =
    12    -4     0     0     0
     7     3     0     0    -8
     0     0     0     0     0
     0     0   -13    11     0
     0     0     2     7    -4
```

This is the equivalent full matrix. Now the following statements test to see if the matrices `A` and `B` are in the sparse class and give the number of non-zeros they contain.

```
>> [issparse(A) issparse(B) nnz(A) nnz(B)]

ans =
     1     0    10    10
```

Clearly, these functions give the expected results. Since `A` is a member of the class of sparse matrices, the value of `issparse(A)` is 1. However, although `B` looks sparse, it is not *stored* as a sparse matrix and hence is not in the class sparse within the MATLAB environment. The next example shows how to generate a large 10,000 × 10,000 sparse matrix and compares the time required to solve a linear system of equations involving this sparse matrix with the time required for the equivalent full matrix. This is provided in script `e4s207.m`.

```
% e4s207.m Generates a sparse triple diagonal matrix
n = 10000;
rowpos = 2:n; colpos = 1:n-1;
values = 2*ones(1,n-1);
Offdiag = sparse(rowpos,colpos,values,n,n);
A = sparse(1:n,1:n,4*ones(1,n),n,n);
A = A+Offdiag+Offdiag.';
%generate full matrix
B = full(A);
%generate arbitrary right side for system of equations
rhs = [1:n].';
tic, x = A\rhs; f1 = toc;
tic, x = B\rhs; f2 = toc;
fprintf('Time to solve sparse matrix = %8.5f\n',f1);
fprintf('Time to solve  full  matrix = %8.5f\n',f2);
```

Script `e4s207.m` provides the results

```
Time to solve sparse matrix =  0.00078
Time to solve  full  matrix =  9.66697
```

In this example there is a massive reduction in the time taken to solve the system when using the sparse class of matrix. We now perform a similar exercise, this time to determine the `lu` decomposition of a 10,000 × 10,000 matrix, using script `e4s208`:

```
% e4s208.m
n = 10000;
offdiag = sparse(2:n,1:n-1,2*ones(1,n-1),n,n);
A = sparse(1:n,1:n,4*ones(1,n),n,n);
A = A+offdiag+offdiag';
%generate full matrix
B = full(A);
%generate arbitrary right side for system of equations
rhs = [1:n]';
tic, lu1 = lu(A); f1 = toc;
tic, lu2 = lu(B); f2 = toc;
fprintf('Time for sparse LU = %8.4f\n',f1);
fprintf('Time for  full  LU = %8.4f\n',f2);
```

The time taken to solve the systems using script `e4s208` are

```
Time for sparse LU =   0.0051
Time for  full  LU =  14.9198
```

Again this provides a considerable reduction in the time taken.

An alternative way to generate sparse matrices is to use the functions `sprandn` and `sprandsym`. These provide random sparse matrices and random sparse symmetric matrices respectively. The call

```
A = sprandn(m,n,d)
```

produces an $m \times n$ random matrix with normally distributed non-zero entries of density `d`. The density is the proportion of the non-zero entries to the total number of entries in the matrix. Thus `d` must be in the range 0 to 1. To produce a symmetric random matrix with normally distributed non-zero entries of density `d`, we use

```
A = sprandsys(n,d)
```

Examples of calls of these functions are given by

```
>> A = sprandn(5,5,0.25)

A =
   (2,1)      -0.4326
   (3,3)      -1.6656
   (5,3)      -1.1465
   (4,4)       0.1253
   (5,4)       1.1909
   (4,5)       0.2877
```

```
>> B = full(A)

B =
         0         0         0         0         0
   -0.4326         0         0         0         0
         0         0   -1.6656         0         0
         0         0         0    0.1253    0.2877
         0         0   -1.1465    1.1909         0

>> As = sprandsym(5,0.25)

As =
   (3,1)        0.3273
   (1,3)        0.3273
   (5,3)        0.1746
   (5,4)       -0.0376
   (3,5)        0.1746
   (4,5)       -0.0376
   (5,5)        1.1892

>> Bs = full(As)

Bs =
         0         0    0.3273         0         0
         0         0         0         0         0
    0.3273         0         0         0    0.1746
         0         0         0         0   -0.0376
         0         0    0.1746   -0.0376    1.1892
```

An alternative call for `sprandsym` is given by

```
A = sprandsym(n,density,r)
```

If `r` is a scalar, then this produces a random sparse symmetric matrix with a condition number equal to $1/r$. Remarkably, if `r` is a vector of length n, a random sparse matrix with eigenvalues equal to the elements of `r` is produced. Eigenvalues are discussed in Section 2.15. A positive definite matrix has all its eigenvalues positive and consequently such a matrix can be generated by choosing each of the n elements of `r` to be positive. An example of this form of call is

```
>> Apd = sprandsym(6,0.4,[1 2.5 6 9 2 4.3])

Apd =
   (1,1)        1.0058
   (2,1)       -0.0294
   (4,1)       -0.0879
```

```
   (1,2)       -0.0294
   (2,2)        8.3477
   (4,2)       -1.9540
   (3,3)        5.4937
   (5,3)       -1.3300
   (1,4)       -0.0879
   (2,4)       -1.9540
   (4,4)        3.1465
   (3,5)       -1.3300
   (5,5)        2.5063
   (6,6)        4.3000

>> Bpd = full(Apd)

Bpd =
    1.0058   -0.0294         0   -0.0879         0         0
   -0.0294    8.3477         0   -1.9540         0         0
         0         0    5.4937         0   -1.3300         0
   -0.0879   -1.9540         0    3.1465         0         0
         0         0   -1.3300         0    2.5063         0
         0         0         0         0         0    4.3000
```

This provides an important method for generating test matrices with required properties since, by providing a list of eigenvalues with a range of values, we can produce positive definite matrices that are very badly conditioned.

We now examine further the value of using sparsity. The reasons for the very high level of improvement in computing efficiency when using the \ operator, illustrated in the example at the beginning of this section, are complex. The process includes a special preordering of the columns of the matrix. This special preordering, called *minimum degree ordering*, is used in the case of the \ operator. This preordering takes different forms depending on whether the matrix is symmetric or non-symmetric. The aim of any preordering is to reduce the amount of *fill-in* from any subsequent matrix operations. Fill-in is the introduction of additional non-zero elements.

We can examine this preordering process using the `spy` function and the function `symamd` which implements *symmetric minimum degree ordering* in MATLAB. The function is automatically applied when working on matrices which belong to the class of sparse matrices for the standard functions and operators of MATLAB. However, if we are required to use this preordering in non-standard applications, then we may use the `symmmd` function. The following examples illustrate the use of this function.

We first consider the simple process of multiplication applied to a full and a sparse matrix. The sparse multiplication uses the minimum degree ordering. The following script `e4s209.m` generates a sparse matrix, obtains a minimum degree ordering for it and then examines the result of multiplying the matrix by itself transposed. This is compared with the same operations carried out on the full matrix, and the time required for each operation is compared.

```
% e4s209.m
% generate a sparse matrix
```

```
n = 5000;
offdiag = sparse(2:n,1:n-1,2*ones(1,n-1),n,n);
offdiag2 = sparse(4:n,1:n-3,3*ones(1,n-3),n,n);
offdiag3 = sparse(n-5:n,1:6,7*ones(1,6),n,n);
A = sparse(1:n,1:n,4*ones(1,n),n,n);
A = A+offdiag+offdiag'+offdiag2+offdiag2'+offdiag3+offdiag3';
A = A*A.';
% generate full matrix
B = full(A);
m_order = symamd(A);
tic
spmult = A(m_order,m_order)*A(m_order,m_order).';
flsp = toc;
tic, fulmult = B*B.'; flful = toc;
fprintf('Time for sparse mult = %6.4f\n',flsp)
fprintf('Time for  full  mult = %6.4f\n',flful)
```

Running this script `e4s209.m` results in the following output:

```
Time for sparse mult = 0.0073
Time for  full  mult = 3.2559
```

We now perform a similar experiment to that of `e4s209.m`, but for a more complex numerical process than multiplication. In the script that follows we examine LU decomposition. We consider the result of using a minimum degree ordering on the LU decomposition process by comparing the performance of the `lu` function with and without preordering. This is illustrated in script `e4s210.m`.

```
% e4s210.m
% generate a sparse matrix
n = 100;
offdiag = sparse(2:n,1:n-1,2*ones(1,n-1),n,n);
offdiag2 = sparse(4:n,1:n-3,3*ones(1,n-3),n,n);
offdiag3 = sparse(n-5:n,1:6,7*ones(1,6),n,n);
A = sparse(1:n,1:n,4*ones(1,n),n,n);
A = A+offdiag+offdiag'+offdiag2+offdiag2'+offdiag3+offdiag3';
A = A*A.';
A1 = flipud(A);
A = A+A1;
n1 = nnz(A)
B = full(A); %generate full matrix
m_order = symamd(A);
lud = lu(A(m_order,m_order));
n2 = nnz(lud)
fullu = lu(B);
n3 = nnz(fullu)
subplot(2,2,1), spy(A,'k');
```

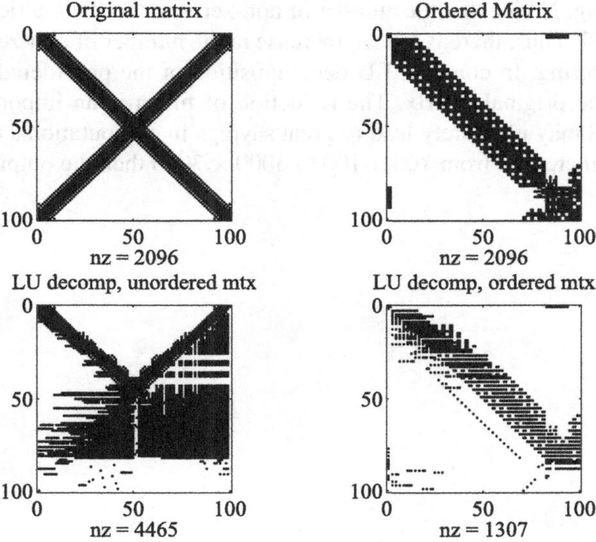

FIGURE 2.7

Effect of minimum degree ordering on LU decomposition. The spy function shows the matrix, the ordered matrix, and LU decomposition with and without preordering.

```
title('Original matrix')
subplot(2,2,2), spy(A(m_order,m_order),'k')
title('Ordered Matrix')
subplot(2,2,3), spy(fullu,'k')
title('LU decomposition,unordered matrix')
subplot(2,2,4), spy(lud,'k')
title('LU decomposition, ordered matrix')
```

Running script e4s210.m gives

```
n1 =
        2096

n2 =
        1307

n3 =
        4723
```

As expected, by using a sparse operation we achieve a reduction in the time taken to determine the LU decomposition. Fig. 2.7 shows the original matrix with 2096 non-zero elements, the reordered matrix (which has the same number of non-zeros), and the LU decomposition structure both with and without minimum degree ordering. Notice that the number of non-zeros in the LU matrices with preordering is 1307 and without is 4465. Thus, there is a large increase in the number of non-zero elements in the LU matrices without preordering. In contrast, LU decomposition of the preordered matrix has produced fewer non-zeros than the original matrix. The reduction of fill-in is an important feature of sparse numerical processes and may ultimately lead to great savings in computational effort. Note that if the size of the matrices are increased from 100×100 to 3000×3000 then the output from `e4s210` is

```
n1 =
       65896

n2 =
       34657

n3 =
      510266

Time for sparse lu = 0.0131
Time for  full  lu = 0.8313
```

Here we obtain a more substantial reduction by using sparse operations.

The MATLAB function `symamd` provides a minimum degree ordering for symmetric matrices. For non-symmetric matrices MATLAB provides the function `colmmd` which gives the column minimum degree ordering for non-symmetric matrices. An alternative ordering, which is used to reduce bandwidth, is the reverse Cuthill–MacGee ordering. This is implemented in MATLAB by the function `symrcm`. The execution of the statement `p = symrcm(A)` provides the permutation vector `p` to produce the required preordering and `A(p,p)` is the reordered matrix.

We have shown that in general taking account of sparsity will provide savings in floating-point operations. However, these savings fall off as the matrices on which we are operating become less sparse as script `e4s211.m` illustrates.

```
% e4s211.m
n = 1000;  b = 1:n;
disp('   density    time_sparse   time_full');
for density = 0.004:0.003:0.039
    A = sprandsym(n,density)+0.1*speye(n);
    density = density+1/n;
    tic, x = A\b'; f1 = toc;
    B = full(A);
    tic, y = B\b'; f2 = toc;
    fprintf('%10.4f %12.4f %12.4f\n',density,f1,f2);
end
```

In script `e4s211` a diagonal of elements has been added to the randomly generated sparse matrix. This is done to ensure that each row of the matrix contains a non-zero element; otherwise the matrix may be singular. Adding this diagonal modifies the density. If the original $n \times n$ matrix has a density of d, then, assuming that this matrix has only zeros on the diagonal, the modified density is $d + 1/n$.

```
density     time_sparse   time_full
 0.0050       0.0204        0.1907
 0.0080       0.0329        0.1318
 0.0110       0.0508        0.1332
 0.0140       0.0744        0.1399
 0.0170       0.0892        0.1351
 0.0200       0.1064        0.1372
 0.0230       0.1179        0.1348
 0.0260       0.1317        0.1381
 0.0290       0.1444        0.1372
 0.0320       0.1516        0.1369
 0.0350       0.1789        0.1404
 0.0380       0.1627        0.1450
```

This output shows that the advantage of using a sparse class of matrix diminishes as the density increases.

Another application where sparsity is important is in solving the least squares problem. This problem is known to be ill-conditioned and hence any saving in computational effort is particularly beneficial. This is directly implemented by using `A\b` where `A` is non-square and sparse. To illustrate the use of the \ operator with sparse matrices and compare its performance when no account is taken of sparsity we use script `e4s212.m`:

```
% e4s212.m
% generate a sparse triple diagonal matrix
n = 2000;
rowpos = 2:n;  colpos = 1:n-1;
values = ones(1,n-1);
offdiag = sparse(rowpos,colpos,values,n,n);
A = sparse(1:n,1:n,4*ones(1,n),n,n);
A = A+offdiag+offdiag';
%Now generate a sparse least squares system
Als = A(:,1:n/2);
%generate full matrix
Cfl = full(Als);
rhs = 1:n;
tic, x = Als\rhs'; f1 = toc;
tic, x = Cfl\rhs'; f2 = toc;
fprintf('Time for sparse least squares solve = %8.4f\n',f1)
fprintf('Time for  full  least squares solve = %8.4f\n',f2)
```

Script `e4s212.m` provides the following results

```
Time for sparse least squares solve =   0.0027
Time for  full  least squares solve =   0.9444
```

Again we see the advantage of using sparsity.

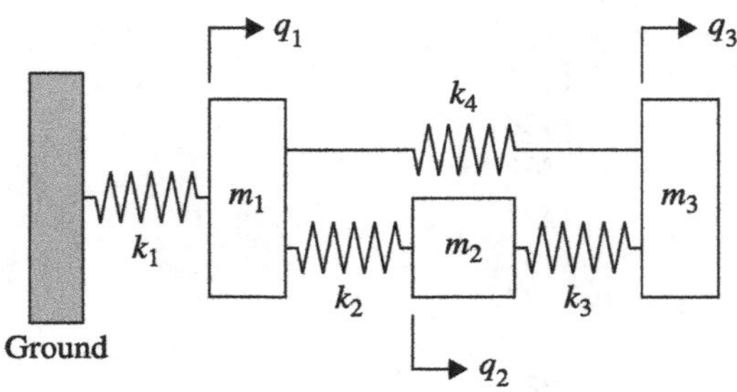

FIGURE 2.8

Mass-spring system with three degrees of freedom.

We have not covered all aspects of sparsity or described all the related functions. However, we hope this section provides a helpful introduction to this difficult but important and valuable feature of MATLAB.

2.15 THE EIGENVALUE PROBLEM

Eigenvalue problems arise in many branches of science and engineering. For example, the vibration characteristics of structures are determined from the solution of an algebraic eigenvalue problem. Here, we consider a particular example of a system of masses and springs shown in Fig. 2.8. The equations of motion for this system are

$$\begin{aligned} m_1\ddot{q}_1 + (k_1 + k_2 + k_4)\,q_1 - k_2q_2 - k_4q_3 &= 0 \\ m_2\ddot{q}_2 - k_2q_1 + (k_2 + k_3)\,q_2 - k_3q_3 &= 0 \\ m_3\ddot{q}_3 - k_4q_1 - k_3q_2 + (k_3 + k_4)\,q_3 &= 0 \end{aligned} \tag{2.35}$$

where m_1, m_2, and m_3 are the system masses and $k_1, \ldots, k_4$ are the spring stiffnesses. If we assume a harmonic solution for each coordinate, then $q_i(t) = u_i \exp(\jmath\omega t)$ where $\jmath = \sqrt{-1}$, for $i = 1$, 2, and 3. Hence $d^2q_i/dt^2 = -\omega^2 u_i \exp(\jmath\omega t)$. Substituting in (2.35) and canceling the common factor $\exp(\jmath\omega t)$ gives

$$\begin{aligned} -\omega^2 m_1u_1 + (k_1 + k_2 + k_4)\,u_1 - k_2u_2 - k_4u_3 &= 0 \\ -\omega^2 m_2u_2 - k_2u_1 + (k_2 + k_3)\,u_2 - k_3u_3 &= 0 \\ -\omega^2 m_3u_3 - k_4u_1 - k_3u_2 + (k_3 + k_4)\,u_3 &= 0 \end{aligned} \tag{2.36}$$

If $m_1 = 10$ kg, $m_2 = 20$ kg, $m_3 = 30$ kg, $k_1 = 10$ kN/m, $k_2 = 20$ kN/m, $k_3 = 25$ kN/m, and $k_4 =$ 15 kN/m, then (2.36) becomes

$$\begin{aligned} -\omega^2 10u_1 + 45{,}000u_1 - 20{,}000u_2 - 15{,}000u_3 &= 0 \\ -\omega^2 20u_1 - 20{,}000u_1 + 45{,}000u_2 - 25{,}000u_3 &= 0 \\ -\omega^2 30u_1 - 15{,}000u_1 - 25{,}000u_2 + 40{,}000u_3 &= 0 \end{aligned}$$

This can be expressed in matrix notation as

$$-\omega^2\mathbf{Mu}+\mathbf{Ku}=\mathbf{0} \tag{2.37}$$

where

$$\mathbf{M}=\begin{bmatrix}10 & 0 & 0\\ 0 & 20 & 0\\ 0 & 0 & 30\end{bmatrix}\text{ kg and }\mathbf{K}=\begin{bmatrix}45 & -20 & -15\\ -20 & 45 & -25\\ -15 & -25 & 40\end{bmatrix}\text{ kN/m} \tag{2.38}$$

Eq. (2.37) can be rearranged in a variety of ways. For example, it can be written

$$\mathbf{Mu}=\lambda\mathbf{Ku}\text{ where }\lambda=\frac{1}{\omega^2} \tag{2.39}$$

This is an algebraic eigenvalue problem and solving it determines values for the eigenvectors **u** and the eigenvalues, λ. MATLAB provides a function `eig` to solve the eigenvalue problem and to illustrate its use we apply it to the solution of (2.37).

```
>> M = [10 0 0;0 20 0;0 0 30];
>> K = 1000*[45 -20 -15;-20 45 -25;-15 -25 40];
>> lambda = eig(M,K).'

lambda =
    0.0002    0.0004    0.0073

>> omega = sqrt(1./lambda)

omega =
   72.2165   52.2551   11.7268
```

This result tells us that the system of Fig. 2.8 vibrates with natural frequencies 11.72, 52.25, and 72.21 rad/s. In this example we have chosen not to determine `u`, the eigenvectors. We will discuss further the use of the function `eig` in Section 2.17.

Having provided an example of an eigenvalue problem, we consider the standard form of this problem thus:

$$\mathbf{Ax}=\lambda\mathbf{x} \tag{2.40}$$

This equation is an algebraic eigenvalue problem where **A** is a given $n\times n$ matrix of coefficients, **x** is an unknown column vector of n elements and λ is an unknown scalar. Eq. (2.40) can alternatively be written as

$$(\mathbf{A}-\lambda\mathbf{I})\mathbf{x}=\mathbf{0} \tag{2.41}$$

Our aim is to discover the values of **x**, called the characteristic vectors or eigenvectors, and the corresponding values of λ, called the characteristic values or eigenvalues. The values of λ that satisfy (2.41) are given by the roots of the equation

$$|\mathbf{A}-\lambda\mathbf{I}|=0 \tag{2.42}$$

These values of λ are such that $(\mathbf{A} - \lambda\mathbf{I})$ is singular. Since (2.41) is a homogeneous equation system, non-trivial solutions exist for these values of λ. Evaluation of the determinant (2.42) leads to an nth degree polynomial in λ which is called the characteristic polynomial. This characteristic polynomial has n roots, some of which may be repeated, giving the n values of λ. In MATLAB we can create the coefficients of the characteristic polynomial using the function `poly`, and the roots of the resulting characteristic polynomial can be found using the function `roots`. For example,

$$\mathbf{A} = \begin{bmatrix} 1 & 2 & 3 \\ 4 & 5 & -6 \\ 7 & -8 & 9 \end{bmatrix}$$

then we have

```
>> A = [1 2 3;4 5 -6;7 -8 9];
>> p = poly(A)

p =
    1.0000  -15.0000  -18.0000  360.0000
```

Hence the characteristic equation is $\lambda^3 - 15\lambda^2 - 18\lambda + 360 = 0$. To find the roots of this equation we use the statement

```
>> roots(p).'

ans =
   14.5343   -4.7494    5.2152
```

We can verify this result using the function `eig`, thus:

```
>> eig(A).'

ans =
   -4.7494    5.2152   14.5343
```

Having obtained the eigenvalues we can substitute back into (2.41) to obtain linear equations for the characteristic vectors thus:

$$(\mathbf{A} - \lambda_i\mathbf{I})\mathbf{x} = \mathbf{0} \text{ for } i = 1, 2, ..., n \tag{2.43}$$

This homogeneous system provides n non-trivial solutions for $\mathbf{x}$. However, the use of (2.42) and (2.43) is not a practical means of solving eigenvalue problems.

If $\mathbf{A}$ is a symmetric real matrix or a Hermitian matrix, (such that $\mathbf{A} = \mathbf{A}^{\mathsf{H}}$), then its eigenvalues are real, but not necessarily positive, and the corresponding eigenvectors are real if $\mathbf{A}$ is real and symmetric, and complex if $\mathbf{A}$ is Hermitian. For such a system, if λ_i, $\mathbf{x}_i$ and λ_j, $\mathbf{x}_j$ satisfy the eigenvalue problem and λ_i and λ_j are distinct, then

$$\mathbf{x}_i^{\mathsf{H}}\mathbf{x}_j = 0 \; i \neq j \tag{2.44}$$

and

$$\mathbf{x}_i^{\mathsf{H}}\mathbf{A}\mathbf{x}_j = 0 \; i \neq j \tag{2.45}$$

Eqs. (2.44) and (2.45) are called the orthogonality relationships. If $\mathbf{A}$ is real and symmetric, then in (2.44) and (2.45) we take the transpose of the vectors denoted by (^{T}), rather than the complex conjugate transpose denoted by (^{H}). If $i = j$, then in general $\mathbf{x}_i^{\mathsf{H}}\mathbf{x}_i$ and $\mathbf{x}_i^{\mathsf{H}}\mathbf{A}\mathbf{x}_i$ are non-zero. The vector $\mathbf{x}_i$ includes an arbitrary scalar multiplier because the vector multiplies both sides of (2.40). Hence the product $\mathbf{x}_i^{\mathsf{H}}\mathbf{x}_i$ must be arbitrary. However, if the arbitrary scalar multiplier is adjusted so that $\mathbf{x}_i^{\mathsf{H}}\mathbf{x}_i = 1$ then the vectors are said to be *normalized*. In fact, the MATLAB function `eig` normalizes the eigenvectors, so that

$$\mathbf{x}_i^{\mathsf{H}}\mathbf{x}_i = 1 \;\text{ for }\; i = 1, 2, \ldots, n \tag{2.46}$$

When the vectors are normalized so that (2.46) is valid, then

$$\mathbf{x}_i^{\mathsf{H}}\mathbf{A}\mathbf{x}_i = \lambda_i \;\text{ for }\; i = 1, 2, \ldots, n \tag{2.47}$$

Note that the eigenvectors are orthogonal, that is (2.44) and (2.45) are valid, then they are orthogonal irrespective of whether or not the eigenvectors are normalized according to (2.46).

Sometimes the eigenvalues are not distinct and the eigenvectors associated with these equal or repeated eigenvalues are not necessarily orthogonal. If $\lambda_i = \lambda_j$ and the other eigenvalues, λ_k, are distinct, then

$$\left.\begin{aligned} \mathbf{x}_i^{\mathsf{H}}\mathbf{x}_k &= 0 \\ \mathbf{x}_j^{\mathsf{H}}\mathbf{x}_k &= 0 \end{aligned}\right\} \quad k = 1,\ 2,\ \ldots, n,\ \ k \neq i,\ \ k \neq j \tag{2.48}$$

For consistency, we can choose to make $\mathbf{x}_i^{\mathsf{H}}\mathbf{x}_j = 0$. When $\lambda_i = \lambda_j$, the eigenvectors $\mathbf{x}_i$ and $\mathbf{x}_j$ are not unique and a linear combination of them, $(\alpha\mathbf{x}_i + \gamma\mathbf{x}_j)$ where α and γ are arbitrary constants, also satisfies the eigenvalue problem.

We can combine all the eigensolutions into a single matrix equation by placing all the normalized eigenvectors into a single matrix, $\mathbf{X}$ and the eigenvalues in a diagonal matrix, $\mathbf{D}$ thus:

$$\mathbf{X} = [\ \mathbf{x}_1 \quad \mathbf{x}_2 \quad \ldots \quad \mathbf{x}_n\ ] \tag{2.49}$$

$$\mathbf{D} = \begin{bmatrix} \lambda_1 & 0 & \ldots & 0 \\ 0 & \lambda_2 & \ldots & 0 \\ \vdots & \vdots & \vdots & \vdots \\ 0 & 0 & \ldots & \lambda_n \end{bmatrix} \tag{2.50}$$

If $\mathbf{A}$ is an Hermitian matrix, then $\mathbf{X}$ is a unitary matrix, that is $\mathbf{X}^{\mathsf{H}} = \mathbf{X}^{-1}$. Its determinant is either $+1$ or -1 and its eigenvalues are complex but lie on a unit circle in the complex plane, that is their amplitudes all equal one but their phases differ. If $\mathbf{A}$ is an $n \times n$ real symmetric matrix, then $\mathbf{X}$ is an orthogonal matrix, that is $\mathbf{X}^{\mathsf{T}} = \mathbf{X}^{-1}$. The determinant of $\mathbf{X}$ is $+1$ and its eigenvalues are complex conjugate pairs if n is even, or complex conjugate pairs plus one real eigenvalue if n is odd. Again, all the eigenvalues have an absolute value of one and thus lie on a unit circle in the complex plane.

The n solutions (2.49) and (2.50) satisfy the eigenvalue problem (2.40). Assembling all the eigen-solutions into one equation, we have:

$$\mathbf{AX} = \mathbf{XD} \tag{2.51}$$

Using similar reasoning to that above, and substituting (2.49) and (2.50) in (2.46) gives

$$\mathbf{X}^{\mathsf{H}}\mathbf{X} = \mathbf{I} \tag{2.52}$$

Pre-multiplying (2.51) by $\mathbf{X}^{\mathsf{H}}$, and noting the relationship given by (2.52) we have:

$$\mathbf{X}^{\mathsf{H}}\mathbf{AX} = \mathbf{D} \tag{2.53}$$

Example 2.7. Determine the eigenvalues of the Hermitian matrix

$$\mathbf{A} = \begin{bmatrix} 1+0\imath & 2+2\imath \\ 2-2\imath & 3+0\imath \end{bmatrix}$$

Using MATLAB function `eig` we can determine both the eigenvalues and eigenvectors of this matrix as follows:

```
>> A = [1 2;2 3]+j*[0 2;-2 0];
>> [X,D] = eig(A)

X =
   0.5774 + 0.5774i   0.4082 + 0.4082i
  -0.5774 + 0.0000i   0.8165 + 0.0000i

D =
   -1.0000         0
         0    5.0000
```

where the diagonal elements of **D** are the eigenvalues and the corresponding eigenvector are given by **X**. We can verify, for this problem, the relationships given by (2.52) and (2.53) as follows:

```
>> X'*A*X

ans =
  -1.0000 + 0.0000i  -0.0000 - 0.0000i
  -0.0000 + 0.0000i   5.0000 + 0.0000i

>> X'*X

ans =
   1.0000 + 0.0000i  -0.0000 - 0.0000i
  -0.0000 + 0.0000i   1.0000 + 0.0000i
```

In summary, if **A** is real, both its eigenvalues and eigenvectors are real, but the eigenvalues are not necessarily positive. If **A** is a Hermitian matrix, its eigenvalues are real, but not necessarily positive, but its eigenvectors are generally complex.

Consider now the case when **A** is a general complex square matrix (that is, not Hermitian) with a set of eigenvalues $\mathbf{x}_i$ and corresponding eigenvalues λ_i. A pair of related eigenvalue problems exist as follows:

$$\mathbf{A}\mathbf{x} = \lambda\mathbf{x} \tag{2.54}$$

and

$$\mathbf{A}^{\mathsf{T}}\mathbf{y} = \beta\mathbf{y} \tag{2.55}$$

The second eigenvalue problem, (2.55), can be transposed to give

$$\mathbf{y}^{\mathsf{T}}\mathbf{A} = \beta\mathbf{y}^{\mathsf{T}} \tag{2.56}$$

The vectors **x** and **y** are called the right and left vectors of **A**, respectively. The equations $|\mathbf{A} - \lambda\mathbf{I}| = 0$ and $|\mathbf{A}^{\mathsf{T}} - \beta\mathbf{I}| = 0$ must have the same solutions for λ and β because the determinant of a matrix and the determinant of its transpose are equal. Thus, the eigenvalues of **A** and $\mathbf{A}^{\mathsf{T}}$ are identical but the eigenvectors **x** and **y** will, in general, differ from each other. The eigenvalues and eigenvectors of a non-symmetric real matrix are either real or pairs of complex conjugates. If λ_i, $\mathbf{x}_i$, $\mathbf{y}_i$ and λ_j, $\mathbf{x}_j$, $\mathbf{y}_j$ are solutions that satisfy the eigenvalue problems of (2.54) and (2.55) and λ_i and λ_j are distinct, then

$$\mathbf{y}_i^{\mathsf{T}}\mathbf{x}_j = 0 \quad i \neq j \tag{2.57}$$

and

$$\mathbf{y}_i^{\mathsf{T}}\mathbf{A}\mathbf{x}_j = 0 \quad i \neq j \tag{2.58}$$

Eqs. (2.57) and (2.58) are called the bi-orthogonal relationships. As with (2.46) and (2.47) if, in these equations, $i = j$, then in general $\mathbf{y}_i^{\mathsf{T}}\mathbf{x}_i$ and $\mathbf{y}_i^{\mathsf{T}}\mathbf{A}\mathbf{x}_i$ are not zero. The eigenvectors $\mathbf{x}_i$ and $\mathbf{y}_i$ include arbitrary scaling factors and so the product of these vectors will also be arbitrary. However, if the vectors are adjusted so that

$$\mathbf{y}_i^{\mathsf{T}}\mathbf{x}_i = 1 \text{ for } i = 1, 2, \ldots, n \tag{2.59}$$

then

$$\mathbf{y}_i^{\mathsf{T}}\mathbf{A}\mathbf{x}_i = \lambda_i \text{ for } i = 1, 2, \ldots, n \tag{2.60}$$

We cannot, in these circumstances, describe either $\mathbf{x}_i$ or $\mathbf{y}_i$ as normalized; the vectors still include an arbitrary scale factor; only their product can be chosen uniquely.

The eigensolutions can be combined us:

$$\mathbf{X} = [\ \mathbf{x}_1 \quad \mathbf{x}_2 \quad \ldots \quad \mathbf{x}_n\ ] \tag{2.61}$$

$$\mathbf{Y} = [\ \mathbf{y}_1 \quad \mathbf{y}_2 \quad \ldots \quad \mathbf{y}_n\ ] \tag{2.62}$$

$$\mathbf{D} = \begin{bmatrix} \lambda_1 & 0 & \dots & 0 \\ 0 & \lambda_2 & \dots & 0 \\ \vdots & \vdots & \vdots & \vdots \\ 0 & 0 & \dots & \lambda_n \end{bmatrix} \tag{2.63}$$

The n solutions (2.61) and (2.63) satisfy the eigenvalue problem (2.54). Assembling all the eigensolutions into one equation, we have:

$$\mathbf{AX} = \mathbf{XD} \tag{2.64}$$

Similarly, noting that $\beta_i = \lambda_i$, the n solutions (2.62) and (2.63) satisfy the eigenvalue problem (2.55). Assembling all the eigensolutions into one equation, we have:

$$\mathbf{A}^{\mathsf{T}}\mathbf{Y} = \mathbf{YD} \text{ or } \mathbf{Y}^{\mathsf{T}}\mathbf{A} = \mathbf{DY}^{\mathsf{T}} \tag{2.65}$$

Using (2.49) and (2.50) we can write (2.59) in matrix form:

$$\mathbf{Y}^{\mathsf{T}}\mathbf{X} = \mathbf{I} \tag{2.66}$$

Pre-multiplying (2.64) by $\mathbf{Y}^{\mathsf{T}}$ and noting the relationship given by (2.66) we have

$$\mathbf{Y}^{\mathsf{T}}\mathbf{AX} = \mathbf{D} \tag{2.67}$$

In summary, if **A** is real (but not symmetric) or complex (but not Hermitian) then both its eigenvalues and eigenvectors will be complex.

Example 2.8. determine the eigenvalues of the following complex matrix,

$$\mathbf{A} = \begin{bmatrix} 1 + 2\imath & 2 + 3\imath \\ 4 - 1\imath & 3 + 1\imath \end{bmatrix}$$

```
>> A = [1 2;4 3]+j*[2 3;-1 1]

A =

   1.0000 + 2.0000i   2.0000 + 3.0000i
   4.0000 - 1.0000i   3.0000 + 1.0000i

>> [X,D] = eig(A)

X =

   0.7518 + 0.0000i   0.4126 + 0.4473i
  -0.6068 + 0.2580i   0.7935 + 0.0000i

D =

  -1.6435 + 0.2649i   0.0000 + 0.0000i
   0.0000 + 0.0000i   5.6435 + 2.7351i
```

```
>> [Y,B] = eig(A.')

Y =
   0.7935 + 0.0000i   0.6068 - 0.2580i
  -0.4126 - 0.4473i   0.7518 + 0.0000i

B =
  -1.6435 + 0.2649i   0.0000 + 0.0000i
   0.0000 + 0.0000i   5.6435 + 2.7351i
```

Note that the eigenvalues given by **B** and **D** are identical but the right and left eigenvalues given by **X** and **Y** are different. We can verify, for this problem, the relationships given by (2.66) and (2.67) as follows:

```
>> p = Y.'*X

p =
   0.9623 + 0.1650i  -0.0000 - 0.0000i
  -0.0000 + 0.0000i   0.9623 + 0.1650i

>> X(:,1) = X(:,1)/p(1,1); X(:,2) = X(:,2)/p(2,2);
>> Y.'*X

ans =
   1.0000 - 0.0000i  -0.0000 - 0.0000i
  -0.0000 + 0.0000i   1.0000 + 0.0000i

>> Y.'*A*X

ans =
  -1.6435 + 0.2649i  -0.0000 - 0.0000i
   0.0000 - 0.0000i   5.6435 + 2.7351i
```

2.15.1 EIGENVALUE DECOMPOSITION

Consider the eigensolution of a Hermitian (or real symmetric matrix) given by (2.51). Then, post-multiplying by $\mathbf{X}^{-1}$ gives

$$\mathbf{A} = \mathbf{X}\mathbf{D}\mathbf{X}^{-1}$$

This is not a particularly useful decomposition of **A** since it involves a matrix inversion. However, from (2.52), we have $\mathbf{X}^{-1} = \mathbf{X}^{\mathsf{H}}$ and hence

$$\mathbf{A} = \mathbf{X}\mathbf{D}\mathbf{X}^{\mathsf{H}} \tag{2.68}$$

Thus **A** has been decomposed into the product of two orthogonal or unitary matrices and a diagonal matrix. This is useful decomposition. For example, to compute the square of a Hermitian matrix. We have

$$\mathbf{A}^2 = \mathbf{AA} = (\mathbf{XDX}^{\mathsf{H}})(\mathbf{XDX}^{\mathsf{H}})$$

and thus

$$\mathbf{A}^2 = \mathbf{XDIDX}^{\mathsf{H}} = \mathbf{XD}^2\mathbf{X}^{\mathsf{H}} \tag{2.69}$$

since $\mathbf{X}^{\mathsf{H}}\mathbf{X} = \mathbf{I}$. Computing $\mathbf{D}^2$ is trivial since $\mathbf{D}$ is diagonal and so each element on the diagonal is squared. In fact, this result can be generalized to

$$\mathbf{A}^p = \mathbf{XD}^p\mathbf{X}^{\mathsf{H}} \tag{2.70}$$

Another useful result is obtained by take in inverse of (2.68) to give

$$\mathbf{A}^{-1} = (\mathbf{XDX}^{\mathsf{H}})^{-1} = \mathbf{X}^{-\mathsf{H}}\mathbf{D}^{-1}\mathbf{X}^{-1} \tag{2.71}$$

Since $\mathbf{X}^{-1} = \mathbf{X}^{\mathsf{H}}$, on inverting $\mathbf{X} = \mathbf{X}^{-\mathsf{H}}$ so (2.71) becomes

$$\mathbf{A}^{-1} = \mathbf{XD}^{-1}\mathbf{X}^{\mathsf{H}} \tag{2.72}$$

This is an efficient method of inverting a matrix since inverting the diagonal matrix $\mathbf{D}$ is trivial; it is obtained by taking the reciprocal of each element on the diagonal. Of course, this assumes that the eigenvalue problem has been solved.

Example 2.9. Eqs. (2.68) to (2.72) are illustrated in MATLAB by continuing **Example 2.7**

```
>> A = [1 2;2 3]+j*[0 2;-2 0];
>> [X D]= eig(A);
>> AA = X*D*X'

AA =
    1.0000 + 0.0000i   2.0000 + 2.0000i
    2.0000 - 2.0000i   3.0000 + 0.0000i

>> Ainv = X*inv(D)*X'

Ainv =
   -0.6000 + 0.0000i   0.4000 + 0.4000i
    0.4000 - 0.4000i  -0.2000 + 0.0000i

>> inv(A)

ans =
   -0.6000 + 0.0000i   0.4000 + 0.4000i
    0.4000 - 0.4000i  -0.2000 + 0.0000i
```

Consider the decomposition of a general, complex matrix. By post-multiplying (2.64) by $\mathbf{X}^{-1}$ we have

$$\mathbf{A} = \mathbf{X}\mathbf{D}\mathbf{X}^{-1} \tag{2.73}$$

Similarly, pre-multiplying the second expression of (2.65) by $\mathbf{Y}^{-\mathsf{T}}$ we have

$$\mathbf{A} = \mathbf{Y}^{-\mathsf{T}}\mathbf{D}\mathbf{Y}^{\mathsf{T}} \tag{2.74}$$

Neither of these are particularly useful transforms since they both require matrix inversion. However, from (2.66), $\mathbf{X}^{-1} = \mathbf{Y}^{\mathsf{T}}$. Substituting in (2.73) gives

$$\mathbf{A} = \mathbf{X}\mathbf{D}\mathbf{Y}^{\mathsf{T}} \tag{2.75}$$

This decomposition avoids the need to invert a matrix, but it requires the solution of two eigenvalue problems for $\mathbf{A}$ and $\mathbf{A}^{\mathsf{T}}$.

Example 2.10. Consider the matrices of **Example 2.8**

```
>> A = [1 2;4 3]+j*[2 3;-1 1];
>> [X,D] = eig(A);
>> [Y,B] = eig(A.');
>> p = Y.'*X;
>> X(:,1) = X(:,1)/p(1,1); X(:,2) = X(:,2)/p(2,2);
>> A1 = X*D*inv(X)

A1 =
   1.0000 + 2.0000i   2.0000 + 3.0000i
   4.0000 - 1.0000i   3.0000 + 1.0000i

>> A2 = inv(Y.')*D*Y.'

A2 =
   1.0000 + 2.0000i   2.0000 + 3.0000i
   4.0000 - 1.0000i   3.0000 + 1.0000i

>> A3 = X*D*Y.'

A3 =
   1.0000 + 2.0000i   2.0000 + 3.0000i
   4.0000 - 1.0000i   3.0000 + 1.0000i
```

Eigenvalue decomposition will fail if the matrix being decomposed is defective. A defective matrix is a square matrix that does not have a full set of linearly independent eigenvectors and consequently can not be diagonalized. $\mathbf{A}$ is a normal matrix if $\mathbf{A}^{\mathsf{H}}\mathbf{A} = \mathbf{A}\mathbf{A}^{\mathsf{H}}$. A normal matrix (which includes Hermitian and real symmetric matrices as special cases) is never defective.

2.15.2 COMPARING EIGENVALUE AND SINGULAR VALUE DECOMPOSITION

Eigenvalue decomposition (EVD) can only decompose square matrices whereas singular value decomposition (SVD) can decompose rectangular or square matrices. There is, however, a relationship between singular values and eigenvalues. Consider a general, $n \times m$ matrix $\mathbf{A}$. The SVD of $\mathbf{A}$ is

$$\mathbf{A} = \mathbf{U}\mathbf{S}\mathbf{V}^\mathsf{T} \tag{2.76}$$

From this SVD we can write

$$\mathbf{A}\mathbf{A}^\mathsf{T} = (\mathbf{U}\mathbf{S}\mathbf{V}^\mathsf{T})(\mathbf{V}\mathbf{S}^\mathsf{T}\mathbf{U}^\mathsf{T}) = \mathbf{U}\mathbf{S}\mathbf{S}^\mathsf{T}\mathbf{U}^\mathsf{T}$$

because $\mathbf{V}^\mathsf{T}\mathbf{V} = \mathbf{I}$. Let the EVD of $\mathbf{A}\mathbf{A}^\mathsf{T}$ be

$$\mathbf{A}\mathbf{A}^\mathsf{T} = \mathbf{X}\mathbf{D}\mathbf{X}^\mathsf{T} \tag{2.77}$$

Comparing (2.76) and (2.77), we see $\mathbf{D} = \mathbf{S}\mathbf{S}^\mathsf{T}$ and $\mathbf{U} = \mathbf{X}$. That is $\mathbf{U}$ is equal to the eigenvector of $\mathbf{A}\mathbf{A}^\mathsf{T}$ and $\mathbf{S}\mathbf{S}^\mathsf{T}$ is equal to the eigenvalues of $\mathbf{A}\mathbf{A}^\mathsf{T}$.

From the SVD of $\mathbf{A}$ we can also write

$$\mathbf{A}^\mathsf{T}\mathbf{A} = (\mathbf{V}\mathbf{S}\mathbf{U}^\mathsf{T})(\mathbf{U}\mathbf{S}^\mathsf{T}\mathbf{V}^\mathsf{T}) = \mathbf{V}\mathbf{S}\mathbf{S}^\mathsf{T}\mathbf{V}^\mathsf{T}$$

since $\mathbf{U}^\mathsf{T}\mathbf{U} = \mathbf{I}$. Let the EVD of $\mathbf{A}^\mathsf{T}\mathbf{A}$ be

$$\mathbf{A}^\mathsf{T}\mathbf{A} = \mathbf{Y}\mathbf{D}\mathbf{Y}^\mathsf{T}$$

Then $\mathbf{V} = \mathbf{Y}$ and again $\mathbf{D} = \mathbf{S}\mathbf{S}^\mathsf{T}$.

The fact that the eigenvalues of $\mathbf{A}^\mathsf{T}\mathbf{A}$ are the square of the non-zero singular values of $\mathbf{A}$ has implications for the condition number $\mathbf{A}$. The condition number of $\mathbf{A}\mathbf{A}^\mathsf{T}$ is the condition number of $\mathbf{A}$ squared. Thus if $\mathbf{A}$ is ill-conditioned, working with $\mathbf{A}^\mathsf{T}\mathbf{A}$ will make the ill-conditioning worse. Thus, in difficult cases singular values are computed more accurately than eigenvalues.

Example 2.11. Determine the eigenvalues and condition number of $\mathbf{A}\mathbf{A}^\mathsf{T}$ and the SVD and condition number of $\mathbf{A}$. Note that $\mathbf{A}$ is a 3×4 matrix.

```
>> A = [4 -2 0 7;-2 7 -5 3;0 -5 12 -2];
>> [X,D] = eig(A*A')

X =

    0.0620    0.9979   -0.0163
    0.8321   -0.0607   -0.5512
    0.5511   -0.0206    0.8342

D =
   20.0386         0         0
         0   69.1435         0
         0         0  239.8179
```

```
>> [U,S,V] = svd(A)

U =
   -0.0163    0.9979    0.0620
   -0.5512   -0.0607    0.8321
    0.8342   -0.0206    0.5511

S =
   15.4861         0         0         0
         0    8.3153         0         0
         0         0    4.4764         0

V =
    0.0670    0.4947   -0.3164   -0.8067
   -0.5164   -0.2787    0.6580   -0.4719
    0.8244    0.0068    0.5478   -0.1423
   -0.2219    0.8231    0.4084    0.3261

>> S*S'

ans =
  239.8179         0         0
         0   69.1435         0
         0         0   20.0386

>> cond(A)

ans =
    3.4595

>> cond(A*A')

ans =
   11.9678
```

We note that $\mathbf{S}^2$ is equal to $\mathbf{D}$ (but the eigenvalues are listed in a different order) and the condition number of $\mathbf{AA}^\mathsf{T}$ is the square of the condition number of $\mathbf{A}$.

There is a special case when $\mathbf{A}$ is a square, symmetric, positive definite matrix, illustrated in the following example.

Example 2.12. The matrix $\mathbf{A}$ is square and symmetric. Determining its eigensolution we have

```
>> A = [4 -2 0;-2 7 -5; 0 -5 12];
>> [X,D] = eig(A)
```

```
X =
   -0.7055   -0.7022    0.0957
   -0.6312    0.5611   -0.5355
   -0.3224    0.4382    0.8391

D =
    2.2107         0         0
         0    5.5980         0
         0         0   15.1913
```

Since all the eigenvalues of the square matrix **A** are positive, it must be a positive definite matrix. Determining its SVD we have

```
>> [U,S,V] = svd(A)

U =
   -0.0957   -0.7022    0.7055
    0.5355    0.5611    0.6312
   -0.8391    0.4382    0.3224

S =
   15.1913         0         0
         0    5.5980         0
         0         0    2.2107

V =
   -0.0957   -0.7022    0.7055
    0.5355    0.5611    0.6312
   -0.8391    0.4382    0.3224
```

This example has illustrated the fact that the eigenvalues and singular values of a square, symmetrical, positive definite matrix are identical. Because the eigenvalue and singular values have been computed in different orders it isn't immediately obvious that **X** is equivalent to **V**. However, if we rearrange **X** so that it becomes equal to $[-\mathbf{x}_3 \;\; \mathbf{x}_2 \;\; -\mathbf{x}_1]$ then **X** does equal **V**. Note that the columns of **X** have been reordered in line with the order of the singular values, and two columns have been multiplied by -1. Recall that each eigenvector can be multiplied by a constant, in this case -1, and still be a solution to the eigenvalue problem.

2.16 ITERATIVE METHODS FOR SOLVING THE EIGENVALUE PROBLEM

The first of two simple iterative procedures described here determines the dominant or largest eigenvalue. The method, which is called the power method or matrix iteration, can be used on both symmetric and non-symmetric matrices. However, for a non-symmetric matrix the user must be alert to the possibility that there is not a single real dominant eigenvalue value but a complex conjugate pair. Under these conditions simple iteration does not converge.

Consider the eigenvalue problem defined by (2.40) and let the vector $\mathbf{u}_0$ be an initial trial solution. The vector $\mathbf{u}_0$ is an unknown linear combination of all the eigenvectors of the system provided they are linearly independent. Thus:

$$\mathbf{u}_0 = \sum_{i=1}^{n} \alpha_i \mathbf{x}_i \tag{2.78}$$

where α_i are unknown coefficients and $\mathbf{x}_i$ are the unknown eigenvectors. Let the iterative scheme be

$$\mathbf{u}_1 = \mathbf{A}\mathbf{u}_0, \;\; \mathbf{u}_2 = \mathbf{A}\mathbf{u}_1, \;\; \ldots, \;\; \mathbf{u}_p = \mathbf{A}\mathbf{u}_{p-1} \tag{2.79}$$

Substituting (2.78) into the sequence (2.79) we have

$$\begin{aligned}
\mathbf{u}_1 &= \sum_{i=1}^{n} \alpha_i \mathbf{A}\mathbf{x}_i = \sum_{i=1}^{n} \alpha_i \lambda_i \mathbf{x}_i \quad \text{since } \mathbf{A}\mathbf{x}_i = \lambda_i \mathbf{x}_i \\
\mathbf{u}_2 &= \sum_{i=1}^{n} \alpha_i \lambda_i \mathbf{A}\mathbf{x}_i = \sum_{i=1}^{n} \alpha_i \lambda_i^2 \mathbf{x}_i \\
&\ldots\ldots\ldots\ldots\ldots\ldots\ldots\ldots \\
\mathbf{u}_p &= \sum_{i=1}^{n} \alpha_i \lambda_i^{p-1} \mathbf{A}\mathbf{x}_i = \sum_{i=1}^{n} \alpha_i \lambda_i^{p} \mathbf{x}_i
\end{aligned} \tag{2.80}$$

The final equation can be rearranged, thus:

$$\mathbf{u}_p = \lambda_1^p \left[\alpha_1 \mathbf{x}_1 + \sum_{i=2}^{n} \alpha_i \left(\frac{\lambda_i}{\lambda_1} \right)^p \mathbf{x}_i \right] \tag{2.81}$$

It is the accepted convention that the n eigenvalues, λ_i of a matrix are numbered such that

$$|\lambda_1| > |\lambda_2| > \ldots > |\lambda_n|$$

Hence

$$\left[\frac{\lambda_i}{\lambda_1} \right]^p$$

tends to zero as p tends to infinity for $i = 2, 3, \ldots, n$. As p becomes large, we have from (2.81)

$$\mathbf{u}_p \Rightarrow \lambda_1^p \alpha_1 \mathbf{x}_1 \tag{2.82}$$

Thus $\mathbf{u}_p$ becomes proportional to $\mathbf{x}_1$ and the ratio between corresponding components of $\mathbf{u}_p$ and $\mathbf{u}_{p-1}$ tends to λ_1.

The algorithm, often referred to as the Power Method, is not usually implemented exactly as described previously because problems could arise due to numeric overflows. Usually, after each iteration, the resulting trial vector is normalized by dividing it by its largest element, thereby reducing the largest element in the vector to unity. This can be expressed mathematically as

$$\left.\begin{aligned} \mathbf{v}_p &= \mathbf{A}\mathbf{u}_p \\ \mathbf{u}_{p+1} &= \left(\frac{1}{\max(\mathbf{v}_p)}\right)\mathbf{v}_p \end{aligned}\right\} \quad p = 0, 1, 2, \ldots \tag{2.83}$$

where $\max(\mathbf{v}_p)$ is the element of $\mathbf{v}_p$ with the maximum modulus. The pair of Eqs. (2.83) are iterated until convergence is achieved. This modification to the algorithm does not affect the rate of convergence of the iteration. In addition to preventing the build up of very large numbers, the modification described before has the added advantage that it is now much easier to decide at what stage the iteration should be terminated. Post-multiplying the coefficient matrix $\mathbf{A}$ by one of its eigenvectors gives the eigenvector multiplied by the corresponding eigenvalue. Thus, when we stop the iteration because $\mathbf{u}_{p+1}$ is sufficiently close to $\mathbf{u}_p$ to ensure convergence, $\max(\mathbf{v}_p)$ will be an estimate of the eigenvalue.

The rate of convergence of the iteration is primarily dependent on the distribution of the eigenvalues; the smaller the ratios $|\lambda_i/\lambda_1|$ where $i = 2, 3, \ldots, n$, the faster the convergence.

The following MATLAB function `eigit` implements the iterative method to find the dominant eigenvalue and the associated eigenvector.

```
function [lam u iter] = eigit(A,tol)
% Solves EVP to determine dominant eigensolution by the power method
% Sample call: [lam u iter]=eigit(A,tol)
% A is a square matrix, tol is the accuracy
% lam is the dominant eigenvalue, u is the associated vector
% iter is the number of iterations required
[n n] = size(A);
err = 100*tol;
u0 = ones(n,1);   iter = 0;
while err>tol
    v = A*u0;
    u1 = (1/max(v))*v;
    err = max(abs(u1-u0));
    u0 = u1;  iter = iter+1;
end
u = u0;   lam = max(v);
```

We now apply the power method to find the dominant eigenvalue and corresponding vector for following eigenvalue problem.

$$\begin{bmatrix} 1 & 2 & 3 \\ 2 & 5 & -6 \\ 3 & -6 & 9 \end{bmatrix}\begin{bmatrix} x_1 \\ x_2 \\ x_3 \end{bmatrix} = \lambda \begin{bmatrix} x_1 \\ x_2 \\ x_3 \end{bmatrix} \tag{2.84}$$

```
>> A = [1 2 3;2 5 -6;3 -6 9];
>> [lam u iterations] = eigit(A,1e-8)
```

```
lam =
   13.4627

u =
    0.1319
   -0.6778
    1.0000

iterations =
    18
```

The dominant eigenvalue, to eight decimal places, is 13.46269899.

The power method can also be used to determine the smallest eigenvalue of a system. The eigenvalue problem $\mathbf{Ax} = \lambda\mathbf{x}$ can be rearranged to give

$$\mathbf{A}^{-1}\mathbf{x} = (1/\lambda)\,\mathbf{x}$$

Here the iterations will converge to the largest value of $1/\lambda$, that is, the smallest value of λ. However, as a general rule, matrix inversion should be avoided, particularly in large systems.

We have seen that direct iteration of $\mathbf{Ax} = \lambda\mathbf{x}$ leads to the largest or dominant eigenvalue. A second iterative procedure, called inverse iteration, provides a powerful method of determining subdominant eigensolutions. Consider again the eigenvalue problem of (2.40). Subtracting $\mu\mathbf{x}$ from both sides of this equation we have

$$(\mathbf{A} - \mu\mathbf{I})\,\mathbf{x} = (\lambda - \mu)\,\mathbf{x} \tag{2.85}$$

$$(\mathbf{A} - \mu\mathbf{I})^{-1}\,\mathbf{x} = \left(\frac{1}{\lambda - \mu}\right)\mathbf{x} \tag{2.86}$$

Consider the iterative scheme that begins with a trial vector $\mathbf{u}_0$. Then using the equivalent of (2.83) to (2.86) we have

$$\left.\begin{aligned} \mathbf{v}_s &= (\mathbf{A} - \mu\mathbf{I})^{-1}\,\mathbf{u}_s \\ \mathbf{u}_{s+1} &= \left(\frac{1}{\max(\mathbf{v}_s)}\right)\mathbf{v}_s \end{aligned}\right\} s = 0, 1, 2, \ldots \tag{2.87}$$

Iteration will lead to the largest value of $1/(\lambda - \mu)$, that is the smallest value of $(\lambda - \mu)$. The smallest value of $(\lambda - \mu)$ implies that the value of λ will be the eigenvalue closest to μ and $\mathbf{u}$ will have converged to the eigenvector $\mathbf{x}$ corresponding to this particular eigenvalue. Thus, by a suitable choice of μ, we have a procedure for finding subdominant eigensolutions.

Iteration is terminated when $\mathbf{u}_{s+1}$ is sufficiently close to $\mathbf{u}_s$. When convergence is complete

$$\frac{1}{\lambda - \mu} = \max(\mathbf{v}_s)$$

Thus, the value of λ nearest to μ is given by

$$\lambda = \mu + \frac{1}{\max(\mathbf{v}_s)} \tag{2.88}$$

The rate of convergence is fast, provided the chosen value of μ is close to an eigenvalue. If μ is equal to an eigenvalue, then $(\mathbf{A} - \mu\mathbf{I})$ is singular. In practice this seldom presents difficulties because it is unlikely that μ would be chosen, by chance, to exactly equal an eigenvalue. However, if $\mathbf{A} - \mu\mathbf{I}$ is singular then we have confirmation that the eigenvalue is known to a very high precision. The corresponding eigenvector can then be obtained by changing μ by adding a small quantity and iterating to determine the eigenvector.

Although inverse iteration can be used to find the eigensolutions of a system about which we have no previous knowledge, it is more usual to use inverse iteration to refine the approximate eigensolution obtained by some other technique. In practice $(\mathbf{A} - \mu\mathbf{I})^{-1}$ is not formed explicitly; instead, $(\mathbf{A} - \mu\mathbf{I})$ is usually decomposed into the product of a lower and an upper triangular matrix; explicit matrix inversion is avoided and is replaced by two efficient substitution procedures. In the simple MATLAB implementation of this procedure shown next, the operator \ is used to avoid matrix inversion.

```
function [lam u iter] = eiginv(A,mu,tol)
% Determines eigenvalue of A closest to mu with a tolerance tol.
% Sample call: [lam u] = eiginv(A,mu,tol)
% lam is the eigenvalue and u the corresponding eigenvector.
[n,n] = size(A);
err = 100*tol;
B = A-mu*eye(n,n);
u0 = ones(n,1);
iter = 0;
while err>tol
    v = B\u0; f = 1/max(v);
    u1 = f*v;
    err = max(abs(u1-u0));
    u0 = u1; iter = iter+1;
end
u = u0; lam = mu+f;
```

We now apply this function to find the eigenvalue of (2.84) nearest to 4 and the corresponding eigenvector.

```
>> A = [1 2 3;2 5 -6;3 -6 9];
>> [lam u iterations] = eiginv(A,4,1e-8)

lam =
    4.1283
```

```
u =
    1.0000
    0.8737
    0.4603

iterations =
     6
```

The eigenvalue closest to 4 is 4.12827017 to eight decimal places. The functions `eigit` and `eiginv` should be used with care when solving large-scale eigenvalue problems since convergence is not always guaranteed and in adverse circumstances may be slow.

2.17 SOLUTION OF THE GENERAL EIGENVALUE PROBLEM

There are many algorithms available to solve the eigenvalue problem. The method chosen is influenced by many factors such as the form and size of the eigenvalue problem, whether or not it is symmetric, whether it is real or complex, whether or not only the eigenvalues are required, whether all or only some of the eigenvalues and vectors are required.

We now describe the algorithms that are generally used to solve eigenvalue problems. The alternative approaches are as follows.

If **A** is a general matrix, it is first reduced to Hessenberg form using Householder's transformation method. A Hessenberg matrix has zeros everywhere below the diagonal except for the first sub-diagonal. If **A** is a symmetric matrix the transform creates a tridiagonal matrix. Then the eigenvalues and eigenvectors of the real upper Hessenberg matrix are found by the iterative application of the QR procedure. The QR procedure involves decomposing the Hessenberg matrix into an upper triangular matrix and a unitary matrix.

The following script, `e4s213.m` uses the MATLAB function `hess` to convert the original matrix to a Hessenberg matrix, followed by the iterative application of the QR decomposition using MATLAB function `qr` to determine the eigenvalues of a symmetric matrix. Note that in this script we have iterated 10 times, rather than use a formal test for convergence since the purpose of the script is merely to illustrate the functioning of the iterative application of the QR procedure.

```
% e4s213.m
A = [5 4 1 1;4 5 1 1; 1 1 4 2;1 1 2 4];
H1 = hess(A);
for i = 1:10
    [Q R] = qr(H1);
    H2 = R*Q;  H1 = H2;
    p = diag(H1)';
    fprintf('%2.0f %8.4f %8.4f',i,p(1),p(2))
    fprintf('%8.4f %8.4f\n',p(3),p(4))
end
```

Running script `e4s213.m` gives

```
 1   1.0000    8.3636  6.2420   2.3944
 2   1.0000    9.4940  5.4433   2.0627
 3   1.0000    9.8646  5.1255   2.0099
 4   1.0000    9.9655  5.0329   2.0016
 5   1.0000    9.9913  5.0084   2.0003
 6   1.0000    9.9978  5.0021   2.0000
 7   1.0000    9.9995  5.0005   2.0000
 8   1.0000    9.9999  5.0001   2.0000
 9   1.0000   10.0000  5.0000   2.0000
10   1.0000   10.0000  5.0000   2.0000
```

The iteration converges to the values 1, 10, 5, and 2 which are the correct. This QR iteration could be applied directly to the full matrix **A** but in general it would be inefficient. We have not given details of how the eigenvectors are computed.

When the arguments in the MATLAB function `eig` are two real or complex matrices, the QZ algorithm is used instead of the QR algorithm. The QZ algorithm (Golub and Van Loan, 1989) has been modified to deal with the complex case. When `eig` is called using a single complex matrix `A` then the algorithm works by applying the QZ algorithm to `eig(A,eye(size(A)))`. The QZ algorithm begins by noting that there exists a unitary **Q** and **Z** such that $\mathbf{Q}^{\mathsf{H}}\mathbf{AZ} = \mathbf{T}$ and $\mathbf{Q}^{\mathsf{H}}\mathbf{BZ} = \mathbf{S}$ and hence both **T** and **S** are both upper triangular. This is called generalized Schur decomposition. Providing s_{kk} is not zero then the eigenvalues are computed from the ratio t_{kk}/s_{kk}, where $k = 1, 2, ..., n$. The script `e4s214.m` demonstrates that the ratios of the diagonal elements of the **T** and **S** matrices give the required eigenvalues.

```
% e4s214.m
A = [10+2i 1 2;1-3i 2 -1;1 1 2];
B = [1 2-2i -2;4 5 6;7+3i 9 9];
[T S Q Z V] = qz(A,B);
r1 = diag(T)./diag(S)
r2 = eig(A,B)
```

Running script `e4s214.m` gives

```
r1 =
   1.6154 + 2.7252i
  -0.4882 - 1.3680i
   0.1518 + 0.0193i


r2 =
   1.6154 + 2.7252i
  -0.4882 - 1.3680i
   0.1518 + 0.0193i
```

Schur decomposition is closely related to the eigenvalue problem. The MATLAB function `schur(a)` produces an upper triangular matrix **T** with real eigenvalues on its diagonal and complex eigenvalues in 2×2 blocks on the diagonal. The matrix **A** decomposed to give

$$\mathbf{A} = \mathbf{UTU}^{\mathsf{H}}$$

where **U** is a unitary matrix such that $\mathbf{U}^H\mathbf{U} = \mathbf{I}$. The output of the script `e4s215` shows the similarity between Schur decomposition given by the MATLAB function `schur` and the eigenvalues determined using `eig`, of a given matrix.

```
% e4s215.m
A = [4 -5 0 3;0 4 -3 -5;5 -3 4 0;3 0 5 4];
T = schur(A), lam = eig(A)
```

Running script `e4s215` gives

```
T =
   12.0000    0.0000   -0.0000   -0.0000
         0    1.0000   -5.0000   -0.0000
         0    5.0000    1.0000   -0.0000
         0         0         0    2.0000

lam =
  12.0000 + 0.0000i
   1.0000 + 5.0000i
   1.0000 - 5.0000i
   2.0000 + 0.0000i
```

We can readily identify the four eigenvalues in the matrix `T`. The real parts of the eigenvalues lie on the diagonal of `T` and any non-zero elements in the sub- and super-diagonal of `T` are the imaginary parts of the eigenvalues. The script `e4s216.m` compares the performance of the `eig` function when solving various classes of problem.

```
% e4s216.m
disp('        real1    realsym1      real2     realsym2      comp1       comp2')
for n = 100:50:500
    A = rand(n); C = rand(n);
    S = A+C*i;
    T = rand(n)+i*rand(n);
    tic, [U,V] = eig(A); f1 = toc;
    B = A+A.'; D = C+C.';
    tic, [U,V] = eig(B); f2 = toc;
    tic, [U,V] = eig(A,C); f3 = toc;
    tic, [U,V] = eig(B,D); f4 = toc;
    tic, [U,V] = eig(S); f5 = toc;
    tic, [U,V] = eig(S,T); f6 = toc;
    fprintf('%12.3f %10.3f %10.3f %10.3f %10.3f %10.3f\n',f1,f2,f3,f4,f5,f6);
end
```

This script, `e4s216.m` gives the time taken (in seconds) to carry out the various operations. The output is as follows:

real1	realsym1	real2	realsym2	comp1	comp2
0.188	0.002	0.027	0.019	0.033	0.142
0.053	0.009	0.060	0.107	0.148	0.190
0.196	0.011	0.138	0.120	0.151	0.209
0.100	0.011	0.153	0.151	0.182	0.276
0.127	0.013	0.244	0.249	0.236	0.423
0.162	0.021	0.391	0.375	0.338	0.709
0.202	0.027	0.653	0.635	0.463	1.066
0.253	0.038	0.886	0.913	0.610	1.696
0.300	0.046	1.250	1.270	0.792	2.220

Note that each row is for a different size of $n \times n$ matrix, from 100 to 500 in steps of 50.

In some circumstances not all the eigenvalues and eigenvectors are required. For example, in a complex engineering structure, modeled with many hundreds of degrees of freedom, we may only require the first 15 eigenvalues, giving the natural frequencies of the model, and the corresponding eigenvectors. MATLAB provides the function `eigs` which finds a selected number of eigenvalues, such as those with the largest amplitude, the largest or smallest real or imaginary part, etc. This function is particularly useful when seeking a small number of eigenvalues of very large sparse matrices. Eigenvalue reduction algorithms are used to reduce the size of eigenvalue problem (for example, Guyan, 1965) but still allow selected eigenvalues to be computed to an acceptable level of accuracy.

MATLAB also includes the facility to find the eigenvalues of a sparse matrix. The script `e4s217.m` compares the number of floating-point operations required to find the eigenvalues of a matrix treated as sparse with the number of floating-point operations required to find the eigenvalues of the corresponding full matrix.

```
% e4s217.m
% generate a sparse triple diagonal matrix
n = 4000;
rowpos = 2:n; colpos = 1:n-1;
values = ones(1,n-1);
offdiag = sparse(rowpos,colpos,values,n,n);
A = sparse(1:n,1:n,4*ones(1,n),n,n);
A = A+offdiag+offdiag.';
% generate full matrix
B = full(A);
tic, eig(A); sptim = toc;
tic, eig(B); futim = toc;
fprintf('Time for sparse eigen solve = %8.6f\n',sptim)
fprintf('Time for  full  eigen solve = %8.6f\n',futim)
```

The results from running script `e4s217.m` are

```
Time for sparse eigen solve = 0.511121
Time for  full  eigen solve = 0.591209
```

Clearly a significant saving in time.

2.18 THE GOOGLE 'PAGERANK' ALGORITHM

An interesting application of eigenvalue analysis is the Google algorithm for determining the most important pages found in a web search.

Search engines use crawlers to continuously search the web, finding new pages and adding them to the vast index of pages. When you undertake a web search, the search engine looks at the index and finds the pages containing the key words used in your search. For example, a Google search for "eigenvalue" finds about 1,670,000 results in 0.54 seconds. The next step is to decide in what order to present these pages: clearly the most relevant pages should be shown first. Different search engines use different ways of ranking the pages. What makes the Google search engine particularly attractive is the 'PageRank' algorithm that decides which are the most important pages. The Google 'PageRank' algorithm was first described by Brin and Page (1998).

The pages containing the key search words can be viewed as a network where a large number of nodes, representing topics or pages, are connected by edges. Such a network is often called a graph. Certain nodes have more connections than others, depending on their importance and relevance. A node which has many other node connections pointing to it, and thus referencing it, may be considered more important than others with fewer references. When searching on the web for specific information amongst the huge numbers of items, it would clearly be useful to rank the pages in terms of their relevance to the user, since this is what the user requires. The Google algorithm provides an objective and efficient page ranking process. It ranks the pages by relevance based on the number of citations a page receives, assigns a page rank to each page, and lists the pages in order of relevance to the user. A key feature of page rank is that a page has high rank if many other pages of high rank point to it, so it is not only the issue of the number of pages which point to a page, the rank of these pages is also important.

So how can an algorithm be built to implement this process? First a model of the network of potentially interesting pages must be devised and then we must produce an objective mathematical process to establish the required page ranking in numerical form. From our previous discussion it is clear the model of the pages could be a connected network or graph and this is the approach adopted. Since we wish to use a numerical process we must then represent the graph numerically. This can be achieved by representing the graph as a matrix.

If we think of the matrix having elements at positions (i, j) where i is the row number of the matrix and j is the column of the matrix then the elements of **A** may be denoted by A_{ij}. In our model of the internet, we may define the elements of the matrix as the probabilities that information will be sent from page j to page i. We can set the probability values for each node i which has a connection to node j, to $1/n_j$ assuming each node j has n_j outward pointing connections to other nodes i. If there is no connection to any other node then the probability is of course zero. Thus the sum of probabilities in each column must equal 1 and probabilities lie in the range zero to one. Such a matrix is a column stochastic matrix, defined as one whose elements are less than or equal to one and the columns of which sum to 1. A stochastic matrix has some specific properties, one of which is that its largest or dominant eigenvalue is equal to one.

The question now arises, how can we use this model of the internet to rank the internet pages? It can be shown that the dominant eigenvector of the stochastic matrix **A** which we have used to model the internet links provides the page ranking we require. The numerical ordering of the individual values of the eigenvector rank the relevance of the specific page on the web to the searcher; the largest being

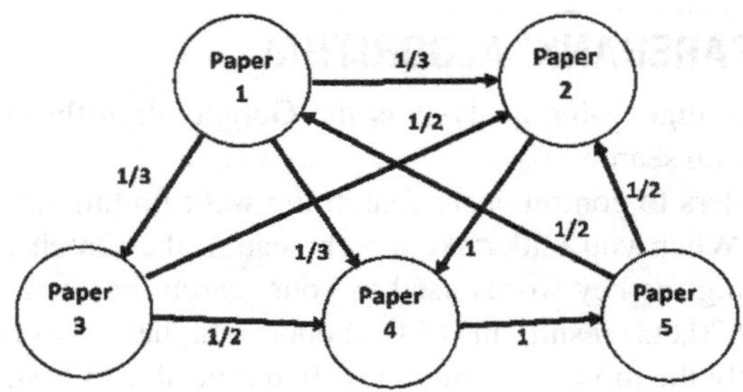

FIGURE 2.9

Connections of different strengths between five pages of the internet.

the most relevant and the smallest being the least relevant. However, this dominant eigenvector of the stochastic matrix **A** is not unique. To ensure uniqueness the Google matrix **G** was introduced as follows:

$$\mathbf{G} = \alpha\mathbf{A} + (1-\alpha)\mathbf{U}\mathbf{P}^{\mathsf{T}} \text{ where } 0 \le \alpha < 1$$

where **U** is a matrix of 1s and **P** is sometimes called a personalization vector such that $\mathbf{P} \geq 0$ and $||\mathbf{P}|| = 1$. This may be taken as the initial probabilities each given by $1/n$ where n is the size of the matrix. This matrix, **G**, the Google matrix, provides a unique dominant eigenvector and is also stochastic. The Google page rank vector **r** may now be defined by the equations:

$$\mathbf{G}\mathbf{r} = \mathbf{r} \text{ where } \mathbf{r} \geq \mathbf{0} \text{ and } ||\mathbf{r}|| = 1$$

This is the standard eigenvalue problem, see (2.40), with the dominant eigenvalue equal to one.

The Google page rank vector **r** may now be used to rank the internet pages and display them in order, largest value first and then in decreasing order of magnitude. The final question is: how can we find the dominant eigenvector of **G**? Fortunately it can be found by using the relatively simple power method, as described in Section 2.16. In the description of the power method in Section 2.16 it is explained that in order to avoid a numeric overflow caused by effectively multiplying the eigenvector by the eigenvalue raised to a power equal to the number of iterations performed, the iteration algorithm must be modified as described by (2.83). This is implemented in the function `eigit`. In determining the eigensolution of a stochastic matrix, since the dominant eigenvalue is close to and tending to unity, this difficulty does not arise and the algorithm can be simplified.

In fact, since we iterate the equations $\mathbf{r} = \mathbf{G}\mathbf{r}_{trial}$ and $\mathbf{r}_{trial} = \mathbf{r}$ this iteration is effectively $\mathbf{r} = \mathbf{G}^p\mathbf{r}_{trial}$, where p is suitably large. Thus irrespective of $\mathbf{r}_{trial}$ the product $\mathbf{G}^p\mathbf{r}_{trial}$ is the same, but scaled differently. This is because every column of $\mathbf{G}^p$ is identical.

Consider the graph shown in Fig. 2.9. The graph representing the connections between five scientific papers on the internet with various links connecting each node. Then this diagram may be represented by the stochastic matrix:

$$\begin{bmatrix} 0 & 0 & 0 & 0 & 1/2 \\ 1/3 & 0 & 1/2 & 0 & 1/2 \\ 1/3 & 0 & 0 & 0 & 0 \\ 1/3 & 1 & 1/2 & 0 & 0 \\ 0 & 0 & 0 & 1 & 0 \end{bmatrix} \tag{2.89}$$

Then the Google matrix is

$$\begin{bmatrix} 0.0300 & 0.0300 & 0.0300 & 0.0300 & 0.4550 \\ 0.3133 & 0.0300 & 0.4550 & 0.0300 & 0.4550 \\ 0.3133 & 0.0300 & 0.0300 & 0.0300 & 0.0300 \\ 0.3133 & 0.8800 & 0.4550 & 0.0300 & 0.0300 \\ 0.0300 & 0.0300 & 0.0300 & 0.8800 & 0.0300 \end{bmatrix} \tag{2.90}$$

Note that the sum of each column is one and for each element G_{ij} in the matrix the condition $0 \leq G_{ij} \leq 1$ is satisfied. Then $\mathbf{G}^{50}$ is

$$\begin{bmatrix} 0.1468 & 0.1468 & 0.1468 & 0.1468 & 0.1468 \\ 0.2188 & 0.2188 & 0.2188 & 0.2188 & 0.2188 \\ 0.0716 & 0.0716 & 0.0716 & 0.0716 & 0.0716 \\ 0.2880 & 0.2880 & 0.2880 & 0.2880 & 0.2880 \\ 0.2748 & 0.2748 & 0.2748 & 0.2748 & 0.2748 \end{bmatrix} \tag{2.91}$$

If $\mathbf{r}_{trial} = [1\ 1\ 1\ 1\ 1]^{\mathrm{T}}$ then

$$\mathbf{G}^{50}\mathbf{r}_{trial} = [0.7340\ 1.0940\ 0.3580\ 1.4400\ 1.3740]$$

In contrast, suppose $\mathbf{r}_{trial} = [1\ -1\ 1\ -1\ 1]^{\mathrm{T}}$ then

$$\mathbf{G}^{50}\mathbf{r}_{trial} = [0.1468\ 0.2188\ 0.0716\ 0.2880\ 0.2748]$$

In both cases, the vectors can be normalized to give

$$\mathbf{r} = [0.5097\ 0.7598\ 0.2486\ 1.0000\ 0.9542]$$

Thus the most important web location or paper is number 4, closely followed by number 5.

To implement these procedures as a formal algorithm a test must be included to determine if the procedure has converged to a sufficiently accurate estimate for **r**.

2.19 SUMMARY

We have described many of the important algorithms related to computational matrix algebra and shown how the power of MATLAB can be used to illustrate the application of these algorithms in a revealing way. We have shown how to solve over- and under-determined systems and eigenvalue problems. We have drawn the attention of the reader to the importance of sparsity in linear systems and demonstrated its significance. The scripts provided should help readers to develop their own applications.

In Chapter 10, we show how the symbolic toolbox can be usefully applied to solve some problems in linear algebra.

2.20 PROBLEMS

2.1. An $n \times n$ Hilbert matrix, $\mathbf{A}$, is defined by

$$a_{ij} = 1/(i + j - 1) \text{ for } i, j = 1, 2, \ldots, n$$

Find the inverse of $\mathbf{A}$ and the inverse of $\mathbf{A}^\mathsf{T}\mathbf{A}$ for $n = 5$. Then, noting that

$$(\mathbf{A}^\mathsf{T}\mathbf{A})^{-1} = \mathbf{A}^{-1}(\mathbf{A}^{-1})^\mathsf{T}$$

find the inverse of $\mathbf{A}^\mathsf{T}\mathbf{A}$ using this result for values of $n = 3, 4, \ldots, 6$. Compare the accuracy of the two results by using the inverse Hilbert function `invhilb` to find the exact inverse using $(\mathbf{A}^\mathsf{T}\mathbf{A})^{-1} = \mathbf{A}^{-1}(\mathbf{A}^{-1})^\mathsf{T}$. *Hint*: Compute norm($\mathbf{P} - \mathbf{R}$) and norm($\mathbf{Q} - \mathbf{R}$) where $\mathbf{P} = (\mathbf{A}^\mathsf{T}\mathbf{A})^{-1}$ and $\mathbf{Q} = \mathbf{A}^{-1}(\mathbf{A}^{-1})^\mathsf{T}$ and $\mathbf{R}$ is the exact value of $\mathbf{Q}$.

2.2. Find the condition number of $\mathbf{A}^\mathsf{T}\mathbf{A}$ where $\mathbf{A}$ is an $n \times n$ Hilbert matrix, defined in Problem 2.1, for $n = 3, 4, \ldots, 6$. How do these results relate to the results of Problem 2.1?

2.3. It can be proved that the series $(\mathbf{I} - \mathbf{A})^{-1} = \mathbf{I} + \mathbf{A} + \mathbf{A}^2 + \mathbf{A}^3 + \ldots$, where $\mathbf{A}$ is an $n \times n$ matrix, converges if the eigenvalues of $\mathbf{A}$ are all less than unity. The following $n \times n$ matrix satisfies this condition if $a + 2b < 1$ and a and b are positive:

$$\begin{bmatrix} a & b & 0 & \ldots & 0 & 0 & 0 \\ b & a & b & \ldots & 0 & 0 & 0 \\ \vdots & \vdots & \vdots & & \vdots & \vdots & \vdots \\ 0 & 0 & 0 & \ldots & b & a & b \\ 0 & 0 & 0 & \ldots & 0 & b & a \end{bmatrix}$$

Experiment with this matrix for various values of n, a, and b to illustrate that the series converges under the condition stated.

2.4. Use the function `eig` to find the eigenvalues of the following matrix:

$$\begin{bmatrix} 2 & 3 & 6 \\ 2 & 3 & -4 \\ 6 & 11 & 4 \end{bmatrix}$$

Then use the `rref` function on the matrix $(\mathbf{A} - \lambda\mathbf{I})$, taking λ equal to any of the eigenvalues. Solve the resulting equations by hand to obtain the eigenvector of the matrix. *Hint*: Note that an eigenvector is the solution of $(\mathbf{A} - \lambda\mathbf{I})\mathbf{x} = \mathbf{0}$ for λ equal to specific eigenvalues. Assume an arbitrary value for x_3.

2.5. For the system given in Problem 2.3, find the eigenvalues, assuming both full and sparse forms with $n = 10:10:30$. Compare your results with the exact solution given by

$$\lambda_k = a + 2b\cos\{k\pi/(n+1)\}, \quad k = 1, 2, \ldots$$

2.6. Find the solution of the over-determined system that follows using `pinv`, `qr`, and the \ operator.

$$\begin{bmatrix} 2.0 & -3.0 & 2.0 \\ 1.9 & -3.0 & 2.2 \\ 2.1 & -2.9 & 2.0 \\ 6.1 & 2.1 & -3.0 \\ -3.0 & 5.0 & 2.1 \end{bmatrix} \begin{bmatrix} x_1 \\ x_2 \\ x_3 \end{bmatrix} = \begin{bmatrix} 1.01 \\ 1.01 \\ 0.98 \\ 4.94 \\ 4.10 \end{bmatrix}$$

2.7. Write a script to generate an $n \times n$ matrix $\mathbf{E}$, where $\mathbf{E} = \{1/(n+1)\}\mathbf{C}$ where

$$\begin{aligned} c_{ij} &= i(n-i+1) && \text{if } i = j \\ &= c_{i,j-1} - i && \text{if } j > i \\ &= c_{ji} && \text{if } j < i \end{aligned}$$

Having generated $\mathbf{E}$ for $n = 5$, solve $\mathbf{Ex} = \mathbf{b}$ where $\mathbf{b} = [1:n]^{\mathsf{T}}$ by
(a) using the \ operator;
(b) using the `lu` function and solving $\mathbf{Ux} = \mathbf{y}$ and $\mathbf{Ly} = \mathbf{b}$.

2.8. Determine the inverse of $\mathbf{E}$ of Problem 2.7 for $n = 20$ and 50. Compare with the exact inverse which is a matrix with 2 along the main diagonal and -1 along the upper and lower sub-diagonals and zero elsewhere.

2.9. Determine the eigenvalues of $\mathbf{E}$ defined in Problem 2.7 for $n = 20$ and 50. The exact eigenvalues for this system are given by $\lambda_k = 1/[2 - 2\cos\{k\pi/(n+1)\}]$ where $k = 1, \ldots, n$.

2.10. Determine the condition number of $\mathbf{E}$ of Problem 2.7 using the MATLAB function `cond`, for $n = 20$ and 50. Compare your results with the theoretical expression for the condition number which is $4n^2/\pi^2$.

2.11. Find the eigenvalues and the left and right eigenvectors using the MATLAB function `eig` for the matrix

$$\mathbf{A} = \begin{bmatrix} 8 & -1 & -5 \\ -4 & 4 & -2 \\ 18 & -5 & -7 \end{bmatrix}$$

2.12. For the following matrix $\mathbf{A}$, using `eigit`, `eiginv`, determine:

a. the largest eigenvalue
b. the eigenvalue nearest 100
c. the smallest eigenvalue

$$\mathbf{A} = \begin{bmatrix} 122 & 41 & 40 & 26 & 25 \\ 40 & 170 & 25 & 14 & 24 \\ 27 & 26 & 172 & 7 & 3 \\ 32 & 22 & 9 & 106 & 6 \\ 31 & 28 & -2 & -1 & 165 \end{bmatrix}$$

2.13. Given that

$$\mathbf{A} = \begin{bmatrix} 1 & 2 & 2 \\ 5 & 6 & -2 \\ 1 & -1 & 0 \end{bmatrix} \text{ and } \mathbf{B} = \begin{bmatrix} 2 & 0 & 1 \\ 4 & -5 & 1 \\ 1 & 0 & 0 \end{bmatrix}$$

and defining **C** by

$$\mathbf{C} = \begin{bmatrix} \mathbf{A} & \mathbf{B} \\ \mathbf{B} & \mathbf{A} \end{bmatrix}$$

verify using `eig` that the eigenvalues of **C** are given by a combination of the eigenvalues of $\mathbf{A}+\mathbf{B}$ and $\mathbf{A}-\mathbf{B}$.

2.14. Write a MATLAB script to generate the matrix

$$\mathbf{A} = \begin{bmatrix} n & n-1 & n-2 & \dots & 2 & 1 \\ n-1 & n-1 & n-2 & \dots & 2 & 1 \\ n-2 & n-2 & n-2 & \dots & 2 & 1 \\ \vdots & \vdots & \vdots & \vdots & \vdots & \vdots \\ 2 & 2 & 2 & \dots & 2 & 1 \\ 1 & 1 & 1 & \dots & 1 & 1 \end{bmatrix}$$

Taking $n = 5$ and $n = 50$ and using the MATLAB function `eig`, find the largest and smallest eigenvalues.
The eigenvalues of this matrix are given by the formula

$$\lambda_i = \frac{1}{2}\left[1 - \cos\frac{(2i-1)\pi}{2n+1}\right]^{-1}, \ i = 1, 2 \dots n$$

Verify your results are correct using this formula.

2.15. Taking $n = 10$, find the eigenvalues of the matrix

$$\mathbf{A} = \begin{bmatrix} 1 & 0 & 0 & \dots & 0 & 1 \\ 0 & 1 & 0 & \dots & 0 & 2 \\ 0 & 0 & 1 & \dots & 0 & 3 \\ \vdots & \vdots & \vdots & \dots & \vdots & \vdots \\ 0 & 0 & 0 & \dots & 1 & n-1 \\ 1 & 2 & 3 & \dots & n-1 & n \end{bmatrix}$$

using `eig`. As an alternative, find the eigenvalues of **A** by first generating the characteristic polynomial using `poly` for the matrix **A** and then using `roots` to find the roots of the resulting polynomial. What conclusions do you draw from these results?

2.16. For the matrix given in Problem 2.12, use `eig` to find the eigenvalues. Then find the eigenvalues of **A** by first generating the characteristic polynomial for **A** using `poly` and then using `roots` to find the roots of the resulting polynomial. Use `sort` to compare the results of the two approaches. What conclusions do you draw from these results?

2.17. For the matrix given in Problem 2.14, taking $n = 10$, show that the trace is equal to the sum of the eigenvalues and the determinant is equal to the product of the eigenvalues. Use the MATLAB functions `det`, `trace`, and `eig`.

2.18. The matrix **A** is defined as follows:

$$\mathbf{A} = \begin{bmatrix} 2 & -1 & 0 & 0 & \cdots & 0 \\ -1 & 2 & -1 & 0 & \cdots & 0 \\ 0 & -1 & 2 & -1 & \cdots & 0 \\ \vdots & \vdots & \vdots & \vdots & \vdots & \vdots \\ 0 & 0 & \cdots & -1 & 2 & -1 \\ 0 & 0 & \cdots & 0 & -1 & 2 \end{bmatrix}$$

The condition number, c, for this matrix is given by $c = pn^q$ where n is the size of the matrix and p and q are constants. By computing the condition number for the matrix **A** for $n = 5:5:50$ using the MATLAB function `cond`, fit the function pn^q to the set of results you produce. *Hint*: Take logs of both sides of the equation for c and solve the system of over-determined equations using the \ operator.

2.19. An approximation for the inverse of $(\mathbf{I} - \mathbf{A})$ where **I** is an $n \times n$ unit matrix and **A** is an $n \times n$ matrix is given by

$$(\mathbf{I} - \mathbf{A})^{-1} = \mathbf{I} + \mathbf{A} + \mathbf{A}^2 + \mathbf{A}^3 + \dots$$

This series only converges and the approximation is valid only if the maximum eigenvalue of **A** is less than 1. Write a MATLAB function `invapprox(A,k)` which obtains an approximation to $(\mathbf{I} - \mathbf{A})^{-1}$ using k terms of the given series. The function must find all eigenvalues of **A** using the MATLAB function `eig`. If the largest eigenvalue is greater than one then a message will be output indicating the method fails. Otherwise the function will compute an approximation to $(\mathbf{I} - \mathbf{A})^{-1}$ using k terms of the series expansion given. Taking $k = 4$, test the function on the matrices:

$$\begin{bmatrix} 0.2 & 0.3 & 0 \\ 0.3 & 0.2 & 0.3 \\ 0 & 0.3 & 0.2 \end{bmatrix} \text{ and } \begin{bmatrix} 1.0 & 0.3 & 0 \\ 0.3 & 1.0 & 0.3 \\ 0 & 0.3 & 1.0 \end{bmatrix}$$

Use the `norm` function to compare the accuracy of the inverse of the matrix $(\mathbf{I} - \mathbf{A})$ found using the MATLAB `inv` function and the function `invapprox(A,k)` for $k = 4, 8, 16$.

2.20. The system of equations $\mathbf{Ax} = \mathbf{b}$, where **A** is a matrix of m rows and n columns, **x** is an n element column vector, **b** an m element column vector, is said to be under-determined if $n > m$. The direct use of the MATLAB `inv` function to solve this system fails since the matrix **A** is not square. However, multiplying both sides of the equation by $\mathbf{A}^\mathsf{T}$ gives:

$$\mathbf{A}^\mathsf{T}\mathbf{Ax} = \mathbf{A}^\mathsf{T}\mathbf{b}$$

$\mathbf{A}^\mathsf{T}\mathbf{A}$ is a square matrix and the MATLAB `inv` function can now be used to solve the system. Write a MATLAB function to use this result to solve under-determined systems. The function should allow the input of the **b** vector and the **A** matrix, form the necessary matrix products and use the MATLAB `inv` function to solve the system. The accuracy of the solution should be checked by using the MATLAB `norm` function to measure the difference between **Ax** and **b**. The function must also include the direct use of the MATLAB \ symbol to solve the same under-determined linear system, again include a check on the accuracy of the solution by using the MATLAB `norm` function to measure the difference between **Ax** and **b**. The function should take the form `udsys(A,b)` and return the solutions given by the different methods and the norms produced by the two methods. Test your program by using it to solve the under-determined system of linear equations **Ax** = **b** where:

$$\mathbf{A} = \begin{bmatrix} 1 & -2 & -5 & 3 \\ 3 & 4 & 2 & -7 \end{bmatrix} \text{ and } \mathbf{b} = \begin{bmatrix} -10 \\ 20 \end{bmatrix}$$

What conclusions do you draw regarding the two methods by comparing the norms that the two methods produce?

2.21. An orthogonal matrix **A** is defined as a square matrix such that the product of the matrix and its transpose equals the unit matrix or

$$\mathbf{A}\mathbf{A}^\mathsf{T} = \mathbf{I}$$

Use MATLAB to verify that the following matrices are orthogonal:

$$\mathbf{B} = \begin{bmatrix} \frac{1}{\sqrt{3}} & \frac{1}{\sqrt{6}} & -\frac{1}{\sqrt{2}} \\ \frac{1}{\sqrt{3}} & \frac{-2}{\sqrt{6}} & 0 \\ \frac{1}{\sqrt{3}} & \frac{1}{\sqrt{6}} & \frac{1}{\sqrt{2}} \end{bmatrix}$$

$$\mathbf{C} = \begin{bmatrix} \cos(\pi/3) & \sin(\pi/3) \\ -\sin(\pi/3) & \cos(\pi/3) \end{bmatrix}$$

2.22. Write MATLAB scripts to implement both the Gauss–Seidel and the Jacobi method and use them to solve, with an accuracy of 0.000005, the equation system **Ax** = **b** where the elements of **A** are

$$a_{ii} = -4$$
$$a_{ij} = 2 \text{ if } |i - j| = 1$$
$$a_{ij} = 0 \text{ if } |i - j| \geq 2 \text{ where } i, j = 1, 2, \ldots, 10$$

and

$$\mathbf{b}^\mathsf{T} = [2\ 3\ 4 \ldots 11]$$

Use initial values $x_i = 0$, where $i = 1, 2, \ldots, 10$. (You might also like to experiment with other initial values.) Check your results by solving the system using the MATLAB \ operator.

CHAPTER

SOLUTION OF NON-LINEAR EQUATIONS

3

Abstract

The problem of solving non-linear equations arises frequently and naturally from the study of a wide range of practical problems. The problem may involve one or a system of non-linear equations in many variables. In this chapter, general methods of solving non-linear equations are presented, together with specific methods for polynomial equations.

3.1 INTRODUCTION

To illustrate our discussion and provide a practical insight into the solution of non-linear equations we shall consider an equation described by Armstrong and Kulesza (1981). These authors report a problem which arises from the study of resistive mixer circuits. Given an applied current and voltage, it is necessary to find the current flowing in part of the circuit. This leads to a simple non-linear equation which after some manipulation may be expressed in the form

$$x - \exp(-x/c) = 0 \quad \text{or equivalently} \quad x = \exp(-x/c) \tag{3.1}$$

Here c is a given constant and x is the variable we wish to determine. The solution of such equations is not obvious but Armstrong and Kulesza provide an approximate solution based on a series expansion which gives a reasonably accurate solution for a large range of values of c. This approximation is given in terms of c by

$$x = cu[1 - \log_e\{(1+c)u\}/(1+u)] \tag{3.2}$$

where $u = \log_e(1 + 1/c)$. This is an interesting and useful result since it is reasonably accurate for values of c in the five-decade range $[10^{-3}, 100]$ and gives a relatively efficient way of finding the solutions of a whole family of equations generated by varying c. Although this result is useful for this particular equation, when we attempt to use this type of *ad hoc* approach for the general solution of non-linear equations, there are significant drawbacks. These are

1. *Ad hoc* approaches to the solutions of equations are rarely as successful as this example in finding a formula for the solution of a given equation, usually it is impossible to obtain such formulae.
2. Even when they exist, such formulae require considerable time and ingenuity to develop.
3. We may require greater accuracy than any *ad hoc* formula can provide.

Numerical Methods. https://doi.org/10.1016/B978-0-12-812256-3.00012-9

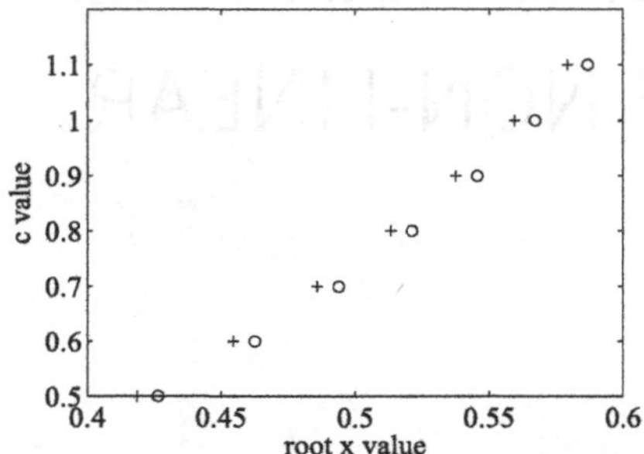

FIGURE 3.1

Solution of $x = \exp(-x/c)$. Results from the function `fzero` are indicated by o and those from the Armstrong and Kulesza formula by +.

To illustrate point 3 consider Fig. 3.1 which is generated by the MATLAB script `e4s301.m`. This figure shows the results obtained using the formula (3.2) together with the results using the MATLAB function `fzero` to solve the same non-linear equation (3.1).

```
% e4s301.m
ro = [ ]; ve = [ ]; x = [ ];
c = 0.5:0.1:1.1; u = log(1+1./c);
x = c.*u.*(1-log((1+c).*u)./(1+u));
% solve equation using MATLAB function fzero
i = 0;
options = optimset('TolFun',0.00005);
for c1 = 0.5:0.1:1.1
    i = i+1;
    ro(i) = fzero(@(x) x-exp(-x/c1),1,options);
end
plot(x,c,'+')
axis([0.4 0.6 0.5 1.2])
hold on
plot(ro,c,'o')
xlabel('Root x value'),  ylabel('c value')
hold off
```

The function `fzero` is discussed in detail in Section 3.10. Note that the call of `fzero` takes the form `fzero(@(x) x-exp(-x/c1),1,options)`. This gives an accuracy of 0.00005 for the roots and uses an initial approximation 1. The function `fzero` provides the root with up to 16-digit accuracy, if required, whereas the formula (3.2) of Armstrong and Kulesza, although faster, gives the result to one or two decimal places only. In fact, the method of Armstrong and Kulesza becomes more accurate for large values of c.

From the preceding discussion, we conclude that, although occasionally ingenious alternatives may be available, in the vast majority of cases we must use algorithms which provide, with reasonable

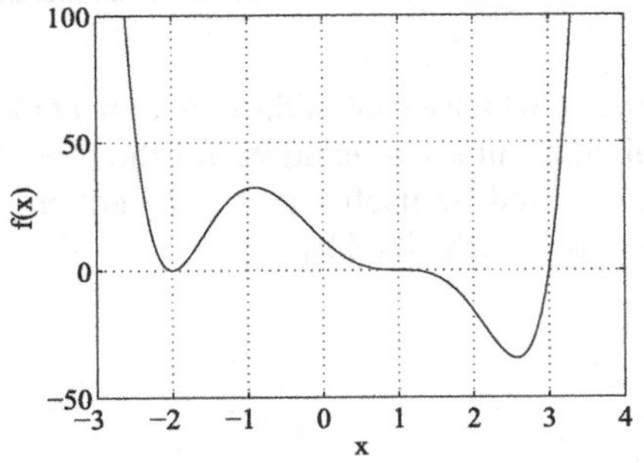

FIGURE 3.2

Plot of the function $f(x) = (x-1)^3(x+2)^2(x-3)$.

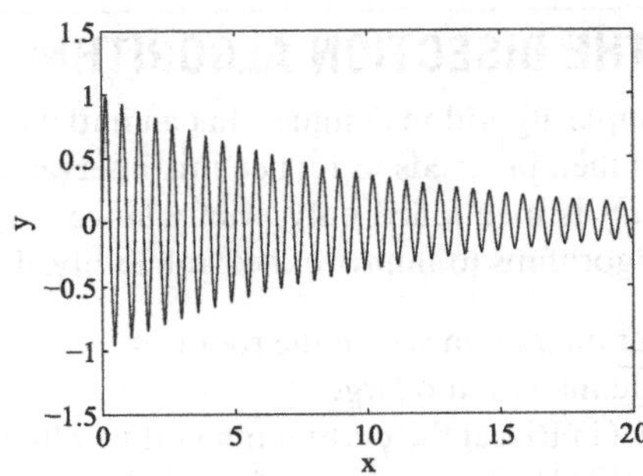

FIGURE 3.3

Plot of $f(x) = \exp(-x/10)\sin(10x)$.

computational effort, the solutions of general problems to any specified accuracy. Before describing the nature of these algorithms in detail, we consider different types of equations and the general nature of their solutions.

3.2 THE NATURE OF SOLUTIONS TO NON-LINEAR EQUATIONS

We illustrate the nature of the solutions to non-linear equations by considering two equations which we wish to solve for the variable x.

$$(a)\ (x-1)^3(x+2)^2(x-3) = 0, \text{ – that is,}$$
$$x^6 - 2x^5 - 8x^4 + 14x^3 + 11x^2 - 28x + 12 = 0$$
$$(b)\ \exp(-x/10)\sin(10x) = 0$$

The first equation is a special type of non-linear equation known as a polynomial equation since it involves only integer powers of the variable x and no other function. Such polynomial equations have the important characteristic that they have n roots where n is the degree of the polynomial. In this example the highest power of x, and hence the degree of the polynomial, is six. The solutions of a polynomial may be complex or real, separate or coincident. Fig. 3.2 illustrates the nature of the solutions of this equation. Although there must be six roots, three are coincident at $x = 1$ and two are coincident at $x = -2$. There is also a single root at $x = 3$. Coincident roots may present difficulties for some algorithms, as do roots which are very close together so it is important to appreciate their existence. The user may require a particular root of the equation or all the roots. In the case of polynomial equations, special algorithms exist to find all the roots simultaneously.

The example (b) is a non-linear equation involving transcendental functions. The task of finding all the roots of this class of non-linear equation is a daunting one, since the number of roots may not be known or there may be an infinity of roots. This situation is illustrated by Fig. 3.3 which shows the graph of the second equation for x in the range [0, 20]. If we extended the range of x, more roots would be revealed.

We now consider some simple algorithms to find a specific root of a given non-linear equation.

3.3 THE BISECTION ALGORITHM

This simple algorithm assumes that an initial interval is known in which a root of the equation $f(x) = 0$ lies and then proceeds to reduce this interval until the required accuracy is achieved for the root. This algorithm is mentioned only briefly since it is not in practice used by itself but in conjunction with other algorithms to improve their reliability. The algorithm may be described by

```
input interval in which the root lies
while interval too large
    (1) Bisect the current interval in which the root lies.
    (2) Determine in which half of the interval the root lies.
end
display root
```

The principles on which this algorithm is based are simple. Given an initial interval in which a specific root lies, the algorithm will provide an improved approximation for the root. However, the requirement that an interval be known is sometimes difficult to achieve, and although the algorithm is reliable it is extremely slow.

Alternative algorithms have been developed which converge more rapidly; this chapter is concerned with describing some of the most important of these. All the algorithms we consider are iterative in character, – that is they proceed by repeating the same sequence of steps until the root approximation is accurate enough to satisfy the user. We now consider the general form of an iterative method, the nature of the convergence of such methods and the problems they encounter.

3.4 ITERATIVE OR FIXED POINT METHODS

We wish to solve the general equation $f(x) = 0$; however, to illustrate iterative methods clearly we consider a simple example. Suppose we wish to solve the quadratic

$$x^2 - x - 1 = 0 \tag{3.3}$$

This equation can be solved by using the standard formula for solving quadratics but we take a different approach. Rearrange (3.3) as follows:

$$x = 1 + 1/x$$

Then rewrite it in iterative form using subscripts as follows:

$$x_{r+1} = 1 + 1/x_r \quad \text{for } r = 0, 1, 2, \ldots \tag{3.4}$$

Assuming we have an initial approximation x_0 to the root we are seeking, we can proceed from one approximation to another using this formula. The iterates we obtain in this way may or may not converge to the solution of the original equation. This is not the only iterative procedure for attempting to solve (3.3); we can generate two others from (3.3) as follows:

$$x_{r+1} = x_r^2 - 1 \quad \text{for } r = 0, 1, 2, \ldots \tag{3.5}$$

Table 3.1 Difference between exact root and iterate for $x^2 - x - 1 = 0$

Iteration (3.4)	Iteration (3.5)	Iteration (3.6)
−0.1180	1.3820	0.1140
0.0486	6.3820	0.0349
−0.0180	61.3820	0.0107
0.0070	3966.3820	0.0033
−0.0026	15,745,021.3820	0.0010

and

$$x_{r+1} = \sqrt{x_r + 1} \quad \text{for } r = 0, 1, 2, \ldots \tag{3.6}$$

Starting from the same initial approximation, these iterative procedures may or may not converge to the same root. Table 3.1 shows what happens when we use the initial approximation $x_0 = 2$ with the iterative procedures (3.4), (3.5), and (3.6). It shows that iterations (3.4) and (3.6) converge but (3.5) does not.

Note that when the root is reached no further improvement is possible and the point remains fixed. Hence the roots of the equation are the *fixed points* of the iteration. To remove the unpredictability of this approach we must be able to find general conditions which determine when such iterative schemes converge, when they do not and the nature of this convergence.

3.5 THE CONVERGENCE OF ITERATIVE METHODS

The procedure described in Section 3.4 can be applied to any equation $f(x) = 0$ and has the general form

$$x_{r+1} = g(x_r) \text{ for } r = 0, 1, 2, \ldots \tag{3.7}$$

It is not our purpose to give the details of the derivation of convergence conditions for this form of iteration, but to point out some of the difficulties which may arise in using them even when this condition is satisfied. The detailed derivation is given in many text books; see, for example, Lindfield and Penny (1989). It can be shown that the approximate relation between the current error ε_{r+1} at the $(r+1)$th iteration and the previous error ε_r is given by

$$\varepsilon_{r+1} = \varepsilon_r g'(t_r)$$

where t_r is a point lying between the exact root and the current approximation to the root. Thus, the error will be decreasing if the absolute value of the derivative at these points is less than 1. However, this does not guarantee convergence from all starting points and the initial approximation must be sufficiently close to the root for convergence to occur.

In the case of the specific iterative procedures (3.4) and (3.5), Table 3.2 shows how the values of the derivatives of the corresponding $g(x)$ vary with the values of the approximations to x_r. This table provides numerical evidence for the theoretical assertion in the case of iterations (3.4) and (3.5).

Table 3.2 Values of the derivatives for iterations given by (3.4) and (3.5)

Iteration (3.4)	Derivative	Iteration (3.5)	Derivative
−0.1180	−0.44	1.3820	6.00
0.0486	−0.36	6.3820	16.00
−0.0180	−0.39	61.3820	126.00
0.0070	−0.38	3966.3820	7936.00

However, the concept of convergence is more complex than this. We need to give some answer to the crucial question: if an iterative procedure converges, how can we classify the rate of convergence? We do not derive this result but refer the reader to Lindfield and Penny (1989) and state the answer to the question. Suppose all derivatives of the function $g(x)$ of order 1 to $p-1$ are zero at the exact root a. Then the relation between the current error ε_{r+1} at the $(r+1)$th iteration and the previous error ε_r is given by

$$\varepsilon_{r+1} = (\varepsilon_r)^p g^{(p)}(t_r)/p! \tag{3.8}$$

where t_r lies between the exact root and the current approximation to the root and $g^{(p)}$ denotes the pth derivative of g. The importance of this result is that it means the current error is proportional to the pth power of the previous error and clearly, on the basis of the reasonable assumption that the errors are much smaller than 1, the higher the value of p, the faster the convergence. Such methods are said to have pth-order convergence. In general it is cumbersome to derive iterative methods of order higher than two or three and second-order methods, where the current error is proportional to the square of the previous error, have proved very satisfactory in practice for solving a wide range of non-linear equations. Note that if the current error is proportional to the previous error it is called linear convergence and if the current error is proportional to the square of the previous error it is called quadratic convergence. This provides a convenient classification for the convergence of iterative methods but avoids the difficult questions: for what range of starting values will the process converge and how sensitive is convergence to changes in the starting values?

3.6 RANGES FOR CONVERGENCE AND CHAOTIC BEHAVIOR

We illustrate some of the problems of convergence by considering a specific example which highlights some of the difficulties. Short (1992) examined the behavior of the iterative process

$$x_{r+1} = -0.5(x_r^3 - 6x_r^2 + 9x_r - 6) \quad \text{for } r = 0, 1, 2, \ldots$$

for solving the equation $(x-1)(x-2)(x-3) = 0$. This iterative procedure clearly has the form

$$x_{r+1} = g(x_r), \quad r = 0, 1, 2, \ldots$$

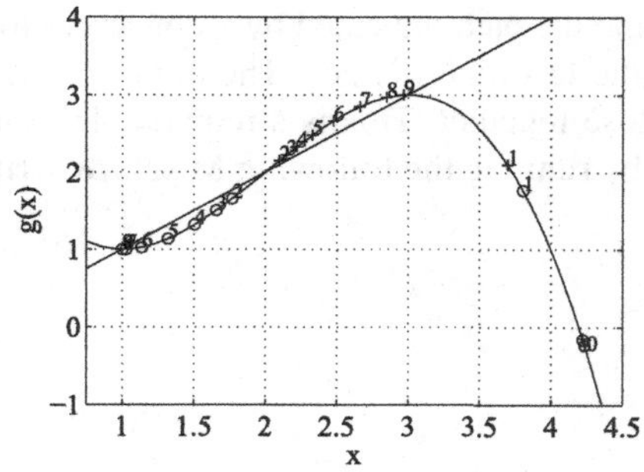

FIGURE 3.4

Iterates in the solution of $(x-1)(x-2)(x-3)=0$ from close but different starting points.

and it is easy to verify that it has the following properties:

$$g'(1)=0 \quad \text{and} \quad g''(1)\neq 0$$
$$g'(2)\neq 0$$
$$g'(3)=0 \quad \text{and} \quad g''(3)\neq 0$$

Thus, by taking $p=2$ in result (3.8), we can expect, for appropriate starting values, quadratic convergence for the roots at $x=1$ and $x=3$ but at best linear convergence for the root at $x=2$. The major problem is, however, to determine the ranges of initial approximation which will converge to the different roots. This is not an easy task but one simple way of doing this is to draw a graph of $y=x$ and $y=g(x)$. The points of intersection provide the roots. The line $y=x$ has a slope of 1, and points where the slope of $g(x)$ is less than this provide a range of initial approximations which converge to one or other of the roots.

This graphical analysis shows that points within the range 1 to 1.43 (approximately) converge to the root 1 and points in the range 2.57 (approximately) to 3 converge to the root 3. This is the obvious part of the analysis. However, Short demonstrates that there are many other ranges of convergence for this iterative procedure, many of them very narrow indeed, which lead to chaotic behavior in the iterative process. He demonstrates for example, that taking $x_0=4.236067968$ will converge to the root $x=3$ whereas taking $x_0=4.236067970$ converges to the root $x=1$, a remarkable change for such a small variation in the initial approximation. This should serve as a warning to the reader that the study of convergence properties is in general not an easy task.

Fig. 3.4 illustrates this point quite strikingly. It shows the graph of x and the graph of $g(x)$ where

$$g(x)=-0.5(x^3-6x^2+9x-6)$$

The x line intersects with $g(x)$ to give the roots of the original equation. The graph also shows iterates starting from $x_0=4.236067968$, indicated by $\circ$, and iterates starting from $x=4.236067970$, indicated by +. The starting points are so close they are of course superimposed on the graph. However, the iterates soon take their separate paths to converge on different roots of the equation. The path indicated by o converges to the root $x = 1$ and the path indicated by + converges to the root $x = 3$. The sequence of numbers on the graph shows the last nine iterates. The point referenced by zero is in fact all the points which are initially very close together. This is a remarkable example and users should verify these phenomena for themselves by running the following MATLAB script:

```
% e4s302.m
x = 0.75:0.1:4.5;
g = -0.5*(x.^3-6*x.^2+9*x-6);
plot(x,g)
axis([.75,4.5,-1,4])
hold on, plot(x,x)
xlabel('x'), ylabel('g(x)'), grid on
ch = ['o','+'];
num = [ '0','1','2','3','4','5','6','7','8','9'];
ty = 0;
for x1 = [4.236067970 4.236067968]
    ty = ty+1;
    for i = 1:19
        x2 = -0.5*(x1^3-6*x1^2+9*x1-6);
        % First ten points very close, so represent by '0'
        if i==10
            text(4.25,-0.2,'0')
        elseif i>10
            text(x1,x2+0.1,num(i-9))
        end
        plot(x1,x2,ch(ty))
        x1 = x2;
    end
end
hold off
```

It is interesting to note that the iterative form

$$x_{r+1} = x_r^2 + c \quad \text{for } r = 0, 1, 2, ...$$

demonstrates strikingly chaotic behavior when the iterates are plotted in the complex plane and for complex ranges of values for c.

We now return to the more mundane task of developing algorithms that work in general for the solution of non-linear equations. In the next section we shall consider a simple method of order 2.

3.7 NEWTON'S METHOD

This method for the solution of the equation $f(x) = 0$ is based on the simple geometric properties of the tangent to the curve $f(x)$. The method requires some initial approximation to the root and that the

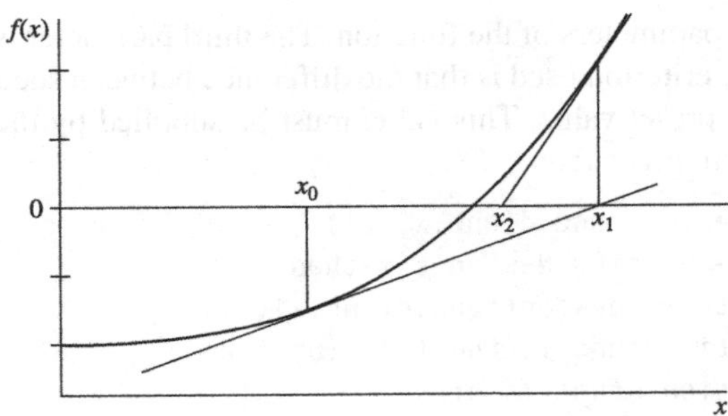

FIGURE 3.5

Geometric interpretation of Newton's method.

derivative of $f(x)$ exists in the range of interest. Fig. 3.5 illustrates the operation of the method. The diagram shows the tangent to the curve at the current approximation x_0. This tangent strikes the x-axis at x_1 and provides us with an improved approximation to the root. Similarly, the tangent at x_1 gives the improved approximation x_2.

The process is repeated until some convergence criterion is satisfied. It is easy to translate this geometrical procedure into a numerical method for finding the root since the tangent of the angle between the x-axis and the tangent equals

$$f(x_0)/(x_1 - x_0)$$

and the slope of this tangent itself equals $f'(x_0)$, the derivative of $f(x)$ at x_0. So we have the equation

$$f'(x_0) = f(x_0)/(x_1 - x_0)$$

Thus the improved approximation, x_1, is given by

$$x_1 = x_0 - f(x_0)/f'(x_0)$$

This may be written in iterative form as

$$x_{r+1} = x_r - f(x_r)/f'(x_r) \quad \text{where } r = 0, 1, 2, \ldots \tag{3.9}$$

We note that this method is of the general iterative form

$$x_{r+1} = g(x_r) \quad \text{where } r = 0, 1, 2, \ldots$$

Consequently, the discussion of Section 3.5 applies to it. On computing $g'(a)$, where a is the exact root, we find it is zero. However, $g''(a)$ is in general non-zero so the method is of order 2 and we expect convergence to be quadratic. For a sufficiently close initial approximation, convergence to the root will be rapid.

A MATLAB function `fnewton` is supplied for Newton's method. The function that forms the left side of the equation we wish to solve *and* its derivative must be supplied by the user as functions; these become the first and second parameters of the function. The third parameter is an initial approximation to the root. The convergence criterion used is that the difference between successive approximations to the root is less than a small preset value. This value must be supplied by the user and is given as the fourth parameter of the function `fnewton`

```
function [res, it] = fnewton(func,dfunc,x,tol)
% Finds a root of f(x) = 0 using Newton's method.
% Example call: [res, it] = fnewton(func,dfunc,x,tol)
% The user defined function func is the function f(x),
% The user defined function dfunc is df/dx.
% x is an initial starting value, tol is required accuracy.
it = 0; x0 = x;
d = feval(func,x0)/feval(dfunc,x0);
while abs(d) > tol
    x1 = x0-d;   it = it+1;   x0 = x1;
    d = feval(func,x0)/feval(dfunc,x0);
end
res = x0;
```

We will now find a root of the equation

$$x^3 - 10x^2 + 29x - 20 = 0$$

To use Newton's method we must define the function and its derivative thus:

```
>> f = @(x) x.^3-10*x.^2+29*x-20;
>> df = @(x) 3*x.^2-20*x+29;
```

We may call the function `fnewton` as follows:

```
>> [x,it] = fnewton(f,df,7,0.00005)

x =

    5.0000


it =

     6
```

The progress of the iterations when solving $x^3 - 10x^2 + 29x - 20 = 0$ by Newton's method is shown in Fig. 3.6.

Table 3.3 gives numerical results for this problem when Newton's method is used to seek a root, starting the iteration at -2. The second column of the table gives the current error ε_r by subtracting the known exact root from the current iterate. The third column contains the value of $2\varepsilon_{r+1}/\varepsilon_r^2$. This value tends to a constant as the process proceeds. From theoretical considerations, this value should approach the second derivative of the right side of the Newton iterative formula. This follows from (3.8) with $p = 2$. The final column contains the value of the second-order derivative of $g(x)$ calculated as follows. From (3.9) we have $g(x) = x - f(x)/f'(x)$. Thus, on differentiating $g(x)$, we have

$$g'(x) = 1 - [\{f'(x)\}^2 - f''(x)f(x)]/[f'(x)]^2 = f''(x)f(x)/[f'(x)]^2$$

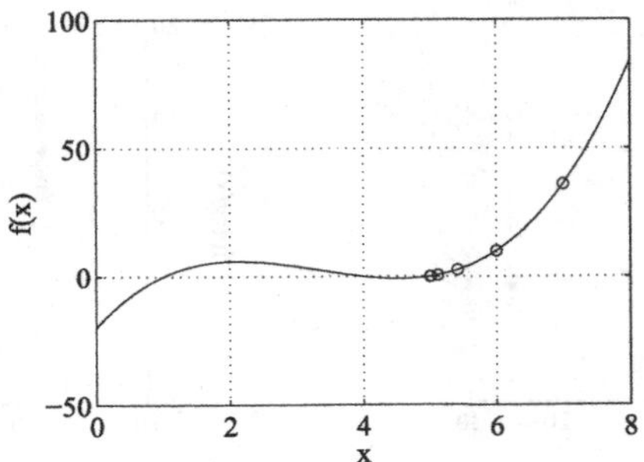

FIGURE 3.6

Plot of $x^3 - 10x^2 + 29x - 20 = 0$ with the iterates of Newton's method shown by o.

Table 3.3 Newton's method to solve $x^3 - 10x^2 + 29x - 20 = 0$ with an initial approximation of −2

x value	error ε_r	$2\varepsilon_{r+1}/\varepsilon_r^2$	Approximate Second derivative of g
−2.000000	3.000000	−0.320988	−0.395062
−0.444444	1.444444	−0.513956	−0.589028
0.463836	0.536164	−0.792621	−0.845260
0.886072	0.113928	−1.060275	−1.076987
0.993119	0.006881	−1.159637	−1.160775
0.999973	0.000027	−1.166639	−1.166643
1.000000	0.000000	−1.166639	−1.166667

On differentiating again,

$$g''(x) = [\{f'(x)\}^2\{f'''(x)f(x) + f''(x)f'(x)\} - 2f'(x)\{f''(x)\}^2 f(x)]/[f'(x)]^4$$

Putting $x = a$, where a is the exact root, since $f(a) = 0$, we have

$$g'(a) = 0 \quad \text{and} \quad g''(a) = f''(a)/f'(a) \tag{3.10}$$

Thus, we have a value for the second derivative of $g(x)$ when $x = a$. We note that as x approaches the root, the final column of Table 3.3, which uses this formula, gives an increasingly accurate approximation to the second derivative of $g(x)$. The table thus verifies our theoretical expectations.

We can find complex roots using Newton's method, providing our initial approximation is complex. For example, consider

$$\cos(x) - x = 0. \tag{3.11}$$

This equation has only one real root which is $x = 0.7391$, but it has an infinity of complex roots. Fig. 3.7 shows the distribution of the roots of (3.11) in the complex plane in the range $-30 < \text{Re}(x) < 30$. Working with complex values presents no additional difficulty in the MATLAB environment since

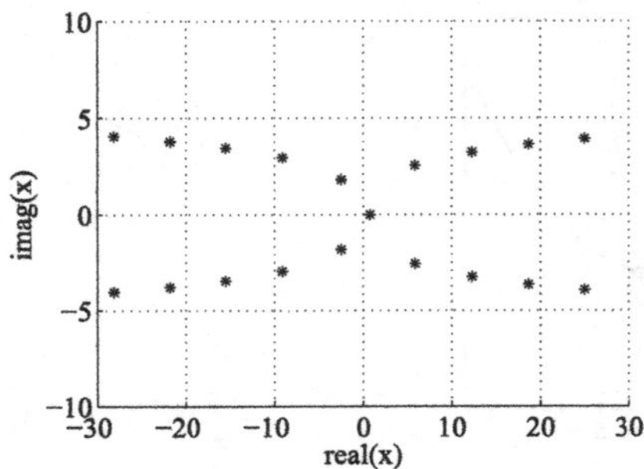

FIGURE 3.7

Plot showing the complex roots of $\cos(x) - x = 0$.

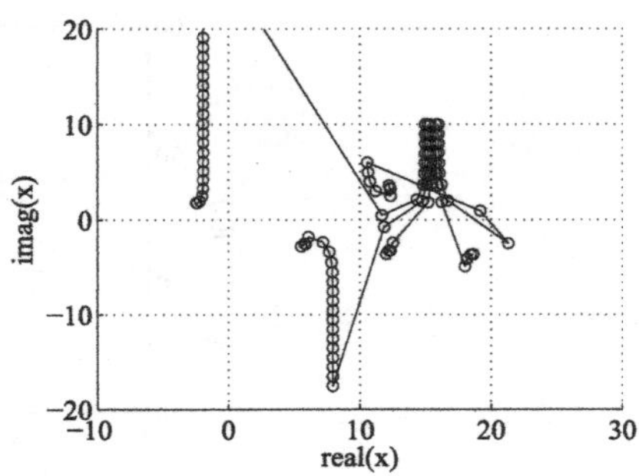

FIGURE 3.8

Plot of the iterates for five complex initial approximations for the solution of $\cos(x) - x = 0$ using Newton's method. Each iterate is shown by o.

MATLAB implements complex arithmetic and so we can use the function `fnewton` without modification to deal with these cases.

Fig. 3.8 illustrates the fact that it is difficult to predict which root we will find from a given starting value. This figure shows that the starting values $15 + \jmath 10$, $15.2 + \jmath 10$, $15.4 + \jmath 10$, $15.8 + \jmath 10$, and $16 + \jmath 10$, which are close together, lead to a sequence of iterations which converge to very different roots. In one case the complete trajectory is not shown because the complex part of the intermediate iterates is well outside the range of the graph.

Newton's method requires the first derivative of $f(x)$ to be supplied by the user. To make the procedure more self-contained we can use a standard approximation to the first derivative which takes the form

$$f'(x_r) = \{f(x_r) - f(x_{r-1})\}/(x_r - x_{r-1}) \tag{3.12}$$

Substituting this result in (3.9) gives the new procedure for calculating the improvements to x as

$$x_{r+1} = [x_{r-1} f(x_r) - x_r f(x_{r-1})]/[f(x_r) - f(x_{r-1})] \tag{3.13}$$

This method does not require the calculation of the first derivative of $f(x)$ but does require that we know two initial approximations to the root, x_0 and x_1. Geometrically, we have simply approximated the slope of the tangent to the curve by the slope of a secant. For this reason the method is known as the *secant method*. The convergence of this method is slower than Newton's method. Another procedure similar to the secant method is called *regula falsi*. In this method two values of x which enclose the root are chosen to start the next iteration rather than the most recent pair of x values as in the secant method.

Newton's method and the secant method work well on a wide range of problems. However, for problems where the roots of an equation are close together or equal, the convergence may be slow. We now consider a simple adjustment to Newton's method which provides good convergence even with multiple roots.

3.8 SCHRODER'S METHOD

In Section 3.2 we described how coincident roots present significant problems for most algorithms. In the case of Newton's method its performance is no longer quadratic for finding a coincident root and the procedure must be modified if it is to maintain this property. The iteration for Schroder's method for finding multiple roots has a form similar to that of Newton's method given in (3.9) except for the inclusion of a multiplying factor m. Thus,

$$x_{r+1} = x_r - mf(x_r)/f'(x_r) \text{ where } r = 0, 1, 2, \ldots \quad (3.14)$$

Here m is an integer equal to the multiplicity of the root to which we are trying to converge. Since the user may not know the value of m, it may have to be found experimentally.

It can be verified by some simple but lengthy algebraic manipulation that for a function $f(x)$ with multiple roots at $x = a$, $g'(a) = 0$. Here $g(x)$ is the right side of (3.14) and a is the exact root. This modification is sufficient to preserve the quadratic convergence of Newton's method

A MATLAB function for Schroder's method, `schroder`, is provided as follows:

```
function [res, it] = schroder(func,dfunc,m,x,tol)
% Finds a multiple root of f(x) = 0 using Schroder's method.
% Example call: [res, it] = schroder(func,dfunc,m,x,tol)
% The user defined function func is the function f(x),
% The user defined function dfunc is df/dx.
% x is an initial starting value, tol is required accuracy.
% function has a root of multiplicity m.
% x is a starting value, tol is required accuracy.
it = 0; x0 = x;
d = feval(func,x0)/feval(dfunc,x0);
while abs(d)>tol
    x1 = x0-m*d; it = it+1; x0 = x1;
    d = feval(func,x0)/feval(dfunc,x0);
end
res = x0;
```

We will now use the function `schroder` to solve $f(x) = (e^{-x} - x)^2 = 0$. In this case we must set the multiplying factor m to 2. We write functions for `f` and its derivative `df` and call the function `schroder` as follows:

```
>> f = @(x) (exp(-x)-x).^2;
>> df = @(x) 2*(exp(-x)-x).*(-exp(-x)-1);
>> [x, it] = schroder(f,df,2,-2,0.00005)

x =

    0.5671


it =

    5
```

It is interesting to note that Newton's method took 17 iterations to solve this problem in contrast to the 5 required by Schroder's method.

When a function $f(x)$ is known to have repeated roots, an alternative to Schroder's approach is to apply Newton's method to the function $f(x)/f'(x)$ rather than to the function $f(x)$ itself. It can be easily shown by direct differentiation that if $f(x)$ has a root of any multiplicity then $f(x)/f'(x)$ will have the same root but with multiplicity 1. Thus, the algorithm has the iterative form (3.9) but modified by replacing $f(x)$ with $f(x)/f'(x)$. The advantage of this approach is that the user does not have to know the multiplicity of the root which is to be found. The considerable disadvantage is that both the first- and second-order derivatives must be supplied by the user.

3.9 NUMERICAL PROBLEMS

We now consider the following problems which arise in solving single-variable non-linear equations:

1. Finding good initial approximations.
2. Ill-conditioned functions.
3. Deciding on the most suitable convergence criteria.
4. Discontinuities in the equation to be solved.

These problems are now examined in detail:

1. Finding an initial approximation can be difficult for some non-linear equations and a graph can be a considerable help in supplying such a value. The advantage of working in a MATLAB environment is that the script for the graph of the function can easily be generated and input can be taken from it directly. The function `plotapp` that is defined here finds an approximation to the root of a function supplied by the user in the range given by the parameters `rangelow` and `rangeup` using a step given by `interval`.

```
function approx = plotapp(func,rangelow,interval,rangeup)
% Plots a function and allows the user to approximate a
% particular root using the cursor.
% Example call: approx = plotapp(func,rangelow,interval,rangeup)
% Plots the user defined function func in the range rangelow to
% rangeup using a step given by interval. Returns approx to root.
approx = [ ];
x = rangelow:interval:rangeup;
plot(x,feval(func,x))
hold on, xlabel('x'), ylabel('f(x)')
title(' ** Place cursor close to root and click mouse ** ')
grid on
% Use ginput to get approximation from graph using mouse
approx = ginput(1);
fprintf('Approximate root is %8.2f\n',approx(1)), hold off
```

The script `e4s303.m` shows how this function may be used with the MATLAB function `fzero` to find a root of $x - \cos(x) = 0$.

```
% e4s303.m
g = @(x) x-cos(x);
approx = plotapp(g,-2,0.1,2);
% Use this approximation  and fzero to find exact root
options = optimset('TolFun',0.00005);
root = fzero(g,approx(1),options);
fprintf('Exact root is %8.5f\n',root)
```

Fig. 3.9 gives the graph of $x - \cos(x) = 0$ generated by `plotapp` and shows the cross-hairs cursor generated by the `ginput` function close to the root. The call `ginput(1)` means only one point is taken. The cursor can be positioned over the intersection of the curve with the $f(x) = 0$ line. This provides a useful initial approximation, the accuracy of which depends on the scale of the graph. In this example an initial approximation was found to be 0.74 and the more exact value was found using `fzero` to be 0.73909.

2. Ill-conditioning in a non-linear equation means that small changes in the coefficients of the equation lead to unexpectedly large errors in the solutions. An interesting example of a very ill-conditioned polynomial is Wilkinson's polynomial. The MATLAB function `poly(v)` generates the coefficients of a polynomial, beginning with the coefficient of the highest power, with roots which are equal to the elements of the vector `v`. Thus, `poly(1:n)` generates the coefficients of the polynomial with the roots $1, 2, ..., n$ which is Wilkinson's polynomial of degree $n - 1$.
3. In the design of any numerical algorithm for the solution of non-linear equations the termination criterion is particularly important. There are two major indicators of convergence: the difference between successive iterates and the value of the function at the current iterate. Taken separately these indicators may be misleading. For example, some non-linear functions are such that small changes in the independent variable value may lead to large changes in the function value. In this case it may be better to monitor both indicators.
4. The function $f(x) = \sin(1/x)$ is particularly difficult to plot, and $\sin(1/x) = 0$ is very difficult to solve since it has an infinite number of roots, all clustered between 1 and -1. The function has a discontinuity at $x = 0$. Fig. 3.10 attempts to illustrate the behavior of this function. In fact, the graph shown does not truly represent the function and this plotting problem is discussed in more detail in Chapter 4. Near a discontinuity the function changes rapidly for small changes in the independent variable and some algorithms may have problems with this.

All the preceding points emphasize that algorithms for solving non-linear equations to be not only fast and efficient but robust as well. The next algorithm combines these properties and is relatively undemanding on the user.

3.10 THE MATLAB FUNCTION fzero AND COMPARATIVE STUDIES

Some problems may present particular difficulties for algorithms which in general work well. For example, algorithms which have fast ultimate convergence may initially diverge. One way to improve the reliability of an algorithm is to ensure that at each stage the root is confined to a known interval and the method of bisection, introduced in Section 3.3, may be used to provide an interval in which the

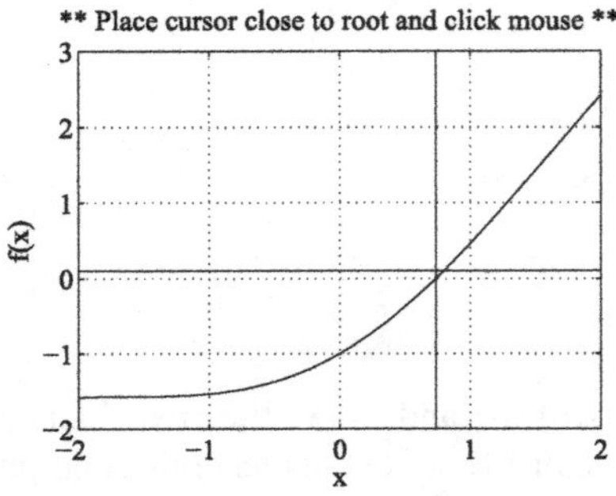

FIGURE 3.9

The cursor is shown close to the position of the root.

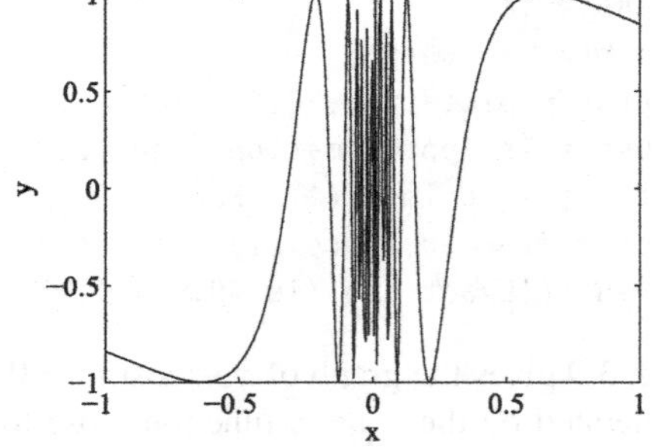

FIGURE 3.10

Plot of graph $f(x) = \sin(1/x)$. This plot is spurious in the range ± 0.2.

root lies. Thus, a method which combines bisection with a rapidly convergent procedure may be able to provide both rapid *and* reliable convergence.

The method of Brent combines inverse quadratic interpolation with bisection to provide a powerful method that has been found to be successful on a wide range of difficult problems. The method is easily implemented and a detailed description of the algorithm may be found in Brent (1971). Similar algorithms of comparable efficiency have been developed by Dekker (1969).

Experience with Brent's algorithm has shown it to be both reliable and efficient on a wide range of problems. A variation of this method is directly available in MATLAB and is called `fzero`. It may be used as follows:

```
x = fzero('funcname',x0,options);
```

where `funcname` is replaced by the name of any system function such as `cos`, `sin`, etc., or the name of a function predefined by the user. The initial approximation is `x0`. The accuracy of the solution is set by `options` using the `optimset`. For example, in the script `e4s304`, `optimset` sets the accuracy to 0.00005.

Only the first two parameters need be given and so an alternative call of this function is given by

```
x = fzero('funcname',x0);
```

To plot the function $(e^x - \cos(x))^3$ and then determine some roots of $(e^x - \cos(x))^3 = 0$ with tolerance 0.00005, initial approximations of 1.65 and -3 and no trace of the iterations, we use `fzero` in script `e4s304.m`:

```
% e4s304.m
f = @(x) (exp(x)-cos(x)).^3;
x = -4:0.02:0.5;
plot(x,f(x)), grid on
xlabel('x'), ylabel('f(x)');
title('f(x) = (exp(x)-cos(x)).^3')
options = optimset('TolFun',0.00005);
```

Table 3.4 Solution of equations (1) through (4) with the same starting point $x = -2$ and accuracy = 0.00005

Function	1	2	3	4
fnewton	Fail	0.999795831	Fail	−1.352673831
fzero	−0.318309886	1.000000000	−1.570796327	−1.352678708

```
root = fzero(f,1.65,options);
fprintf('A root of this equation is %6.4f\n',root)
root = fzero(f,-3,0.00005);
fprintf('A root of this equation is %6.4f\n',root)
```

The output and plot generated by this script are not given. However, the script is provided for reader experimentation.

Before we deal with the problem of finding many roots of a polynomial equation simultaneously, we present a comparative study of the MATLAB function `fzero` with the function `fnewton`. The following functions are considered:

1. $\sin(1/x) = 0$
2. $(x-1)^5 = 0$
3. $x - \tan x = 0$
4. $\cos\{(x^2+5)/(x^4+1)\} = 0$

The results of these comparative studies are given in Table 3.4. We see that `fnewton` is less reliable than `fzero` and that `fzero` produces more accurate answers.

3.11 METHODS FOR FINDING ALL THE ROOTS OF A POLYNOMIAL

The problem of solving polynomial equations is a special one in that these equations contain only combinations of integer powers of x and no other functions. Because of their special structure, algorithms have been developed to find all of the roots of a polynomial equation simultaneously. The function `roots` is provided in MATLAB. This function sets up the companion matrix for the polynomial and determines its eigenvalues, which can be shown to be the roots of the polynomial. For a description of the companion matrix, see Appendix A.

In the following sections, we describe the methods of Bairstow and Laguerre but do not give a detailed theoretical justification of them. We provide a MATLAB function for Bairstow's method.

3.11.1 BAIRSTOW'S METHOD

Consider the polynomial

$$a_0x^n + a_1x^{n-1} + a_2x^{n-2} + \ldots + a_n = 0 \tag{3.15}$$

Since this is a polynomial equation of degree n, it has n roots. A common approach for locating the roots of a polynomial is to find all its quadratic factors. These will have the form

$$x^2 + ux + v \tag{3.16}$$

where u and v are the constants we wish to determine. Once all the quadratic factors are found it is easy to solve the quadratics to find all the roots of the equation. We now outline the major steps used in Bairstow's method for finding these quadratic factors.

If $R(x)$ is the remainder after the division of polynomial (3.15) by the quadratic factor (3.16), then there will clearly exist constants $b_0, b_1, b_2, \ldots$ such that the equality (3.17) holds

$$\begin{aligned}&(x^2+ux+v)(b_0x^{n-2}+b_1x^{n-3}+b_2x^{n-4}+\ldots+b_{n-2})+R(x)\\&=x^n+a_1x^{n-1}+a_2x^{n-2}+\ldots+a_n\end{aligned} \tag{3.17}$$

where a_0 is set at 1 and $R(x)$ will have the form $rx+s$. To ensure that x^2+ux+v is an exact factor of the polynomial (3.15), the remainder $R(x)$ must be zero. For this to be true both r and s must be zero and we must adjust u and v until this is true. Thus since both r and s depend on u and v, the problem reduces to solving the simultaneous equations

$$\begin{aligned}r(u,v)&=0\\s(u,v)&=0\end{aligned}$$

To solve these equations we use an iterative method which assumes some initial approximations u_0 and v_0. Then we require improved approximations u_1 and v_1 where $u_1=u_0+\Delta u_0$ and $v_1=v_0+\Delta v_0$ such that

$$\begin{aligned}r(u_1,v_1)&=0\\s(u_1,v_1)&=0\end{aligned}$$

or r and s are as close to zero as possible.

Now we wish to find the changes Δu_0 and Δv_0 which will result in this improvement. Consequently, we must expand the two equations

$$\begin{aligned}r(u_0+\Delta u_0, v_0+\Delta v_0)&=0\\s(u_0+\Delta u_0, v_0+\Delta v_0)&=0\end{aligned}$$

using a Taylor's series expansion and neglecting higher powers of Δu_0 and Δv_0. This leads to two approximating linear equations for Δu_0 and Δv_0:

$$\begin{aligned}r(u_0,v_0)+(\partial r/\partial u)_0\Delta u_0+(\partial r/\partial v)_0\Delta v_0&=0\\s(u_0,v_0)+(\partial s/\partial u)_0\Delta u_0+(\partial s/\partial v)_0\Delta v_0&=0\end{aligned} \tag{3.18}$$

The subscript 0 denotes that the partial derivatives are calculated at the point u_0, v_0. Once the corrections are found, the iteration can be repeated until r and s are sufficiently close to zero. The method we have used here is a two-variable form of Newton's method which will be described in Section 3.12.

Clearly, this method requires the first-order partial derivatives of r and s with respect to u and v. The form of these is not obvious; however, they may be determined using recurrence relations derived from equating coefficients in (3.17) and then differentiating them. The details of this derivation are not given here but a clear description of the process is given by Froberg (1969). Once the quadratic factor is found, the same process is applied to the residual polynomial with the coefficients b_i to obtain the remaining quadratic factors. The details of this derivation are not provided here but a MATLAB function bairstow follows.

```
function [rts,it] = bairstow(a,n,tol)
% Bairstow's method for finding the roots of a polynomial of degree n.
% Example call: [rts,it] = bairstow(a,n,tol)
% a is a row vector of REAL coefficients so that the
% polynomial is x^n+a(1)*x^(n-1)+a(2)*x^(n-2)+...+a(n).
% The accuracy to which the polynomial is satisfied is given by tol.
% The output is produced as an (n x 2) matrix rts.
% Cols 1 & 2 of rts contain the real & imag part of root respectively.
% The number of iterations taken is given by it.
it = 1;
while n>2
    %Initialise for this loop
    u = 1; v = 1; st = 1;
    while st>tol
        b(1) = a(1)-u; b(2) = a(2)-b(1)*u-v;
        for k = 3:n
            b(k) = a(k)-b(k-1)*u-b(k-2)*v;
        end
        c(1) = b(1)-u; c(2) = b(2)-c(1)*u-v;
        for k = 3:n-1
            c(k) = b(k)-c(k-1)*u-c(k-2)*v;
        end
        %calculate change in u and v
        c1 = c(n-1); b1 = b(n); cb = c(n-1)*b(n-1);
        c2 = c(n-2)*c(n-2); bc = b(n-1)*c(n-2);
        if n>3, c1 = c1*c(n-3); b1 = b1*c(n-3); end
        dn = c1-c2;
        du = (b1-bc)/dn; dv = (cb-c(n-2)*b(n))/dn;
        u = u+du; v = v+dv;
        st = norm([du dv]); it = it+1;
    end
    [r1,r2,im1,im2] = solveq(u,v,n,a);
    rts(n,1:2) = [r1 im1]; rts(n-1,1:2) = [r2 im2];
    n = n-2;
    a(1:n) = b(1:n);
end
% Solve last quadratic or linear equation
u = a(1); v = a(2);
[r1,r2,im1,im2] = solveq(u,v,n,a);
rts(n,1:2) = [r1 im1];
if n==2
    rts(n-1,1:2) = [r2 im2];
```

```
end
% -----------------------------------------------------------
function [r1,r2,im1,im2] = solveq(u,v,n,a);
% Solves x^2 + ux + v = 0 (n ~= 1) or x + a(1) = 0 (n = 1).
% Example call: [r1,r2,im1,im2] = solveq(u,v,n,a)
% r1, r2 are real parts of the roots,
% im1, im2 are the imaginary parts of the roots.
% Called by function bairstow.
if n==1
    r1 = -a(1); im1 = 0; r2 = 0; im2 = 0;
else
    d = u*u-4*v;
    if d<0
        d = -d;
        im1 = sqrt(d)/2; r1 = -u/2; r2 = r1; im2 = -im1;
    elseif d>0
        r1 = (-u+sqrt(d))/2; im1 = 0; r2 = (-u-sqrt(d))/2; im2 = 0;
    else
        r1 = -u/2; im1 = 0; r2 = -u/2; im2 = 0;
    end
end
```

Note that the MATLAB function `solveq` is nested within the function `bairstow`. The function is not stored separately and so it can only be accessed by `bairstow`. We may now use `bairstow` to solve the specific polynomial equation

$$x^5 - 3x^4 - 10x^3 + 10x^2 + 44x + 48 = 0$$

In this case we take the coefficient vector as `c` where `c = [-3 -10 10 44 48]` and if we require accuracy of four decimal places we take `tol` as 0.00005. The script `e4s305.m` uses `bairstow` to solve the given polynomial.

```
% e4s305.m
c = [-3 -10 10 44 48];
[rts, it] = bairstow(c,5,0.00005);
for i = 1:5
    fprintf('\nroot%3.0f Real part=%7.4f',i,rts(i,1))
    fprintf(' Imag part=%7.4f',rts(i,2))
end
fprintf('\n')
```

Note how `fprintf` is used to provide a clearer output from the matrix `rts`.

```
root  1 Real part= 4.0000 Imag part= 0.0000
root  2 Real part=-1.0000 Imag part=-1.0000
root  3 Real part=-1.0000 Imag part= 1.0000
```

Table 3.5 Time required to obtain all roots (in seconds)

Polynomial:	MATLAB function `roots`	MATLAB function `bairstow`
p1	7	33
p2	6	19
p3	6	14
p4	10	103
p5	11	37

```
root  4 Real part=-2.0000 Imag part= 0.0000
root  5 Real part= 3.0000 Imag part= 0.0000
```

As we have indicated, MATLAB provides a function `roots` to determine the roots of a polynomial. It is interesting to compare this function with Bairstow's method. Table 3.5 gives the results of this comparison applied to specific polynomials. The problems p1 through p5 are the polynomials tested:

$$\begin{aligned}
\text{p1}: &\quad x^5 - 3x^4 - 10x^3 + 10x^2 + 44x + 48 = 0 \\
\text{p2}: &\quad x^3 - 3.001x^2 + 3.002x - 1.001 = 0 \\
\text{p3}: &\quad x^4 - 6x^3 + 11x^2 + 2x - 28 = 0 \\
\text{p4}: &\quad x^7 + 1 = 0 \\
\text{p5}: &\quad x^8 + x^7 + x^6 + x^5 + x^4 + x^3 + x^2 + x + 1 = 0
\end{aligned}$$

The results for these problems are given in Table 3.5. Both methods determine the correct roots for all problems, although the function `roots` is more efficient.

3.11.2 LAGUERRE'S METHOD

Laguerre's method provides a rapidly convergent procedure for locating the roots of a polynomial. The algorithm is interesting and for this reason it is described in this section. The method is applied to a polynomial in the form

$$p(x) = x^n + a_1 x^{n-1} + a_2 x^{n-2} + \ldots + a_n$$

Starting with an initial approximation x_1, we apply the iterative formula (3.19) to the polynomial $p(x)$

$$x_{i+1} = x_i - np(x_i)/[p'(x_i) \pm \surd\{h(x_i)\}] \text{ for } i = 1, 2, \ldots \tag{3.19}$$

where

$$h(x_i) = (n-1)[(n-1)\{p'(x_i)\}^2 - np(x_i)p''(x_i)]$$

and n is the degree of the polynomial. The sign taken in (3.19) is determined so that it is the same as the sign of $p'(x_i)$.

It is important to give some justification for using a formula with such a complex structure. The reader will notice that if the square root term were not present in (3.19), the iterative form would be similar to that of Newton's method, (3.9), and identical to that of Schroder's method, (3.14). Thus, we would have a method with quadratic convergence for the roots of the polynomial. In fact, the more complex structure of (3.19) provides third-order convergence since the error is proportional to the cube of the previous error and consequently provides faster convergence than Newton's method. Thus, given an initial approximation, the method will converge rapidly to a root of the polynomial which we can denote by r.

To obtain the other roots of the polynomial we divide the polynomial $p(x)$ by the factor $(x - r)$ which provides another polynomial of degree $n - 1$. We can then apply iteration (3.19) to this polynomial and repeat the whole procedure again. This is repeated until all roots are found to the required accuracy. The process of dividing by $(x - r)$ is known as deflation and can be performed in a simple and efficient way, described as follows.

Since we have a known factor $(x - r)$, then

$$\begin{aligned} &a_0x^n + a_1x^{n-1} + a_2x^{n-2} + \ldots + a_n \\ &= (x - r)(b_0x^{n-1} + b_1x^{n-2} + b_2x^{n-3} + \ldots + b_{n-1}) \end{aligned} \tag{3.20}$$

On equating coefficients of the powers of x on both sides, we have

$$\begin{aligned} b_0 &= a_0 \\ b_i &= a_i + rb_{i-1} \text{ for } i = 1, 2, \ldots, n - 1 \end{aligned} \tag{3.21}$$

This process is known as synthetic division. Care must be taken here, particularly if the root is found to low accuracy, since ill-conditioning can magnify the effect of small errors in the coefficients of the deflated polynomial.

This completes the description of the method but a few important points should be noted. Assuming sufficient accuracy can be maintained in calculations, the method of Laguerre will converge for any value of the initial approximation. Convergence to complex roots and multiple roots can be achieved but at a slower rate because the convergence rate is linear. In the case of a complex root the value of the function $h(x_i)$ becomes negative and consequently the algorithm must be adjusted to deal with this situation. A key feature that should be considered is that the derivatives of the polynomial can be found efficiently by synthetic division.

To summarize the important features of the algorithm:

1. The algorithm is third order, thus providing rapid convergence to individual roots.
2. All roots of the polynomial can be found by using synthetic division.
3. Derivatives can be calculated efficiently using synthetic division.

3.12 SOLVING SYSTEMS OF NON-LINEAR EQUATIONS

The methods considered so far have been concerned with finding one or all the roots of a non-linear algebraic equation with one independent variable. We now consider methods for solving systems of non-linear algebraic equations in which each equation is a function of a specified number of variables.

We can write such a system in the form

$$f_i(x_1, x_2, \ldots, x_n) = 0 \text{ for } i = 1, 2, 3, \ldots, n \tag{3.22}$$

A simple method for solving this system of non-linear equations is based on Newton's method for the single equation. To illustrate this procedure we first consider a system of two equations in two variables:

$$\begin{aligned} f_1(x_1, x_2) &= 0 \\ f_2(x_1, x_2) &= 0 \end{aligned} \tag{3.23}$$

Given initial approximations x_1^0 and x_2^0 for x_1 and x_2, we may find new approximations x_1^1 and x_2^1 as follows:

$$\begin{aligned} x_1^1 &= x_1^0 + \Delta x_1^0 \\ x_2^1 &= x_2^0 + \Delta x_2^0 \end{aligned} \tag{3.24}$$

These approximations should be such that they drive the values of the functions closer to zero, so that

$$\begin{aligned} f_1(x_1^1, x_2^1) &\approx 0 \\ f_2(x_1^1, x_2^1) &\approx 0 \end{aligned}$$

or

$$\begin{aligned} f_1(x_1^0 + \Delta x_1^0,\ x_2^0 + \Delta x_2^0) &\approx 0 \\ f_2(x_1^0 + \Delta x_1^0,\ x_2^0 + \Delta x_2^0) &\approx 0 \end{aligned} \tag{3.25}$$

Applying a two-dimensional Taylor's series expansion to (3.25) gives

$$\begin{aligned} f_1(x_1^0, x_2^0) + \{\partial f_1/\partial x_1\}^0 \Delta x_1^0 + \{\partial f_1/\partial x_2\}^0 \Delta x_2^0 + \ldots &\approx 0 \\ f_2(x_1^0, x_2^0) + \{\partial f_2/\partial x_1\}^0 \Delta x_1^0 + \{\partial f_2/\partial x_2\}^0 \Delta x_2^0 + \ldots &\approx 0 \end{aligned} \tag{3.26}$$

If we neglect terms involving powers of Δx_1^0 and Δx_2^0 higher than one, then (3.26) represents a system of two linear equations in two unknowns. The zero superscript means that the function is to be calculated at the initial approximation and Δx_1^0 and Δx_2^0 are the unknowns we wish to find. Having solved (3.26) we can obtain our new improved approximations and then repeat the process until we have obtained the accuracy we require. A common convergence criterion is to continue iterations until

$$\sqrt{(\Delta x_1^r)^2 + (\Delta x_2^r)^2} < \varepsilon$$

where r denotes the iteration number and ε is a small positive quantity preset by the user.

It is a simple step to generalize this procedure for any number of variables and equations. We may write the general system of equations as

$$\mathbf{f}(\mathbf{x}) = \mathbf{0}$$

where $\mathbf{f}$ denotes the column vector of n components $(f_1, f_2, ..., f_n)^\mathsf{T}$ and $\mathbf{x}$ is a column vector of n components $(x_1, x_2, ..., x_n)^\mathsf{T}$. Let $\mathbf{x}^{r+1}$ denote the value of $\mathbf{x}$ at the $(r+1)$th iteration; then

$$\mathbf{x}^{r+1} = \mathbf{x}^r + \Delta\mathbf{x}^r \quad \text{for } r = 0, 1, 2, \ldots$$

If $\mathbf{x}^{r+1}$ is an improved approximation to $\mathbf{x}$, then

$$\mathbf{f}(\mathbf{x}^{r+1}) \approx \mathbf{0}$$

or

$$\mathbf{f}(\mathbf{x}^r + \Delta\mathbf{x}^r) \approx \mathbf{0} \tag{3.27}$$

Expanding (3.27) by using an n-dimensional Taylor's series expansion gives

$$\mathbf{f}(\mathbf{x}^r + \Delta\mathbf{x}^r) = \mathbf{f}(\mathbf{x}^r) + \nabla\mathbf{f}(\mathbf{x}^r)\Delta\mathbf{x}^r + \ldots \tag{3.28}$$

where ∇ is a vector operator of partial derivatives with respect to each of the n components of $\mathbf{x}$. If we neglect higher-order terms in $(\Delta\mathbf{x}^r)^2$, this gives, by virtue of (3.27),

$$\mathbf{f}(\mathbf{x}^r) + \mathbf{J}_r\Delta\mathbf{x}^r \approx \mathbf{0} \tag{3.29}$$

where $\mathbf{J}_r = \nabla\mathbf{f}(\mathbf{x}^r)$. $\mathbf{J}_r$ is called the Jacobian matrix. The subscript r denotes that the matrix is evaluated at the point $\mathbf{x}^r$ and it can be written in component form as

$$\mathbf{J}_r = [\partial f_i(\mathbf{x}^r)/\partial x_j] \quad \text{for } i = 1, 2, \ldots, n \text{ and } j = 1, 2, \ldots, n$$

On solving (3.29) we have the improved approximation

$$\mathbf{x}^{r+1} = \mathbf{x}^r - \mathbf{J}_r^{-1}\mathbf{f}(\mathbf{x}^r) \quad \text{for } r = 1, 2, ...$$

The matrix $\mathbf{J}_r$ may be singular and in this situation the inverse, $\mathbf{J}_r^{-1}$, cannot be calculated.

This is the general form of Newton's method. However, there are two major disadvantages with this method:

1. The method may not converge unless the initial approximation is a good one.
2. The method requires the user to provide the derivatives of each function with respect to each variable. The user must therefore provide n^2 derivatives and any computer implementation must evaluate the n functions and the n^2 derivatives at each iteration.

The MATLAB function `newtonmv` given here implements this method.

```
function [xv,it] = newtonmv(x,f,jf,n,tol)
% Newton's method for solving a system of n nonlinear equations
% in n variables.
% Example call: [xv,it] = newtonmv(x,f,jf,n,tol)
% Requires an initial approximation column vector x. tol is
% required accuracy. User must define functions f (system equations)
```

```
% and jf (partial derivatives). xv is the solution vector, the it
% parameter is number of iterations taken.
% WARNING. The method may fail, for example if initial estimates are poor.
it = 0; xv = x;
fr = feval(f,xv);
while norm(fr) > tol
    Jr = feval(jf,xv);  xv = xv-Jr\fr;
    fr = feval(f,xv);  it = it+1;
end
```

Fig. 3.11 illustrates the following system of two equations in two variables:

$$\begin{aligned} x^2 + y^2 &= 4 \\ xy &= 1 \end{aligned} \tag{3.30}$$

To solve the system (3.30) we define the MATLAB function by `f` and its Jacobian by `Jf` and then call `newtonmv` using initial approximations for the roots $x = 3$ and $y = -1.5$ and a tolerance of 0.00005 as follows:

```
>> f = @(v) [v(1)^2+v(2)^2-4; v(1)*v(2)-1];
>> Jf = @(v) [2*v(1) 2*v(2); v(2) v(1)];
>> [rootvals,iter] = newtonmv([3 -1.5]',f,Jf,2,0.00005)
```

This results in the MATLAB output

```
rootvals =
    1.9319
    0.5176


iter =
    5
```

The solution is $x = 1.9319$ and $y = 0.5176$. Clearly the user must supply a large amount of information for this function. The next section attempts to deal with this problem.

3.13 BROYDEN'S METHOD FOR SOLVING NON-LINEAR EQUATIONS

The method of Newton described in Section 3.12 does not provide a practical procedure for solving any but the smallest systems of non-linear equations. As we have seen, the method requires the user to provide not only the function definitions but also the definitions of the n^2 partial derivatives of the functions. Thus, for a system of 10 equations in 10 unknowns, the user must provide 110 function definitions!

To deal with this problem a number of techniques have been proposed but the group of methods that appear most successful is the class known as the quasi-Newton methods. The quasi-Newton methods avoid the calculation of the partial derivatives by obtaining approximations to them involving only the

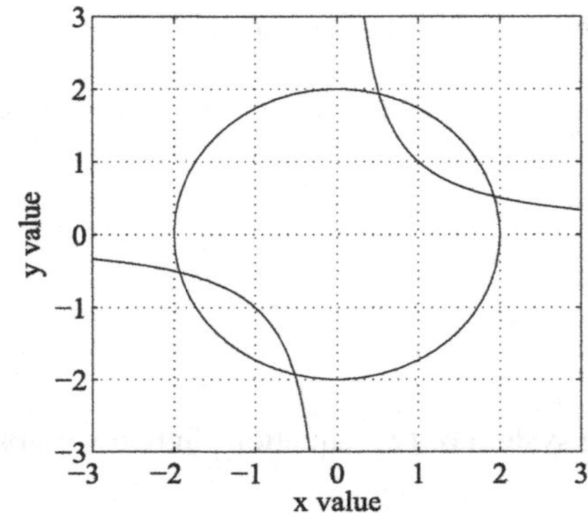

FIGURE 3.11

Plot of system (3.30). Intersections show roots.

function values. The set of derivatives of the functions evaluated at any point $\mathbf{x}^r$ may be written in the form of the Jacobian matrix

$$\mathbf{J}_r = [\partial f_i(\mathbf{x}^r)/\partial x_j] \quad \text{for } i = 1, 2, \ldots, n \text{ and } j = 1, 2, \ldots, n \tag{3.31}$$

The quasi-Newton methods provide an updating formula which gives successive approximations to the Jacobian for each iteration. Broyden and others have shown that under specified circumstances these updating formulae provide satisfactory approximations to the inverse Jacobian. The structure of the algorithm suggested by Broyden is:

1. Input an initial approximation to the solution. Set the counter r to zero.
2. Calculate or assume an initial approximation to the inverse Jacobian $\mathbf{B}^r$.
3. Calculate $\mathbf{p}^r = -\mathbf{B}^r\mathbf{f}^r$ where $\mathbf{f}^r = \mathbf{f}(\mathbf{x}^r)$.
4. Determine the scalar parameter t_r such that $||\mathbf{f}(\mathbf{x}^r + t_r\mathbf{p}^r)|| < ||\mathbf{f}^r||$ where the symbols $||\quad||$ denote that the norm of the vector is to be taken.
5. Calculate $\mathbf{x}^{r+1} = \mathbf{x}^r + t_r\mathbf{p}^r$.
6. Calculate $\mathbf{f}^{r+1} = \mathbf{f}(\mathbf{x}^{r+1})$. If $||\mathbf{f}^{r+1}|| < \varepsilon$ (where ε is a small preset positive quantity), then exit. If not continue with step (7).
7. Use the updating formula to obtain the required approximation to the Jacobian $\mathbf{B}^{r+1} = \mathbf{B}^r - (\mathbf{B}^r\mathbf{y}^r - \mathbf{p}^r)(\mathbf{p}^r)^\mathsf{T}\mathbf{B}^r/\{(\mathbf{p}^r)^\mathsf{T}\mathbf{B}^r\mathbf{y}^r\}$ where $\mathbf{y}^r = \mathbf{f}^{r+1} - \mathbf{f}^r$.
8. Set $i = i + 1$ and return to step (3).

The initial approximation to the inverse Jacobian $\mathbf{B}$ is usually taken as a scalar multiple of the unit matrix. The success of this algorithm depends on the nature of the functions to be solved and on the closeness of the initial approximation to the solution. In particular, step (4) may present major problems. It may be very expensive in computer time and to avoid this t_r is sometimes set as a constant, usually one or smaller. This may reduce the stability of the algorithm but speeds it up.

It should be noted that other updating formulae have been suggested and it is fairly easy to replace the Broyden formula by others in the preceding algorithm. In general, the problem of solving a system of non-linear equations is a very difficult one. There is no algorithm that is guaranteed to work for all systems of equations. For large systems of equations the available algorithms tend to require large amounts of computer time to obtain accurate solutions.

The MATLAB function `broyden` implements Broyden's method. It should be noted that this avoids the difficulty of implementing step (4) by taking $t_r = 1$.

```
function [xv,it] = broyden(x,f,n,tol)
% Broyden's method for solving a system of n nonlinear equations
% in n variables.
% Example call: [xv,it] = broyden(x,f,n,tol)
% Requires an initial approximation column vector x. tol is required
% accuracy. User must define function f.
% xv is the solution vector, parameter it is number of iterations
% taken. WARNING. Method may fail, for example, if initial estimates
% are poor.
fr = zeros(n,1); it = 0; xv = x;
Br = eye(n); %Set initial Br
fr = feval(f, xv);
while norm(fr)>tol
    it = it+1; pr = -Br*fr; tau = 1;
    xv = xv+tau*pr;
    oldfr = fr; fr = feval(f,xv);
    % Update approximation to Jacobian using Broyden's formula
    y = fr - oldfr; oldBr = Br;
    oyp = oldBr*y-pr; pB = pr'*oldBr;
    for i = 1:n
        for j = 1:n
            M(i,j) = oyp(i)*pB(j);
        end
    end
    Br = oldBr-M./(pr'*oldBr*y);
end
```

To solve the system (3.30) using Broyden's method we call `broyden` as follows:

```
>> f = @(v) [v(1)^2+v(2)^2-4; v(1)*v(2)-1];
>> [x, iter] = broyden([3 -1.5]',f,2,0.00005)
```

This results in

```
x =
    0.5176
    1.9319


iter =
    36
```

This is a correct root of system (3.30) but it is not the same root as that found by Newton's method, even though the starting values for the iteration are the same.

As a second example we consider the following system of equations which are taken from the MATLAB User's Guide (1989):

$$\begin{aligned} \sin x + y^2 + \log_e z &= 7 \\ 3x + 2y - z^3 &= -1 \\ x + y + z &= 5 \end{aligned} \tag{3.32}$$

The function g, which implements (3.32), is given here:

```
>> g = @(p) [sin(p(1))+p(2)^2+log(p(3))-7; 3*p(1)+2^p(2)-p(3)^3+1;
            p(1)+p(2)+p(3)-5];
```

The result of solving (3.32) is given next. The starting values used are $x = 0$, $y = 2$, and $z = 2$.

```
>> x = broyden([0 2 2]',g,3,0.00005)

x =
    0.5991
    2.3959
    2.0050
```

We can now verify this result by substituting this solution in g(p) thus:

```
>> g(x)'

ans =
   1.0e-05 *
   -0.2867    -0.5105     0.0068
```

The residue should, of course, be zero, but here the residue is very small, being of the order of 10^{-5}. This shows that the method is successful for two problems and does not require the evaluation of the partial derivatives. The reader may be interested in applying the function newtonmv to this problem. Nine first-order partial derivatives will be required.

3.14 COMPARING THE NEWTON AND BROYDEN METHODS

We end our discussion of the solution of non-linear systems of equations by comparing the performance of the functions broyden and newtonmv, developed in Sections 3.12 and 3.13 when solving the system (3.30). The following script calls both functions and provides the number of iterations required for convergence.

```
>> f = @(v) [v(1)^2+v(2)^2-4; v(1)*v(2)-1];
>> [x,it] = broyden([3 -1.5]',f,2,0.00005)
```

```
x =
    0.5176
    1.9319

it =
    36

>> J = @(v) [2*v(1) 2*v(2);v(2) v(1)];
>> [x,it] = newtonmv([3,-1.5]',f,J,2,0.00005)

x =
    1.9319
    0.5176

it =
     5
```

Note that although a correct solution is found in each case, it is a different root.

The first-order partial derivatives are required for the Newton method and this requires a considerable effort on the part of the user. Solving the problem using Broyden's method demonstrates that the relatively simple form of the function `broyden` is attractive since it relieves the user of this effort.

In Sections 3.12 and 3.13 two relatively simple algorithms were provided for the solution of a very difficult problem. They cannot always be guaranteed to work and for large problems will converge only slowly.

3.15 SUMMARY

The user wishing to solve non-linear equations will find that this is an area which can present particular difficulties. It is always possible to devise or meet problems which particular algorithms either cannot solve or take a long time to solve. For example, it is just not possible for many algorithms to find the roots of the apparently trivial problem $x^{20} = 0$ very accurately. However, the algorithms described, if used with care, provide ways of solving a wide range of problems. MATLAB is well suited for this study because it allows interactive experimentation and graphical insights into the behavior of methods and functions. The reader is referred to Chapter 10, Section 10.6 for applications of the symbolic toolbox for solving non-linear equations. The algorithms `solve`, `fnewtsym`, `newtmvsym` are described and applied in that section.

3.16 PROBLEMS

3.1. Omar Khayyam (who lived in the twelfth century) solved, by geometric means, a cubic equation with the form

$$x^3 - cx^2 + b^2x + a^3 = 0$$

The positive roots of this equation are the x coordinates of points of intersection in the first quadrant of the circle and parabola given in the following:

$$x^2 + y^2 - (c - a^3/b^2)x + 2by + b^2 - ca^3/b^2 = 0$$
$$xy = a^3/b$$

For $a = 1$, $b = 2$, and $c = 3$ use MATLAB to plot these two functions and note the x coordinates of the points of intersection. Using the MATLAB function `fzero`, solve the cubic equation and hence verify Omar Khayyam's method. *Hint*: You may find it helpful to use the MATLAB function `ginput`.

3.2. Use the MATLAB function `fnewton` to find a root of

$$x^{1.4} - \sqrt{x} + 1/x - 100 = 0$$

given an initial approximation 50. Use an accuracy of 10^{-4}.

3.3. Find the two real roots of $|x^3| + x - 6 = 0$ using the MATLAB function `fnewton`. Use initial approximations -1 and 1 and an accuracy of 10^{-4}. Plot the function using MATLAB to verify that the equation has only two real roots. *Hint:* Take care in finding the derivative of the function.

3.4. Explain why it is relatively difficult to find the root of $\tan x - c = 0$ when c is a large. Use the MATLAB function `fnewton`, with initial approximations 1.3 and 1.4 and accuracy 10^{-4}, to find a root of this equation when $c = 5$ and $c = 10$. Compare the number of iterations required in both cases. *Hint*: A MATLAB `plot` will be useful.

3.5. Find a root of the polynomial $x^5 - 5x^4 + 10x^3 - 10x^2 + 5x - 1 = 0$ correct to four decimal places by using the MATLAB function `schroder` with $m = 5$ and a starting value $x_0 = 2$. Use MATLAB function `fnewton` to solve the same problem. Compare the result and the number of iterations using both methods. Use an accuracy of 5×10^{-7}.

3.6. Use the simple iterative method to solve the equation $x^{10} = e^x$. Express the equation in the form $x = f(x)$ in different ways and start the iterations with the initial approximation $x = 1$. Compare the efficiency of the formulae you have devised and check your answer(s) using the MATLAB function `fnewton`.

3.7. The historic Kepler's equation has the form $E - e \sin E = M$. Solve this equation for $e = 0.96727464$, the eccentricity of Halley's comet, and $M = 4.527594 \times 10^{-3}$. Use the MATLAB function `fnewton`, with an accuracy of 0.00005 and a starting value of 1.

3.8. Examine the performance of the function `fzero` for solving $x^{11} = 0$ with an initial value of -1.5 and also 1. Use an accuracy of 1×10^{-5}.

3.9. The smallest positive root of the equation

$$1 - x + x^2/(2!)^2 - x^3/(3!)^2 + x^4/(4!)^2 - \ldots = 0$$

is 1.4458. By considering in turn only the first four, five, and six terms in the series, show that a root of the truncated series approaches this result. Use the MATLAB function `fzero` to derive these results, with an initial value of 1 and an accuracy of 10^{-4}.

3.10. Reduce the following system of equations to one equation in terms of x and solve the resulting equation using the MATLAB function `fnewton`.

$$e^{x/10} - y = 0$$
$$2\log_e y - \cos x = 2$$

Use the MATLAB function `newtonmv` to solve these equations directly and compare your results. Use an initial approximation $x = 1$ for `fnewton` and approximations $x = 1$, $y = 1$ for `newtonmv` and accuracy 10^{-4} in both cases.

3.11. Solve the pair of equations that follow using the MATLAB function `broyden`, with the starting point $x = 10$, $y = -10$ and accuracy 10^{-4}.

$$2x = \sin\{(x + y)/2\}$$
$$2y = \cos\{(x - y)/2\}$$

3.12. Solve the two equations that follow using the MATLAB functions `newtonmv` and `broyden` with the starting point $x = 1$ and $y = 2$ and accuracy 10^{-4}.

$$x^3 - 3xy^2 = 1/2$$
$$3x^2y - y^3 = \sqrt{3}/2$$

3.13. The polynomial equation

$$x^4 - (13 + \varepsilon)x^3 + (57 + 8\varepsilon)x^2 - (95 + 17\varepsilon)x + 50 + 10\varepsilon = 0$$

has roots 1, 2, 5, $5 + \varepsilon$. Use the functions `bairstow` and `roots` to find all the roots of this polynomial for $\varepsilon = 0.1, 0.01$, and 0.001. What happens as ε becomes smaller? Use an accuracy of 10^{-5}.

3.14. Employ the MATLAB function `bairstow` to find all the roots of the following polynomial using an accuracy requirement of 10^{-4}.

$$x^5 - x^4 - x^3 + x^2 - 2x + 2 = 0$$

3.15. Use the MATLAB function `roots` to find all the roots of the equation

$$t^3 - 0.5 - \sqrt{3/2}\imath = 0 \quad \text{where } \imath = \sqrt{-1}$$

Compare with the exact solution

$$\cos\{(\pi/3 + 2\pi k)/3\} + \imath\sin\{(\pi/3 + 2\pi k)/3\} \quad \text{for } k = 0, 1, 2.$$

Use an accuracy of 10^{-4}.

3.16. An outline algorithm for the Illinois method for finding a root of $f(x) = 0$ (Dowell and Jarrett, 1971) is as follows:

For $k = 0, 1, 2, \ldots$
$x_{k+1} = x_k - f_k / f[x_{k-1}, x_k]$
if $f_k f_{k+1} > 0$ set $x_k = x_{k-1}$ and $f_k = g f_{k-1}$
where $f_k = f(x_k)$, $f[x_{k-1}, x_k] = (f_k - f_{k-1})/(x_k - x_{k-1})$
and $g = 0.5$.

Write a MATLAB function to implement this method. Note that the regula falsi method is similar but differs in that g is taken as one.

3.17. The following iterative formulae can be used to solve the equation $x^2 - a = 0$:

$$x_{k+1} = (x_{k+1} + a/x_k)/2, \quad k = 0, 1, 2, \ldots$$

and

$$x_{k+1} = (x_{k+1} + a/x_k)/2 - (x_k - a/x_k)^2/(8x_k), \quad k = 0, 1, 2, \ldots$$

These iterative formulae are second- and third-order methods, respectively, for solving this equation. Write a MATLAB script to implement them and compare the number of iterations required to obtain the square root of 100.112 to five decimal places. For the purpose of illustration, use an initial approximation of 1000.

3.18. Show how MATLAB can be used to study chaotic behavior by considering the iteration

$$x_{k+1} = g(x_k) \quad \text{for } k = 0, 1, 2, \ldots$$

where

$$g(x) = cx(1 - x)$$

for different values of the constant c. This simple iteration arises from an attempt to solve a simple quadratic equation. However, its behavior is complex and for some values of c is chaotic. Write a MATLAB script to plot the value of the iterates against the iterate number for this function and study the behavior of the iterations for $c = 2.8$, 3.25, 3.5, and 3.8. Use an initial value of $x_0 = 0.7$.

3.19. For the functions solved in Problems 3.2, 3.3 and 3.7, use the MATLAB function `plotapp`, given in Section 3.9, to find approximate solutions for these functions.

3.20. It can be shown that the cubic polynomial equation

$$x^3 - px - q = 0$$

will have real roots if the inequality $p^3/q^2 > 27/4$ is satisfied. Select five pairs of values for p and q for which this inequality is satisfied and hence, using the MATLAB function `roots`, verify in each case that the roots of the equation are real.

3.21. In the 16th century the mathematician Ioannes Colla suggested the following problem: Divide 10 into three parts such that they shall be in continued proportion to each other and the product of the first two shall be 6. Taking x, y, and z as three parts, this problem can be stated as:

$$x + y + z = 10, \; x/y = y/z, \; xy = 6$$

Now by simple manipulation these equations can be expressed in terms of the specific variable y as:

$$y^4 + 6y^2 - 60y + 36 = 0$$

Clearly if we can solve this equation for y then we can easily find the other variables x and z from the original equations. Use the MATLAB function `roots` to find values for y and hence solve Colla's problem.

3.22. The natural frequencies of a simply supported beam are given by the roots of the equation

$$c_1^2 - x^4 c_3^2 = 0$$

where

$$c_1 = (\sinh(x) + \sin(x))/(2x)$$

and

$$c_3 = (\sinh(x) - \sin(x))/(2x^3)$$

Substituting for c_1 and c_3 gives

$$((\sinh(x) + \sin(x))/(2x))^2 - x^4((\sinh(x) - \sin(x))/(2x^3))^2 = 0$$

When searching for the roots of this equation no difficulty is found in determining a root for trial values of x, providing x is small (say $x < 10$). For values of $x > 25$ the process becomes erratic. The roots of this equation are actually $x = k\pi$ where k is a positive integer. Use the MATLAB function `fzero` with initial approximations $x = 5$ and $x = 30$ to obtain a solution close to these initial approximations for this equation. For the purpose of this exercise, do not simplify this equation.

Why are the results so poor? If you simplify the preceding equation, which equation do you obtain and what is its solution?

3.23. Use the MATLAB function `roots` to solve the cubic equation:

$$x^3 + 3px - 2q = 0$$

with $p = 3$ and $q = 4$. One exact solution of this equation is give by Cardano's formula:

$$x = \sqrt[3]{q + \sqrt[2]{q^2 + p^3}} + \sqrt[3]{q - \sqrt[2]{q^2 + p^3}} = r_1 + r_2$$

The complex roots are obtained using $w_1 r_1 + w_2 r_2$ and $w_2 r_1 + w_1 r_2$ where $w_1 = (-1 + \jmath\sqrt{3})/2$ and $w_2 = (-1 - \jmath\sqrt{3})/2$.

Check your result using the MATLAB function `roots`. Hint. To find the real root using the formula you should use the MATLAB function `nthroot`.

3.24. If a, b, and c are real, and $a < b < c$ then the equation:

$$\frac{1+ax}{x-a}+\frac{1+bx}{x-b}+\frac{1+cx}{x-c}+d=0$$

has three real roots. Taking $a = 1$, $b = 2$, $c = 3$ and rearranging this equation as a cubic. If you have access to the Symbolic Toolbox you may wish to check your cubic equation using the symbolic function `collect`. Then solve this equation using the MATLAB function `roots` and verify the statement regarding the roots of this equation.

3.25. Determine the roots of the equation

$$A\sin(3\theta)+B\cos^2(\theta)+C\sin(\theta)+1=0$$

where $A = 0.157$, $B = -0.940$, and $C = -0.900$. *Hint*: Expand $\sin(3\theta)$ and $\cos^2(\theta)$ in terms of powers of $\sin(\theta)$ and solve the resultant polynomial using the MATLAB function `roots` to obtain values for $\sin(\theta)$, and hence θ.

CHAPTER

DIFFERENTIATION AND INTEGRATION

4

Abstract

Differentiation and integration are the fundamental operations of differential calculus and occur in almost every field of mathematics, science, and engineering. Determining the derivative of a function analytically may be tedious but is relatively straightforward. The inverse of this process, that of analytically determining the integral of a function, can often be difficult or impossible. These difficulties have encouraged the development of many numerical procedures for determining approximately the value of definite integrals. In this chapter we introduce a wide range of algorithm for both numerical differentiation and integration.

4.1 INTRODUCTION

In the next section of this chapter, we show how the derivative of a function may be estimated for a particular value of the independent variable. The numerical approximations for derivatives require only function values. These approximations can be used to great advantage when derivatives are required in a program. Their application saves the program user the task of determining the analytical expressions for these derivatives. In Section 4.3 and beyond, we introduce the reader to a range of numerical integration methods, including methods suitable for infinite ranges of integration. Generally, numerical integration works well, but there are pathological integrals that will defeat the best numerical algorithms.

4.2 NUMERICAL DIFFERENTIATION

In this section, we present a range of approximations for first- and higher-order derivatives. Before we derive these approximations in detail, we give a simple example which illustrates the dangers of the careless or naive use of such derivative approximations. The simplest approximation for the first-order derivative of a given function $f(x)$ may be developed from the formal definition of the derivative:

$$\frac{df}{dx} = \lim_{h \to 0} \left(\frac{f(x+h) - f(x)}{h} \right) \tag{4.1}$$

One interpretation of (4.1) is that the derivative of a function $f(x)$ is the slope of the tangent to the function at the point x. For small h we obtain the approximation to the derivative:

$$\frac{df}{dx} \approx \left(\frac{f(x+h) - f(x)}{h} \right) \tag{4.2}$$

Numerical Methods. https://doi.org/10.1016/B978-0-12-812256-3.00013-0

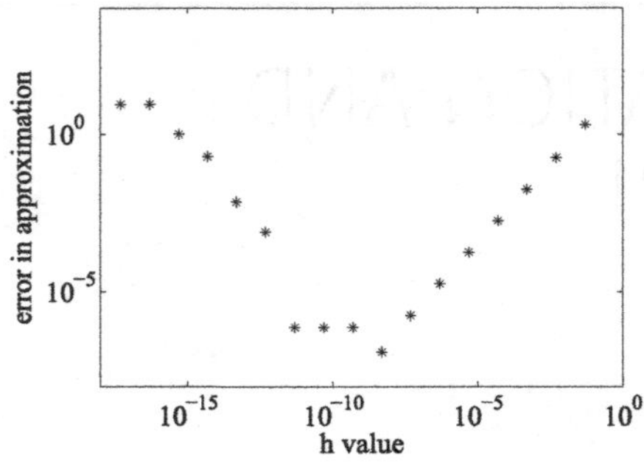

FIGURE 4.1

A log–log plot showing the error in a simple derivative approximation.

The error is of $\mathcal{O}(h)$ and would appear to imply that the smaller the value of h, the better the value of our approximation given by (4.2). The following MATLAB script plots Fig. 4.1 which shows the error for various values of h.

```
% e4s401.m
g = @(x) x.^9;
x = 1; h(1) = 0.5;
hvals = [ ]; dfbydx = [ ];
for i = 1:17
    h = h/10;
    b = g(x); a = g(x+h);
    hvals = [hvals h];
    dfbydx(i) = (a-b)/h;
end;
exact = 9;
axes('position',[0.30 0.30 0.50 0.50])
loglog(hvals,abs(dfbydx-exact),'*')
axis([1e-18 1 1e-8 1e4])
xlabel('h value'), ylabel('Error in approximation')
```

Fig. 4.1 shows that for large values of h the error is large but falls rapidly as h is decreased. However, when h becomes less than about 10^{-9}, rounding errors dominate and the approximation becomes much worse. Clearly care must be taken in the choice of h. With this warning in mind we develop methods of differing accuracies for any order derivative.

We have seen how a simple approximate formula for the first derivative can be easily obtained from the formal definition of the derivative. However, it is difficult to approximate higher derivatives and deduce more accurate formulae in this way; instead we will use the Taylor series expansion of the function $y = f(x)$. To determine a more refined approximation for the derivative of this function at x_i

we expand $f(x_i + h)$ thus:

$$f(x_i + h) = f(x_i) + hf'(x_i) + (h^2/2!)f''(x_i) + (h^3/3!)f'''(x_i) + (h^4/4!)f^{(iv)}(x_i) + \ldots \tag{4.3}$$

We evaluate $f(x)$ at points a distance h apart and write $x_i + h$ as x_{i+1}, etc. We will also write $f(x_i)$ as f_i and $f(x_{i+1})$ as f_{i+1}. Thus,

$$f_{i+1} = f_i + hf'(x_i) + (h^2/2!)f''(x_i) + (h^3/3!)f'''(x_i) + (h^4/4!)f^{(iv)}(x_i) + \ldots \tag{4.4}$$

Similarly

$$f(x_i - h) = f_{i-1} = f_i - hf'(x_i) + (h^2/2!)f''(x_i) - (h^3/3!)f'''(x_i) + (h^4/4!)f^{(iv)}(x_i) - \ldots \tag{4.5}$$

We can find an approximation to the first derivative as follows. Subtracting (4.5) from (4.4) gives

$$f_{i+1} - f_{i-1} = 2hf'(x_i) + 2\left(h^3/3!\right) f'''(x_i) + \ldots$$

Thus, neglecting terms in h^3 and higher, since we may assume h is small, we have

$$f'(x_i) = (f_{i+1} - f_{i-1})/2h \text{ with errors of } O\left(h^2\right) \tag{4.6}$$

This is referred to as the central difference approximation and differs from (4.2) which is a forward difference approximation. Eq. (4.6) is more accurate than (4.2) but in the limit as h approaches zero, the two are identical.

To determine an approximation for the second derivative we add (4.4) and (4.5) to obtain

$$f_{i+1} + f_{i-1} = 2f_i + 2\left(h^2/2!\right) f''(x_i) + 2\left(h^4/4!\right) f^{(iv)}(x_i) + \ldots$$

Thus, neglecting terms in h^4 and higher, we have

$$f''(x_i) = (f_{i+1} - 2f_i + f_{i-1})/h^2 \text{ with errors of } O\left(h^2\right) \tag{4.7}$$

By taking more terms in the Taylor series, together with the Taylor series for $f(x + 2h)$ and $f(x - 2h)$, etc., and performing similar manipulations, we can obtain higher derivatives and more accurate approximations if required. Table 4.1 gives examples of these formulae.

The following MATLAB function `diffgen` computes the first, second, third, and fourth derivative of a given function with errors of $O(h^4)$ for a specified value of x using data from the table.

```
function q = diffgen(func,n,x,h)
% Numerical differentiation.
% Example call: q = diffgen(func,n,x,h)
```

Table 4.1 Derivative approximations

	Multipliers for $f_{i-3} \cdots f_{i+3}$							
	f_{i-3}	f_{i-2}	f_{i-1}	f_i	f_{i+1}	f_{i+2}	f_{i+3}	order of error
$2hf'(x_i)$	0	0	−1	0	1	0	0	h^2
$h^2 f''(x_i)$	0	0	1	−2	1	0	0	h^2
$2h^3 f'''(x_i)$	0	−1	2	0	−2	1	0	h^2
$h^4 f^{(iv)}(x_i)$	0	1	−4	6	−4	1	0	h^2
$12hf'(x_i)$	0	1	−8	0	8	−1	0	h^4
$12h^2 f''(x_i)$	0	−1	16	−30	16	−1	0	h^4
$8h^3 f'''(x_i)$	1	−8	13	0	−13	8	−1	h^4
$6h^4 f^{(iv)}(x_i)$	−1	12	−39	56	−39	12	−1	h^4

```
% Provides nth order derivatives, where n = 1 or 2 or 3 or 4
% of the user defined function func at the value x, using a step h.
if (n==1)|(n==2)|(n==3)|(n==4)
    c = zeros(4,7);
    c(1,:) = [ 0 1 -8 0 8 -1 0];
    c(2,:) = [ 0 -1 16 -30 16 -1 0];
    c(3,:) = [1.5 -12 19.5 0 -19.5 12 -1.5];
    c(4,:) = [ -2 24 -78 112 -78 24 -2];
    y = feval(func,x+[-3:3]*h);
    q = c(n,:)*y.';  q = q/(12*h^n);
else
    disp('n must be 1, 2, 3 or 4'), return
end
```

For example,

```
result = diffgen('cos',2,1.2,0.01)
```

determines the second derivative of $\cos(x)$ for $x = 1.2$ with $h = 0.01$ and gives `-0.3624` for the result. The following script calls the function `diffgen` four times to determine the first four derivatives of $y = x^7$ when $x = 1$:

```
% e4s402.m
g = @(x) x.^7;
h = 0.5;
disp('     h       1st deriv  2nd deriv    3rd deriv     4th deriv');
while h>=1e-5
    t1 = h;
    t2 = diffgen(g, 1, 1, h);
    t3 = diffgen(g, 2, 1, h);
    t4 = diffgen(g, 3, 1, h);
    t5 = diffgen(g, 4, 1, h);
```

```
    fprintf('%10.5f %10.5f %10.5f %11.5f %12.5f\n',t1,t2,t3,t4,t5);
    h = h/10;
end
```

The output from the script `e4s402` is

```
     h       1st deriv   2nd deriv   3rd deriv     4th deriv
  0.50000     1.43750    38.50000   191.62500    840.00000
  0.05000     6.99947    41.99965   209.99816    840.00000
  0.00500     7.00000    42.00000   210.00000    840.00001
  0.00050     7.00000    42.00000   210.00000    839.95625
  0.00005     7.00000    42.00000   209.98669   -189.47806
```

Note that as h is decreased the estimates for the first and second derivatives steadily improve, but when $h = 5 \times 10^{-4}$ the estimate for the fourth derivative begins to deteriorate. When $h = 5 \times 10^{-5}$ the estimate for the third derivative also begins to deteriorate and the fourth derivative is very inaccurate. In general we cannot predict when this deterioration will begin. It should be noted that different computer platforms may give different results for this value.

The function `diffgen` numerically differentiates the mathematical function provided. This function is then sampled. The method `diffgen` could be configured to take a vector of data instead of a mathamatical function. Thus `diffgen` could be modified as follows:

```
function q = diffgend(x,n,k,h)
```

where `x` is a vector of equispaced data of increment `h`, `n` is the order of the required derivative and `k` is the index of `x`, so that the derivatives are determined for the point `x(k)`.

4.3 NUMERICAL INTEGRATION

We begin by examining the definite integral

$$I = \int_a^b f(x)\,dx \tag{4.8}$$

The evaluation of such integrals is often called quadrature and we will develop methods for both finite and infinite values of a and b.

The definite integral (4.8) is a summation process but it may also be interpreted as the area under the curve $y = f(x)$ from a to b. Any areas above the x-axis are counted positive; any areas below the x-axis are counted as negative. Integration is a smoothing process and errors in the approximation tend to cancel each other. Many numerical methods for integration are based on using this interpretation to derive approximations to the integral. Typically the interval $[a, b]$ is divided into a number of smaller subintervals, and by making simple approximations to the curve $y = f(x)$ in the subinterval, the area of the subinterval may be obtained. The areas of all the subintervals are then summed to give an approximation to the integral in the interval $[a, b]$. Variations of this technique are developed by taking groups of subintervals and fitting different degree polynomials to approximate $y = f(x)$ in each of these groups. The simplest of these methods is the trapezoidal rule.

The trapezoidal rule is based on the idea of approximating the function $y = f(x)$ in each subinterval by a straight line so that the shape of the area in the subinterval is trapezoidal. Clearly, as the number of subintervals used increases, the straight lines will approximate the function more closely. Dividing the interval from a to b into n subintervals of width h (where $h = (b - a)/n$) we can calculate the area of each subinterval since the area of a trapezium is its base times the mean of its heights. These heights are f_i and f_{i+1} where $f_i = f(x_i)$. Thus, the area of the trapezium is

$$h(f_i + f_{i+1})/2 \text{ for } i = 0, 1, 2, \ldots, n - 1$$

Summing all the areas of the trapezia gives the composite trapezoidal rule for approximating (4.8) thus:

$$I \approx h\{(f_0 + f_n)/2 + f_1 + f_2 + \ldots + f_{n-1}\} \tag{4.9}$$

The truncation error, E_n, which is the error due to the implicit approximation in the trapezoidal rule, is

$$E_n \leq (b - a)h^2 M/12 \tag{4.10}$$

where M is the upper bound for $|f''(t)|$ and t must be in the range a to b. The MATLAB function `trapz` implements this procedure and we use it in Section 4.4 to compare the performance of the trapezoidal rule with the more accurate Simpson's rule.

The level of accuracy obtained from a numerical integration procedure is dependent on three factors. The first two are the nature of the approximating function and the number of intervals used. These are controlled by the user and give rise to the truncation error, that is, the error inherent in the approximation. The third factor influencing accuracy is the rounding error, the error caused by the fact that practical computation has limited precision. For a particular approximating function the truncation error will decrease as the number of subintervals increases. Integration is a smoothing process and rounding errors do not present a major problem. However, when many intervals are used the time to solve the problem becomes more significant because of the increased amount of computation. This problem may be reduced by writing the script efficiently.

4.4 SIMPSON'S RULE

Simpson's rule is based on using a quadratic polynomial approximation to the function $f(x)$ over a pair of subintervals; it is illustrated in Fig. 4.2. If we integrate the quadratic polynomial passing through the points (x_0, f_0); (x_1, f_1); (x_2, f_2), where $f_1 = f(x_1)$, etc., the following formula is obtained:

$$\int_{x_0}^{x_2} f(x)\,dx = \frac{h}{3}(f_0 + 4f_1 + f_2) \tag{4.11}$$

This is Simpson's rule for one pair of intervals. Applying the rule to all pairs of intervals in the range a to b and adding the results produces the following expression, known as the composite Simpson's rule:

$$\int_a^b f(x)\,dx = \frac{h}{3}\{f_0 + 4(f_1 + f_3 + f_5 + \ldots + f_{2n-1}) + 2(f_2 + f_4 + \ldots + f_{2n-2}) + f_{2n}\} \tag{4.12}$$

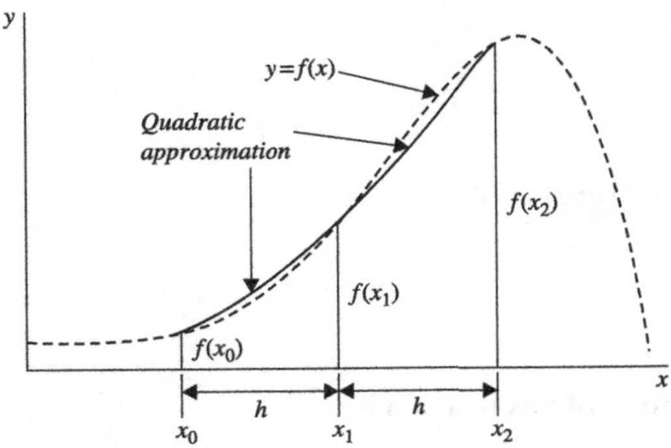

FIGURE 4.2

Simpson's rule, using a quadratic approximation over two intervals.

Here n indicates the number of pairs of intervals and $h = (b - a)/(2n)$. The composite rule may be also be written as a vector product thus:

$$\int_a^b f(x)\,dx = \frac{h}{3}\left(\mathbf{c}^\mathsf{T}\mathbf{f}\right) \tag{4.13}$$

where $\mathbf{c} = [1\ 4\ 2\ 4\ 2\ \ldots\ 2\ 4\ 1]^\mathsf{T}$ and $\mathbf{f} = [f_1\ f_2\ f_3\ \cdots\ f_{2n}]^\mathsf{T}$.

The error arising from the approximation, called the truncation error, is approximated by

$$E_n = (b - a)h^4 f^{(iv)}(t)/180$$

where t lies between a and b. An upper bound for the error is given by

$$E_n \leq (b - a)h^4 M/180 \tag{4.14}$$

where M is an upper bound for $|f^{(iv)}(t)|$. The upper bound for the error in the simpler trapezoidal rule, (4.10), is proportional to h^2 rather than h^4. This makes Simpson's rule superior to the trapezoidal rule in terms of accuracy at the expense of more function evaluations.

To illustrate different ways of implementing Simpson's rule we provide two alternatives, `simp1` and `simp2`. The function `simp1` creates a vector of coefficients `v` and a vector of function values `y` and multiplies the two vectors together. Function `simp2` provides a more conventional implementation of Simpson's rule. In each case the user must provide the definition of the function to be integrated, the lower and upper limits of integration and the number of subintervals to be used. The number of subintervals must be an even number since the rule fits a function to a *pair* of subintervals.

```
function q = simp1(func,a,b,m)
% Implements Simpson's rule using vectors.
% Example call: q = simp1(func,a,b,m)
% Integrates user defined function func from a to b, using m divisions
if (m/2)~=floor(m/2)
```

```
    disp('m must be even'); return
end
h = (b-a)/m; x = a:h:b;
y = feval(func,x);
v = 2*ones(m+1,1);  v2 = 2*ones(m/2,1);
v(2:2:m) = v(2:2:m)+v2;
v(1) = 1;  v(m+1) = 1;
q = (h/3)*y*v;
```

The second non-vectorized form of this function is

```
function q = simp2(func,a,b,m)
% Implements Simpson's rule using for loop.
% Example call: q = simp2(func,a,b,m)
% Integrates user defined function
% func from a to b, using m divisions
if (m/2) ~= floor(m/2)
    disp('m must be even'); return
end
h = (b-a)/m;
s = 0; yl = feval(func,a);
for j = 2:2:m
    x = a+(j-1)*h;  ym = feval(func,x);
    x = a+j*h;  yh = feval(func,x);
    s = s+yl+4*ym+yh;  yl = yh;
end
q = s*h/3;
```

The script `e4s403` calls either `simp1` or `simp2`. These functions can be used to demonstrate the effect on accuracy of the number of pairs of intervals used. The script `e4s403` evaluates the integral of x^7 in the range 0 to 1.

```
% e4s403.m
n = 4; i = 1;
disp('    n integral value')
while n < 1025
    simpval = simp1(@(x) x.^7,0,1,n); % or simpval = simp2(etc.);
    fprintf('%5.0f %15.12f\n',n,simpval)
    n = 2*n; i = i+1;
end
```

The output from script `e4s403` using `simp1` is shown thus:

```
n integral value
    4  0.129150390625
    8  0.125278472900
   16  0.125017702579
   32  0.125001111068
   64  0.125000069514
  128  0.125000004346
  256  0.125000000272
  512  0.125000000017
 1024  0.125000000001
```

On running this script, but using `simp2`, we obtain the same values for the integral. However, we would expect that the vectorized version, `simp1`, would be faster than `simp2`.

Eq. (4.14) shows that the truncation error will decrease rapidly for values of h smaller than 1. The preceding results illustrate this. The rounding error in Simpson's rule is due to evaluating the function $f(x)$ and the subsequent multiplications and additions.

We now evaluate the same integral using the MATLAB function `trapz`. To call this function the user must provide a vector of function values `f`. The function `trapz(f)` estimates the integral of the function assuming unit spacing between the data points. Thus, to determine the integral we multiply `trapz(f)` by the increment `h`.

```
% e4s404.m
n = 4; i = 1; f = @(x) x.^7;
disp('    n  integral value')
while n<1025
    h = 1/n; x = 0:h:1;
    trapval = h*trapz(f(x));
    fprintf('%5.0f %15.12f\n',n,trapval)
    n = 2*n; i = i+1;
end
```

Running script `e4s404` gives

```
    n  integral value
    4  0.160339355469
    8  0.134043693542
   16  0.127274200320
   32  0.125569383381
   64  0.125142397981
  128  0.125035602755
  256  0.125008900892
  512  0.125002225236
 1024  0.125000556310
```

These results illustrate the fact that the trapezoidal rule is less accurate than the Simpson rule.

As with numerical differentiation, we have chosen to design the MATLAB functions `simp1` and `simp2` to integrate a mathematical function. Sometimes, instead of a function, we have a set data representing the function and `simp1` could be modified to accept equispaced data rather than a function as follows:

```
function q = simpd(y,x0,xm)
```

where `y` is a vector of an even number of equispaced data, `x0` and `xm` are the values of `x` corresponding to `y(1)` and `y(m)`.

4.5 NEWTON–COTES FORMULAE

Simpson's rule is an example of a Newton–Cotes formula for integration. Other examples of these formulae can be obtained by fitting higher-degree polynomials through the appropriate number of points. In general we fit a polynomial of degree n through $n+1$ points. The resulting polynomial can then be integrated to provide an integration formula. Here are some examples of Newton–Cotes formulae together with estimates of their truncation errors. It is important for the reader to note that in this section the truncation error approximations provided are for the range of integration of the specific rule only. This is in contrast to the truncation error given for Simpson's rule which is for the composite rule applied to the whole integral range.

For $n = 3$ we have

$$\int_{x_0}^{x_3} f(x)\,dx = \frac{3h}{8}(f_0 + 3f_1 + 3f_2 + f_3) + \text{truncation error } \frac{3h^5}{80} f^{iv}(t) \tag{4.15}$$

where t lies in the interval x_0 to x_3.

For $n = 4$ we have

$$\int_{x_0}^{x_4} f(x)\,dx = \frac{2h}{45}(7f_0 + 32f_1 + 12f_2 + 32f_3 + 7f_4) + \text{truncation error } \frac{8h^7}{945} f^{(vi)}(t) \tag{4.16}$$

where t lies in the interval x_0 to x_4. Composite rules for a specific number of intervals can be generated using either (4.15) or (4.16).

The truncation errors indicate that some improvement in accuracy may be obtained by using these rules rather than Simpson's rule. However, the rules are more complex; consequently, greater computational effort is involved and rounding errors may become a more significant problem.

4.6 ROMBERG INTEGRATION

A major problem that arises with the Simpson's or Newton–Cotes rules is that the number of intervals required to provide the required accuracy is initially unknown. Clearly one approach to this problem is to double successively the number of intervals used and compare the results of applying a particular rule, as illustrated by the examples in Section 4.4. Romberg's method provides an organized approach to this problem and utilizes the results obtained by applying Simpson's rule with different interval sizes to reduce the truncation error.

Romberg integration may be formulated as follows. Let I be the exact value of the integral and T_i the approximate value of the integral obtained using Simpson's rule with i intervals. Consequently, we may write an approximation for the integral I which includes contributions from the truncation error as follows (note that the error terms are expressed in powers of h^4):

$$I = T_i + c_1 h^4 + c_2 h^8 + c_3 h^{12} + \ldots \tag{4.17}$$

If we double the number of intervals, h is halved, giving

$$I = T_{2i} + c_1 (h/2)^4 + c_2 (h/2)^8 + c_3 (h/2)^{12} + \ldots \tag{4.18}$$

We can eliminate the terms in h^4 by subtracting (4.17) from 16 times (4.18), giving

$$I = (16T_{2i} - T_i)/15 + k_2 h^8 + k_3 h^{12} + \ldots \tag{4.19}$$

Notice that the dominant or most significant term in the truncation error is now of order h^8. In general this will provide a significantly improved approximation to I. For the remainder of this discussion it is advantageous to use a double subscript notation. If we generate an initial set of approximations by successively halving the interval we may represent them by $T_{0,k}$ where $k = 0, 1, 2, 3, 4, \ldots$. These results may be combined in a similar manner to that described in (4.19) by using the general formula

$$\begin{aligned} &T_{r,k} = \left(16^r T_{r-1,k+1} - T_{r-1,k}\right) / \left(16^r - 1\right) \\ &\text{for } k = 0,\ 1,\ 2,\ 3\ \ldots \text{ and } r = 1,\ 2,\ 3,\ \ldots \end{aligned} \tag{4.20}$$

Here r represents the current set of approximations we are generating. The calculations may be tabulated as follows:

$$\begin{array}{lllll} T_{0,0} & T_{0,1} & T_{0,2} & T_{0,3} & T_{0,4} \\ T_{1,0} & T_{1,1} & T_{1,2} & T_{1,3} & \\ T_{2,0} & T_{2,1} & T_{2,2} & & \\ T_{3,0} & T_{3,1} & & & \\ T_{4,0} & & & & \end{array}$$

In this case, the interval has been halved four times to generate the first five values in the table denoted by $T_{0,k}$. The formula (4.20) for $T_{r,k}$ is used to calculate the remaining values in the table and at each stage the order of the truncation error is increased by four. A common alternative is to write the preceding table with the rows and columns interchanged.

At each stage the interval size is given by

$$(b-a)/2^k \text{ for } k = 0, 1, 2, \ldots \tag{4.21}$$

Romberg integration is implemented in the following MATLAB function, romb:

```
function [W T] = romb(func,a,b,d)
% Implements Romberg integration.
% Example call: W = romb(func,a,b,d)
% Integrates user defined function func from a to b, using d stages.
T = zeros(d+1,d+1);
for k = 1:d+1
    n = 2^k;  T(1,k) = simp1(func,a,b,n);
end
for p = 1:d
    q = 16^p;
    for k = 0:d-p
        T(p+1,k+1) = (q*T(p,k+2)-T(p,k+1))/(q-1);
    end
end
W = T(d+1,1);
```

We now apply function romb to the evaluation of $x^{0.1}$ in the range 0 to 1. The call of the function romb with five stages is

```
>> [integral table] = romb(@(x) x.^0.1,0,1,5)
```

Calling this function gives the following output. Note that the best estimate is the non-zero value in the last row of the table.

```
integral =
    0.9066

table =
    0.7887    0.8529    0.8829    0.8969    0.9034    0.9064
    0.8572    0.8849    0.8978    0.9038    0.9066         0
    0.8850    0.8978    0.9038    0.9066         0         0
    0.8978    0.9038    0.9066         0         0         0
    0.9038    0.9066         0         0         0         0
    0.9066         0         0         0         0         0
```

This integral is a surprisingly difficult one and obtaining an accurate result presents a significant problem. The exact solution to four decimal places is 0.9090 so the application of the Romberg method gives only two places of accuracy. However, taking $d = 10$ does give the answer correct to four places thus:

```
>> integral = romb(@(x) x.^0.1,0,1,10)

integral =
    0.9090
```

Generally the Romberg method is very efficient and accurate.

An interesting exercise for the reader is to convert function `romb` to work with the MATLAB function `trapz` instead of `simp1`.

4.7 GAUSSIAN INTEGRATION

The common feature of the methods considered so far is that the integrand is evaluated at equal intervals within the range of integration. In contrast, Gaussian integration requires the evaluation of the integrand at specified, but unequal, intervals. For this reason Gaussian integration cannot be applied to data values that are sampled at equal intervals of the independent variable. The general form of the rule is

$$\int_{-1}^{1} f(x)\,dx = \sum_{i=1}^{n} A_i f(x_i) \tag{4.22}$$

The parameters A_i and x_i are chosen so that, for a given n, the rule is exact for polynomials up to and including degree $2n-1$. It should be noticed that the range of integration is required to be from -1 to 1. This does not restrict the integrals to which Gaussian integration can be applied since if $f(x)$ is to be integrated in the range a to b, then it can be replaced by the function $g(t)$ integrated from -1 to 1 where

$$t = (2x - a - b)/(b - a)$$

Note that in the preceding formula, when $x = a$, $t = -1$ and when $x = b$, $t = 1$.

We now determine the four parameters A_i and x_i for $n = 2$ in (4.22). Thus (4.22) now becomes

$$\int_{-1}^{1} f(x)\,dx = A_1 f(x_1) + A_2 f(x_2) \tag{4.23}$$

This integration rule will be exact for polynomials up to and including degree 3 by ensuring that the rule is exact for the polynomials 1, x, x^2, and x^3 in turn. Thus four equations are obtained, as follows:

$$\begin{aligned}
f(x) = 1 &\text{ gives } \textstyle\int_{-1}^{1} 1\,dx = 2 = A_1 + A_2,\\
f(x) = x &\text{ gives } \textstyle\int_{-1}^{1} x\,dx = 0 = A_1x_1 + A_2x_2,\\
f(x) = x^2 &\text{ gives } \textstyle\int_{-1}^{1} x^2\,dx = 2/3 = A_1x_1^2 + A_2x_2^2,\\
f(x) = x^3 &\text{ gives } \textstyle\int_{-1}^{1} x^3\,dx = 0 = A_1x_1^3 + A_2x_2^3
\end{aligned} \tag{4.24}$$

Solving these equations gives

$$x_1 = -1/\sqrt{3},\ \ x_2 = 1/\sqrt{3},\ \ A_1 = 1,\ \ A_2 = 1.$$

Thus,

$$\int_{-1}^{1} f(x)\,dx = f\left(-\frac{1}{\sqrt{3}}\right) + f\left(\frac{1}{\sqrt{3}}\right) \tag{4.25}$$

Notice that this rule, like Simpson's rule, is exact for cubic equations but requires fewer function evaluations.

A general procedure for obtaining the values of A_i and x_i is based on the fact that in the range of integration it can be shown that $x_1, x_2, ..., x_n$ are the roots of the Legendre polynomial of degree n. The values of A_i can then be obtained from an expression involving the Legendre polynomial of degree n, evaluated at x_i. Tables have been produced for the values of x_i and A_i for various values of n; see Abramowitz and Stegun (1965) and Olver et al. (2010). Abramowitz and Stegun provide an excellent reference not only for these functions but for a very extensive range of mathematical functions. However, this classic work is now becoming outdated and a newer handbook of mathematical functions for the twenty-first century by Olver et al. has been published with many improvements, for example, clearer, colored graphics. However, this new text contains far fewer tables of functions, since most can now be rapidly computed on a personal computer.

Function `fgauss` performs Gaussian integration. It includes a substitution so that integration in the range a to b is converted to an integration in the range -1 to 1.

```
function q = fgauss(func,a,b,n)
% Implements Gaussian integration.
% Example call: q = fgauss(func,a,b,n)
% Integrates user defined function func from a to b, using n divisions
% n must be 2 or 4 or 8 or 16.
if (n==2)|(n==4)|(n==8)|(n==16)
    c = zeros(8,4);  t = zeros(8,4);
    c(1,1) = 1;
    c(1:2,2) = [.6521451548; .3478548451];
    c(1:4,3) = [.3626837833; .3137066458; .2223810344; .1012285362];
    c(:,4 )= [.1894506104; .1826034150; .1691565193; .1495959888; ...
             .1246289712; .0951585116; .0622535239; .0271524594];
    t(1,1) = .5773502691;
    t(1:2,2) = [.3399810435; .8611363115];
    t(1:4,3) = [.1834346424; .5255324099; .7966664774; .9602898564];
    t(:,4) = [.0950125098; .2816035507; .4580167776; .6178762444; ...
             .7554044084; .8656312023; .9445750230; .9894009350];
    j = 1;
    while j<=4
        if 2^j==n; break;
        else
            j = j+1;
        end
    end
    s = 0;
    for k = 1:n/2
```

```
            x1 = (t(k,j)*(b-a)+a+b)/2;
            x2 = (-t(k,j)*(b-a)+a+b)/2;
            y = feval(func,x1)+feval(func,x2);
            s = s+c(k,j)*y;
        end
        q = (b-a)*s/2;
    else
        disp('n must be equal to 2, 4, 8 or 16'); return
    end
```

The script e4s405 calls the function fgauss to integrate $x^{0.1}$ from 0 to 1.

```
% e4s405.m
disp('  n  integral value');
for j = 1:4
    n = 2^j;
    int = fgauss(@(x) x.^0.1,0,1,n);
    fprintf('%3.0f %14.9f\n',n,int)
end
```

The output from script e4s405.m is

```
  n  integral value
  2     0.916290737
  4     0.911012914
  8     0.909561226
 16     0.909199952
```

Gaussian integration with $n = 16$ gives a better result than that obtained by Romberg's method with five divisions of the interval.

4.8 INFINITE RANGES OF INTEGRATION

Other formulae of the Gauss type are available to allow us to deal with integrals having a special form and infinite ranges of integration. These are the Gauss–Laguerre and Gauss–Hermite formula and take the following form.

4.8.1 GAUSS–LAGUERRE FORMULA

This method is developed from (4.26) as follows:

$$\int_0^\infty e^{-x} g(x)\, dx = \sum_{i=1}^{n} A_i g(x_i) \tag{4.26}$$

The parameters A_i and x_i are chosen so that, for a given n, the rule is exact for polynomials up to and including degree $2n-1$. Considering the case when $n=2$, we have

$$\begin{aligned}
g(x) &= 1 \text{ gives } \textstyle\int_0^\infty e^{-x}\,dx = 1 = A_1 + A_2 \\
g(x) &= x \text{ gives } \textstyle\int_0^\infty xe^{-x}\,dx = 1 = A_1x_1 + A_2x_2 \\
g(x) &= x^2 \text{ gives } \textstyle\int_0^\infty x^2e^{-x}\,dx = 2 = A_1x_1^2 + A_2x_2^2 \\
g(x) &= x^3 \text{ gives } \textstyle\int_0^\infty x^3e^{-x}\,dx = 6 = A_1x_1^3 + A_2x_2^3
\end{aligned} \tag{4.27}$$

Having evaluated the integrals on the left-hand side of Eqs. (4.27) we may solve for the four unknowns x_1, x_2, A_1, and A_2 so that (4.26) becomes

$$\int_0^\infty e^{-x}g(x)\,dx = \frac{2+\sqrt{2}}{4}g(2-\sqrt{2}) + \frac{2-\sqrt{2}}{4}g(2+\sqrt{2})$$

It can be shown that the x_i are the roots of the nth-order Laguerre polynomial and the coefficients A_i can be calculated from an expression involving the derivative of an nth-order Laguerre polynomial evaluated at x_i.

In general, we wish to evaluate integrals of the form

$$\int_0^\infty f(x)\,dx$$

We may write this integral as

$$\int_0^\infty e^{-x}\{e^x f(x)\}\,dx$$

Thus, using (4.26), we have

$$\int_0^\infty f(x)\,dx = \sum_{i=1}^{n} A_i \exp(x_i) f(x_i) \tag{4.28}$$

Eq. (4.28) allows integrals to be evaluated over an infinite range, assuming that the value of the integral is finite.

The Gauss–Laguerre method is implemented by the MATLAB function `galag` thus:

```
function s = galag(func,n)
% Implements Gauss-Laguerre integration.
% Example call: s = galag(func,n)
% Integrates user defined function func from 0 to inf
% using n divisions. n must be 2 or 4 or 8.
if (n==2)|(n==4)|(n==8)
    c = zeros(8,3);   t = zeros(8,3);
    c(1:2,1) = [1.533326033; 4.450957335];
    c(1:4,2) = [.8327391238; 2.048102438; 3.631146305; 6.487145084];
    c(:,3) = [.4377234105; 1.033869347; 1.669709765; 2.376924702;...
             3.208540913; 4.268575510; 5.818083368; 8.906226215];
```

```
        t(1:2,1) = [.5857864376; 3.414213562];
        t(1:4,2) = [.3225476896; 1.745761101; 4.536620297; 9.395070912];
        t(:,3) = [.1702796323; .9037017768; 2.251086630; 4.266700170;...
                  7.045905402; 10.75851601; 15.74067864; 22.86313174];
        j = 1;
        while j<=3
            if 2^j==n; break
            else
                j = j+1;
            end
        end
        s = 0;
        for k = 1:n
            x = t(k,j); y = feval(func,x);
            s = s+c(k,j)*y;
        end
    else
        disp('n must be 2, 4 or 8'); return
    end
```

Sample values x_i and the product $A_i \exp(x_i)$ are given by `c` and `t` respectively in the MATLAB function. A more complete list of coefficients may be found in Abramowitz and Stegun (1965) and Olver et al. (2010).

We now evaluate the integral $\log_e(1+e^{-x})$ from zero to infinity. The script `e4s406.m` evaluates the integral using the function `galag`.

```
% e4s406.m
disp(' n   integral value');
for j = 1:3
    n = 2^j;
    int = galag(@(x) log(1+exp(-x)),n);
    fprintf('%3.0f%14.9f\n',n,int)
end
```

The output is as follows:

```
  n   integral value
  2    0.822658694
  4    0.822358093
  8    0.822467051
```

Note that the exact result is $\pi^2/12 = 0.82246703342411$. The eight-point integration formula is accurate to six decimal places!

4.8.2 GAUSS–HERMITE FORMULA

This method is developed from (4.29) as follows:

$$\int_{-\infty}^{\infty} \exp(-x^2) g(x)\, dx = \sum_{i=1}^{n} A_i g(x_i) \tag{4.29}$$

Again the parameters A_i and x_i are chosen so that, for a given n, the rule is exact for polynomials up to and including degree $2n-1$. For the case $n=2$ we have

$$\begin{aligned}
g(x) = 1 \text{ gives } & \textstyle\int_{-\infty}^{\infty} \exp(-x^2)\, dx = \sqrt{\pi} = A_1 + A_2 \\
g(x) = x \text{ gives } & \textstyle\int_{-\infty}^{\infty} x \exp(-x^2)\, dx = 0 = A_1 x_1 + A_2 x_2 \\
g(x) = x^2 \text{ gives } & \textstyle\int_{-\infty}^{\infty} x^2 \exp(-x^2)\, dx = \frac{\sqrt{\pi}}{2} = A_1 x_1^2 + A_2 x_2^2 \\
g(x) = x^3 \text{ gives } & \textstyle\int_{-\infty}^{\infty} x^3 \exp(-x^2)\, dx = 0 = A_1 x_1^3 + A_2 x_2^3
\end{aligned} \tag{4.30}$$

We have evaluated the integrals on the left-hand side of Eqs. (4.30) and may now solve for the four unknowns x_1, x_2, A_1, and A_2 so that (4.29) becomes

$$\int_{-\infty}^{\infty} \exp(-x^2) g(x)\, dx = \frac{\sqrt{\pi}}{2} g\left(-\frac{1}{\sqrt{2}}\right) + \frac{\sqrt{\pi}}{2} g\left(\frac{1}{\sqrt{2}}\right)$$

An alternative approach is to note that x_i are the roots of the nth-order Hermite polynomial $H_n(x)$. The coefficients A_i can then be determined from an expression involving the derivative of the nth-order Hermite polynomial evaluated at x_i.

In general, we wish to evaluate integrals of the form

$$\int_{-\infty}^{\infty} f(x)\, dx$$

We may write this integral as

$$\int_{-\infty}^{\infty} \exp(-x^2)\{\exp(x^2) f(x)\}\, dx$$

and using (4.29) we have

$$\int_{-\infty}^{\infty} f(x)\, dx = \sum_{i=1}^{n} A_i \exp(x_i^2) f(x_i) \tag{4.31}$$

Again, care must be taken to apply (4.31) only to functions that have a finite integral in the range $-\infty$ to ∞. Extensive tables of x_i and A_i are given in Abramowitz and Stegun (1965) and Olver et al. (2010). The MATLAB function `gaherm` implements Gauss–Hermite integration thus:

```
function s = gaherm(func,n)
% Implements Gauss-Hermite integration. Chkd 13/01/11
% Example call: s = gaherm(func,n)
```

```
% Integrates user defined function func from -inf to +inf,
% using n divisions. n must be 2 or 4 or 8 or 16
if (n==2)|(n==4)|(n==8)|(n==16)
    c = zeros(8,4);  t = zeros(8,4);
    c(1,1) = 1.461141183;
    c(1:2,2) = [1.059964483; 1.240225818];
    c(1:4,3) = [.7645441286; .7928900483; .8667526065; 1.071930144];
    c(:,4) = [.5473752050; .5524419573; .5632178291; .5812472754; ...
              .6097369583; .6557556729; .7382456223; .9368744929];
    t(1,1) = .7071067811;
    t(1:2,2) = [.5246476233; 1.650680124];
    t(1:4,3) = [.3811869902; 1.157193712; 1.981656757; 2.930637420];
    t(:,4) = [.2734810461; .8229514491; 1.380258539; 1.951787991; ...
              2.546202158; 3.176999162; 3.869447905; 4.688738939];
    j = 1;
    while j<=4
        if 2^j==n; break;
        else
            j = j+1;
        end
    end
    s=0;
    for k = 1:n/2
        x1 = t(k,j); x2 = -x1;
        y = feval(func,x1)+feval(func,x2);
        s = s+c(k,j)*y;
    end
else
    disp('n must be equal to 2, 4, 8 or 16'); return
end
```

We now evaluate the integral

$$\int_{-\infty}^{\infty} \frac{dx}{(1+x^2)^2}$$

by the Gauss–Hermite method. The script `e4s407` uses `gaherm` to evaluate this integral.

```
% e4s407.m
disp(' n   integral value');
for j = 1:4
    n = 2^j;
    int = gaherm(@(x) 1./(1+x.^2).^2,n);
    fprintf('%3.0f%14.9f\n',n,int)
end
```

The results from running script `e4s407` are

```
 n   integral value
 2    1.298792163
 4    1.482336098
 8    1.550273058
16    1.565939612
```

The exact value of this integral is $\pi/2 = 1.570796...$

4.9 GAUSS–CHEBYSHEV FORMULA

We now consider two interesting cases where the sample points x_i and weights w_i are known in a closed or analytical form. The two integrals together with their closed forms are

$$\int_{-1}^{1} \frac{f(x)}{\sqrt{1-x^2}}\,dx = \frac{\pi}{n}\sum_{k=1}^{n} f(x_k) \text{ where } x_k = \cos\left(\frac{(2k-1)\pi}{2n}\right) \tag{4.32}$$

$$\int_{-1}^{1} \sqrt{1-x^2} f(x)\,dx = \frac{\pi}{n+1}\sum_{k=1}^{n} \sin^2\left(\frac{k\pi}{n+1}\right) f(x_k) \tag{4.33}$$

$$\text{where } x_k = \cos\left(\frac{k\pi}{n+1}\right)$$

These expressions are members of the Gauss family, in this case Gauss–Chebyshev formula. Clearly it is extremely easy to use these formulae for integrands of the required form which have a specified $f(x)$. It is simply a matter of evaluating the function at the specified points, multiplying by the appropriate factor and summing these products. A MATLAB script or function can easily be developed and is left as an exercise for the reader. (See Problem 4.11.)

4.10 GAUSS–LOBATTO INTEGRATION

Lobatto integration or quadrature (Abramowitz and Stegun, 1965), is named after Dutch mathematician Rehuel Lobatto. It is similar to Gaussian quadrature, which we discussed in Section 4.7, but the integration points include the end points of the integration interval. This has an advantage when the procedure is used in a subinterval because data can be shared between consecutive subintervals. However, Lobatto quadrature is less accurate than the Gaussian formula.

Lobatto quadrature of function $f(x)$ on interval $[-1\ 1]$ is given by the formula

$$\int_{-1}^{1} f(x)\,dx = \frac{2}{n(n-1)}\left[f(1)+f(-1)\right] + \sum_{i=2}^{n-1} w_i f(x_i) + R_n$$

Here the points x_i are the roots of the Legendre polynomial $P_{n-1}(x) = 0$. The weights for $f(1)$ and $f(-1)$ are both equal $2/(n(n-1))$. The remaining weights, w_i, are calculated from the formula:

$$w_i = \frac{2}{n(n-1)[P_{n-1}(x_i)]^2} \quad (x_i \neq \pm 1)$$

Clearly from this description it is an easy matter to calculate the weights required if the roots of the derivative of the Legendre polynomial are found.

The coefficients of any order Legendre polynomial can be found using Bonnet's recursion formula thus

$$(n+1)P_{n+1}(x) = (2n+1)xP_n(x) - nP_{n-1}(x)$$

where $P_0(x) = 1$, $P_1(x) = x$, and $P_n(x)$ is the nth Legendre polynomial. Alternatively a recurrence relation for the polynomials can be found using the differential equation definition of the Legendre function.

The following MATLAB function is based on generating the polynomial coefficients using a recurrence formula and then finding the roots of the derivative of this polynomial using the MATLAB function `roots`. The range has been converted to any range a to b.

```
function Iv = lobattof(func,a,b,n)
% Implementation of Lobattos method
% func is the function to be integrated from the a to b
% using n points.
% Generate Legendre polynomials based on recurrence relation
% derived from the differetial equation which the legendre polynomial
% satisfies.
% Obtain derivitive of that polynomial
% The roots of this polynomial give the Lobatto nodes
% From the nodes calculate the weights using standard algorithm
lc = [ ];
for k = 0:n-1
    if n>=2*k
        fnk = factorial(2*n-2*k);
        fnp = 2^n*factorial(k)*factorial(n-k)*factorial(n-2*k);
        lc(n-2*k+1) = (-1)^k*fnk/fnp;
    end
end
% Find coefficients of derivitive of the polynomial
lcd = [ ];
for k = 0:n-1
    if n>=2*k
        lcd(n-2*k+1) = (n-2*k)*lc(n-2*k+1);
    end
end
lcd(n) = 0;
```

```
% Obtain Lobatto points
x = roots(fliplr(lcd(2:n+1)));
x1 = sort(x,'descend');
pv = zeros(size(x));
% Calculate Lobatto weights
for k = 1:n+1
    pv = pv+lc(k)*x.^(k-1);
end
n = n+1;
w = 2./(n*(n-1)*pv.^2);
w = [2/(n*(n-1)); w; 2/(n*(n-1))];
% Transform to range a to b
x1 = (x*(b-a)+(a+b))/2;
pts = [a; x1; b];
% Implement rule for integration
Iv = (b-a)*w'*feval(func,pts)/2;
```

To test the function the MATLAB script `e4s408.m` is used

```
% e4s408.m
g = @(x) exp(5*x).*cos(2*x); a = 0; b = pi/2;
for n = [2 4 8 16 32 64]
    Iv = lobattof(g,a,b,n);
    fprintf('%3.0f%19.9f\n',n,real(Iv))
end
exact = -5*(exp(2.5*pi)+1)/29;
fprintf('\n Exact %15.9f\n',exact)
```

The script `e4s408.m` gives the following results:

```
  2     -674.125699610
  4     -443.869707406
  8     -444.305258004
 16     -444.305258038
 32     -444.305439477
 64      -15.828963820

 Exact  -444.305258034
```

Note that as the number of points used is increased up to 16, the integration becomes more accurate. However, above this value the accuracy decreases. This is because the function `lobattof` determines the abscissae weights by finding the roots of a polynomial. This becomes less accurate as n increases.

An alternative approach to determine the value of an integral is to subdivide the range of integration into a specific number of subintervals, m and then apply a Lobatto rule with a small number of points to each subinterval. The following function allows the user to choose the number of points in the Lobatto integration and the number of subintervals in which the integration is applied.

```
function s = lobattomp(func,a,b,n,m)
% n is the number of points in the Labatto quadrature
% m is the number of subintervals of the range of the integration.
h = (b-a)/m; s = 0;
for panel = 0:m-1
    a0 =a+panel*h; b0 = a+(panel+1)*h;
    s = s+lobattof(func,a0,b0,n);
end
```

The following script evaluates the error in the integration of $e^{5x}\cos(2x)$ over the range 0 to $\pi/2$. The script considers a 4, 5, ..., 8 point Lobatto integration rule applied to subintervals, the number of subintervals ranging from 2, 4, 8 to 256.

```
% e4s409.m
g = @(x) exp(5*x).*cos(2*x); a = 0; b = pi/2;
format short e
m = 2; k = 0;
while m<512
    % m is number of panels, k is the index
    k = k+1;
    p = 0;
    for n = 4:8
        % n number of Labotto points, p is index
        p = p+1;
        Integral_err(k,p) = real(lobattomp(g,a,b,n,m))+5*(exp(2.5*pi)+1)/29;
    end
    m = 2*m;
end
Integral_err
```

Running script `e4s409` gives the following output. Each row gives the value of the error for the specified number of panels, beginning at the first row with `m = 2`, and doubling each time until `m = 256`. Each column, from left to right, gives the value of the error for the specified number of points in the Lobatto integration, for `n = 4` in steps of one to `n = 8`.

```
Integral_err =
   1.5122e-02   2.6320e-04   1.6910e-06   1.8365e-09  -3.5470e-11
   1.0050e-04   3.5484e-07   4.4213e-10  -3.9790e-13  -9.0949e-13
   4.4719e-07   3.7176e-10   2.8422e-13  -2.2737e-13  -7.3896e-13
   1.8038e-09   1.1369e-13   1.1369e-13  -6.2528e-13  -9.6634e-13
   7.0486e-12  -1.7053e-13   2.2737e-13  -6.8212e-13  -8.5265e-13
            0  -2.2737e-13   2.8422e-13  -6.2528e-13  -9.0949e-13
            0  -2.2737e-13   2.8422e-13  -6.2528e-13  -9.0949e-13
  -5.6843e-14  -2.8422e-13   2.8422e-13  -5.6843e-13  -8.5265e-13
```

It is evident that increasing the number of subintervals (m) and increasing the number of points in the Lobatto integration (n) reduces the error in the integration. However, when the number of points in the Lobatto integration and the number of subintervals increase beyond a certain values the accuracy of the integration begins to decrease. The values of m and n at which this happens is problem dependent.

A further disadvantage of the Gauss formula is that the location and weight of the abscissae change as their number increases. For example, suppose we have evaluated an integral using an n point Gauss quadrature rule. To increase the accuracy we could now increase the number of points and use the Gauss rule again, but all the points would be at a new location. An alternative strategy is to keep the existing n points and add to them $n+1$ points located at the best positions. This is the Kronrod method (Kronrod, 1965). Thus, a three point Gauss method can be extended by keeping the three points and adding four more to give a seven point rule. The MATLAB function `quadgk` implements adaptive Gauss–Kronrod quadrature.

A discussion of the family of Gaussian quadrature methods is given by Thompson (2010).

4.11 FILON'S SINE AND COSINE FORMULAE

These formulae can be applied to integrals of the form

$$\int_a^b f(x)\cos kx\,dx \quad \text{and} \quad \int_a^b f(x)\sin kx\,dx \tag{4.34}$$

The formulae are generally more efficient than standard methods for this form of integral. To derive the Filon formulae we first consider an integral of the form

$$\int_0^{2\pi} f(x)\cos kx\,dx$$

By the method of undetermined coefficients we can obtain an approximation to this integrand as follows. Let

$$\int_0^{2\pi} f(x)\,\cos x\,dx = A_1 f(0) + A_2 f(\pi) + A_3 f(2\pi) \tag{4.35}$$

Requiring that this should be exact for $f(x) = 1, x$, and x^2, we have

$$\begin{aligned} 0 &= A_1 + A_2 + A_3 \\ 0 &= A_2\pi + A_3 2\pi \\ 4\pi &= A_2\pi^2 + A_3 4\pi^2 \end{aligned}$$

Thus $A_1 = 2/\pi$, $A_2 = -4/\pi$, and $A_3 = 2/\pi$. Hence,

$$\int_0^{2\pi} f(x)\,\cos x\,dx = \frac{1}{\pi}[2f(0) - 4f(\pi) + 2f(2\pi)] \tag{4.36}$$

More general results can be developed as follows:

$$\int_0^{2\pi} f(x)\cos kx\,dx = h[A\{f(x_n)\sin kx_n - f(x_0)\sin kx_0\} + BC_e + DC_o]$$

$$\int_0^{2\pi} f(x)\sin kx\,dx = h[A\{f(x_0)\cos kx_0 - f(x_n)\cos kx_n\} + BS_e + DS_o]$$

where $h = (b-a)/n$, $q = kh$, and

$$A = \left(q^2 + q\sin 2q/2 - 2\sin^2 q\right)/q^3 \tag{4.37}$$

$$B = 2\left\{q\left(1+\cos^2 q\right) - \sin 2q\right\}/q^3 \tag{4.38}$$

$$D = 4\left(\sin q - q\cos q\right)/q^3 \tag{4.39}$$

$$C_o = \sum_{i=1,3,5\ldots}^{n-1} f(x_i)\cos kx_i$$

$$C_e = \frac{1}{2}\{f(x_0)\cos kx_0 + f(x_n)\cos kx_n\} + \sum_{i=2,4,6\ldots}^{n-2} f(x_i)\cos kx_i$$

It can be seen that C_o and C_e are odd and even sums of cosine terms. S_o and S_e are similarly defined with respect to sine terms.

It is important to note that Filon's method, when applied to functions of the form given in (4.34), usually gives better results than Simpson's method for the same number of intervals.

Approximations may be used for the expressions for A, B, and D given in (4.37), (4.38), and (4.39) by expanding them in series of ascending powers of q. This leads to the following results:

$$A = 2q^2\left(q/45 - q^3/315 + q^5/4725 - \ldots\right)$$

$$B = 2\left(1/3 + q^2/15 - 2q^4/105 + q^6/567 - \ldots\right)$$

$$D = 4/3 - 2q^2/15 + q^4/210 - q^6/11340 + \ldots$$

When the number of intervals becomes very large, h and hence q become small. As q tends to zero, A tends to zero, B tends to 2/3 and D tends to 4/3. Substituting these values into the formula for Filon's method, it can be shown that it becomes equivalent to Simpson's rule. However, in these circumstances the accuracy of Filon's rule may be worse than Simpson's rule owing to the additional complexity of the calculations.

The MATLAB function `filon` implements Filon's method for the evaluation of appropriate integrals. In the parameter list, function `func` defines $f(x)$ of (4.34) and this is multiplied by $\cos kx$ when `cas = 1` or $\sin kx$ when `cas ~= 1`. The parameters `l` and `u` specify the lower and upper limit of the integral and `n` specifies the number of divisions required. The script incorporates a modification to the standard Filon method such that the series approximation is used if q is less than 0.1 rather than (4.37) to (4.39). The justification for this is that as q becomes small, the accuracy of series approximation is sufficient and easier to compute.

```
function int = filon(func,cas,k,l,u,n)
% Implements filon's integration.
% Example call: int = filon(func,cas,k,l,u,n)
% If cas = 1, integrates cos(kx)*f(x) from l to u using n divisions.
% If cas ~= 1, integrates sin(kx)*f(x) from l to u using n divisions.
% User defined function func defines f(x).
if (n/2)~=floor(n/2)
    disp('n must be even'); return
else
    h = (u-l)/n; q = k*h;
    q2 = q*q; q3 = q*q2;
    if q<0.1
        a = 2*q2*(q/45-q3/315+q2*q3/4725);
        b = 2*(1/3+q2/15+2*q2*q2/105+q3*q3/567);
        d = 4/3-2*q2/15+q2*q2/210-q3*q3/11340;
    else
        a = (q2+q*sin(2*q)/2-2*(sin(q))^2)/q3;
        b = 2*(q*(1+(cos(q))^2)-sin(2*q))/q3;
        d = 4*(sin(q)-q*cos(q))/q3;
    end
    x = l:h:u;
    y = feval(func,x);
    yodd = y(2:2:n); yeven = y(3:2:n-1);
    if cas == 1
        c = cos(k*x);
        codd = c(2:2:n); co = codd*yodd';
        ceven = c(3:2:n-1);
        ce = (y(1)*c(1)+y(n+1)*c(n+1))/2;
        ce = ce+ceven*yeven';
        int = h*(a*(y(n+1)*sin(k*u)-y(1)*sin(k*l))+b*ce+d*co);
    else
        s = sin(k*x);
        sodd = s(2:2:n); so = sodd*yodd';
        seven = s(3:2:n-1);
        se = (y(1)*s(1)+y(n+1)*s(n+1))/2;
        se = se+seven*yeven';
        int = h*(-a*(y(n+1)*cos(k*u)-y(1)*cos(k*l))+b*se+d*so);
    end
end
```

We now test the function `filon` by integrating $\sin x/x$ in the range 1×10^{-10} to 1. The lower limit is set at 1×10^{-10} to avoid the singularity at zero.

The script `e4s410` uses `filon` and `filonmod` to evaluate the integral. The function `filonmod` removes the ability to switch to the series formula in `filon`. Note that from (4.34) we use the function $f(x) = 1/x$ for this particular problem.

```
% e4s410.m
n = 4;
g = @(x) 1./x;
disp('  n  Filon no switch  Filon with switch');
while n<=4096
    int1 = filonmod(g,2,1,1e-10,1,n);
    int2 = filon(g,2,1,1e-10,1,n);
    fprintf('%4.0f %17.8e %17.8e\n',n,int1,int2)
    n = 2*n;
end
```

Running this script gives

```
   n  Filon no switch  Filon with switch
   4    1.72067549e+006    1.72067549e+006
   8    1.08265940e+005    1.08265940e+005
  16    6.77884667e+003    6.77884667e+003
  32    4.24742208e+002    4.24742207e+002
  64    2.74361110e+001    2.74361124e+001
 128    2.60175423e+000    2.60175321e+000
 256    1.04956252e+000    1.04956313e+000
 512    9.52549009e-001    9.52550585e-001
1024    9.46489412e-001    9.46487290e-001
2048    9.46109716e-001    9.46108334e-001
4096    9.46085291e-001    9.46084649e-001
```

The exact value of the integral is 0.9460831.

In this particular problem, the switch occurs when $n = 16$. The output from script e4s410 shows that the values of the integral obtained with the switch are marginally more accurate. However, it should be noted that experiments carried out by us have shown that for a lower accuracy of computation than that supplied in the MATLAB environment, the accuracy of Filon's method, including the switch, is significantly better. The reader may find it interesting to experiment with the value of q at which the switch occurs. This is currently set at 0.1.

Finally we choose a function which is appropriate for Filon's method and compare the results with Simpson's rule. The function is $\exp(-x/2)\cos(100x)$ integrated between 0 and 2π.

The MATLAB script e4s411 that implements this comparison is

```
% e4s411.m
n = 4;
disp('  n   Simpsons value   Filons value');
g1 = @(x) exp(-x/2);
g2 = @(x) exp(-x/2).*cos(100*x);
while n<=2048
    int1 = filon(g1,1,100,0,2*pi,n);
    int2 = simp1(g2,0,2*pi,n);
    fprintf('%4.0f %17.8e %17.8e\n',n,int2,int1)
```

```
    n = 2*n;
end
```

The results of this comparison are

```
   n    Simpsons value    Filons value
   4    1.91733833e+000    4.55229440e-005
   8   -5.73192992e-001    4.72338540e-005
  16    2.42801799e-002    4.72338540e-005
  32    2.92263624e-002    4.76641931e-005
  64   -8.74419731e-003    4.77734109e-005
 128    5.55127202e-004    4.78308678e-005
 256   -1.30263888e-004    4.78404787e-005
 512    4.53408415e-005    4.78381786e-005
1024    4.77161559e-005    4.78381120e-005
2048    4.78309107e-005    4.78381084e-005
```

The exact value of the integral to 10 significant digits is $4.783810813 \times 10^{-5}$. In this particular problem the switch to the series approximations does not take place because of the high value of the coefficient k. The output shows that using 2048 intervals, Filon's method is accurate to eight significant digits. In contrast, Simpson's rule is accurate to only five significant digits and its behavior is highly erratic. However, timing the evaluation of this integral shows that Simpson's method is about 25% faster than Filon's method.

4.12 ADAPTIVE INTEGRATION

Some functions represent significant difficulties for integration because of their rapid change over a narrow range of their independent variable or variables. In some cases the magnitude of these changes may be such that the integrand approaches a discontinuity. To deal with this type of change the interval of integration may be changed adaptively according to some measure of the behavior of the function in the specific region. The basic approach to this is to divide the integral into two parts, with equal ranges by setting

$$\int_a^b f(x)\,dx = \int_a^p f(x)\,dx + \int_p^b f(x)\,dx \tag{4.40}$$

where $p = (b - a)/2$. These two integrals are evaluated separately and then the interval is subdivided and evaluated recursively, repeating the subdivision as long as there are significant differences between the values of successive integrals. Usually the interval is halved at each stage. Although this is the basic method much depends on the integration rule used for the individual integral evaluation; for example the Simpson's rule or the Lobatto rule.

There are two problems here; deciding the measure of the rate of change of the function and introducing the extra computation required efficiently. These are important decisions since the extra computation may slow down the computation process significantly. Adaptive integration using recursion is a true, tried and tested method and has been implemented in MATLAB, for example, in the implementation of the function `quad` which is based on an adaptive Simpson's rule and lately the `quadl` and `integral` functions which are based on the more accurate Lobatto and Kronrod rules.

When using an adaptive integration routine, the user may well meet with the warning "recursion level reached", this was particularly true for the earlier implementation of adaptive integration such as MATLAB `quad`. This message is useful in that it gives a clear warning of the nature of the problem. The recursion level, that was preset in MATLAB, was not adequate for some problems, since it could not reach the level of accuracy required by the user.

In an outstanding paper by Gander and Gautschi (2000) they considered this problem and examined in detail how it could be resolved. They provided an algorithm which dealt with the problems effectively. This algorithm is an adaptive Lobatto algorithm and the coefficients supplied by this method give a high level of accuracy for each recursion. Here we give a description of the key concepts which are used in its derivation.

Gander and Gautschi began the development of their algorithm with a discussion of the stopping criteria which should be used for adaptive integration. They pointed out that the stopping criteria used at that time in the MATLAB function `quad` failed for some problems. The algorithm compared successive estimates of the integral and terminated if the difference of these estimates were less than a set tolerance, these estimates being obtained by halving the interval used. MATLAB also included a check on the recursion level and terminated with a warning when this had been reached. In an example which Gander and Gautschi cite the integration of $\sqrt{x}$ for x in the range 0 to 1 using MATLAB at that time. They found that the evaluation only provide 6 of the 12 digits required. The software also output a recursion level failure warning.

Consequently they introduced an additional requirement for the stopping criteria which is that the difference between the current approximations to the integral is negligible compared to the current best value of the integral. This presents problems since it requires a good estimate of the integral. Thus an estimate of the integral, say I_e, must be determined. Gander and Gautschi initial used a Monte Carlo estimate for the integral. Letting the two current estimates of the integral obtained by halving the interval be I_1 and I_2, Gander and Gautschi propose the stopping criteria should be formulated as: stop when $I_e + (I_1 - I_2)$ is equal to I_e, to the required, or machine accuracy. They indicate that this guarantees the termination of the algorithm. This allows the integral to be computed to machine accuracy. If a calculation to a tolerance tol is required they use $I_e \times tol/eps$ rather than I_e. A final amendment to the criteria is added which ensures that the values may include the end points. Gander and Gautschi illustrate these points by providing a recursive MATLAB program using the standard Simpson's rule.

However, the main element of their paper is to refine the method for the individual recursions of the integral evaluations and perform this using the Lobatto method and the Kronrod extension. We do not describe the Lobatto method here since this is described in Section 4.10. Essentially all of these techniques supply constant coefficients for the basic integration process which is used in the recursive programs, different sets of coefficients providing different levels of accuracy. For example the method used for the implementation of the adaptive Simpson's method is the basic Simpson's rule. They introduce the Lobatto Quadrature method and generate a particular set of coefficients for this program.

Gander and Gautschi noted that to test the accuracy of their adaptive Lobatto quadrature method another accurate method is required. They introduce the Kronrod extension. This provides another set of coefficients which are an alternative and more accurate quadrature rule. To obtain an estimate of the improved accuracy of the Kronrod rule a further development of the Kronrod process is used. This

Table 4.2 Evaluation of test integrals using different algorithms

function	quadl	adaptsim	integral
Integral (4.41)	$9.99000999 \times 10^{-1}$	$9.99000999 \times 10^{-1}$	$9.99000999 \times 10^{-1}$
Integral (4.42)	$1.34924856 \times 10^{-2}$	$1.34924856 \times 10^{-2}$	$1.34924856 \times 10^{-2}$
Integral (4.43)	1.47763348×10	0	1.47763348×10

extension provides a more accurate method and a different set of coefficients. The MATLAB script they developed on these principles uses the Simpson's and Kronrod rules to provide the values of I_1 and I_2 and the more accurate Kronrod extension to an overall estimate of the integral. These rules are then applied recursively to provide the quadrature program.

Currently, the MATLAB user is advised to employ the adaptive integration function `integral` which is based on the work of Gander and Gautschi. This function is called in a way similar to that of the `quad` and `quadl` as follows:

```
integral(fun,xmin,xmax)
```

where `fun`, `xmin`, `xmax` are the function to be integrated and the limits of integration respectively. Other parameters may be included in this list if further options are required in the process of integration.

To illustrate this discussion of these integration functions, we compare the performance of the MATLAB functions `quadl`, `integral` and the function of Gander and Gautschi, given in their 2000 paper, which they name `adaptsim` for a set of test problems, as follows

$$\int_0^1 x^{0.001}\, dx = 1000/1001 = 0.999000999\ldots \tag{4.41}$$

$$\int_0^1 \frac{dx}{1+(230x-30)^2} = (\tan^{-1} 200 + \tan^{-1} 30)/230 = 0.0134924856495 \tag{4.42}$$

$$\int_0^4 x^2(x-1)^2(x-2)^2(x-3)^2(x-4)^2\, dx = 10240/693 = 14.776334776 \tag{4.43}$$

Fig. 4.3 shows plots of the integrands in the range of integration. It can be seen that each function, at some point, changes rapidly with small changes of the independent variable, making such functions extremely difficult to integrate numerically if a high degree of accuracy is required.

The results of Table 4.2 show that the integrals of the functions given by (4.41) and (4.42) are accurate to the number of decimal places given for each method. Only in the case of the function given by (4.43) do we see that the result is poor for the `adaptsim` program but good for both functions `quadl` and `integral`. In fairness the `adaptsim` routine uses only Simpson's rule adaptively, whereas the other two methods use the highly accurate Lobatto rule as described in Gander and Gautschi (2000). The results for the MATLAB function `quadl` and its replacement the MATLAB function `integral` are the same as we would expect. The MATLAB function `integral` uses adaptive Lobatto quadrature.

As a further example of the use of `integral` the script `e4s412` evaluates the integral e^x from 0 to n where $n = 2.5 : 2.5 : 25$ and outputs the errors. We have chosen to compare `integral` with the basic Simpson rule `simp1` (called with both 1024 and 4096 panels). The results show the superiority of the MATLAB function `integral`.

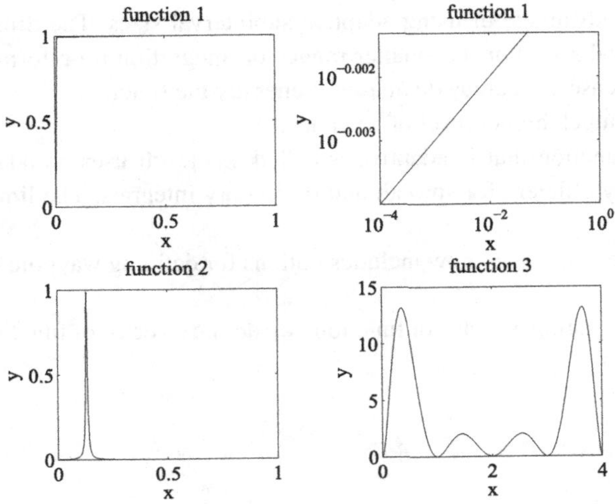

FIGURE 4.3

Plots of functions defined by (4.41), (4.42), and (4.43).

```
% e4s412.m
err = zeros(10,3);
for n = 1:10; n1 = 2.5*n;
    ext = exp(n1)-1;
    err(n,1) = simp1('exp',0,n1,1024)-ext;
    err(n,2) = simp1('exp',0,n1,4096)-ext;
    err(n,3) = integral(@(x) exp(x),0,n1)-ext;
end
err
```

Running script `e4s412.m` gives the following:

```
err =
   2.2080e-12   8.8818e-15            0
   4.6555e-10   1.8190e-12            0
   2.8889e-08   1.1300e-10  -2.2737e-13
   1.1129e-06   4.3474e-09            0
   3.3101e-05   1.2928e-07  -5.8208e-11
   8.3618e-04   3.2666e-06  -4.6566e-10
   1.8872e-02   7.3716e-05            0
   3.9221e-01   1.5322e-03  -1.1921e-07
   7.6535e+00   2.9898e-02  -3.8147e-06
   1.4211e+02   5.5515e-01  -4.5776e-05
```

These results show the advantage of using adaptive subinterval sizes. The Simpson rule (columns 1 and 2) has a fixed interval size. For the smaller ranges of integration it performs very well but as the range of integration increases, accuracy decreases. Generally the function `integral` (column 3), which is adaptive, maintains a much higher level of accuracy.

Another MATLAB function that is adaptive is called `quadgk` It uses an adaptive Gauss–Kronrod rule which is particularly efficient for smooth and oscillatory integrals. The limits of integration may be infinite.

The MATLAB function `integral` now includes options for defining waypoints and vector functions which we now illustrate.

As an example of integrating a vector of functions we define a vector of functions which are: $\cos(x)$, $\cos(x^2)$, $\cos(x^3)$, $\cos(x^4)$

```
>> fvector = @(x) cos(x.^[1  2  3   4])

fvector =
  function_handle with value:
   @(x)cos(x.^[1,2,3,4])

>> integral(fvector,0,1,'arrayvalued', true)

ans =
     0.8415     0.9045     0.9317     0.9468
```

Notice the use of the array valued parameter in the function set at true. A single function may be evaluated in the usual way by replacing the vector by a scalar in the function definition.

Using the parameter `waypoints` ensures that specified waypoints are used in the integration process. For example if we have an integral in the overall range 0 to 4, the pair: `'waypoints', [1, 3]` ensures integration between 0 and 1 and then from 1 to 3 and then 3 to 4. Thus:

```
>> integral(@(x) cos(x),0,4, 'waypoints', [1  3])

ans =
   -0.7568

>> integral(@(x) cos(x),0,4)

ans =
   -0.7568
```

Waypoints are sometimes helpful in dealing singularities.

4.13 PROBLEMS IN THE EVALUATION OF INTEGRALS

The methods outlined in the previous sections are based on the assumption that the function to be integrated is well behaved. If this is not so, then the numerical methods may give poor, or totally useless, results. Problems may occur if:

1. The function is continuous in the range of integration but its derivatives are discontinuous or singular.
2. The function is discontinuous in the range of integration.
3. The function has singularities in the range of integration.
4. The range of integration is infinite.

It is vital that these conditions are identified because in most cases these problems cannot be dealt with directly by numerical techniques. Consequently, some preparation of the integrand is required before the integral can be evaluated by the appropriate numerical method. Case 1 is the least serious condition but since the derivatives of polynomials are continuous, polynomials cannot accurately represent functions with discontinuous derivatives. Ideally, the discontinuity or singularity in the derivative should be located and the integral split into a sum of two or more integrals. The procedure is the same in Case 2; the position of the discontinuities must be found and the integral split into a sum of two or more integrals, the ranges of which avoid the discontinuities. The MATLAB `integral` option `waypoints` can be used for multiple discontinuities. Case 3 can be dealt with in various ways: using a change of variable, integration by parts and splitting the integral. In Case 4 we must use a method suitable for an infinite range of integration (see Section 4.8) or make a substitution.

The following integral, taken from Fox and Mayers (1968), is an example of Case 4:

$$I = \int_1^{\infty} \frac{dx}{x^2 + \cos(x^{-1})} \tag{4.44}$$

This integral can be estimated either by using function `galag` (using the substitution $y = x - 1$ to give a lower limit of zero) or by substituting $z = 1/x$. Thus, $dz = -dx/x^2$ and (4.44) may be transformed as follows:

$$I = -\int_1^{0} \frac{dz}{1 + z^2\cos(z)} \quad \text{or } I = \int_0^{1} \frac{dz}{1 + z^2\cos(z)} \tag{4.45}$$

The integral (4.45) can easily be evaluated by any standard method.

We have discussed a number of techniques for numerical integration. It must be said that even the best methods have difficulty with functions which change very rapidly for small changes in the independent variable. An example of this type of function is $\sin(1/x)$. A MATLAB plot of this function is shown in Section 3.8. However, this plot does not give a true representation of the function in the range -0.1 to 0.1 because in this range the function is changing very rapidly and the number of plotting points and the screen resolution are inadequate. Indeed, as x tends to zero the frequency of the oscillations of the function tends to infinity. A further difficulty is that the function has a singularity at $x = 0$. If we decrease the range of x, then a small section of the function can be plotted and displayed. For example, in the range $x = 2 \times 10^{-4}$ to 2.05×10^{-4} there are approximately 19 cycles of the function $\sin(1/x)$, as shown in Fig. 4.4, and in this limited range the function can be effectively sampled and plotted. Summarizing, the value of this function can change from an extreme positive to an extreme

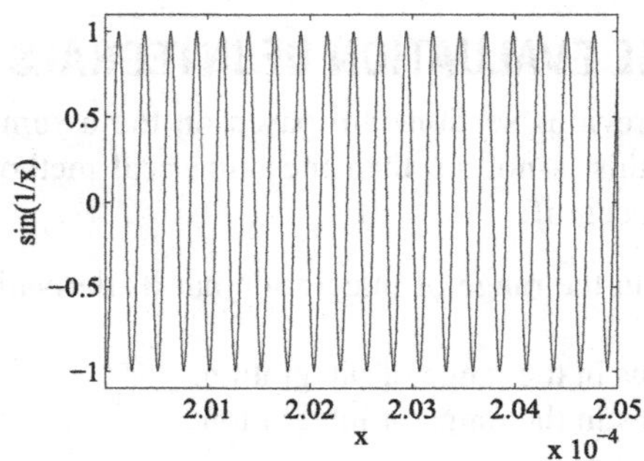

FIGURE 4.4

Function $\sin(1/x)$ in the range $x = 2 \times 10^{-4}$ to 2.05×10^{-4}. Nineteen cycles of the function are displayed.

negative value for a relatively small change in x. The consequence of this is that when estimating the integral of the function, a great number of divisions of the range of integration are needed to provide the required level of accuracy, particularly for smaller values of x.

4.14 TEST INTEGRALS

We now compare the Gauss and Simpson methods of integration with the MATLAB function `integral` using the difficult integrals given in (4.41), (4.42), and (4.43) and repeated here for convenience.

$$\int_0^1 x^{0.001}\,dx = 1000/1001 = 0.999000999\ldots$$

$$\int_0^1 \frac{dx}{1+(230x-30)^2} = (\tan^{-1} 200 + \tan^{-1} 30)/230 = 0.0134924856495$$

$$\int_0^4 x^2(x-1)^2(x-2)^2(x-3)^2(x-4)^2\,dx = 10240/693 = 14.776334776$$

The integrals (4.41), (4.42), and (4.43) are difficult to evaluate, see Fig. 4.3. To generate the comparative results we define function `ftable` thus:

```
function y = ftable(fname,lowerb,upperb)
% Generates table of results.
intg = fgauss(fname,lowerb,upperb,16);
ints = simp1(fname,lowerb,upperb,2048);
intq = integral(fname,lowerb,upperb,'AbsTol',1e-6);
fprintf('%19.8e %18.8e %18.8e \n',intg,ints,intq)
```

The script `e4s413` applies this function to the three test integrals:

```
% e4s413.m
clear
disp('function       Gauss              Simpson               integral')
fprintf('Func 1'), ftable(@(x) x.^0.001,0,1)
fprintf('Func 2'), ftable(@(x) 1./(1+(230*x-30).^2),0,1)
g = @(x) (x.^2).*((1-x).^2).*((2-x).^2).*((3-x).^2).*((4-x).^2);
fprintf('Func 3'), ftable(g,0,4)
```

The output from script e4s413 is

```
function      Gauss              Simpson            integral
Func 1      9.99003302e-01     9.98839883e-01     9.99000999e-01
Func 2      1.46785776e-02     1.34924856e-02     1.34924856e-02
Func 3      1.47763348e+01     1.47763348e+01     1.47763348e+01
```

4.15 REPEATED INTEGRALS

In this section, we confine ourselves to a discussion of repeated integrals using two variables. It is important to note that there is a significant difference between double integrals and repeated integrals. However, it can be shown that if the integrand satisfies certain requirements then double integrals and repeated integrals are equal in value. A detailed discussion of this result is given in Jeffrey (1979).

We have considered in this chapter various techniques for evaluating single integrals. The extension of these methods to repeated integrals can present considerable scripting difficulties. Furthermore, the number of computations required for the accurate evaluation of a repeated integral can be enormous. While many algorithms for the evaluation of single integrals can be extended to repeated integrals, here only extensions to the Simpson and Gauss methods with two variables are presented. These have been chosen as the best compromise between programming simplicity and efficiency.

An example of a repeated integral is

$$\int_{a_1}^{b_1} dx \int_{a_2}^{b_2} f(x, y)\, dy \tag{4.46}$$

In this notation the function is integrated with respect to x from a_1 to b_1 and with respect to y from a_2 to b_2. Here the limits of integration are constant but in some applications they may be variables.

4.15.1 SIMPSON'S RULE FOR REPEATED INTEGRALS

We now apply Simpson's rule to the repeated integral (4.46) by applying it first in the y direction and then in the x direction. Consider three equispaced values of y which are y_0, y_1, and y_2. On applying Simpson's rule, (4.11), to integration with respect to y in (4.46) we have

$$\int_{x_0}^{x_2} dx \int_{y_0}^{y_2} f(x, y)\, dy \approx \int_{x_0}^{x_2} k\{f(x, y_0) + 4f(x, y_1) + f(x, y_2)\}\, dx \tag{4.47}$$

where $k = y_2 - y_l = y_1 - y_0$.

Consider now three equispaced values of x: x_0, x_1, and x_2. Applying Simpson's rule again to integration with respect to x, from (4.47) we have

$$I \approx hk\left[f_{0,0} + f_{0,2} + f_{2,0} + f_{2,2} + 4\left\{f_{0,1} + f_{1,0} + f_{1,2} + f_{2,1}\right\} + 16f_{1,1}\right]/9 \tag{4.48}$$

where $h = x_2 - x_1 = x_1 - x_0$ and, for example, $f_{1,2} = f(x_1, y_2)$.

This is Simpson's rule in two variables. By applying this rule to each group of nine points on the surface $f(x, y)$ and summing, the composite Simpson's rule is obtained. The MATLAB function `simp2v` evaluates repeated integrals in two variables by making direct use of the composite rule.

```
function q = simp2v(func,a,b,c,d,n)
% Implements 2 variable Simpson integration.
% Example call: q = simp2v(func,a,b,c,d,n)
% Integrates user defined 2 variable function func.
% Range for first variable is a to b, and second variable, c to d
% using n divisions of each variable.
if (n/2)~=floor(n/2)
    disp('n must be even'); return
else
    hx = (b-a)/n; x = a:hx:b; nx = length(x);
    hy = (d-c)/n; y = c:hy:d; ny = length(y);
    [xx,yy] = meshgrid(x,y);
    z = feval(func,xx,yy);
    v = 2*ones(n+1,1);  v2 = 2*ones(n/2,1);
    v(2:2:n) = v(2:2:n)+v2;
    v(1) = 1;  v(n+1) = 1;
    S = v*v';  T = z.*S;
    q = sum(sum(T))*hx*hy/9;
end
```

We now apply the function `simp2v` to evaluate the integral Gauss–Lobatto Integration

$$\int_0^{10} dx \int_0^{10} y^2 \sin x \, dy$$

The graph of the function $y^2 \sin x$ is given in Fig. 4.5. The script `e4s414` integrates this function.

```
% e4s414.m
z = @(x,y) y.^2.*sin(x);
disp(' n      integral value');
n = 4; j = 1;
while n<=256
    int = simp2v(z,0,10,0,10,n);
    fprintf('%4.0f %17.8e\n',n,int)
    n = 2*n; j = j+1;
end
```

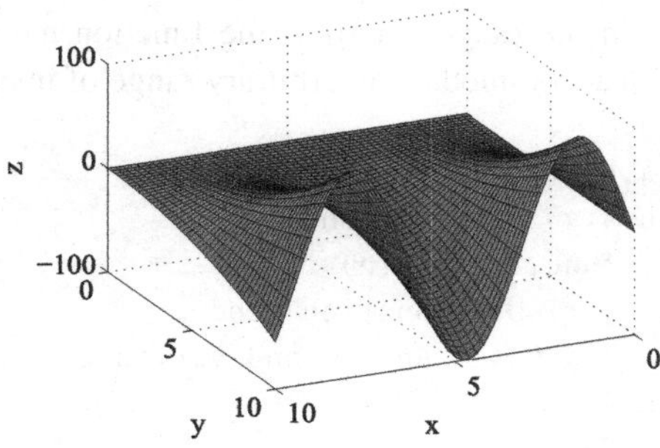

FIGURE 4.5

Graph of $z = y^2 \sin x$.

Running script e4s414 gives the following results:

```
   n      integral value
   4    1.02333856e+003
   8    6.23187046e+002
  16    6.13568708e+002
  32    6.13056704e+002
  64    6.13025879e+002
 128    6.13023970e+002
 256    6.13023851e+002
```

The value of this integral exact to four decimal places is 613.0238 and the number of floating-point operations tends to $7n^2$. It can be proved, – see Salvadori and Baron (1961) – that when Simpson's rule is adapted to evaluate repeated integrals, the error is still of order h^4 and thus it is possible to use an extrapolation scheme similar to the Romberg method of Section 4.6.

4.15.2 GAUSSIAN INTEGRATION FOR REPEATED INTEGRALS

The Gaussian method can be developed to evaluate repeated integrals with constant limits of integration. In Section 4.7 it was shown that for single integrals the integrand must be evaluated at specified points. Thus, if

$$I = \int_{-1}^{1} dx \int_{-1}^{1} f(x, y)\, dy$$

then

$$I \approx \sum_{i=1}^{n} \sum_{j=1}^{m} A_i A_j f(x_i, y_j)$$

The rules for calculating x_i, y_j, and A_i are given in Section 4.7. The MATLAB function gauss2v evaluates integrals using this technique. Because the values of x and y are chosen on the assumption that the integration takes place in the range −1 to 1, the function `gauss2v` includes the necessary manipulations to adjust it so as to accommodate an arbitrary range of integration.

```
function q = gauss2v(func,a,b,c,d,n)
% Implements 2 variable Gaussian integration.
% Example call: q = gauss2v(func,a,b,c,d,n)
% Integrates user defined 2 variable function func.
% Range for first variable is a to b, and second variable, c to d
% using n divisions of each variable.
% n must be 2 or 4 or 8 or 16.
if (n==2)|(n==4)|(n==8)|(n==16)
    co = zeros(8,4); t = zeros(8,4);
    co(1,1) = 1;
    co(1:2,2) = [.6521451548; .3478548451];
    co(1:4,3) = [.3626837833; .3137066458; .2223810344; .1012285362];
    co(:,4) = [.1894506104; .1826034150; .1691565193; .1495959888; ...
               .1246289712;.0951585116; .0622535239; .0271524594];
    t(1,1) = .5773502691;
    t(1:2,2) = [.3399810435; .8611363115];
    t(1:4,3) = [.1834346424; .5255324099; .7966664774; .9602898564];
    t(:,4) = [.0950125098; .2816035507; .4580167776; .6178762444; ...
              .7554044084; .8656312023; .9445750230; .9894009350];
    j = 1;
    while j<=4
        if 2^j==n; break;
        else
            j = j+1;
        end
    end
    s = 0;
    for k = 1:n/2
        x1 = (t(k,j)*(b-a)+a+b)/2;  x2 = (-t(k,j)*(b-a)+a+b)/2;
        for p = 1:n/2
            y1 = (t(p,j)*(d-c)+d+c)/2;  y2 = (-t(p,j)*(d-c)+d+c)/2;
            z = feval(func,x1,y1)+feval(func,x1,y2)+feval(func,x2,y1);
            z = z+feval(func,x2,y2);
            s = s+co(k,j)*co(p,j)*z;
        end
    end
    q = (b-a)*(d-c)*s/4;
else
    disp('n must be equal to 2, 4, 8 or 16'), return
end
```

We now consider the problem of evaluating the following integral:

$$\int_{x^2}^{x^4} dy \int_1^2 x^2 y\,dx \tag{4.49}$$

Integrals of this form cannot be estimated directly by the MATLAB functions `gauss2v` or `simp2v` because neither of these functions was developed to work with variable limits of integration. However, a transformation may be carried out in order to make the limits of integration constant. Let

$$y = (x^4 - x^2)z + x^2 \tag{4.50}$$

Thus, when $z = 1$, $y = x^4$ and when $z = 0$, $y = x^2$ as required. Differentiating (4.50), we have

$$dy = (x^4 - x^2)dz$$

Substituting for y and dy in (4.49), we have

$$\int_0^1 dz \int_1^2 x^2 \left\{(x^4 - x^2)z + x^2\right\} (x^4 - x^2)\,dx \tag{4.51}$$

This integral is now in a form that can be integrated using both `gauss2v` and `simp2v`. However, we must define a MATLAB function thus:

```
w = @(x,z) x.^2.*((x.^4-x.^2).*z+x.^2).*(x.^4-x.^2);
```

This function is used with the functions `simp2v` and `gauss2v` in the script `e4s415`.

```
% e4s415.m
disp('    n   Simpson value      Gauss value')
w = @(x,z) x.^2.*((x.^4-x.^2).*z+x.^2).*(x.^4-x.^2);
n = 2; j = 1;
while n<=16
    in1 = simp2v(w,1,2,0,1,n);
    in2 = gauss2v(w,1,2,0,1,n);
    fprintf('%4.0f%17.8e%17.8e\n',n,in1,in2)
    n = 2*n; j = j+1;
end
```

Running script `e4s415` gives

```
   n   Simpson value      Gauss value
   2  9.54248047e+001  7.65255915e+001
   4  8.48837042e+001  8.39728717e+001
   8  8.40342951e+001  8.39740259e+001
  16  8.39778477e+001  8.39740259e+001
```

The exact integral is equal to 83.97402597 (= 6466/77). This output shows that in general Gaussian integration is superior to Simpson's rule.

4.16 MATLAB FUNCTIONS FOR DOUBLE AND TRIPLE INTEGRATION

Recent versions of MATLAB now provide the functions `integral2` and `integral3` for repeated integration. In this section, we consider these functions and their parameters and provide examples of their use.

For double integration which is repeated integration over two dimensions, the `integral2` function may be used and has the general form:

```
IV2 = integral2(fname,xl,xu,yl,yu,acc)
```

where `fname` is the name of the two-variable function being integrated, which must be defined by the user; `xl` and `xu` the lower and upper limits of the x range of integration; and similarly `yl` and `yu` the lower and upper limits for the y range of integration. The value `acc` provides the required accuracy of the integration and is optional.

The use of `integral2` is illustrated by evaluating the integral:

$$I = \int_0^1 dx \int_0^1 \frac{1}{1-xy}\, dy$$

We may solve this using the MATLAB function `integral2`. It is required that the user predefine the function to be integrated; to do this we use an anonymous function placed directly in the function parameter list. Using `integral2` we have

```
>> I = integral2(@(x,y) 1./(1-x.*y),0,1-1e-6,0,1-1e-6)

I =
    1.6449
```

If we try to integrate this function numerically over the exact range $x = 0$ to 1 and $y = 0$ to 1 then MATLAB gives warnings because of the singularity when $x = y = 1$ but gives the same result.

For triple integration which is repeated integration over three dimensions the `integral3` function may be used and has the general form:

```
IV3 = integral3(fname,xl,xu,yl,yu,zl,zu,acc)
```

where `fname` is the name of the three variable function being integrated, `xl` and `xu` are the lower and upper limits of the x range of integration. Similarly `yl`, `yu` and `zl`, `zu` are the limits for the y and z range of integration.

The use of `integral3` is illustrated by the following example.

$$\int_0^1 dx \int_0^1 dy \int_0^1 64xy(1-x)^2 z\, dz$$

```
>> I3 = integral3(@(x,y,z) 64*x.*y.*(1-x).^2.*z,0,1,0,1,0,1)

I3 =
    1.3333
```

MATLAB provides the function quad2d which allows the user to integrate a function of two variables (say x and y) like the function integral2 but additionally allows the limits in y to be functions of x.

Consider the integral (4.49) repeated here

$$\int_1^2 dx \int_{x^2}^{x^4} x^2 y \, dy$$

Using quad2d we have

```
>> IV = quad2d(@(x,y) x.^2.*y,1,2, @(x) x.^2,@(x) x.^4)

IV =
   83.9740
```

In this example, the anonymous function @(x,y) x.^2.*y is the function to be integrated, 1 and 2 are the lower and upper limits of integration in the x variable, and @(x) x.^2 and @(x) x.^4 are anonymous functions defining lower and upper limits in the y range of integration.

4.17 SUMMARY

In this chapter, we have described simple methods for obtaining the approximate derivatives of various orders for specified functions at given values of the independent variable. The results indicate that these methods, although easy to program, are very sensitive to small changes in key parameters and should be used with considerable care. In addition, we have given a range of methods for integration. For integration, error generation is not such an unpredictable problem but we must be careful to choose the most efficient method for the integral we wish to evaluate. We have also described adaptive integration.

The reader is referred to Sections 10.8, 10.9, and 10.10 for the application of the Symbolic Toolbox to integration and differentiation problems.

4.18 PROBLEMS

4.1. Use the function diffgen to find the first and second derivatives of the function $x^2 \cos x$ at $x = 1$ using $h = 0.1$ and $h = 0.01$.

4.2. Evaluate the first derivative of $\cos x^6$ for $x = 1, 2$, and 3 using the function diffgen and taking $h = 0.001$.

4.3. Write a MATLAB function to differentiate a given function using formulae (4.6) and (4.7). Use it to solve Problems 4.1 and 4.2.

4.4. Find the gradient of $y = \cos x^6$ at $x = 3.1, 3.01, 3.001$, and 3 using the function diffgen with $h = 0.001$. Compare your results with the exact result.

4.5. The approximations for partial derivatives may be defined as

$$\partial f/\partial x \approx \{f(x+h, y) - f(x-h, y)\}/(2h)$$
$$\partial f/\partial y \approx \{f(x, y+h) - f(x, y-h)\}/(2h)$$

Write a function to evaluate these derivatives. The function call should have the form

```
[pdx,pdy] = pdiff('func',x,y,h)
```

Determine the partial derivatives of $\exp(x^2 + y^3)$ at $x = 2$, $y = 1$ using this function with $h = 0.005$.

4.6. In a letter sent to Hardy, the Indian mathematician Ramanujan proposed that the number of numbers between a and b which are either squares or sums of two squares is given approximately by the integral

$$0.764 \int_a^b \frac{dx}{\sqrt{\log_e x}}$$

Test this proposition for the following pairs of values of a and b: $(1, 10)$, $(1, 17)$, and $(1, 30)$. You should use the MATLAB function `fgauss` with 16 points to evaluate the integrals required.

4.7. Verify the equality

$$\int_0^\infty \frac{dx}{(1+x^2)(1+r^2x^2)(1+r^4x^2)} = \frac{\pi(r^2+r+1)}{2(r^2+1)(r+1)^2}$$

for the values of $r = 0, 1, 2$. This result was proposed by Ramanujan. You should use the MATLAB function `galag` for your investigations, using eight points.

4.8. Raabe established the result that

$$\int_a^{a+1} \log_e \Gamma(x)\, dx = a \log_e a - a + \log_e \sqrt{2\pi}$$

Verify this result for $a = 1$ and $a = 2$. Use the MATLAB function `simp1` with 32 divisions to evaluate the integrals required and the MATLAB function `gamma` to set up the integrand.

4.9. Use the MATLAB function `fgauss` with 16 points to evaluate the integral

$$\int_0^1 \frac{\log_e x\, dx}{1+x^2}$$

Explain why the function `fgauss` is appropriate for this problem but `simp1` is not.

4.10. Use the MATLAB function `fgauss` with 16 points to evaluate the integral

$$\int_0^1 \frac{\tan^{-1} x}{x}\, dx$$

Note: Integration by parts shows the integrals in Problems 4.9 and 4.10 to be the same value except for a sign.

4.11. Write a MATLAB function to implement the formulae (4.32) and (4.33) given in Section 4.9 and use your function to evaluate the following integrals using 10 points for the formula. Compare your results with the Gauss 16-point rule.

$$\textbf{(a)} \int_{-1}^{1} \frac{e^x}{\sqrt{1-x^2}}\,dx, \quad \textbf{(b)} \int_{-1}^{1} e^x\sqrt{1-x^2}\,dx$$

4.12. Use the MATLAB function `simp1` to evaluate the Fresnel integrals

$$C(1) = \int_0^1 \cos\left(\frac{\pi t^2}{2}\right) dt \text{ and } S(1) = \int_0^1 \sin\left(\frac{\pi t^2}{2}\right) dt$$

Use 32 intervals. The exact values, to seven decimal places, are $C(1) = 0.7798934$ and $S(1) = 0.4382591$.

4.13. Use the MATLAB function `filon`, with 64 intervals, to evaluate the integral

$$\int_0^{\pi} \sin x \cos kx\,dx$$

for $k = 0$, 4, and 100. Compare your results with the exact answer, $2/(1-k^2)$ if k is even and 0 if k is odd.

4.14. Solve Problem 4.13 for $k = 100$ using Simpson's rule with 1024 divisions and Romberg's methods with 9 divisions.

4.15. Evaluate the following integral using the eight point Gauss–Laguerre method:

$$\int_0^{\infty} \frac{e^{-x}dx}{x+100}$$

Compare your answer with the exact solution 9.9019419×10^{-3} (103/10,402).

4.16. Evaluate the integral

$$\int_0^{\infty} \frac{e^{-2x} - e^{-x}}{x}\,dx$$

using 8-point Gauss–Laguerre integration. Compare your result with the exact answer which is $-\log_e 2 = -0.6931$.

4.17. Evaluate the following integral using the 16-point Gauss–Hermite method.

$$\int_{-\infty}^{\infty} \exp(-x^2) \cos x\,dx$$

Compare your answer with the exact solution $\sqrt{\pi}\exp(-1/4)$.

4.18. Evaluate the following integrals, using Simpson's rule for repeated integrals, MATLAB function `simp2v`, using 64 divisions in each direction.

$$\textbf{(a)} \int_{-1}^{1} dy \int_{-\pi}^{\pi} x^4 y^4 dx, \quad \textbf{(b)} \int_{-1}^{1} dy \int_{-\pi}^{\pi} x^{10} y^{10} dx$$

4.19. Evaluate the following integrals, using `simp2v`, with 64 divisions in each direction.

$$\text{(a)} \int_0^3 dx \int_1^{\sqrt{x/3}} \exp(y^3)\,dy, \quad \text{(b)} \int_0^2 dx \int_0^{2-x} (1+x+y)^{-3} dy$$

4.20. Evaluate part (b) in Problems 4.18 and 4.19 using Gaussian integration, MATLAB function `gauss2v`. *Note*: To use this function the range of integration must be constant.

4.21. The definition of the sine integral Si(z) is

$$\mathrm{Si}(z) = \int_0^z \frac{\sin t}{t}\,dt$$

Evaluate this integral using the 16-point Gauss method for $z = 0.5$, 1, and 2. Why does the Gaussian method work and yet the Simpson and Romberg methods fail?

4.22. Evaluate the following double integral using Gaussian integration for two variables.

$$\int_0^1 dy \int_0^1 \frac{1}{1-xy}\,dx$$

Compare your result with the exact answer, $\pi^2/6 = 1.6449$.

4.23. The probability, P, that a certain type of gas turbine engine will fail within a period of time of T hours is given by the equation:

$$P(x < T) = \int_0^T \frac{ab^a}{(x+b)^{a+1}}\,dx$$

where $a = 3.5$ and $b = 8200$.

By evaluating this integral for values of $T = 500 : 100 : 2000$, draw a graph of P against T in this range. What proportion of the number of gas turbines of this type fail within 1600 hours. For more information on the probability of failure, see Percy (2011).

4.24. Consider the following integral:

$$\int_0^1 \frac{x^p - x^q}{\log_e(x)} x^r\,dx = \log_e\left(\frac{p+r+1}{q+r+1}\right)$$

Use the MATLAB function `integral` to verify this result for $p = 3, q = 4, r = 2$.

4.25. Consider the three integrals:

$$A = -\int_0^1 \frac{\log_e x\,dx}{1+x^2}, \quad B = \int_0^1 \frac{\tan^{-1} x}{x}\,dx, \quad C = \int_0^\infty \frac{xe^{-x}}{1+e^{-2x}}\,dx$$

Use the MATLAB function `integral` to evaluate the two integrals A and B and hence verify that they are equal.

Use 8-point Gauss–Laguerre integration to verify that the integral C is also equal to A and B.

4.26. Use 16-point Gauss–Hermite integration to verify that the integral I, where

$$I = \int_{-\infty}^{\infty} \frac{\sin x}{1+x^2} dx$$

and show that its value is approximately equal to zero.

4.27. Use 16-point Gauss–Hermite integration to evaluate the following integral I

$$I = \int_{-\infty}^{\infty} \frac{\cos x}{1+x^2} dx$$

Check your answer by comparing with the exact answer, π/e.

4.28. Use 8-point Gauss–Laguerre integration to find the value of the following integral:

$$I = \int_{0}^{\infty} \frac{x^{\alpha-1}}{1+x^{\beta}} dx$$

for values of $(\alpha, \beta) = (2, 3)$ and $(3, 4)$. You can verify your answers using the exact value of the integral which is $\pi/(\beta \sin(\alpha\pi/\beta))$.

4.29. An interesting relationship between the Riemann zeta function, ζ, and the integral

$$S_3 = -\int_{0}^{\infty} \log_e(x)^3 e^{-x}\, dx$$

is given by:

$$S_3 = \gamma^3 + \frac{1}{2}\gamma\pi^2 + 2\zeta(3)$$

where $\gamma = 0.57722$.
Use the MATLAB function `quadgk` to evaluate the integral and show that it is a good estimate of S_3.

4.30. A value for the total resistance of a certain network of unit resistors has been shown to be given by $R(m, n)$, where:

$$R(m, n) = \frac{1}{\pi^2} \int_{0}^{\pi} dx \int_{0}^{\pi} \frac{1 - \cos mx\ \cos ny}{2 - \cos x - \cos y} dy$$

Evaluate this integral for $R(50, 100)$ using the MATLAB functions `integral2` and `simp2v`. Use a lower limit close to zero, say 0.0001. If a lower limit equal to zero is used, then the denominator of the integrand will be zero. For large values of m and n an approximation for this integral is given by

$$R(m, n) = \frac{1}{\pi}\left(\gamma + \frac{3}{2}\log_e 2 + \frac{1}{2}\log_e(m^2 + n^2)\right)$$

where γ is Euler's constant and can be obtained from the MATLAB expression `-psi(1)`. The function `psi(x)` is called the digamma function. Use this to check your result.

4.31. A value for the total resistance of a cubic network of unit resistors has been shown to be given by $R(s,m,n)$ where:

$$R(s,m,n) = \frac{1}{\pi^3}\int_0^{\pi} dx \int_0^{\pi} dy \int_0^{\pi} \frac{1-\cos sx\,\cos my\,\cos nz}{3-\cos x-\cos y-\cos z}\,dz$$

Evaluate this integral using the MATLAB function for `integral3` using the values $s=2$, $m=1$, $n=3$. The lower limit should be set at a small non-zero value, say 0.0001.

4.32. In calculations of certain moments of inertia the following two equations must be satisfied by choosing appropriate values for the constants c and γ.

$$\int_0^{\sqrt{c-\gamma}} \frac{1}{\sqrt{p^2-1}}\,dy = 1, \quad \int_0^{\sqrt{c-\gamma}} \frac{p}{\sqrt{p^2-1}}\,dy = \frac{\pi}{2}$$

where $p=(y^2+\gamma)/c$. It can be shown that the values $c=-1.035$ and $\gamma=-2.290$ satisfy these equations. Verify using the MATLAB function `integral` that these values do satisfy these equations. See Taylor (2018).

4.33. The following integral provides the mean distance (M) between two planets, assuming their orbits are circular with radius a and b:

$$M = \frac{1}{2\pi}\int_0^{2\pi} \sqrt{a^2+b^2-2ab\cos\theta}\,d\theta$$

Using this formula calculate M for the orbits of Venus and Mars where $a=0.723AU$ and $b=1.524AU$ approximately. An Astronomical Unit (AU) is defined as the mean distance from the Earth to the Sun.

4.34. The velocity of a body falling vertically under gravity, but subject to resistance is given by the differential equation:

$$mv\frac{dv}{dx} = mg - mk(x)v^2$$

where the resistance to travel is given by the equation

$$k(x) = \exp(-\beta+\alpha x)$$

and x is the downward distance from the starting point. Note when $x=0$ then

$$k(\theta) = \exp(-\beta)$$

It can be shown that the solution to this equation is given by

$$v^2 = \frac{2g}{\alpha\exp(2k(x)/\alpha)}\left(\mathrm{Ei}(2k(x)/\alpha) - \mathrm{Ei}(2k(0)/\alpha)\right)$$

Write a MATLAB script to calculate values showing how v, the velocity, varies with the altitude x and use these results to draw a graph of velocity against distance fallen, given initially

$x = 0.287986$, $\alpha = 1.556362 \times 10^{-4}$, and $\beta = 11.2988350$. Perform your calculations using the MATLAB function `expint`. Increment x by 0.2 and perform 25 steps.
The function $\mathrm{Ei}(x)$ is defined by:

$$\mathrm{Ei}(x) = \int_{-\infty}^{x} \frac{\exp(t)}{t}\, dt$$

Note that this function is available in MATLAB by the name `expint` where

$$\mathrm{Ei}(x) = \mathrm{real}(-\texttt{expint(-x)}), \text{ for real } x > 0.$$

For an interesting description of this problem see Mahoney (2014).

4.35. Evaluate the integral

$$\int_{0}^{\pi/2} \exp(-\sin(2x)) \cos(x)\, dx$$

using Filon's method for cosine function. Use sufficient divisions to obtain an accuracy of 6 decimal places. Note that this integral is equal to the sum to infinity of the series given in Problem 1.39.

4.36. The following functions are the first 4 Legendre polynomials $P_0 = 1$; $P_1 = x$; $P_2 = (3x^2 - 1)/2$; $P_3 = (5x^3 - 3x)/2$. Use Gaussian integration to evaluate the integral

$$I_{m,n} = \int_{-1}^{1} P_m(x) P_n(x)\, dx$$

for $m = 0$ to 3 and for $n = 0$ to 3. What can you deduce about the properties of the first 4 Legendre polynomials.

CHAPTER

SOLUTION OF DIFFERENTIAL EQUATIONS

5

Abstract

Many practical problems involve the study of how rates of change in two or more variables are interrelated. Often the independent variable is time. These problems give rise naturally to differential equations which enable us to understand how the real-world works and how it changes dynamically. In this chapter, we describe a range of widely used algorithms to solve differential equations.

5.1 INTRODUCTION

To illustrate how a differential equation can model a physical situation we will examine a relatively simple problem. Consider the way a hot object cools, – for example, a saucepan of milk, the water in a bath or molten iron. Each of these will cool in a different way dependent on the environment but we shall abstract only the most important features that are easy to model. To model this process by a simple differential equation we use Newton's law of cooling which states that the *rate* at which these objects lose heat as time passes is dependent on the difference between the current temperature of the object and the temperature of its surroundings. This leads to the differential equation

$$dy/dt = K(y - s) \tag{5.1}$$

where y is the current temperature at time t, s is the temperature of the surroundings, and K is a negative constant for the cooling process. In addition we require the initial temperature, y_0, to be specified at time $t = 0$ when the observations begin. This fully specifies our model of the cooling process. We only need values for y_0, K, and s to begin our study. This type of first-order differential equation is called an *initial value problem* because we have an initial value given for the dependent variable y at time $t = 0$.

The solution of (5.1) is easily obtained analytically and will be a function of t and the constants of the problem. However, there are many differential equations which have no analytic solution or the analytic solution does not provide an explicit relation between y and t. In this situation we use numerical methods to solve the differential equation. This means that we approximate the continuous solution with an approximate discrete solution giving the values of y at specified time steps between the initial value of time and some final time value. Thus we compute values of y, which we denote by y_i, for values of t denoted by t_i where $t_i = t_0 + ih$ for $i = 0, 1, ..., n$. Fig. 5.1 illustrates the exact solution and an approximate solution of (5.1) where $K = -0.1$, $s = 10$, and $y_0 = 100$. This figure is generated using the standard MATLAB function for solving differential equations, `ode23`, from time 0 to 60 and plotting the values of y using the symbol "+". The values of the exact solution are plotted on the same graph using the symbol "o".

Numerical Methods. https://doi.org/10.1016/B978-0-12-812256-3.00014-2

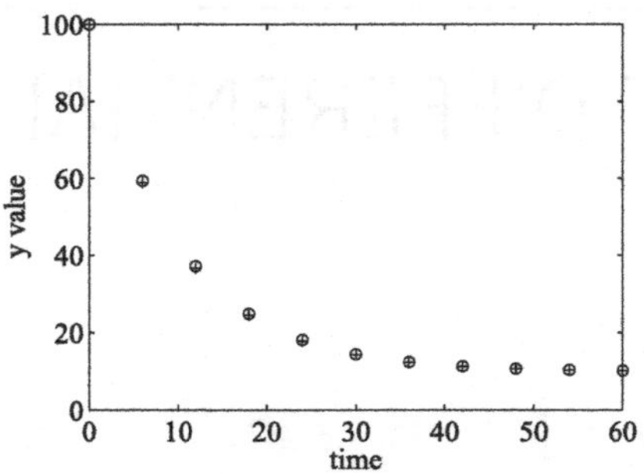

FIGURE 5.1

Exact o and approximate + solution for $dy/dt = -0.1(y - 10)$.

To use `ode23` to solve (5.1) we begin by writing a function `yprime` which defines the right-hand side of (5.1). Then `ode23` is called in the following script and requires the initial and final values of t, 0, and 60 which must be placed in a row vector; a starting value for y of 100; and a low relative tolerance of 0.5. This tolerance is set using the `odeset` function which allows tolerances and other parameters to be set as required.

```
% e4s501.m
yprime = @(t,y) -0.1*(y-10); %RH of diff equn.
options = odeset('RelTol',0.5);
[t y] = ode23(yprime,[0 60],100,options);
plot(t,y,'+')
xlabel('Time'), ylabel('y value'),
hold on
plot(t,90*exp(-0.1.*t)+10,'o'), % Exact solution.
hold off
```

This type of step by step solution is based on computing the current y_i value from a single or combination of functions of previous y values. If the value of y is calculated from a combination of more than one previous value, it is called a *multi-step* method. If only one previous value is used it is called a *single-step* method. We shall now describe a simple *single-step* method known as Euler's method.

5.2 EULER'S METHOD

The dependent variable y and the independent variable t, which we used in the preceding section, can be replaced by any variable names. For example, many textbooks use y as the dependent variable and x as the independent variable. However, for some consistency with MATLAB notation we generally use y to represent the dependent variable and t to represent the independent variable. Clearly initial value problems are not restricted to the time domain although, in most practical situations, they are.

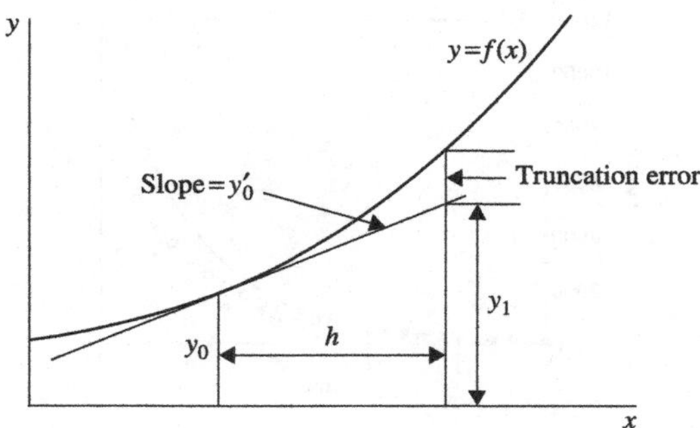

FIGURE 5.2

Geometric interpretation of Euler's method.

Consider the differential equation

$$dy/dt = y \tag{5.2}$$

One of the simplest approaches for obtaining the numerical solution of a differential equation is the method of Euler. This employs Taylor's series but uses only the first two terms of the expansion. Consider the following form of Taylor's series in which the third term is called the remainder term and represents the contribution of all the terms not included in the series.

$$y(t_0 + h) = y(t_0) + y'(t_0)h + y''(\theta)h^2/2 \tag{5.3}$$

where θ lies in the interval (t_0, t_1). For small values of h we may neglect the terms in h^2, and setting $t_1 = t_0 + h$ in (5.3) leads to the formula

$$y_1 = y_0 + hy_0'$$

where the prime denotes differentiation with respect to t and $y_i' = y'(t_i)$. In general,

$$y_{n+1} = y_n + hy_n' \text{ for } n = 0, 1, 2, \ldots$$

By virtue of (5.2) this may be written

$$y_{n+1} = y_n + hf(t_n, y_n) \text{ for } n = 0, 1, 2, \ldots \tag{5.4}$$

This is known as Euler's method and it is illustrated geometrically in Fig. 5.2. This is an example of the use of a single function value to determine the next step. From (5.3) we can see that the local truncation error, i.e., the error for individual steps is of order h^2.

The method is simple to script and is implemented in the MATLAB function `feuler` as follows.

```
function [tvals, yvals] = feuler(f,tspan, startval,step)
% Euler's method for solving
% first order differential equation dy/dt = f(t,y).
```

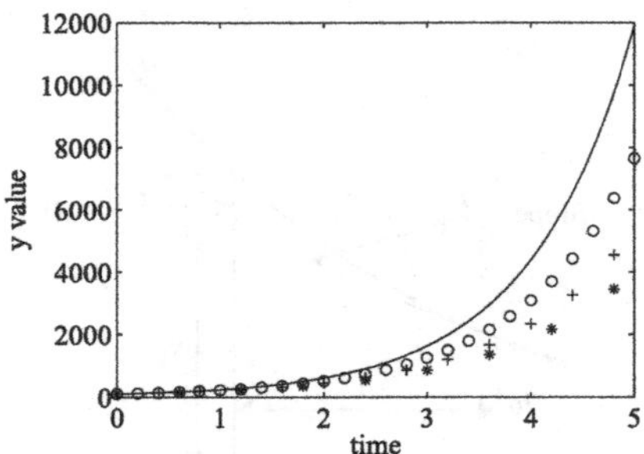

FIGURE 5.3

Points from the Euler solution of $dy/dt = y - 20$ given that $y = 100$ when $t = 0$. Approximate solutions for $h = 0.2$, 0.4, and 0.6 are plotted using o, +, and $*$ respectively. The exact solution is given by the solid line.

```
% Example call: [tvals, yvals]=feuler(f,tspan,startval,step)
% Initial and final value of t are given by tspan = [start finish].
% Initial value of y is given by startval, step size is given by step.
% The function f(t,y) must be defined by the user.
steps = (tspan(2)-tspan(1))/step+1;
y = startval; t = tspan(1);
yvals = startval; tvals = tspan(1);
for i = 2:steps
    y1 = y+step*feval(f,t,y); t1 = t+step;
    %collect values together for output
    tvals = [tvals, t1]; yvals = [yvals, y1];
    t = t1; y = y1;
end
```

Applying this function to the differential equation (5.1) with $K = 1$, $s = 20$, and an initial value of $y = 100$, gives Fig. 5.3, which illustrates how the approximate solution varies for different values of h. The exact value computed from the analytical solution is given for comparison purposes by the solid line. Clearly, in view of the very large errors shown in Fig. 5.3, the Euler method, although simple, requires a very small step h to provide reasonable levels of accuracy. If the differential equation must be solved for a wide range of values of t, the method becomes very expensive in terms of computer time because of the very large number of small steps required to span the interval of interest. In addition, the errors made at each step may accumulate in an unpredictable way. This is a crucial issue, and we discuss this in the next section.

5.3 THE PROBLEM OF STABILITY

To ensure that errors do not accumulate we require that the method for solving the differential equation be stable. We have seen that the error at each step in Euler's method is of order h^2. This error is known as the local truncation error since it tells only how accurate the individual step is, not what the error is for a sequence of steps. The error for a sequence of steps is difficult to find since the error from one step affects the accuracy of the next in a way that is often complex. This leads us to the issue of absolute and relative stability. We now discuss these concepts and examine their effects in relation to a simple equation and explain how the results for this equation may be extended to differential equations in general.

Consider the differential equation

$$dy/dt = Ky \tag{5.5}$$

Since $f(t, y) = Ky$, Euler's method will have the form

$$y_{n+1} = y_n + hKy_n \tag{5.6}$$

Thus, using this recursion repeatedly and assuming that there are no errors in the computation from stage to stage we obtain

$$y_{n+1} = (1 + hK)^{n+1} y_0 \tag{5.7}$$

For small enough h, it is easily shown that this value will approach the exact value e^{Kt}.

To obtain some understanding of how errors propagate when using Euler's method, let us assume that y_0 is perturbed. This perturbed value of y_0 may be denoted by y_0^a where $y_0^a = (y_0 - e_0)$ and e_0 is the error. Thus (5.7) becomes, on using this approximate value instead of y_0,

$$y_{n+1}^a = (1 + hK)^{n+1} y_0^a = (1 + hK)^{n+1}(y_0 - e_0) = y_{n+1} - (1 + hK)^{n+1} e_0$$

Consequently, the initial error will be magnified if $|1 + hK| \geq 1$. After many steps this initial error will grow and may dominate the solution. This is the characteristic of instability and in these circumstances Euler's method is said to be unstable. If, however, $|1 + hK| < 1$, then the error dies away and the method is said to be absolutely stable. Rewriting this inequality leads to the condition for absolute stability:

$$-2 < hK < 0 \tag{5.8}$$

This condition may be too demanding and we may be content if the error does not increase as a proportion of the y values. This is called relative stability. Notice that Euler's method is not absolutely stable for any positive value of K.

The condition for absolute stability can be generalized to an ordinary differential equation of the form of (5.2). It can be shown that the condition becomes

$$-2 < h\partial f/\partial y < 0 \tag{5.9}$$

This inequality implies that, since $h > 0$, $\partial f/\partial y$ must be negative for absolute stability. Fig. 5.4 and Fig. 5.5 give a comparison of the absolute and relative error for $h = 0.1$, for the differential equation $dy/dt = y$ where $y = 1$ when $t = 0$. Fig. 5.4 shows that the error is increasing rapidly and the errors are large for even relatively small step sizes. Fig. 5.5 shows that the error is becoming an increasing proportion of the solution values. Thus the relative error is increasing linearly and so the method is neither relatively stable nor absolutely stable for this problem.

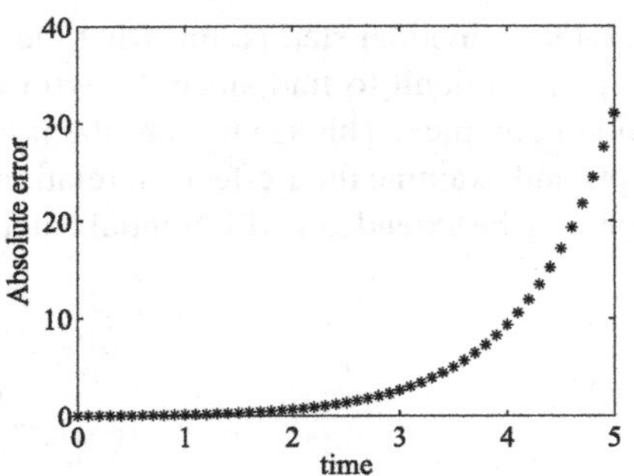

FIGURE 5.4

Absolute errors in the solution of $dy/dt = y$ where $y = 1$ when $t = 0$, using Euler's method with $h = 0.1$.

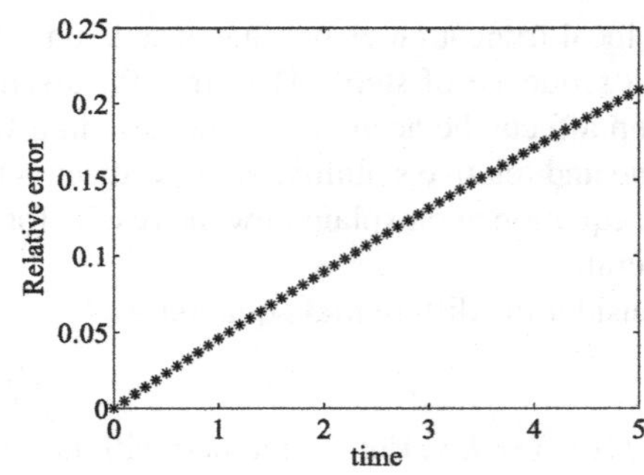

FIGURE 5.5

Relative errors in the solution of $dy/dt = y$ where $y = 1$ when $t = 0$, using Euler's method with $h = 0.1$.

We have seen that Euler's method may be unstable for some values of h. For example, if $K = -100$, then Euler's method is only absolutely stable for $0 < h < 0.02$. Clearly, if we are required to solve the differential equation between 0 and 10, we would require 500 steps. We now consider an improvement to this method called the trapezoidal method which has improved stability features although it is similar in principle to Euler's method.

5.4 THE TRAPEZOIDAL METHOD

The trapezoidal method has the form

$$y_{n+1} = y_n + h\{f(t_n, y_n) + f(t_{n+1}, y_{n+1})\}/2 \text{ for } n = 0, 1, 2, \ldots \quad (5.10)$$

Applying the error analysis of Section 5.3 to this problem gives us, from (5.5), that

$$y_{n+1} = y_n + h(Ky_n + Ky_{n+1})/2 \text{ for } n = 0, 1, 2, \ldots \quad (5.11)$$

Thus expressing y_{n+1} in terms of y_n gives

$$y_{n+1} = (1 + hK/2)/(1 - hK/2)y_n \text{ for } n = 0, 1, 2, \ldots \quad (5.12)$$

Using this result recursively for $n = 0, 1, 2, \ldots$ leads to the result

$$y_{n+1} = \{(1 + hK/2)/(1 - hK/2)\}^{n+1} y_0 \quad (5.13)$$

Now, as in Section 5.3, we can obtain some understanding of how error propagates by assuming that y_0 is perturbed by the error e_0 so that it is replaced by $y_0^a = (y_0 - e_0)$. Hence using the same procedure (5.13) becomes

$$y_{n+1} = \{(1 + hK/2)/(1 - hK/2)\}^{n+1}(y_0 - e_0)$$

This leads directly to the result

$$y_{n=1}^{a} = y_{n+1} - \{(1 + hK/2)/(1 - hK/2)\}^{n+1} e_0$$

Thus, we conclude from this that the influence of the error term which involves e_0 will die away if its multiplier is less than unity in magnitude, i.e.,

$$|(1 + hK/2)/(1 - hK/2)| < 1$$

If K is negative, then for all h the method is absolutely stable. For positive K it is not absolutely stable for any h.

This completes the error analysis of this method. However, we note that the method requires a value for y_{n+1} before we can start. An estimate for this value can be obtained by using Euler's method, that is

$$y_{n+1} = y_n + hf(t_n, y_n) \text{ for } n = 0, 1, 2, \ldots$$

This value can now be used in the right-hand side of (5.10) as an estimate for y_{n+1}. This combined method is often known as the Euler–trapezoidal method. The method can be written formally as:

1. Start with n set at zero where n indicates the number of steps taken.
2. Calculate $y_{n+1}^{(1)} = y_n + hf(t_n, y_n)$.
3. Calculate $f(t_{n+1}, y_{n+1}^{(1)})$ where $t_{n+1} = t_n + h$.
4. For $k = 1, 2, \ldots$ calculate

$$y_{n+1}^{(k+1)} = y_n + h\{f(t_n, y_n) + f(t_{n+1}, y_{n+1}^{(k)})\}/2 \tag{5.14}$$

At step 4, when the difference between successive values of y_{n+1} is sufficiently small, increment n by 1 and repeat steps 2, 3, and 4. This method is implemented in MATLAB function `eulertp` thus:

```
function [tvals, yvals] = eulertp(f,tspan,startval,step)
% Euler trapezoidal method for solving
% first order differential equation dy/dt = f(t,y).
% Example call: [tvals, yvals] = eulertp(f,tspan,startval,step)
% Initial and final value of t are given by tspan = [start finish].
% Initial value of y is given by startval, step size is given by step.
% The function f(t,y) must be defined by the user.
steps = (tspan(2)-tspan(1))/step+1;
y = startval; t = tspan(1);
yvals = startval; tvals = tspan(1);
for i = 2:steps
    y1 = y+step*feval(f,t,y);
    t1 = t+step;
    loopcount = 0; diff = 1;
    while abs(diff)>0.05
        loopcount = loopcount+1;
        y2 = y+step*(feval(f,t,y)+feval(f,t1,y1))/2;
```

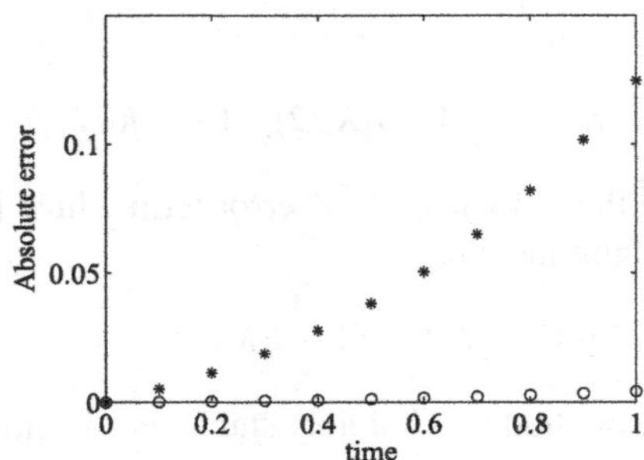

FIGURE 5.6

Solution of $dy/dt = y$ using Euler (∗) and trapezoidal method, o. Step $h = 0.1$ and $y_0 = 1$ at $t = 0$.

```
        diff = y1-y2; y1 = y2;
    end
    %collect values together for output
    tvals = [tvals, t1]; yvals = [yvals, y1];
    t = t1; y = y1;
end
```

We use `eulertp` to study the performance of this method compared with Euler's method for solving $dy/dt = y$. The results are given in Fig. 5.6 which shows graphs of the absolute errors of the two methods. The difference is clear but although the Euler–trapezoidal method gives much greater accuracy for this problem, in other cases the difference may be less marked. In addition, the Euler–trapezoidal method takes longer.

An important feature of this method is the number of iterations that are required to obtain convergence in step 4. If this is high, the method is likely to be inefficient. However, for the example we just solved, a maximum of two iterations at step 4 was required. This algorithm may be modified to use only one iteration at step 4 in (5.14). This is called Heun's method.

Finally, we examine theoretically how the error in Heun's method compares with Euler's method. By considering the Taylor series expansion of y_{n+1} we can obtain the order of the error in terms of the step size h thus:

$$y_{n+1} = y_n + hy_n' + h^2 y_n''/2! + h^3 y_n'(\theta)/3! \tag{5.15}$$

where θ lies in the interval (t_n, t_{n+1}). It can be shown that y_n'' may be approximated by

$$y_n'' = (y_{n+1}' - y_n')/h + O(h) \tag{5.16}$$

Substituting this expression for y_n'' in (5.15) gives

$$\begin{aligned} y_{n+1} &= y_n + hy_n' + h(y_{n+1}' - y_n')/2! + O(h^3) \\ &= y_n + h(y_{n+1}' + y_n')/2! + O(h^3) \end{aligned}$$

This shows that the local truncation error is of order h^3 so there is a significant improvement in accuracy over the basic Euler method which has a truncation error of order h^2.

We now describe a range of methods which will be considered under the collective title of Runge–Kutta methods.

5.5 RUNGE–KUTTA METHODS

The Runge–Kutta methods comprise a large family of methods having a common structure. Heun's method, described by (5.14) but with only one iteration of the corrector, can be recast in the form of a simple Runge–Kutta method. We set

$$k_1 = hf(t_n, y_n) \text{ and } k_2 = hf(t_{n+1}, y_{n+1})$$

since

$$y_{n+1} = y_n + hf(t_n, y_n)$$

We have

$$k_2 = hf(t_{n+1}, y_n + hf(t_n, y_n))$$

Hence from (5.10) we have Heun's method in the form: for $n = 0, 1, 2, \ldots$

$$k_1 = hf(t_n, y_n)$$
$$k_2 = hf(t_{n+1}, y_n + k_1)$$

and

$$y_{n+1} = y_n + (k_1 + k_2)/2$$

This is a simple form of a Runge–Kutta method.

The most commonly used Runge–Kutta method is the classical one; it has the form for each step $n = 0, 1, 2, \ldots$

$$\begin{aligned} k_1 &= hf(t_n, y_n) \\ k_2 &= hf(t_n + h/2, y_n + k_1/2) \\ k_3 &= hf(t_n + h/2, y_n + k_2/2) \\ k_4 &= hf(t_n + h, y_n + k_3) \end{aligned} \tag{5.17}$$

and

$$y_{n+1} = y_n + (k_1 + 2k_2 + 2k_3 + k_4)/6$$

It has a global error of order h^4. The next Runge–Kutta method we consider is a variation on the formula (5.17). It is due to Gill (1951) and takes the form for each step $n = 0, 1, 2, \ldots$

$$
\begin{aligned}
k_1 &= hf(t_n, y_n) \\
k_2 &= hf(t_n + h/2, y_n + k_1/2) \\
k_3 &= hf(t_n + h/2, y_n + (\sqrt{2} - 1)k_1/2 + \left(2 - \sqrt{2}\right) k_2/2) \\
k_4 &= hf(t_n + h, y_n - \sqrt{2}k_2/2 + (1 + \sqrt{2}/2)k_3)
\end{aligned}
\tag{5.18}
$$

and

$$y_{n+1} = y_n + \left\{k_1 + (2 - \sqrt{2})k_2 + (2 + \sqrt{2})k_3 + k_4\right\}/6$$

Again this method is fourth-order and has a local truncation error of order h^5 and a global error of order h^4.

A number of other forms of the Runge–Kutta method have been derived which have particularly advantageous properties. The equations for these methods will not be given but their important features are as follows:

1. *Merson–Runge–Kutta method* (Merson, 1957). This method has an error term of order h^5 and in addition allows an estimate of the local truncation error to be obtained at each step in terms of known values.
2. *The Ralston–Runge–Kutta method* (Ralston, 1962). We have some degree of freedom in assigning the coefficients for a particular Runge–Kutta method. In this formula the values of the coefficients are chosen so as to minimize the truncation error.
3. *The Butcher–Runge–Kutta method* (Butcher, 1964). This method provides higher accuracy at each step, the error being of order h^6.

Runge–Kutta methods have the general form for each step $n = 0, 1, 2, \ldots$

$$
\begin{aligned}
k_1 &= hf(t_n, y_n) \\
k_i &= hf(t_n + hd_i, y_n + \sum_{j=1}^{i-1} c_{ij}k_j), \quad i = 2, 3, \ldots, p
\end{aligned}
\tag{5.19}
$$

$$y_{n+1} = y_n + \sum_{j=1}^{p} b_j k_j \tag{5.20}$$

The order of this general method is p.

The derivation of the various Runge–Kutta methods is based on the expansion of both sides of (5.20) as a Taylor's series and equating coefficients. This is a relatively straightforward idea but involves lengthy algebraic manipulation.

We now discuss the stability of the Runge–Kutta methods. Since the instability which may arise in the Runge–Kutta methods can usually be reduced by a step size reduction, it is known as partial instability. To avoid repeated reduction of the value of h and re-running the method, an estimate of the value of h which will provide stability for the fourth-order Runge–Kutta methods is given by the inequality

$$-2.78 < h\partial f/\partial y < 0$$

In practice $\partial f/\partial y$ may be approximated using the difference of successive values of f and y.

Finally, it is interesting to see how we can provide an elegant MATLAB function for the general Runge–Kutta method given by (5.20) and (5.19). We define two vectors **d** and **b**, where **d** contains the coefficients d_i in (5.19) and **b** contains the coefficients b_j in (5.20), and a matrix **c** which contains the coefficients c_{ij} in (5.19). If the computed values of the k_j are assigned to a vector **k**, then the MATLAB statements that generate the values of the function and the new value of y are relatively simple; they will have the form

```
k(1) = step*feval(f,t,y);
for i = 2:p
    k(i)=step*feval(f,t+step*d(i),y+c(i,1:i-1)*k(1:i-1)');
end
y1 = y+b*k';
```

This is of course repeated for each step. A MATLAB function, `rkgen`, based on this is given below. Since `c` and `d` are easily changed in the script, any form of Runge–Kutta method can be implemented using this function and it is useful for experimenting with different techniques.

```
function[tvals,yvals] = rkgen(f,tspan,startval,step,method)
% Runge Kutta methods for solving
% first order differential equation dy/dt = f(t,y).
% Example call:[tvals,yvals]=rkgen(f,tspan,startval,step,method)
% The initial and final values of t are given by tspan = [start finish].
% Initial y is given by startval and step size is given by step.
% The function f(t,y) must be defined by the user.
% The parameter method (1, 2 or 3) selects
% Classical, Butcher or Merson RK respectively.
b = [ ]; c = [ ]; d = [ ];
switch method
    case 1
        order = 4;
        b = [ 1/6 1/3 1/3 1/6]; d = [0 .5 .5 1];
        c=[0 0 0 0;0.5 0 0 0;0 .5 0 0;0 0 1 0];
        disp('Classical method selected')
    case 2
        order = 6;
        b = [0.07777777778 0 0.355555556 0.13333333 ...
             0.355555556 0.0777777778];
        d = [0 .25 .25 .5 .75 1];
        c(1:4,:) = [0 0 0 0 0 0;0.25 0 0 0 0 0;0.125 0.125 0 0 0 0; ...
                    0 -0.5 1 0 0 0];
```

```
            c(5,:) = [.1875 0 0 0.5625 0 0];
            c(6,:) = [-.4285714 0.2857143 1.714286 -1.714286 1.1428571 0];
            disp('Butcher method selected')
        case 3
            order = 5;
            b = [1/6 0 0 2/3 1/6];
            d = [0 1/3 1/3 1/2 1];
            c = [0 0 0 0 0;1/3 0 0 0 0;1/6 1/6 0 0 0;1/8 0 3/8 0 0; ...
                1/2 0 -3/2 2 0];
            disp('Merson method selected')
        otherwise
            disp('Invalid selection')
end
steps = (tspan(2)-tspan(1))/step+1;
y = startval; t = tspan(1);
yvals = startval; tvals = tspan(1);
for j = 2:steps
    k(1) = step*feval(f,t,y);
    for i = 2:order
        k(i) = step*feval(f,t+step*d(i),y+c(i,1:i-1)*k(1:i-1)');
    end
    y1 = y+b*k'; t1 = t+step;
    %collect values together for output
    tvals = [tvals, t1]; yvals = [yvals, y1];
    t = t1; y = y1;
end
```

A further issue that needs to be considered is that of the adaptive step size adjustment. Where a function is relatively smooth in the area of interest, a large step may be used throughout the region. If the region is such that rapid changes in y occur for small changes in t, then a small step size is required. However, for functions where both these regions exist, then rather than use a small step in the whole region, adaptive step size adjustment would be more efficient. The details of producing this step adjustment are not provided here; however, for an elegant discussion see Press et al. (1990). This type of procedure is implemented for Runge–Kutta methods in the MATLAB functions `ode23` and `ode45`. The basic algorithm for these functions is the Runge–Kutta–Fehlberg algorithm with adaptive step size. However the difference between the two functions is that `ode23` uses second and third order formulae while `ode45` uses fourth and fifth order formulae.

Fig. 5.7 plots the relative errors in the solution of the specific differential equation $dy/dt = -y$ by the classical, Merson and Butcher–Runge–Kutta methods using the MATLAB script: `e4s502.m`:

```
% e4s502.m
yprime = @(t,y) -y;
char = 'o*+';
for meth = 1:3
    [t, y] = rkgen(yprime,[0 3],1,0.25,meth);
```

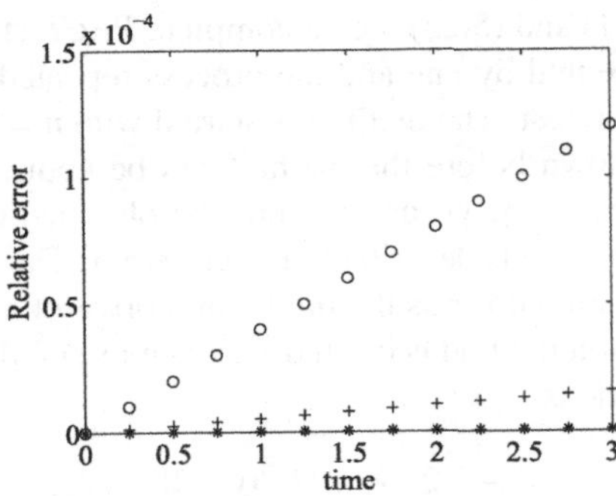

FIGURE 5.7

Solution of $dy/dt = -y$. The $*$ represents Butcher's method, + Merson's method, and o the classical method.

```
    re = (y-exp(-t))./exp(-t);
    plot(t,re,char(meth))
    hold on
end
hold off, axis([0 3 0 1.5e-4])
xlabel('Time'), ylabel('Relative error')
```

It is clear from the graphs that Butcher's method is the best and both Butcher's and Merson's methods are significantly more accurate than the classical method.

5.6 PREDICTOR–CORRECTOR METHODS

The trapezoidal method, which has already been described, is a simple example of both a Runge–Kutta method and a predictor–corrector method with a truncation error of order h^3. The predictor–corrector methods we consider now have much smaller truncation errors. As an initial example we consider the Adams–Bashforth–Moulton method. This method is based on the following equations:

$$\begin{aligned} y_{n+1} &= y_n + h(55y_n' - 59y_{n-1}' + 37y_{n-2}' - 9y_{n-3}')/24 \quad (P) \\ y_{n+1}' &= f(t_{n+1}, y_{n+1}) \quad (E) \end{aligned} \tag{5.21}$$

and

$$\begin{aligned} y_{n+1} &= y_n + h\left(9y_{n+1}' + 19y_n' - 5y_{n-1}' + y_{n-2}'\right)/24 \quad (C) \\ y_{n+1}' &= f(t_{n+1}, y_{n+1}) \quad (E) \end{aligned} \tag{5.22}$$

where $t_{n+1} = t_n + h$. In (5.21) we use the predictor equation (P), followed by a function evaluation (E). Then in (5.22) we use the corrector equation (C), followed by a function evaluation (E). The truncation error for both the predictor and corrector is $O(h^5)$. The first equation in the system (5.21) requires a number of initial values to be known before y can be calculated.

After each application of (5.21) and (5.22), i.e., a complete *PECE* step, the independent variable t_n is incremented by h, n is incremented by one and the process repeated until the differential equation has been solved in the range of interest. The method is started with $n = 3$ and consequently the values of y_3, y_2, y_1, and y_0 must be known before the method can be applied. For this reason, it is called a *multi-point method*. In practice y_3, y_2, y_1, and y_0 must be obtained using a self-starting procedure such as one of the Runge–Kutta methods described in Section 5.5. The self-starting procedure chosen should have the same order truncation error as the predictor–corrector method.

The Adams–Bashforth–Moulton method is often used since its stability is relatively good. Its range of absolute stability in *PECE* mode is

$$-1.25 < h\partial f/\partial y < 0$$

Apart from the need for initial starting values, the Adams–Bashforth–Moulton method in the *PECE* mode requires less computation at each step than the fourth-order Runge–Kutta method. For a true comparison of these methods, however, it is necessary to consider how they behave over a range of problems since applying any method to some differential equations results, at each step, in a growth of error that ultimately swamps the calculation since the step is outside the range of absolute stability.

The Adams–Bashforth–Moulton method is implemented by the function `abm`. It should be noted that errors arise from the choice of starting procedure, in this case the classical Runge–Kutta method. It is, however, easy to amend this function to include the option of entering highly accurate initial values.

```
function [tvals, yvals] = abm(f,tspan,startval,step)
% Adams Bashforth Moulton method for solving
% first order differential equation dy/dt = f(t,y).
% Example call: [tvals, yvals]=abm(f,tspan,startval,step)
% The initial and final values of t are given by tspan = [start finish].
% Initial y is given by startval and step size is given by step.
% The function f(t,y) must be defined by the user.
% 3 steps of Runge-Kutta are required so that ABM method can start.
% Set up matrices for Runge-Kutta methods
b = [ ]; c = [ ]; d = [ ]; order = 4;
b = [1/6 1/3 1/3 1/6]; d = [0 .5 .5 1];
c = [0 0 0 0;0.5 0 0 0;0 .5 0 0;0 0 1 0];
steps = (tspan(2)-tspan(1))/step+1;
y = startval; t = tspan(1); fval(1) = feval(f,t,y);
ys(1) = startval; yvals = startval; tvals = tspan(1);
for j = 2:4
    k(1) = step*feval(f,t,y);
    for i = 2:order
        k(i) = step*feval(f,t+step*d(i),y+c(i,1:i-1)*k(1:i-1)');
    end
    y1 = y+b*k'; ys(j) = y1; t1 = t+step;
    fval(j) = feval(f,t1,y1);
    %collect values together for output
```

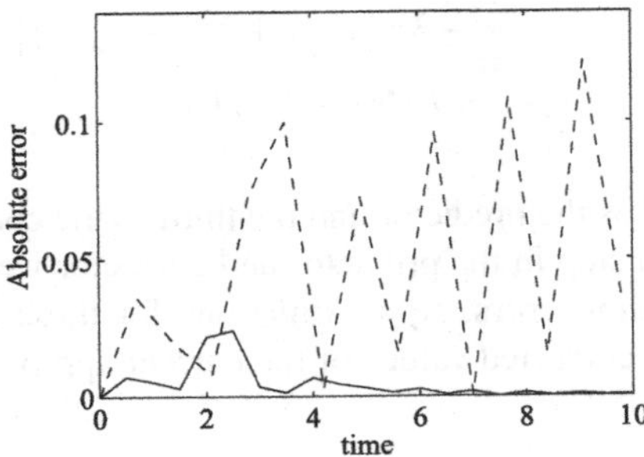

FIGURE 5.8

Absolute error in solution of $dy/dt = -2y$ using the Adams–Bashforth–Moulton method. The *solid line* plots the errors with a step size of 0.5. The *dot-dashed line* plots the errors with step size 0.7.

```
    tvals = [tvals,t1]; yvals = [yvals,y1];
    t = t1; y = y1;
end
%ABM now applied
for i = 5:steps
    y1 = ys(4)+step*(55*fval(4)-59*fval(3)+37*fval(2)-9*fval(1))/24;
    t1 = t+step; fval(5) = feval(f,t1,y1);
    yc = ys(4)+step*(9*fval(5)+19*fval(4)-5*fval(3)+fval(2))/24;
    fval(5) = feval(f,t1,yc);
    fval(1:4) = fval(2:5);
    ys(4) = yc;
    tvals = [tvals,t1]; yvals = [yvals,yc];
    t = t1; y = y1;
end
```

Fig. 5.8 illustrates the behavior of the Adams–Bashforth–Moulton method when applied to the specific problem $dy/dt = -2y$ where $y = 1$ when $t = 0$, using a step size equal to 0.5 and 0.7 in the interval 0 to 10. It is interesting to note that for this problem, since $\partial f/\partial y = -2$, the range of steps for absolute stability is $0 \le h \le 0.625$. For $h = 0.5$, a value inside the range of absolute stability, the plot shows that the absolute error does die away. However, for $h = 0.7$, a value outside the range of absolute stability, the plot shows that the absolute error increases.

5.7 HAMMING'S METHOD AND THE USE OF ERROR ESTIMATES

The method of Hamming (1959) is based on the following pair of predictor–corrector equations:

$$\begin{aligned} y_{n+1} &= y_{n-3} + 4h\left(2y'_n - y'_{n-1} + 2y'_{n-2}\right)/3 \quad (P) \\ y'_{n+1} &= f\left(t_{n+1}, y_{n+1}\right) \quad (E) \end{aligned} \tag{5.23}$$

$$y_{n+1} = \{9y_n - y_{n-2} + 3h\left(y'_{n+1} + 2y'_n - y'_{n-1}\right)\}/8 \quad (C)$$
$$y'_{n+1} = f\left(t_{n+1}, y_{n+1}\right) \quad (E)$$

where $t_{n+1} = t_n + h$.

The first equation (P) is used as the predictor and the third as the corrector (C). To obtain a further improvement in accuracy at each step in the predictor and corrector we modify these equations using expressions for the local truncation errors. Approximations for these local truncation errors can be obtained using the predicted and corrected values of the current approximation to y. This leads to the equations

$$y_{n+1} = y_{n-3} + 4h\left(y'_n - y'_{n-1} + 2y'_{n-2}\right)/3 \quad (P) \tag{5.24}$$

$$(y^M)_{n+1} = y_{n+1} - 112(Y_P - Y_C)/121 \tag{5.25}$$

In this equation Y_P and Y_C represent the predicted and corrected value of y at the nth step.

$$y^*_{n+1} = \left\{9y_n - y_{n-2} + 3h((y^M)'_{n+1} + 2y'_n - y'_{n-1})\right\}/8 \quad (C) \tag{5.26}$$

In this equation $(y^M)'_{n+1}$ is the value of y'_{n+1} calculated using the modified value of y_{n+1} which is $(y^M)_{n+1}$.

$$y_{n+1} = y^*_{n+1} + 9(y_{n+1} - y^*_{n+1}) \tag{5.27}$$

Eq. (5.24) is the predictor and (5.25) modifies the predicted value by using an estimate of the truncation error. Eq. (5.26) is the corrector which is modified by (5.27) using an estimate of the truncation error. The equations in this form are each used only once before n is incremented and the steps repeated again. This method is implemented as MATLAB function `fhamming` thus:

```
function [tvals, yvals] = fhamming(f,tspan,startval,step)
% Hamming's method for solving
% first order differential equation dy/dt = f(t,y).
% Example call: [tvals, yvals]=fhamming(f,tspan,startval,step)
% The initial and final values of t are given by tspan = [start finish].
% Initial y is given by startval and step size is given by step.
% The function f(t,y) must be defined by the user.
% 3 steps of Runge-Kutta are required so that hamming can start.
% Set up matrices for Runge-Kutta methods
b = [ ]; c = [ ]; d = [ ]; order = 4;
b = [1/6 1/3 1/3 1/6]; d = [0 0.5 0.5 1];
c = [0 0 0 0;0.5 0 0 0;0 0.5 0 0;0 0 1 0];
steps = (tspan(2)-tspan(1))/step+1;
y = startval; t = tspan(1);
fval(1) = feval(f,t,y);
ys(1) = startval;
yvals = startval; tvals = tspan(1);
for j = 2:4
```

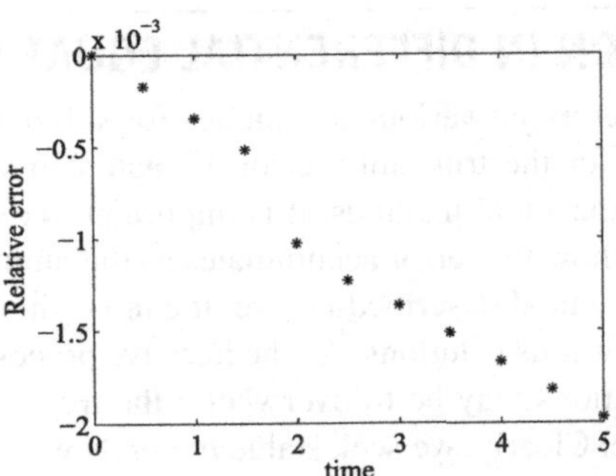

FIGURE 5.9

Relative error in the solution of $dy/dt = y$ where $y = 1$ when $t = 0$, using Hamming's method with a step size of 0.5.

```
    k(1) = step*feval(f,t,y);
    for i = 2:order
        k(i) = step*feval(f,t+step*d(i),y+c(i,1:i-1)*k(1:i-1)');
    end
    y1 = y+b*k'; ys(j) = y1; t1 = t+step; fval(j) = feval(f,t1,y1);
    %collect values together for output
    tvals = [tvals, t1]; yvals = [yvals, y1]; t = t1; y = y1;
end
%Hamming now applied
for i = 5:steps
    y1 = ys(1)+4*step*(2*fval(4)-fval(3)+2*fval(2))/3;
    t1 = t+step; y1m = y1;
    if i>5, y1m = y1+112*(c-p)/121; end
    fval(5) = feval(f,t1,y1m);
    yc = (9*ys(4)-ys(2)+3*step*(2*fval(4)+fval(5)-fval(3)))/8;
    ycm = yc+9*(y1-yc)/121;
    p = y1; c = yc;
    fval(5) = feval(f,t1,ycm); fval(2:4) = fval(3:5);
    ys(1:3) = ys(2:4); ys(4) = ycm;
    tvals = [tvals, t1]; yvals = [yvals, ycm];
    t = t1;
end
```

The choice of h must be made carefully so that the error does not increase without bound. Fig. 5.9 shows Hamming's method used to solve the equation $dy/dt = y$. This is the problem used in Section 5.6.

5.8 ERROR PROPAGATION IN DIFFERENTIAL EQUATIONS

In the preceding sections, we described various techniques for solving differential equations and the order, or a specific expression, for the truncation error at each step was given. As we discussed in Section 5.3 for the Euler and trapezoidal methods, it is important to examine not only the magnitude of the error at each step but also how that error accumulates as the number of steps taken increases.

For the predictor–corrector method described before, it can be shown that the predictor–corrector formulae introduce additional spurious solutions. As the iterative process proceeds, for some problems the effect of these spurious solutions may be to overwhelm the true solution. In these circumstances the method is said to be unstable. Clearly, we seek stable methods where the error does not develop in an unpredictable and unbounded way.

It is important to examine each numerical method to see if it is stable. In addition, if it is not stable for all differential equations we should provide tests to determine when it can be used with confidence. The theoretical study of stability for differential equations is a major undertaking and it is not intended to include a detailed analysis here. In Section 5.9 we summarize the stability characteristics of specific methods and compare the performance of the major methods considered on a number of example differential equations.

5.9 THE STABILITY OF PARTICULAR NUMERICAL METHODS

A good discussion of the stability of many of the numerical methods for solving first-order differential equations is given by Ralston and Rabinowitz (1978) and Lambert (1973). Some of the more significant features, assuming all variables are real, are as follows:

1. Euler and trapezoidal methods. For a detailed discussion see Sections 5.3 and 5.4.
2. Runge–Kutta methods: Runge–Kutta methods do not introduce spurious solutions but instability may arise for some values of h. This may be removed by reducing h to a sufficiently small value. We have already described how the Runge–Kutta methods are less efficient than the predictor–corrector methods because of the greater number of function evaluations that may be required at each step. If h is reduced too far, the number of function evaluations required may make the method uneconomic. The restriction on the size of the interval required to maintain stability may be estimated from the inequality $M < h\partial f/\partial y < 0$ where M is dependent on the particular Runge–Kutta method being used and may be estimated. Clearly this emphasizes the need for careful step size adjustment during the solution process. This is efficiently implemented in the functions `ode23` and `ode45` so that this question does not present a problem when applying these MATLAB functions.
3. Adams–Bashforth–Moulton method: In *PECE* mode the range of absolute stability is given by $-1.25 < h\partial f/\partial y < 0$, implying that $\partial f/\partial y$ must be negative for absolute stability.
4. Hamming's method: In *PECE* mode the range of absolute stability is given by $-0.5 < h\partial f/\partial y < 0$, again implying that $\partial f/\partial y$ must be negative for absolute stability.

Notice that the formulae given for estimating the step size can be difficult to use if f is a general function of y and t. However, in some cases the derivative of f is easily calculated, for example, when $f = Cy$ where C is a constant.

We now give some results of applying the methods discussed in previous sections to solve more general problems. The script `e4s503.m` solves the three examples that follow by setting `example` equal to 1, 2, or 3 in the first line of the script.

```
% e4s503.m
example = 1;
switch example
    case 1
        yprime = @(t,y) 2*t*y;
        sol = @(t) 2*exp(t^2);
        disp('Solution of dy/dt = 2yt')
        t0 = 0; y0 = 2;
    case 2
        yprime = @(t,y) (cos(t)-2*y*t)/(1+t^2);
        sol = @(t) sin(t)/(1+t^2);
        disp('Solution of (1+t^2)dy/dt = cos(t)-2yt')
        t0 = 0; y0 = 0;
    case 3
        yprime = @(t,y) 3*y/t;
        disp('Solution of dy/dt = 3y/t')
        sol = @(t) t^3;
        t0 = 1; y0 = 1;
end
tf = 2; tinc = 0.25; steps = floor((tf-t0)/tinc+1);
[t,x1] = abm(yprime,[t0 tf],y0,tinc);
[t,x2] = fhamming(yprime,[t0 tf],y0,tinc);
[t,x3] = rkgen(yprime,[t0 tf],y0,tinc,1);
disp('t         abm         Hamming      Classical      Exact')
for i = 1:steps
    fprintf('%4.2f%12.7f%12.7f',t(i),x1(i),x2(i))
    fprintf('%12.7f%12.7f\n',x3(i),sol(t(i)))
end
```

Example 5.1. Solve

$$dy/dt = 2yt \text{ where } y = 2 \text{ when } t = 0.$$

Exact solution: $y = 2\exp(t^2)$. Running the script `e4s503` with `example = 1` produces the following output:

```
Solution of dy/dt = 2yt
t         abm         Hamming      Classical      Exact
0.00   2.0000000   2.0000000   2.0000000   2.0000000
0.25   2.1289876   2.1289876   2.1289876   2.1289889
0.50   2.5680329   2.5680329   2.5680329   2.5680508
0.75   3.5099767   3.5099767   3.5099767   3.5101093
```

```
1.00    5.4340314    5.4294215    5.4357436    5.4365637
1.25    9.5206761    9.5152921    9.5369365    9.5414664
1.50   18.8575896   18.8690552   18.9519740   18.9754717
1.75   42.1631012   42.2832017   42.6424234   42.7618855
2.00  106.2068597  106.9045567  108.5814979  109.1963001
```

Example 5.2. Solve

$$(1+t^2)dy/dt + 2ty = \cos t \text{ where } y = 0 \text{ when } t = 0$$

Exact solution: $y = (\sin t)/(1+t^2)$. Running script `e4s503` with `example = 2` produces the following output:

```
Solution of (1+t^2)dy/dt = cos(t)-2yt
t          abm         Hamming     Classical     Exact
0.00    0.0000000    0.0000000    0.0000000    0.0000000
0.25    0.2328491    0.2328491    0.2328491    0.2328508
0.50    0.3835216    0.3835216    0.3835216    0.3835404
0.75    0.4362151    0.4362151    0.4362151    0.4362488
1.00    0.4181300    0.4196303    0.4206992    0.4207355
1.25    0.3671577    0.3705252    0.3703035    0.3703355
1.50    0.3044513    0.3078591    0.3068955    0.3069215
1.75    0.2404465    0.2432427    0.2421911    0.2422119
2.00    0.1805739    0.1827267    0.1818429    0.1818595
```

Example 5.3. Solve

$$dy/dt = 3y/t \text{ where } y = 1 \text{ when } t = 1$$

Exact solution: $y = t^3$. Running script `e4s503` with `example = 3` produces the following output:

```
Solution of dy/dt = 3y/t
t          abm         Hamming     Classical     Exact
1.00    1.0000000    1.0000000    1.0000000    1.0000000
1.25    1.9518519    1.9518519    1.9518519    1.9531250
1.50    3.3719182    3.3719182    3.3719182    3.3750000
1.75    5.3538346    5.3538346    5.3538346    5.3593750
2.00    7.9916917    7.9919728    7.9912355    8.0000000
```

Examples 5.2 and 5.3 appear to show that there is little difference between the three methods considered and they are all fairly successful for the step size $h = 0.25$ in this range. Example 5.1 is a relatively difficult problem in which the classical Runge–Kutta method performs well.

For a further comparison, we now use the MATLAB function `ode113`. This employs a predictor–corrector method based on the PECE approach described in Section 5.6 for the Adams–Bashforth–Moulton method. However the method implemented in the `ode113` is of variable order. The standard call of the function takes the form

```
[t,y] = ode113(f,tspan,y0,options);
```

where `f` is the name of the function providing the right-hand sides of the system of differential equations; `tspan` is the range of solution for the differential equation, given as a vector `[to tfinal]`; `y0` is the vector of initial values for the differential equation at time $t = 0$; `options` is an optional parameter providing additional settings for the differential equation such as accuracy. To illustrate the use of this function we consider the example

$$dy/dt = 2yt \text{ with initial condition } y = 2 \text{ when } t = 0$$

The call to solve this differential equation is

```
>> options = odeset('RelTol', 1e-5,'AbsTol',1e-6);
>> [t,yy] = ode113(@(t,x) 2*t*x,[0,2],[2],options); y = yy', time = t'
```

The result of executing these statements is

```
y =
  Columns 1 through 7
    2.0000    2.0000    2.0000    2.0002    2.0006    2.0026    2.0103
  Columns 8 through 14
    2.0232    2.0414    2.0650    2.0943    2.1707    2.2731    2.4048
  Columns 15 through 21
    2.5703    2.7755    3.0279    3.3373    3.7161    4.1805    4.7513
  Columns 22 through 28
    5.4557    6.3290    7.4177    8.7831   10.5069   15.5048   22.7912
  Columns 29 through 32
   34.6321   54.3997   88.3328  109.1944

time =
  Columns 1 through 7
         0    0.0022    0.0045    0.0089    0.0179    0.0358    0.0716
  Columns 8 through 14
    0.1073    0.1431    0.1789    0.2147    0.2862    0.3578    0.4293
  Columns 15 through 21
    0.5009    0.5724    0.6440    0.7155    0.7871    0.8587    0.9302
  Columns 22 through 28
    1.0018    1.0733    1.1449    1.2164    1.2880    1.4311    1.5599
  Columns 29 through 32
    1.6887    1.8175    1.9463    2.0000
```

Although a direct comparison between each step is not possible, because `ode113` uses a variable step size, we can compare the result for $t = 2$ with the results given for Example 5.1 above. This shows that the final y value given by `ode113` is better than those given by the other methods.

5.10 SYSTEMS OF SIMULTANEOUS DIFFERENTIAL EQUATIONS

The numerical techniques we have described for solving a single first-order differential equation can be applied, after simple modification, to solve systems of first-order differential equations. Systems of differential equations arise naturally from mathematical models of the physical world. In this section we consider systems of differential equations. The first is the Zeeman model and the second is the predator–prey problem.

5.10.1 ZEEMAN MODEL

Zeeman introduced a simplified model of the heart and incorporates ideas from catastrophe theory. The model is described briefly here but more detail is given in the excellent text of Beltrami (1987). The resulting system of differential equations will be solved using the MATLAB function `ode23` and the graphical facilities of MATLAB will help to clarify the interpretation of the results.

The starting point for this model of the heart is Van der Pol's equation which may be written in the form

$$dx/dt = u - \mu(x^3/3 - x)$$
$$du/dt = -x$$

This is a system of two simultaneous equations. The choice of this differential equation reflects our wish to imitate the beat of the heart. The fluctuation in the length of the heart fiber, as the heart contracts and dilates subject to an electrical stimulus, thus pumping blood through the system, may be represented by this pair of differential equations. The fluctuation has certain subtleties which our model should allow for. Starting from the relaxed state, the contraction begins with the application of the stimulus slowly at first and then becomes faster, so giving a sufficient final impetus to the blood. When the stimulus is removed, the heart dilates slowly at first and then more rapidly until the relaxed state is again reached and the cycle can begin again.

To follow this behavior, the Van der Pol equation requires some modification so that the x variable represents the length of heart fiber and the u variable can be replaced by one which represents the stimulus applied to the heart. This is achieved by making the substitution $s = -u/\mu$, where s represents the stimulus and μ is a constant. Since ds/dt is equal to $(-du/dt)/\mu$, it follows that $du/dt = -\mu ds/dt$. Hence we obtain

$$dx/dt = \mu(-s - x^3/3 + x)$$
$$ds/dt = x/\mu$$

If these simultaneous differential equations are solved for s and x for a range of time values, we find that s and x oscillate in a manner representing the fluctuations in the heart fiber length and stimulus. However, Zeeman proposed the introduction into this model of a tension factor p, where $p > 0$, in an attempt to account for the effects of increased blood pressure in terms of increased tension on the heart fiber. The model he suggested has the form

$$dx/dt = \mu(-s - x^3/3 + px)$$
$$du/dt = x/\mu$$

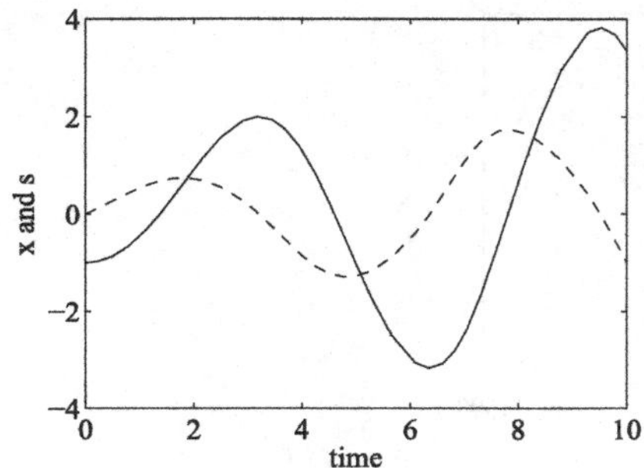

FIGURE 5.10

Solution of Zeeman's model with $p = 1$ and accuracy 0.005. The solid line represents s and the dashed line represents x.

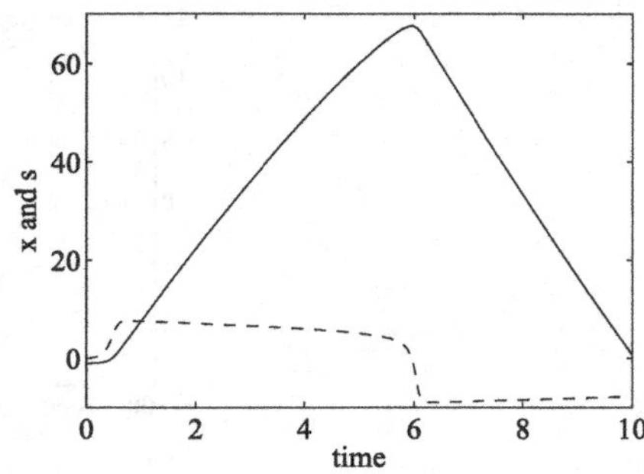

FIGURE 5.11

Solution of Zeeman's model with $p = 20$ and accuracy 0.005. The *solid line* represents s and the *dashed line* represents x.

Although the motivation for such a modification is plausible, the effects of these changes are by no means obvious.

This problem provides an interesting opportunity to apply MATLAB to simulate the heart beat in an experimental environment that allows us to monitor its changes under the effects of differing tension values. The following script solves the differential equations and draws various graphs.

```
% e4s504.m Solving Zeeman's Catastrophe model of the heart
clear all
p = input('enter tension value ');
simtime = input('enter runtime ');
acc = input('enter accuracy value ');
xprime = @(t,x) [0.5*(-x(2)-x(1)^3/3+p*x(1)); 2*x(1)];
options = odeset('RelTol',acc);
initx = [0 -1]';
[t x] = ode23(xprime,[0 simtime],initx,options);
% Plot results against time
plot(t,x(:,1),'--',t,x(:,2),'-')
xlabel('Time'), ylabel('x and s')
```

In the preceding function definition, $\mu = 0.5$. Fig. 5.10 shows graphs of the fiber length x and the stimulus s against time for a relatively small tension factor set at 1. The graphs show that a steady periodic oscillation of fiber length for this tension value is achieved for small stimulus values. However, Fig. 5.11 plots x and s against time with the tension set at 20. This shows that the behavior of the oscillation is clearly more labored and that much larger values of stimulus are required to produce the fluctuations in fiber length for the much higher tension value. Thus, the graphs show the deterioration in the beat with increasing tension. The results parallel the expected physical effects and also give some degree of experimental support to the validity of this simple model.

A further interesting study can be made. The interrelation of the three parameters x, s, and p can be represented by a three-dimensional surface called the cusp catastrophe surface. This surface can be

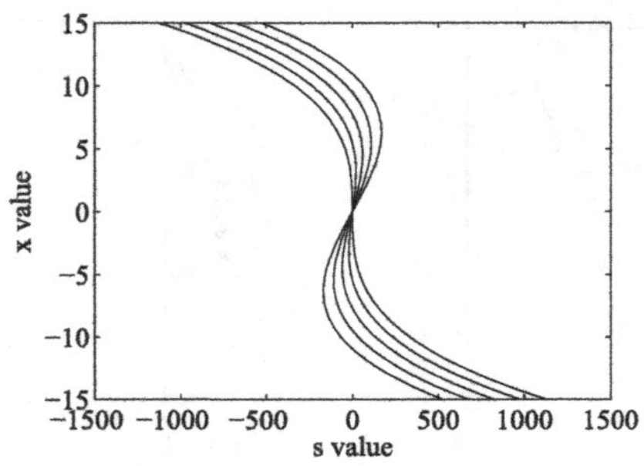

FIGURE 5.12

Sections of the cusp catastrophe curve in Zeeman's model for $p = 0:10:40$.

shown to have the form

$$-s - x^3/3 + px = 0$$

See Beltrami (1987) for a more detailed explanation. Fig. 5.12 shows a series of sections of the cusp catastrophe curve for $p = 0:10:40$. The curve has a pleat which becomes increasingly pronounced in the direction of increasing p. High tension or high p value consequently corresponds to movement on the sharply pleated part of this surface and consequently provides smaller changes in the heart fiber length relative to the stimulus.

5.10.2 THE PREDATOR–PREY PROBLEM

A system of differential equations which models the interaction of competing or predator–prey populations is based on the Volterra equations and may be written in the form

$$\begin{aligned} dP/dt &= K_1 P - CPQ \\ dQ/dt &= -K_2 Q + DPQ \end{aligned} \tag{5.28}$$

together with the initial conditions

$$Q = Q_0 \text{ and } P = P_0 \text{ at time } t = 0$$

The variables P and Q give the size of the prey and predator populations, respectively, at time t. These two populations interact and compete. K_1, K_2, C, and D are positive constants. K_1 relates to the rate of growth of the prey population P, and K_2 relates to the rate of decay of the predator population Q. It seems reasonable to assume that the number of encounters of predator and prey is proportional to P multiplied by Q and that a proportion C of these encounters will be fatal to members of the prey population. Thus the term CPQ gives a measure of the decrease in the prey population and the unrestricted growth in this population, which could occur assuming ample food, must be modified by the subtraction of this term. Similarly the decrease in the population of the predator must be modified by the addition of the term DPQ since the predator population gains food from its encounters with its prey and therefore more of the predators survive.

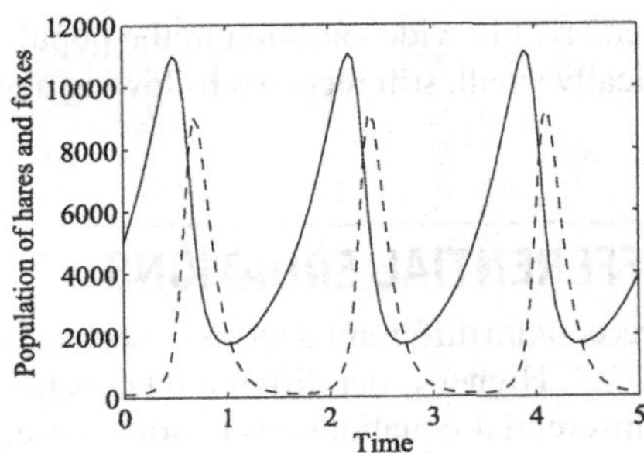

FIGURE 5.13

Variation in the population of lynxes (*dashed line*) and hares (*solid line*) against time, beginning with 5000 hares and 100 lynxes. Accuracy 0.005.

The solution of the differential equation depends on the specific values of the constants and will often result in nature in a stable cyclic variation of the populations. This is because as the predators continue to eat the prey, the prey population will fall and become insufficient to support the predator population which itself then falls. However, as the predator population falls, more of the prey survive and consequently the prey population will then increase. This in turn leads to an increase in the predator population since it has more food and the cycle begins again. This cycle maintains the predator and prey populations between certain upper and lower limits. The Volterra differential equations can be solved directly but this solution does not provide a simple relation between the size of the predator and prey populations; therefore, numerical methods of solution should be applied. An interesting description of this problem is given by Simmons (1972).

We now use MATLAB to study the behavior of a system of equations of the form (5.28) applied to the interaction of the lynx and its prey, the hare. The choice of the constants K_1, K_2, C, and D is not a simple matter if we wish to obtain a stable situation where the populations of the predator and prey never die out completely but oscillate between upper and lower limits. The MATLAB script below uses $K_1 = 2$, $K_2 = 10$, $C = 0.001$, and $D = 0.002$, and considers the interaction of a population of lynxes and hares where it is assumed that this interaction is the crucial feature in determining the size of the two populations. With an initial population of 5000 hares and 100 lynxes, the script `e4s505.m` uses these values to produce the graph in Fig. 5.13.

```
% e4s505.m
% x(1) and x(2) are hare and lynx populations.
simtime = input('enter runtime ');
acc = input('enter accuracy value ');
fv = @(t,x) [2*x(1)-0.001*x(1)*x(2); -10*x(2)+0.002*x(1)*x(2)];
initx = [5000 100]';
options = odeset('RelTol',acc);
[t x] = ode23(fv,[0 simtime],initx,options);
plot(t,x(:,1),'k',t,x(:,2),'k--')
xlabel('Time'), ylabel('Population of hares and lynxes')
```

For these parameters, there is a remarkably wide variation in the populations of hares and lynxes. The lynx population, although periodically small, still recovers following a recovery of the hare population.

5.11 HIGHER-ORDER DIFFERENTIAL EQUATIONS

Higher-order differential equations contain differential terms of second order or above, for example, the third-order differential term d^3y/dx^3. Higher-order differential equations can be solved by converting them to a system of first-order differential equations, or in some case, solved by direct methods. In the following subsections, we describe the process of transforming from a higher-order equation to set of first-order differential equations, and also describe one direct method for solving second-order differential equations.

5.11.1 CONVERSION TO A SET OF SIMULTANEOUS FIRST-ORDER DIFFERENTIAL EQUATIONS

The process of converting a higher-order differential equation to a set of first-order simultaneous equations is best described by an example. To illustrate this, consider the third-order differential equation

$$d^3u/dx^3 - cd^2u/dx^2 + (df/du)(du/dx) - (b/c)u = 0 \tag{5.29}$$

where $f = u(1-u)(u-a)$, $0 < a < 1$, and $b > 0$.

This equation describes a simplified model of the behavior of pulses in a squid axon (Iqbal, 1999). We require the derivative of f. Thus

$$df/du = (1-u)(u-a) + u(-1)(u-1) + u(1-u) = -3u^2 + 2(a+1)u - a$$

If we substitute $v = du/dx$, $w = dv/dx = d^2u/dx^2$ then (5.29) becomes

$$\begin{aligned} du/dx &= v \\ dv/dx &= w \\ dw/dx &= cw + 3u^2 - 2(a+1) + av + (b/c)u \end{aligned} \tag{5.30}$$

Thus, the third-order differential equation has been replaced by a system of three first-order differential equations.

If we have an nth-order differential equation of the form

$$a_n d^n y/dt^n + a_{n-1} d^{n-1} y/dt^{n-1} + \ldots + a_0 y = f(t, y) \tag{5.31}$$

by making the substitutions

$$P_0 = y \text{ and } dP_{i-1}/dt = P_i \text{ for } i = 1, 2, \ldots, n-1 \tag{5.32}$$

(5.31) becomes

$$a_n dP_{n-1}/dt = f(t, y) - a_{n-1}P_{n-1} - a_{n-2}P_{n-2} - \ldots - a_0 P_0 \tag{5.33}$$

Now (5.32) and (5.33) together constitute a system of n first-order differential equations. Initial values will be given for (5.31) in terms of the various order derivatives P_i for $i = 1, 2, \ldots, n-1$ at some initial value t_0 and these can easily be translated into initial conditions for the system of Eqs. (5.32) and (5.33). In general, the solutions of the original nth order differential equation and the system of first-order differential equations, (5.32) and (5.33), are the same. In particular, the numerical solution will provide the values of y for a specified range of t. An excellent discussion of the equivalence of the solutions of the two problems is given in Simmons (1972). We can see from this description that any order differential equation of the form (5.31) with given initial values can be reduced to solving a system of first-order differential equations. This argument is easily extended to the more general nth-order differential equation by making exactly the same substitutions as above in

$$d^n y/dt^n = f(t, y, y', \ldots y^{(n-1)})$$

where $y^{(n-1)}$ denotes the $(n-1)$th-order derivative of y.

5.11.2 NEWMARK'S METHOD

Newmark's method, (Newmark, 1959), allows the direct solution of a second-order differential equation or a system of second-order differential equations without the need for the transformation to a pair of simultaneous first-order differential equations. The method may be applied in various fields of engineering, in particular to dynamic response systems. We begin the derivation of Newmark's method by considering a system of p second-order differential equations:

$$\mathbf{M}\frac{d^2\mathbf{u}}{dt^2} + \mathbf{C}\frac{d\mathbf{u}}{dt} + \mathbf{K}\mathbf{u} = \mathbf{f}(t)$$

or using dashed notation:

$$\mathbf{M}\mathbf{u}'' + \mathbf{C}\mathbf{u}' + \mathbf{K}\mathbf{u} = \mathbf{f}(t) \tag{5.34}$$

together with the initial conditions that $\mathbf{u} = \mathbf{u}(0)$ and $\mathbf{u}' = \mathbf{u}'(0)$ at time $t = 0$. Here, if $\mathbf{M}$, $\mathbf{C}$, and $\mathbf{K}$ are arrays of constant coefficients and the function $\mathbf{f}(t)$ is a function of time only, then the equations are linear differential equations, otherwise they are non-linear differential equations. We will deal with linear differential equations case only. The initial development of Newmark's method uses the mean value theorem for u, one element of $\mathbf{u}$, and which may be stated as follows. There exists a scalar value β such that:

$$u(t + \Delta t) - u(t) = u'(t + \beta \Delta t)\Delta t$$

where the dash denotes the derivative of u and Δt is the time increment.

Newmark used the standard Taylor's series expansion of the functions $u(t + \Delta t)$ and $u'(t + \Delta t)$ as far as the third-order derivative of u. Denoting $u(t + \Delta t)$ by u_{n+1} and $u(t)$ by u_n, these are given by:

$$u_{n+1} = u_n + u'_n \Delta t + u''_n \frac{(\Delta t)^2}{2!} + u'''_n \frac{(\Delta t)^3}{3!} + \ldots \tag{5.35}$$

$$u'_{n+1} = u'_n + u''_n \Delta t + u'''_n \frac{(\Delta t)^2}{2!} \ldots \tag{5.36}$$

Applying the generalized mean value theorem we may write these equations as:

$$u_{n+1} = u_n + u'_n \Delta t + u''_n \frac{(\Delta t)^2}{2!} + \beta u'''_n (\Delta t)^3 + \ldots \tag{5.37}$$

$$u'_{n+1} = u'_n + u''_n \Delta t + \gamma u'''_n (\Delta t)^2 + \ldots \tag{5.38}$$

where there exit $0 \le \gamma \le 1$ and $0 \le \beta \le 1$ such that these equations are true.

The third derivative u''', if linear, may be written:

$$u'''_{n+1} = (u''_{n+1} - u''_n)/\Delta t$$

So far we have considered the third derivative of a single variables u. However, if we are considering a system of p differential equations, then u must be generalized to a column vector of p variable, $\mathbf{u}$. Thus the previous equation becomes

$$\mathbf{u}'''_{n+1} = (\mathbf{u}''_{n+1} - \mathbf{u}''_n)/\Delta t$$

Substituting for each element of $\mathbf{u}'''$ in (5.38) in turn, and collecting the values of $\mathbf{u}'_{n+1}$ into a column vector, gives:

$$\mathbf{u}'_{n+1} = \mathbf{u}'_n + \Delta t(1-\gamma)\mathbf{u}''_n + \Delta t \gamma \mathbf{u}''_{n+1} \tag{5.39}$$

and $0 \le \gamma \le 1$.

Now, substituting for each element of $\mathbf{u}'''_{n+1}$ in (5.37) and collecting the values of $\mathbf{u}_{n+1}$ into a column vector gives

$$\mathbf{u}_{n+1} = \mathbf{u}_n + \Delta t \mathbf{u}'_n + (1-2\beta)\frac{(\Delta t)^2}{2}\mathbf{u}''_n + \beta(\Delta t)^2 \mathbf{u}''_{n+1} \tag{5.40}$$

As above, we place a constraint on the values of β, $0 \le 2\beta \le 1$ and so $0 \le \beta \le 1/2$

We now have formulae for velocities, (5.39) and displacements (5.40). We can obtain a formula for accelerations from the original differential equation, (5.34), thus:

$$\mathbf{u}''_{n+1} = \mathbf{M}^{-1}(\mathbf{f}(t_{n+1}) - \mathbf{C}\mathbf{u}'_{n+1} - \mathbf{K}\mathbf{u}_{n+1}) \tag{5.41}$$

Thus the equations give formulae for displacement, velocity and acceleration. However since these are implicit they must be reformulated so that we obtain expressions for the updated values in terms of the previously known values at step n. This can be achieved by algebraic manipulation. Rearranging Eq. (5.40) to provide an expression for $\mathbf{u}''_{n+1}$ gives:

$$\beta(\Delta t)^2 \mathbf{u}''_{n+1} = \mathbf{u}_{n+1} - \mathbf{u}_n - \Delta t \mathbf{u}'_n - (1-2\beta)((\Delta t)^2/2)\mathbf{u}''_n$$

Now dividing this equation by $\beta(\Delta t)^2$ gives:

$$\mathbf{u}''_{n+1} = (\mathbf{u}_{n+1} - \mathbf{u}_n)/(\beta(\Delta t)^2) - \mathbf{u}'_n/(\beta\Delta t) - (1/(2\beta) - 1)\mathbf{u}''_n \tag{5.42}$$

Inserting the expression for $\mathbf{u}''_{n+1}$ from (5.42) into (5.39) gives:

$$\mathbf{u}'_{n+1} = \gamma(\mathbf{u}_{n+1} - \mathbf{u}_n)/(\beta\Delta t) + \mathbf{u}'_n(1-\gamma/\beta) + \Delta t \mathbf{u}''_n(1-\gamma/(2\beta)) \tag{5.43}$$

We can now insert the explicit expressions for $\mathbf{u}''_{n+1}$ and $\mathbf{u}'_{n+1}$ from (5.42) and (5.43) into the original second-order differential equation (5.34) to obtain:

$$\begin{aligned}
&\mathbf{M}\left[(\mathbf{u}_{n+1}-\mathbf{u}_n)/(\beta(\Delta t)^2)-\mathbf{u}'_n/(\beta\Delta t)-(1/(2\beta)-1)\mathbf{u}''_n\right]\dots\\
&+\mathbf{C}\left[(\mathbf{u}_{n+1}-\mathbf{u}_n)\gamma/(\beta\Delta t)+\mathbf{u}'_n(1-\gamma/\beta)+\Delta t\mathbf{u}''_n(1-\gamma/(2\beta))\right]\dots\\
&+\mathbf{K}\mathbf{u}_{n+1}=\mathbf{f}(t_{n+1})
\end{aligned}\tag{5.44}$$

Now rearranging (5.44) as an equation for u_{n+1} we obtain the equation:

$$\mathbf{A}\mathbf{u}_{n+1}=\mathbf{B}_n\tag{5.45}$$

and thus $\mathbf{u}_{n+1}=\mathbf{A}^{-1}\mathbf{B}_n$.

This is an *explicit* expression for $\mathbf{u}_{n+1}$ in terms of values at an earlier cycle. Where $\mathbf{A}$ and $\mathbf{B}_n$ are defined as follows:

$$\mathbf{A}=\frac{\mathbf{M}}{\beta(\Delta t)^2}+\frac{\gamma\mathbf{C}}{\beta\Delta t}+\mathbf{K}\tag{5.46}$$

$$\begin{aligned}
\mathbf{B}_n=\mathbf{f}(t_{n+1})&+\mathbf{M}\left[\frac{\mathbf{u}_n}{\beta(\Delta t)^2}+\frac{\mathbf{u}'_n}{\beta\Delta t}+\left(\frac{1}{2\beta}-1\right)\mathbf{u}''_n\right]\dots\\
&+\mathbf{C}\left[\frac{\gamma\mathbf{u}_n}{\beta\Delta t}-\mathbf{u}'_n\left(1-\frac{\gamma}{\beta}\right)-\Delta t\mathbf{u}''_n\left(1-\frac{\gamma}{2\beta}\right)\right]
\end{aligned}\tag{5.47}$$

Note that once β and Δt are given specific numerical values, $\mathbf{A}$ is an array of constant values for a particular set of differential equations. This, of course, is not the case for $\mathbf{B}_n$ which changes with each iteration. Having determined $\mathbf{u}_{n+1}$ we use Eqs. (5.42) and (5.43) to determine $\mathbf{u}''_{n+1}$ and $\mathbf{u}'_{n+1}$.

Knowing $\mathbf{u}''_{n+1}$ we can determine $\mathbf{u}'_{n+1}$ the iterative process can then be continued. We now proceed in this way for the number of iterations required.

It remains to choose values for γ and β. An undamped system (that is $\mathbf{C}=0$) is conditionally stable if $\gamma\geq 1/2$, and unconditionally stable if $\gamma\geq 1/2$ and $\beta\geq\frac{1}{4}\left(\gamma+\frac{1}{2}\right)^2$. However, $\gamma>1/2$ introduces artificial damping into the solution, and so Newmark chose $\gamma=1/2$ and $\beta=1/4$. This is sometimes called the average acceleration method and it is unconditionally stable, meaning that the method will converge for all time increments. If β is chosen to be $1/6$ the method is called the linear acceleration method, but it is only conditionally stable.

It is important to consider the stability of Newmark's method, i.e. the conditions under which the algorithm will converge to the correct solution. In certain circumstances Newmark's method is unconditionally stable, this means that the method will converge for all time increments.

This is the case if the following condition is satisfied:

$$2\beta\geq\gamma\geq 1/2\tag{5.48}$$

For example $\beta=1/2$ and $\gamma=1/2$ satisfy this condition. But it is important to note that certain values of γ do introduce significant errors in the computation process. The method is called conditionally convergent when convergence is guaranteed only for a limited range of the step size. This condition is given by:

$$\Delta t \leq \frac{1}{\omega_{max}\sqrt{\frac{\gamma}{2}-\beta}} \tag{5.49}$$

where $\beta \leq 1/2$ and $\gamma \geq 1/2$.

The value ω_{max} can be calculated from the characteristic equation or the eigenvalue problem of the differential equations.

The following MATLAB function, `newmark`, implements this procedure.

```
function [tp,x] = newmark(M,C,K,F,tspan,n,x0,xd0)
% M, C, K are matrices multiplying xddot, xdot and x repectively
% F is column vector of exciations. x0, xd0 are initial x0 and xd vectors
% tspan = [t_initial t_final]; n is tspan/t_increment
dt = (tspan(2)-tspan(1))/n;
tp(1) = tspan(1);
x(:,1) = x0';
xd(:,1) = xd0';
gamma = 1/2;    beta = 1/4;
A = (1/(beta*dt^2))*M+(gamma/(beta*dt))*C+K; invA = inv(A);
xdd(:,1) = inv(M)*(F(:,1)-C*xd(:,1)-K*x(:,1));
for i = 1:n
    B = (F(:,i+1)+...
        M*((1/(beta*dt^2))*x(:,i)+(1/(beta*dt))*xd(:,i)+ ...
(1/(2*beta)-1)*xdd(:,i))+C*((gamma/(beta*dt))*x(:,i)+...
(gamma/beta-1)*xd(:,i)+(gamma/beta-2)*(dt/2)*xdd(:,i)));
    x(:,i+1) = invA*B;
    xdd(:,i+1) = (1/(beta*dt^2))*(x(:,i+1)-x(:,i))...
-(1/(beta*dt))*xd(:,i)-((1/(2*beta))-1)*xdd(:,i);
    xd(:,i+1) = xd(:,i)+(1-gamma)*dt*xdd(:,i)+gamma*dt*xdd(:,i+1);
    tp(i+1) = tp(i)+dt;
end
x = x';
```

We now illustrate the application of Newmark's method to solving the system shown in Fig. 2.8. The equations of motion for this system are given in (2.35) and these equations can be written in matrix form as

$$\mathbf{M}\ddot{\mathbf{x}} + \mathbf{K}\mathbf{x} = \mathbf{0}$$

For the purposes of this illustration, we will add an external force to the system given by $\mathbf{f}(t)$ and a damping matrix, $\mathbf{C}$, so that the above equation becomes

$$\mathbf{M}\ddot{\mathbf{x}} + \mathbf{C}\dot{\mathbf{x}} + \mathbf{K}\mathbf{x} = \mathbf{f}(t)$$

M and **K** are given by (2.38) and repeated here for convenience.

$$\mathbf{M} = \begin{bmatrix} 10 & 0 & 0 \\ 0 & 20 & 0 \\ 0 & 0 & 30 \end{bmatrix} \text{ kg and } \mathbf{K} = \begin{bmatrix} 45 & -20 & -15 \\ -20 & 45 & -25 \\ -15 & -25 & 40 \end{bmatrix} \text{ kN/m}$$

x lists the displacements of the three coordinates. We will make $\mathbf{C} = 3 \times 10^{-2}\mathbf{K}$ and apply a half sine pulse of 0.29 s duration with a peak value of 50 N at coordinate x_3. The MATLAB script `e4s506.m` calls both `newmark` and `ode45` to solve this problem.

```
%e4s506.m
clear all
t_init = 0;   t_final = 3;
t_incr = 0.005;
n = (t_final-t_init)/t_incr;
tt = t_init:t_incr:t_final;
trange = [t_init t_final];
M = [10 0 0;0 20 0;0 0 30];
K = 1e3*[45 -20 -15;-20 45 -25;-15 -25 40];
C = 3e-2*K;
F(1,:) = 0*tt;
F(2,:) = 0*tt;
omega = pi/.29;
F(3,:) = 50*sin(omega*tt);
for j = 1:n
    pulse(j) = sin(omega*tt(j));
    if tt(j) > pi/omega
        pulse(j) = 0;
    end
end
F(3,1:n) = 50*pulse;
x0 = [0 0 0];
xd0= [0 0 0];
[tp,x1] = newmark(M,C,K,F,trange,n,x0,xd0);
x1 = 1000*x1;
figure(1), plot(tp,x1(:,1),'k',tp,x1(:,2),'k',tp,x1(:,3),'k',tp,0*tp,'k')
xlabel('Time, s')
ylabel('x, mm')
axis([0 3 -5 10])
hold on
figure(1), plot(tt(1:60),5*pulse(1:60),'.k')
hold off
[t,x2] = ode45('f10',[t_init:t_incr:t_final],[0 0 0 0 0 0]);
x2 = 1000*x2;
d(:,1) = (x1(:,1)-x2(:,1));
```

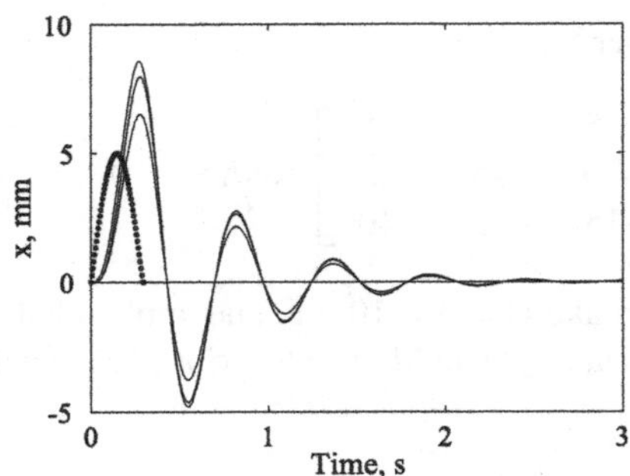

FIGURE 5.14

Graph showing the three coordinate responses of a mass-spring-damper system, shown by full lines, when excited by a half sine pulse, shown by a dotted line.

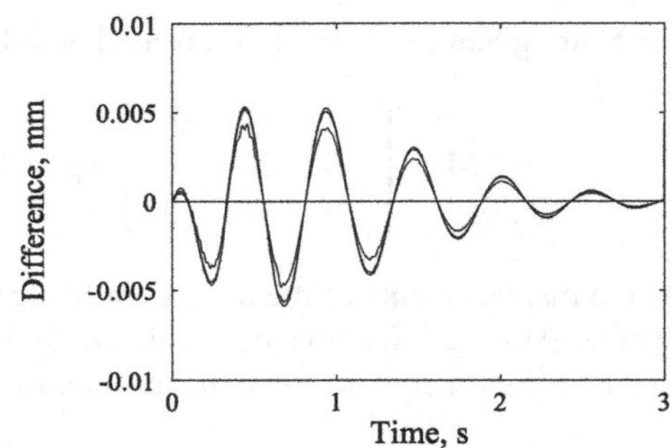

FIGURE 5.15

Plot showing the difference between the Newmark and 4th-order Runge–Kutta method solutions for the three coordinates.

```
d(:,2) = (x1(:,2)-x2(:,2));
d(:,3) = (x1(:,3)-x2(:,3));
figure(2), plot(tp,d(:,1),'k',tp,d(:,2),'k',tp,d(:,3),'k',tp,0*tp,'k')
xlabel('Time, s')
ylabel('Difference, mm')
axis([0 3 -1e-2 1e-2])
```

The function `f10`, required by function `ode45`, is as follows:

```
function yprime = f10(t,y)
m = [10 0 0;0 20 0;0 0 30];
k = 1e3*[45 -20 -15;-20 45 -25;-15 -25 40];
c = 3e-2*k;   f =[0; 0; 50];
omega = pi/.29;
pulse = sin(omega*t);
if t > pi/omega
    pulse = 0;
end
A = [zeros(3,3) eye(3,3); -m\k -m\c];
b = [zeros(3,1); m\f*pulse];
yprime = A*y+b;
```

Running script `e4s506.m` generates Figs. 5.14 and 5.15. Fig. 5.14 shows the response of the three coordinates and the half sine pulse. The pulse is scaled a factor of 10 in order to display on the graph. Fig. 5.15 shows the difference between the output of `newmark` and `ode45`. It is seen that the maximum difference is 0.006 mm.

5.12 CHAOTIC SYSTEMS

Chaos arises in several disciplines, including meteorology, physics, environmental science, engineering, economics, biology and ecology. Because of this it is perhaps surprising that chaotic motion was not properly recognized until the 1960s.

Chaotic systems are difficult to define. Essentially they are dynamic systems that are highly sensitive to small differences in their initial conditions and also to rounding errors in numerical computation. These differences produce widely diverging outcomes for such systems. In 1972 Edward Lorenz presented a talk to the American Association for the Advancement of Science entitled "Predictability: Does the Flap of a Butterfly's Wings in Brazil Set a Tornado in Texas?". The theme of this presentation was that small changes in the state of the atmosphere can result in large differences in its later states. The term *butterfly effect* has entered popular culture as a symbol of chaos.

Chaos occurs even though such systems are deterministic, meaning that their future behavior is fully determined by their initial conditions, with no random elements involved. Given *exactly* the same initial conditions, the same result is obtained. Thus, despite its "random" appearance, chaos is a deterministic outcome.

A simple example of a chaotic system quoted by Lorenz (1993) is a pin ball machine. At the start of the game, the ball is given an initial velocity. The ball then rolls through an array of pins. Striking a pin changes the ball's direction and the ball subsequently strikes other pins in the array. The slightest change in the initial velocity of the ball will cause it to strike different pins and take a totally different route through the array of pins.

5.12.1 THE LORENZ EQUATIONS

As an example of a system of three simultaneous equations, we consider the Lorenz system. This system has a number of important applications including weather forecasting. The system has the form

$$dx/dt = s\,(y - x)$$

$$dy/dt = rx - y - xz$$

$$dz/dt = xy - bz$$

subject to appropriate initial conditions. As the parameters s, r, and b are varied through various ranges of values, the solutions of this system of differential equations vary in form. In particular, for certain values of the parameters the system exhibits chaotic behavior. To provide more accuracy in the computation process we use the MATLAB function `ode45` rather than `ode23`. The MATLAB script `e4s507.m` for solving this problem is given below.

```
% e4s507.m  Solution of the Lorenz equations
r = input('enter a value for the constant r ');
simtime = input('enter runtime ');
acc = input('enter accuracy value ');
xprime = @(t,x) [10*(x(2)-x(1)); r*x(1)-x(2)-x(1)*x(3); ...
            x(1)*x(2)-8*x(3)/3];
```

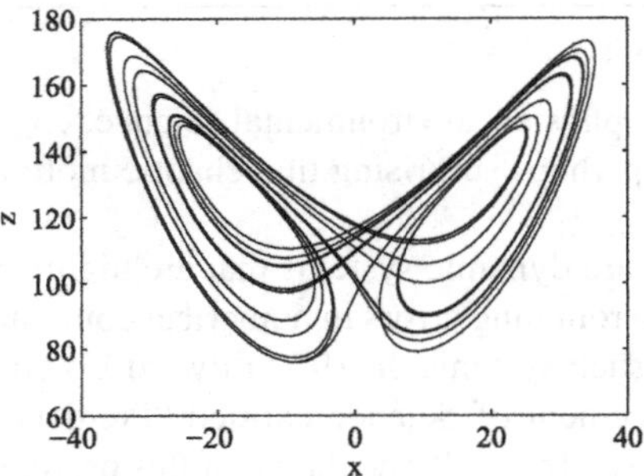

FIGURE 5.16

Solution of Lorenz equations for $r = 126.52$, $s = 10$, and $b = 8/3$ using an accuracy of 0.000005 and terminating at $t = 8$.

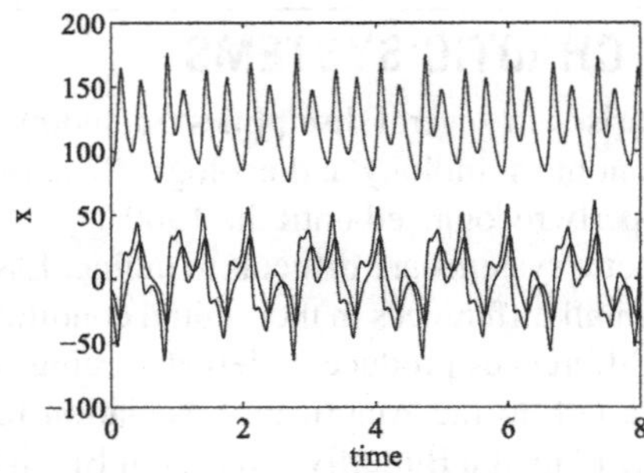

FIGURE 5.17

Solution of Lorenz equations where each variable is plotted against time. Conditions are the same as those used to generate Fig. 5.16. Note the unpredictable nature of the solutions.

```
initx = [-7.69 -15.61 90.39]';
tspan = [0 simtime];
options = odeset('RelTol',acc);
[t x] = ode45(xprime,tspan,initx,options);
% Plot results against time
figure(1), plot(t,x,'k')
xlabel('Time'), ylabel('x')
figure(2), plot(x(:,1),x(:,3),'k')
xlabel('x'), ylabel('z')
```

The results of running script `e4s507.m` are given in Figs. 5.16 and 5.17. Fig. 5.16 is characteristic of the Lorenz equations and shows the complexity of the relationship between x and z. Fig. 5.17 shows how x, y, and z change with time.

For $r = 126.52$ and for other large values of r the behavior of this system is chaotic. In fact for $r > 24.7$ most orbits exhibit chaotic wandering. The trajectory passes around two points of attraction, called *strange attractors*, switching from one to another in an apparently unpredictable fashion. This appearance of apparently random behavior is remarkable considering the clearly deterministic nature of the problem. However, for other values of r the behavior of the trajectories is simple and stable.

We now consider the case of the Lorenz equations with $s = 10$, $b = 8/3$, and $r = 28$. Fig. 5.18 shows the solution of the Lorenz equations with initial conditions of $\mathbf{x} = [5\ 5\ 5]$ and $\mathbf{x} = [5.0091\ 4.9997\ 5.0060]$. Fig. 5.19 shows the solution of Lorenz equations with the initial conditions $\mathbf{x} = [5\ 5\ 5]$ but with two different accuracies; the default accuracy of the MATLAB function `ode45`; namely an absolute tolerance of 10^{-6} and a relative tolerance of 10^{-3}, and an increased accuracy with an absolute tolerance of 10^{-10} and a relative tolerance of 10^{-8}.

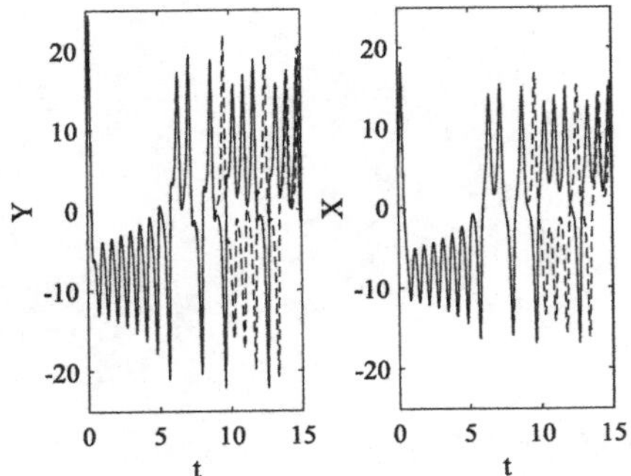

FIGURE 5.18

Solution of Lorenz equations for $r = 28$, $s = 10$, and $b = 8/3$. Initial conditions $\mathbf{x} = [5\ 5\ 5]$ shown by the full line, and $\mathbf{x} = [5.0091\ 4.9997\ 5.0060]$ shown by the dashed line. Note the sudden divergence of the two solutions from each other and unpredictable nature of the solutions.

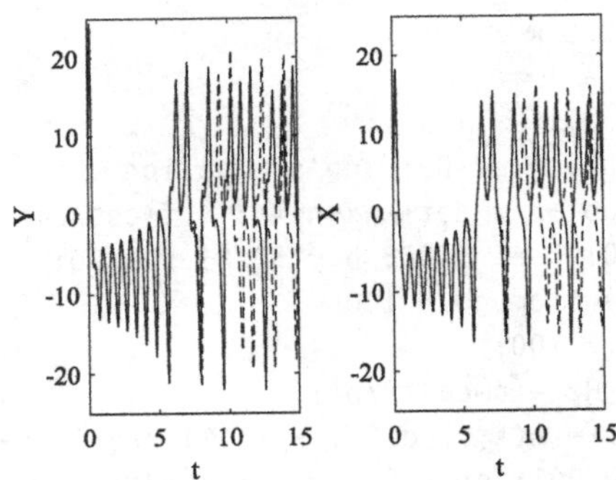

FIGURE 5.19

Solution of Lorenz equations for $r = 28$, $s = 10$, and $b = 8/3$. The full line shows the solution using the default accuracy of the MATLAB Runge–Kutta 4/5 function. The dashed line shows a higher accuracy solution. Note the sudden divergence of the two solutions from each other and unpredictable nature of the solutions.

5.12.2 DUFFING'S EQUATION

The Duffing oscillator is described by Duffing's equation: a second-order differential equation of the form

$$m\ddot{x} + c\dot{x} + kx + hx^3 = f(t)$$

This equation might represent, for example, an electric circuit or a mechanical oscillator. In a mechanical oscillator, m is the system mass, c is the viscous damping coefficient, k and h are stiffness coefficients and $f(t)$ is the force applied to the system. This force is a function of time, t. Duffing's equation is, of course, a non-linear differential equation since it contains a term in x^3. The equation can be solved approximately by seeking a series solution and considering only a small number of terms in the series. These approximate solutions can be very revealing and provide information about the free vibration of the oscillator by setting $f(t)$ to zero, or the system response to a non-zero force $f(t)$. Here we seek a numerical solution using the MATLAB function `ode45`. In order to use `ode45` we must transform the second-order differential equation into a pair of first-order differential equations thus

$$m\dot{u} = f(t) - cu - kx - hx^3 \text{ and } \dot{x} = u$$

We consider two cases. In each case we assume $m = 1$ kg, $c = 5$ Ns/m, $k = -5000$ N/m, $h = 50 \times 10^6$ N/m^3, and $f = 50\sin(\omega t)$ N. In each case only the frequency of the applied force and the initial conditions are changed. The following script `e4s508.m` solves Duffing's equation and generates Figs. 5.20–5.25.

```
% e4s508.m
clear all
close all
% Solution of Duffing's equation
% x_ddot+c*x_dot+k*x+h*x^3 = f*cos(omega*t)
c = 10; k = -5000; h = 50e6; f = 30;
% Case 1 omega = 100
omega = 100;
omega_Hz = omega/2/pi;
qprime = @(t,q)[q(2); -c*q(2)-k*q(1)-h*q(1)^3+f*cos(omega*t)];
dt = 0.001; nt = 8192; T = nt*dt;
t = 0:dt:(nt-1)*dt;
[t q] = ode45(qprime,[0:dt:(nt-1)*dt],[0 0]);
figure(1), plot(t,1000*q(:,1),'k',t,5*cos(omega*t),'k--')
axis([2 3 -25 25])
xlabel('Time, s')
ylabel('Displacement, mm')
grid
% Case 2 omega = 120
omega = 120;
omega_Hz = omega/2/pi;
qprime = @(t,q)[q(2); -c*q(2)-k*q(1)-h*q(1)^3+f*cos(omega*t)];
[t q11] = ode45(qprime,[0:0.001:1],[0 0]);
[t q12] = ode45(qprime,[0:0.001:1],[0.001 1]);
figure(2), plot(t,1000*q11(:,1),'k',t,1000*q12(:,1),'k--')
grid
xlabel('Time, s')
ylabel('Displacement, mm')
[t4 q4] = ode45(qprime,[0:0.001:2],[0.001000 0]);
[t5 q5] = ode45(qprime,[0:0.001:2],[0.001001 0]);
[t6 q6] = ode45(qprime,[0:0.001:2],[0.001002 0]);
figure(3), plot(t4,1000*q4(:,1),'k',t5,1000*q5(:,1),'k--',t6,1000*q6(:,1),'k-.')
grid
axis([0 2 -20 20])
xlabel('Time, s')
ylabel('Displacement, mm')
[t q] = ode45(qprime,[15:0.001:20],[0 0]);
x = 1000*q(:,1); v = q(:,2); theta = omega*t;
figure(4); plot(x,v,'k')
grid
xlabel('Displacement, mm')
ylabel('Velocity, m/s')
[t1 q1] = ode45(qprime,[0:2*pi/omega:400],[0 0]);
figure(5), plot(1000*q1(300:end,1),q1(300:end,2),'k.')
```

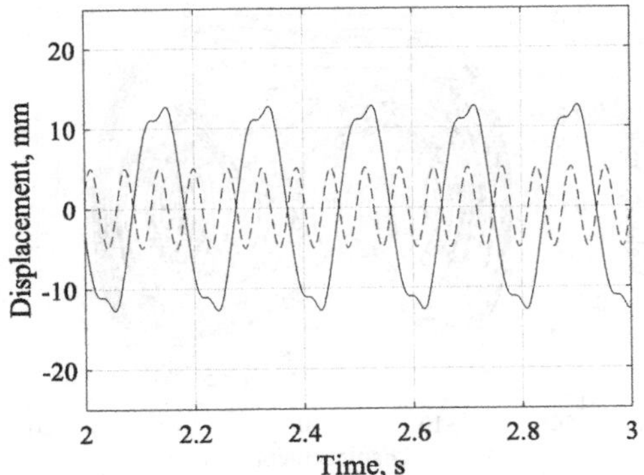

FIGURE 5.20

Case 1: The full line is the output from Duffing oscillator. $\omega = 100$ rad/s (15.92 Hz). Zero initial conditions. The dashed line is the input force, arbitrarily scaled in amplitude.

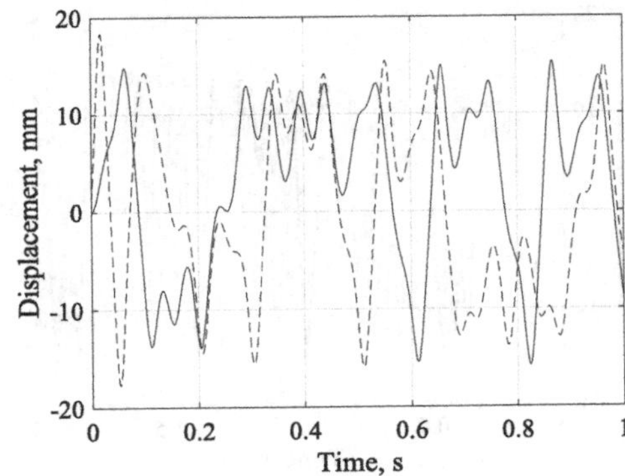

FIGURE 5.21

Output from Duffing oscillator. $\omega = 120$ rad/s. Full line gives output with zero initial conditions. Dashed line give output with an initial displacement of 1 mm and an initial velocity of 1 m/s.

```
axis([-20 20 -1.1 1.1])
xlabel('Displacement, mm')
ylabel('Velocity, m/s')
[t1 q1] = ode45(qprime,[0:2*pi/omega:400],[0 0]);
[t2 q2] = ode45(qprime,[0:2*pi/omega:400],[0.001 1]);
figure(6)
plot(1000*q2(775:25:2000,1),q2(775:25:2000,2),'ko',1000*q1(775:25:2000,1),...
q1(775:25:2000,2),'k+')
axis([-20 20 -1.1 1.1])
xlabel('Displacement, mm')
ylabel('Velocity, m/s')
```

Case 1. Input frequency $\omega = 100$ rad/s (15.92 Hz). With the values of the two initial conditions equal to zero, the oscillator output is shown in Fig. 5.20. Also shown in the figure is the input force, its amplitude arbitrarily scaled to fit the graph. It can be seen that the output frequency is about one third of the input frequency. This can be confirmed by using the discrete Fourier transform (see Chapter 8). The Fourier transform shows the major frequency component in the output is $\omega/3$ and there are also smaller frequency components at ω and $5\omega/3$. However, introducing an initial velocity of 1 m/s causes the $\omega/3$ and $5\omega/3$ frequency components in the output to vanish so that the ω frequency component is dominant. There is then also a small 2ω frequency component present.

Case 2. Input frequency $\omega = 120$ rad/s (19.10 Hz). At this excitation frequency the character of the response has changed completely, as shown in Fig. 5.21. The output appears to be unpredictable. Furthermore, the output with initial conditions of velocity and displacement equal to zero, is strikingly different from the output with an initial displacement of 1 mm and an initial velocity of 1 m/s except, perhaps, in the overall amplitude. Even more remarkable are the three plots shown in Fig. 5.22. In each case the initial velocity is zero but the three initial displacements are 1 mm, 1.001 mm and

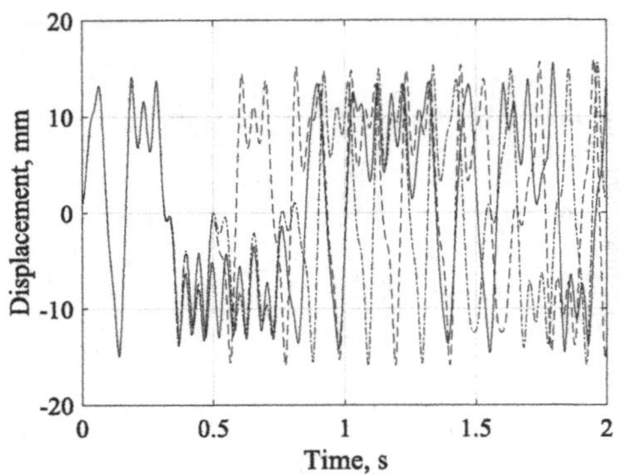

FIGURE 5.22

Output from Duffing oscillator. $\omega = 120$ rad/s. Solution with zero initial velocity and initial displacements of 1, 1.001, and 1.002 mm. (Shown by full, dashed and dot-dashed lines respectively.)

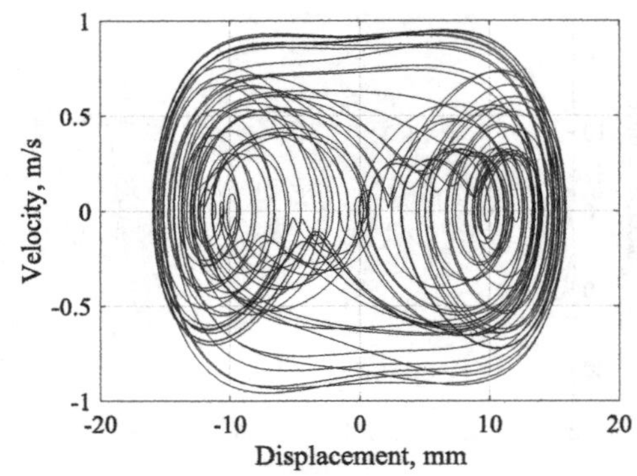

FIGURE 5.23

Output from Duffing oscillator. Phase plane plot. $\omega = 120$ rad/s.

1.002 mm. The outputs differ from each other even though the difference in the initial displacements are only 1 and 2 μm. Clearly, this is an example of chaotic motion. If we plot the output displacement against the output velocity for a short period, Fig. 5.22 is obtained. It is difficult to draw any useful conclusions from this plot, and so we plot the so-called Poincaré map or section to clarify the situation. The output displacement and velocity are sampled once per cycle of the excitation or input signal frequency, for a large number of cycles and the resulting large number of sample points is plotted in the displacement–velocity plane. Thus if $T = 2\pi/\omega$ then the samples are taken at multiples of T, that is $T_s = 0,\ T,\ 2T,\ \ldots,\ nT$. The Poincaré map for this system is shown in Fig. 5.23. We see that there are areas where points cluster and other areas devoid of values. If we change the initial displacement from zero to 1 mm and the initial velocity from zero to 1 m/s the Poincaré map appears unchanged. How can this be since the output is so influenced by the initial condition, as shown in Fig. 5.21? The explanation is that if we plot a small number of the output points at the same instances of time but with the two different sets of initial conditions, the points on the Poincaré map are in different locations but, none the less, they are confined in the same regions. This is shown in Fig. 5.24. A large number of points from either output will create the same Poincaré map but the points generated from the two different sets of initial conditions will not be coincident with each other in the Poincaré map. The Poincaré map helps to clarify the relationship between displacement and velocity.

5.13 DIFFERENTIAL EQUATIONS APPLIED TO NEURAL NETWORKS

Different types of neural networks have been used to solve a wide range of problems. Neural networks often consist of several layers of "neurons" that are "trained" by fixing a set of weights. These weights are found by minimizing the sum of squares of the difference between actual and required outputs. Once trained, the networks can be used to classify a range of inputs. However, here we consider

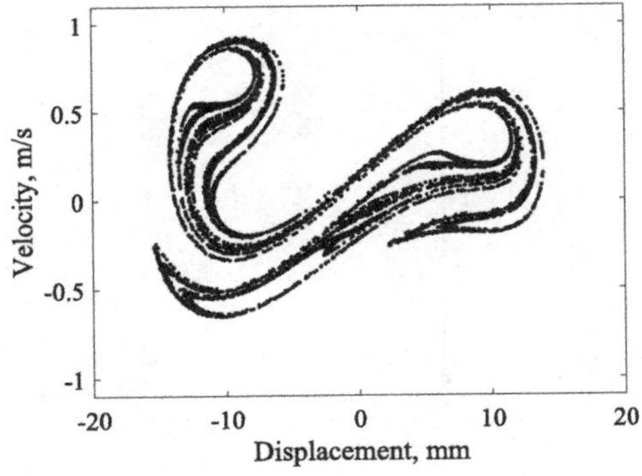

FIGURE 5.24

Poincaré map showing output from Duffing oscillator. $\omega = 120$ rad/s.

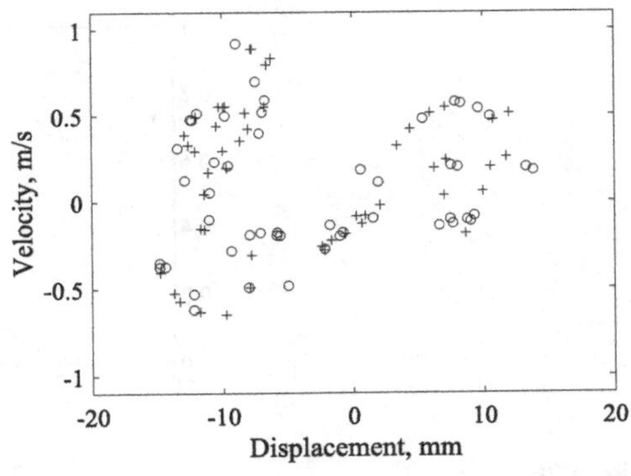

FIGURE 5.25

Output from Duffing oscillator showing where points from two solutions lie on a Poincaré map. + and o indicate points generated from two different initial conditions with $\omega = 120$ rad/s.

a different approach that uses a neural network that may be based directly on considering a system of differential equations. This approach is described by Hopfield and Tank (1985, 1986), who demonstrated the application of neural networks to solving specific numerical problems. It is not our intention to provide the full details or proofs of this process here.

Hopfield and Tank, in their 1985 and 1986 papers, utilized a system of differential equations which take the form

$$\frac{du_i}{dt} = \frac{-u_i}{\tau} + \sum_{j=0}^{n-1} T_{ij} V_j + I_i \text{ for } i = 0, 1, \dots n-1 \tag{5.50}$$

where τ a constant usually taken as 1. This system of differential equations represents the interaction of a system of n neurons, and each differential equation is a simple model of a single biological neuron. (This is only one of a number of possible models of a neural network.) Clearly, to establish a network of such neurons, they must be able to interact with each other and this interaction must be represented in the differential equations. The T_{ij} provide the strengths of the interconnections between the ith and jth neuron and the I_i provide the externally applied current to the ith neuron. These I_i may be viewed as inputs to the system. The V_j values provide the outputs from the system and are directly related to the u_j so that we may write $V_j = g(u_j)$. The function g, called a sigmoidal function, may be specified, for example, by

$$V_j = (1 + \tanh u_j)/2 \text{ for all } j = 0, 1, \dots, n-1$$

A plot of this function is given in Fig. 5.26.

Having provided such a model of a neural network, the question still remains: how can we show that it can be used to solve specific problems? This is the key issue and a significant problem in itself. Before we can solve a given problem using a neural network we must first reformulate our problem so that it can be solved by this approach.

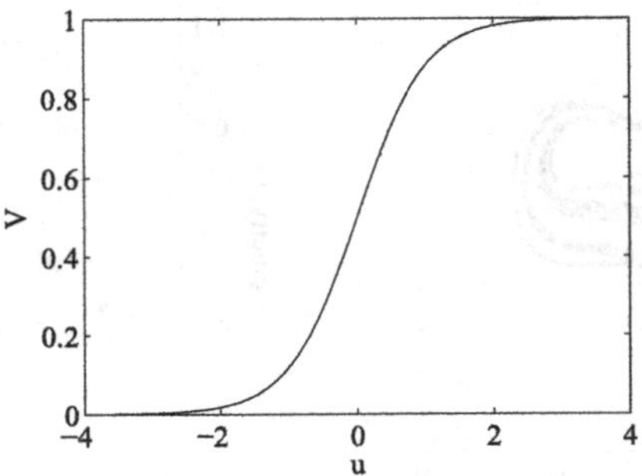

FIGURE 5.26

Plot of sigmoid function $V = (1 + \tanh u)/2$.

To illustrate this process, Hopfield and Tank chose as an example the simple problem of binary conversion. That is, to find the binary equivalent of a given decimal number. Since there is no obvious and direct relationship between this problem and the system of differential equations (5.50) which model the neural network, a more direct link has to be established.

Hopfield and Tank have shown that the stable state solution of (5.50), in terms of the V_j, is given by the minima of the energy function:

$$E = -\frac{1}{2}\sum_{i=0}^{n-1}\sum_{j=0}^{n-1} T_{ij} V_i V_j - \sum_{j=0}^{n-1} I_j V_j \tag{5.51}$$

It is an easy matter to link the solution of the binary conversion problem to the minimization of the function (5.51).

Hopfield and Tank consider the energy function

$$E = \frac{1}{2}\left\{x - \sum_{j=0}^{n-1} V_j 2^j\right\}^2 + \sum_{j=0}^{n-1} 2^{2j-1} V_j \left(1 - V_j\right) \tag{5.52}$$

Now the minimum of (5.52) will be attained when $x = \Sigma V_j 2^j$ and $V_j = 0$ or 1. Clearly the first term ensures that the required binary representation is achieved while the second term provides that the V_j take either 0 or 1 values when the value of E is minimized. On expanding this energy function (5.52) and comparing it with the general energy function (5.51) we find that if we make

$$T_{ij} = -2^{i+j} \text{ for } i \neq j \text{ and } T_{ij} = 0 \text{ when } i = j$$

$$I_j = -2^{2j-1} + 2^j x$$

then the two energy functions are equivalent, apart from a constant. Thus the minimum of one gives the minimum of the other. Solving the binary conversion problem expressed in this way is thus equivalent to solving the system of differential equations (5.50) with this special choice of values for T_{ij} and I_i.

In fact, using an appropriate choice of T_{ij} and I_i, a range of problems can be represented by a neural network in the form of the system of differential equations (5.50). Hopfield and Tank have extended this process from the simple example considered above to attempting to solve the very challenging traveling salesman problem. The details of this are given in Hopfield and Tank (1985, 1986).

In MATLAB we may use `ode23` or `ode45` to solve this problem. The crucial part of this exercise is to define the function which gives the right-hand sides of the differential equation system for the neural network. This can be done very simply using the function `hopbin` below. This function gives the right-hand side for the differential equations which solve the binary conversion problem. In the definition of function `hopbin`, `sc` is the decimal value we wish to convert.

```
function neurf = hopbin(t,x)
global n sc
% Calculate synaptic current
I = 2.^[0:n-1]*sc-0.5*2.^(2.*[0:n-1]);
% Perform sigmoid transformation
V = (tanh(x/0.02)+1)/2;
% Compute interconnection values
p = 2.^[0:n-1].*V';
% Calculate change for each neuron
neurf = -x-2.^[0:n-1]'*sum(p)+I'+2.^(2.*[0:n-1])'.*V;
```

This function `hopbin` is called by the script `e4s509.m` to solve the system of differential equations which define the neural network and hence simulate its operation.

```
% e4s509.m
% Hopfield and Tank neuron model for binary conversion problem
global n sc
n = input('enter number of neurons ');
sc = input('enter number to be converted to binary form ');
simtime = 0.2; acc = 0.005;
initx = zeros(1,n)';
options = odeset('RelTol',acc);
%Call ode45 to solve equation
[t x] = ode45('hopbin',[0 simtime],initx,options);
V = (tanh(x/0.02)+1)/2;
bin = V(end,n:-1:1);
for i = 1:n
    fprintf('%8.4f', bin(i))
end
fprintf('\n\n')
plot(t,V,'k')
xlabel('Time'), ylabel('Binary values')
```

Running this script to convert the decimal number 5 gives

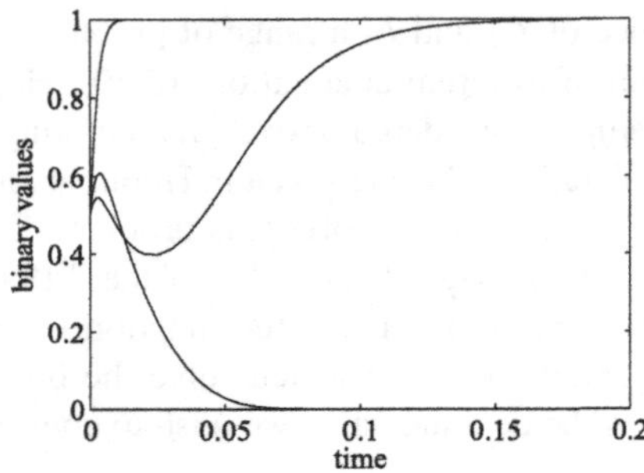

FIGURE 5.27

Neural network finds the binary equivalent of 5 using 3 neurons and an accuracy of 0.005. The three curves show the convergence to the binary digits 1, 0, and 1.

```
enter number of neurons 3
enter number to be converted to binary form 5
  1.0000   0.0000   0.9993
```

together with Fig. 5.27. This plot shows how the neural network model converges to the required results, that is, `V(1) = 1`, `V(2) = 0` and `V(3) = 1` or binary number 101. As a further example, we convert the decimal number 59 using 7 neurons as follows:

```
enter number of neurons 7
enter number to be converted to binary form 59
  0.0000   1.0000   1.0000   1.0000   0.0000   1.0000   0.9999
```

Again, a correct result.

This is an application of neural networks to a trivial problem. A real test for neural computing is the traveling salesman problem. The MATLAB neural network toolbox provides a range of functions to solve neural network problems.

5.14 STIFF EQUATIONS

When the solution of a system of differential equations contains components which change at significantly different rates for given changes in the independent variable, the equation system is said to be "stiff". When this phenomenon is present, a particularly careful choice of the step size must be made if stability is to be achieved.

We will now consider how the stiffness phenomenon arises in an apparently simple system of differential equations. Consider the system given below.

$$\begin{aligned} dy_1/dt &= -by_1 - cy_2 \\ dy_2/dt &= y_1 \end{aligned} \tag{5.53}$$

This system may be written in matrix form as

$$d\mathbf{y}/dt = \mathbf{A}\mathbf{y} \tag{5.54}$$

The solution of (5.54) is

$$\begin{aligned} y_1 &= A\exp(r_1 t) + B\exp(r_2 t) \\ y_2 &= C\exp(r_1 t) + D\exp(r_2 t) \end{aligned} \tag{5.55}$$

where A, B, C, and D are constants set by the initial conditions. It can easily be verified that r_1 and r_2 are the eigenvalues of the matrix $\mathbf{A}$.

If a numerical procedure is applied to solve these systems of differential equations, the success of the method will depend crucially on the eigenvalues of the matrix $\mathbf{A}$ and in particular the ratio of the smallest and largest eigenvalues.

As an example of a matrix with widely spaced eigenvalues we can take the 8×8 Rosser matrix; this is available in MATLAB as `rosser`. The sequence of statements

```
>> a = rosser; lambda = eig(a);
>> eigratio = max(abs(lambda))/min(abs(lambda))

eigratio =
   3.2285e+15
```

produces a matrix with eigenvalue ratios of order 10^{16}. Thus a system of ordinary first-order differential equations involving this matrix would be pathologically difficult to solve. The significance of the eigenvalue ratio in relation to the required step size can be generalized to systems of many equations. Consider the system of n equations

$$d\mathbf{y}/dt = \mathbf{A}\mathbf{y} + \mathbf{P}(t) \tag{5.56}$$

where $\mathbf{y}$ is an n component column vector, $\mathbf{P}(t)$ is an n component column vector of functions of t and $\mathbf{A}$ is an $n \times n$ matrix of constants. It can be shown that the solution of this system takes the form

$$\mathbf{y}(t) = \sum_{i=1}^{n} v_i \, \mathbf{d}_i \exp(r_i t) + \mathbf{s}(t) \tag{5.57}$$

Here r_1, r_2, ... are the eigenvalues and $\mathbf{d}_1$, $\mathbf{d}_2$, ... the eigenvectors of $\mathbf{A}$. The vector function $\mathbf{s}(t)$ is the particular integral of the system, sometimes called the steady-state solution since for negative eigenvalues the exponential terms should die away with increasing t. If it is assumed that the $r_k < 0$ for $k = 1, 2, 3, \ldots$ and we require the steady-state solution of system (5.56), then any numerical method applied to solve this problem may face significant difficulties, as we have seen. We must continue the integration until the exponential components have been reduced to negligible levels and yet we must take sufficiently small steps to ensure stability, thus requiring many steps over a large interval. This is the most significant effect of stiffness.

The definition of stiffness can be extended to any system of the form (5.56). The stiffness ratio is defined as the ratio of the largest and smallest eigenvalues of $\mathbf{A}$ and gives a measure of the stiffness of the system.

The methods used to solve stiff problems must be based on stable techniques. The MATLAB function `ode23s` uses continuous step size adjustment and therefore is able to deal with such problems, although the solution process may be slow. If we use a predictor–corrector method, not only must this method be stable but the corrector must also be iterated to convergence. An interesting discussion of this topic is given by Ralston and Rabinowitz (1978). Specialized methods have been developed for solving stiff problems, and Gear (1971) has provided a number of techniques which have been reported to be successful. Note that although explicit methods are commonly used for stiff equations, implicit methods can be a good choice for this type of problem.

5.15 SPECIAL TECHNIQUES

A further set of predictor–corrector equations may be generated by making use of an interpolation formula due to Hermite. An unusual feature of these equations is that they contain second-order derivatives. It is usually the case that the calculation of second-order derivatives is not particularly difficult and consequently this feature does not add a significant amount of work to the solution of the problem. However, it should be noted that in using a computer program for this technique the user has to supply not only the function on the right-hand side of the differential equation but its derivative as well. To the general user this may be unacceptable.

The equations for Hermite's method take the form

$$\begin{aligned} y_{n+1}^{(1)} &= y_n + h(y_n' - 3y_{n-1}')/2 + h^2(17y_n'' + 7y_{n-1}'')/12 \\ y_{n+1}^{*\,(1)} &= y_{n+1}^{(1)} + 31(y_n - y_n^{(1)})/30 \\ y_{n+1}^{\prime\,(1)} &= f(t_{n+1}, y_{n+1}^{*\,(1)}) \end{aligned} \tag{5.58}$$

For $k = 1, 2, 3, \ldots$

$$y_{n+1}^{(k+1)} = y_n + h(y_{n+1}^{\prime\,(k)} + y_n')/2 + h^2(-y_{n+1}^{\prime\prime\,(k)} + y_n'')/12$$

This method is stable and has a smaller truncation error at each step than Hamming's method. Thus it may be worthwhile accepting the additional effort required by the user. We note that since we have

$$dy/dt = f(t, y)$$

then

$$d^2y/dt^2 = df/dt$$

and thus y_n'', etc., are easily calculated as the first derivative of f. The MATLAB function `fhermite` implements this method, and the script is given below. Note that in this function, the function `f` must provide both the first and second derivatives of y.

```
function [tvals, yvals] = fhermite(f,tspan,startval,step)
% Hermite's method for solving
% first order differential equation dy/dt = f(t,y).
```

```
% Example call: [tvals, yvals] = fhermite(f,tspan,startval,step)
% The initial and final values of t are given by tspan = [start finish].
% Initial value of y is given by startval, step size is given by step.
% The function f(t,y) and its derivative must be defined by the user.
% 3 steps of Runge-Kutta are required so that hermite can start.
% Set up matrices for Runge-Kutta methods
b = [ ]; c = [ ]; d = [ ];
order = 4;
b = [1/6 1/3 1/3 1/6]; d = [0 0.5 0.5 1];
c = [0 0 0 0;0.5 0 0 0;0 0.5 0 0;0 0 1 0];
steps = (tspan(2)-tspan(1))/step+1;
y = startval; t = tspan(1);
ys(1) = startval; w = feval(f,t,y); fval(1) = w(1); df(1) = w(2);
yvals = startval; tvals = tspan(1);
for j = 2:2
    k(1) = step*fval(1);
    for i = 2:order
        w = feval(f,t+step*d(i),y+c(i,1:i-1)*k(1:i-1)');
        k(i) = step*w(1);
    end
    y1 = y+b*k'; ys(j) = y1; t1 = t+step;
    w = feval(f,t1,y1); fval(j) = w(1); df(j) = w(2);
    %collect values together for output
    tvals = [tvals, t1]; yvals = [yvals, y1];
    t = t1; y = y1;
end
%hermite now applied
h2 = step*step/12; er = 1;
for i = 3:steps
    y1 = ys(2)+step*(3*fval(1)-fval(2))/2+h2*(17*df(2)+7*df(1));
    t1 = t+step; y1m = y1; y10 = y1;
    if i>3, y1m = y1+31*(ys(2)-y10)/30; end
    w = feval(f,t1,y1m); fval(3) = w(1); df(3)=w(2);
    yc = 0; er = 1;
    while abs(er)>0.0000001
        yp = ys(2)+step*(fval(2)+fval(3))/2+h2*(df(2)-df(3));
        w = feval(f,t1,yp); fval(3) = w(1); df(3) = w(2);
        er = yp-yc; yc = yp;
    end
    fval(1:2) = fval(2:3); df(1:2) = df(2:3);
    ys(2) = yp;
    tvals = [tvals, t1]; yvals = [yvals, yp];
    t = t1;
end
```

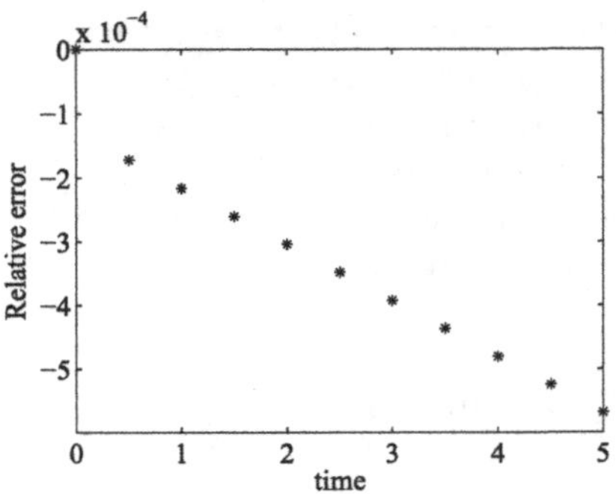

FIGURE 5.28

Relative error in the solution of $dy/dt = y$ using Hermite's method. Initial condition $y = 1$ when $t = 0$ and a step of 0.5.

Fig. 5.28 gives the error when solving the specific equation $dy/dt = y$ using the same step size and starting point as for Hamming's method – see Fig. 5.9. For this particular problem Hermite's method performs better than Hamming's method.

Finally, we compare the Hermite, Hamming, and Adams–Bashforth–Moulton methods for the difficult problem

$$dy/dt = -10y \text{ given } y = 1 \text{ when } t = 0$$

The script `e4s510.m` implements these comparisons.

```
% e4s510.m
vg = @(t,x) [-10*x 100*x];
v = @(t,x) -10*x;
disp('Solution of dx/dt = -10x')
t0 = 0; y0 = 1;
tf = 1; tinc = 0.1; steps = floor((tf-t0)/tinc+1);
[t,x1] = abm(v,[t0 tf],y0,tinc);
[t,x2] = fhamming(v,[t0 tf],y0,tinc);
[t,x3] = fhermite(vg,[t0 tf],y0,tinc);
disp('t          abm          Hamming      Hermite      Exact');
for i = 1:steps
    fprintf('%4.2f%12.7f%12.7f',t(i),x1(i),x2(i))
    fprintf('%12.7f%12.7f\n',x3(i),exp(-10*(t(i))))
end
```

Note that for the function `fhermite`, we must supply both the first and second derivatives of y with respect to t. For the first derivative, we have directly $dy/dt = -10y$ but the second derivative d^2y/dt^2 is given by $-10dy/dt = -10(-10y) = 100y$. Consequently, the function takes the form

```
vg = @(t,x) [-10*x 100*x];
```

The functions `abm` and `fhamming` require only the first derivative of y with respect to t and we define function thus:

```
v = @(t,x) -10*x;
```

Running the above script provides the following results, demonstrating the superiority of the Hermite method.

```
Solution of dx/dt = -10x
t          abm        Hamming     Hermite     Exact
0.00   1.0000000   1.0000000   1.0000000   1.0000000
0.10   0.3750000   0.3750000   0.3750000   0.3678794
0.20   0.1406250   0.1406250   0.1381579   0.1353353
0.30   0.0527344   0.0527344   0.0509003   0.0497871
0.40  -0.0032654   0.0109440   0.0187528   0.0183156
0.50  -0.0171851   0.0070876   0.0069089   0.0067379
0.60  -0.0010598   0.0131483   0.0025454   0.0024788
0.70   0.0023606   0.0002607   0.0009378   0.0009119
0.80  -0.0063684   0.0006066   0.0003455   0.0003355
0.90  -0.0042478   0.0096271   0.0001273   0.0001234
1.00   0.0030171  -0.0065859   0.0000469   0.0000454
```

One feature which may be used to improve many of the methods discussed above is step size adjustment. This means that we adjust the step size h according to the progress of the iteration. One criterion for adjusting h is to monitor the size of the truncation error. If the truncation error is smaller than the accuracy requirement, we can increase h; however, if the truncation error is too large, we can reduce h. Step size adjustment can lead to considerable additional work; for example, if a predictor–corrector method is used, new initial values must be calculated. The following method is an interesting alternative to this kind of procedure.

5.16 EXTRAPOLATION TECHNIQUES

The extrapolation method described in this section is based on a similar procedure to that used in Romberg integration, introduced in Chapter 4. The procedure begins by obtaining successive initial approximations for y_{n+1} using a modified mid-point method. The interval sizes used for obtaining these approximations are calculated from

$$h_i = h_{i-1}/2 \text{ for } i = 1,\ 2,\ \ldots \tag{5.59}$$

with the initial value h_0 given.

Once these initial approximations have been obtained, we can use (5.60), the extrapolation formula, to obtain improved approximations.

$$T_{m,\,k} = (4^m T_{m-1,\,k+1} - T_{m-1,\,k})/(4^m - 1) \quad \text{for } m = 1, 2, \ldots \text{ and } k = 1, 2, \ldots\ s - m \tag{5.60}$$

The calculations are set out in an array in much the same way as the calculations for Romberg's method for integration described in Chapter 4. When $m = 0$, the values of $T_{0,k}$ for $k = 0, 1, 2, ..., s$ are taken as the successive approximations to the values of y_{n+1} using the h_i values obtained from (5.59).

The formula for calculating the approximations used for the initial values $T_{0,k}$ in the above array are computed using the following equations.

$$\begin{aligned} y_1 &= y_0 + hy_0' \\ y_{n+1} &= y_{n-1} + 2hy_n' \text{ for } n = 1,\ 2\ \dots,\ N_k \end{aligned} \tag{5.61}$$

Here $k = 1, 2, ...$ and N_k is the number of steps taken in the range of interest, so that $N_k = 2^k$ as the size of the interval is halved each time. The independent variable increment between y_{n+1} and y_{n-1} is $2h$. Values of this increment may lead to significant variations in the magnitude of the error. Because of this, instead of using the final value of y_{n+1} given by (5.61), Gragg (1965) has suggested that at the final step these values be smoothed using the intermediate value y_n. This leads to the values for $T_{0,k}$

$$T_{0,\,k} = (y_{N-1}^k + 2y_N^k + y_{N+1}^k)/4$$

where the superscript k denotes the value at the kth division of the interval.

Alternatives to the method of Gragg are available for finding the initial values in the function `rombergx` and various combinations of predictor–correctors may be used. It should be noted, however, that if the corrector is iterated until convergence is achieved, this will improve the accuracy of the initial values but at considerable computational expense for smaller step sizes, i.e., for larger N values. The MATLAB function `rombergx` implements the extrapolation method and is given below.

```
function [v W] = rombergx(f,tspan,intdiv,inity)
% Solves dy/dt = f(t,y) using Romberg's method.
% Example call: [v W] = rombergx(f,tspan,intdiv,inity)
% The initial and final values of t are given by tspan = [start finish].
% Initial value of y is given by inity.
% The number of interval divisions is given by intdiv.
% The function f(t,y) must be defined by the user.
W = zeros(intdiv-1,intdiv-1);
for index = 1:intdiv
    y0 = inity; t0 = tspan(1);
    intervals = 2^index;
    step = (tspan(2)-tspan(1))/intervals;
    y1 = y0+step*feval(f,t0,y0);
    t = t0+step;
    for i = 1:intervals
        y2 = y0+2*step*feval(f,t,y1);
        t = t+step;
        ye2 = y2; ye1 = y1; ye0 = y0; y0 = y1; y1 = y2;
    end
    tableval(index) = (ye0+2*ye1+ye2)/4;
end
```

```
for i = 1:intdiv-1
    for j = 1:intdiv-i
        table(j) = (tableval(j+1)*4^i-tableval(j))/(4^i-1);
        tableval(j) = table(j);
    end
    tablep = table(1:intdiv-i);
    W(i,1:intdiv-i) = tablep;
end
v = tablep;
```

We can now call this function to solve $dx/dt = -10x$ with $x = 1$ at $t = 0$. The following MATLAB statement solves this differential equation when $t = 0.5$:

```
>> [fv P] = rombergx(@(t,x) -10*x,[0 0.5],7,1)

fv =

    0.0067

P =

   -2.5677    0.2277    0.1624    0.0245    0.0080    0.0068
    0.4141    0.1580    0.0153    0.0069    0.0067         0
    0.1539    0.0131    0.0068    0.0067         0         0
    0.0125    0.0068    0.0067         0         0         0
    0.0068    0.0067         0         0         0         0
    0.0067         0         0         0         0         0
```

The final value, 0.0067, is better than any of the results achieved for this problem by other methods presented in this chapter. It must be noted that only the final value is found; other values in a given interval can be obtained if intermediate ranges are considered.

This completes our discussion of those types of differential equations known as initial value problems. In Chapter 6 we consider a different type of differential equation known as a boundary value problem.

5.17 SIMULINK

Simulink is part of MATLAB and can only be used when MATLAB is installed. Simulink is used to model, simulate, and analyze dynamic systems. The systems can be linear or non-linear, modeled in continuous or sampled time, or a mixture of both. Simulink provides a graphical user interface that allows the user to drag and drop the appropriate blocks in order to create the required block diagram to model the system. The wide range of blocks in the block library comprise sources (inputs to the system), sinks (outputs from the system) and both linear and non-linear process blocks, and connectors. Simulink also allows the user to modify or create blocks of their own. The parameters for each block are set by the user to be appropriate for their specific problem.

It is reasonable to ask why Simulink was developed. MATLAB provides a range of functions to solve differential equation and hence to simulate and analyze dynamic systems. Where Simulink proves its value is in simulating more complicated dynamic systems, perhaps combining the dynamics of two or more systems. A further advantage of Simulink is that it is possible to visualize a system in terms of process block diagrams (as control engineers often do) rather than in terms of systems of differential equations. A further advantage is the ease with which non-linear phenomena such as dead zones, limits, hysteresis, and so on can be incorporated into the model.

In this brief introduction to Simulink we provide some examples of the way block diagrams can be developed to simulate some simple systems. We do not give details of how the process blocks are selected, dragged, and dropped into the model nor how connections are made between blocks, which, in any case, is quite intuitive. The purpose of this introduction is to show how simple problems can be represented by Simulink models. We do not provide the output from the models because in this context the output data is only of secondary interest.

We now provide a series of Simulink examples.

Example 5.4. Suppose we wish to simulate the dynamic system described by the following second-order differential equation:

$$m\ddot{x} + c\dot{x} + kx = f_0 \sin(\omega t) \tag{5.62}$$

We begin by isolating the highest derivative of the equation to obtain

$$\ddot{x} = (f_0/m)\sin(\omega t) - (c/m)\dot{x} - (k/m)x \tag{5.63}$$

To create the Simulink model we begin by dragging and dropping two integration process blocks, labeled internally $\frac{1}{s}$, see Fig. 5.29. The integration process blocks perform numerical integration. The input to the process block (Integrator 1) is labeled x ddot and its output is x dot. This output becomes the input to Integrator 2 and the output of this integrator is x. A connection is made from x dot to one of the gain process block, labeled Gain c/m, set to a gain of c/m. Gain means that the signal is increased or multiplied by this quantity to give an output (c/m)x dot. Similarly x is connected to the other gain process block, labeled Gain k/m, with a gain of k/m to give an output of (k/m)x. At the left of the Simulink diagram is a process block that generates a sine wave, labeled Sine Wave. The output from this block is a sine wave of amplitude f/m, and is connected to a summing unit. Also connected to the summing unit is the negatives of (c/m)x dot and (k/m)x. From (5.63), $(f_0/m)\sin(\omega t) - (c/m)\dot{x} + (k/m)x$ is equal to $\ddot{x}$, so the output of the summing block is x ddot which is connected to the input of integrator 1. Finally, x and x dot are connected to a Scope block, labeled Scope 1, which, as the name suggests, displays x and $\dot{x}$ against the simulation time.

Before we run the model, we must open the Sine Wave block and set the frequency and amplitude of the input; we must also open the integrators to specify any initial conditions required, in this case the initial velocity at Integrator 1 and initial the displacement at Integrator 2, and open the Gain blocks to set the required gains. Finally, we open the Model Simulation Parameters and set the start and end times of the simulation, the ODE solver to be used (ode45 in this case), relative and absolute accuracy, step size, and so on.

Rearranging (5.62) to isolate the $\ddot{x}$ terms is not the only arrangement we might consider for a Simulink model. We could arrange (5.62) to isolate the x term as thus:

$$x = (f_0/k)\sin(\omega t) - (m/k)\ddot{x} - (c/k)\dot{x}$$

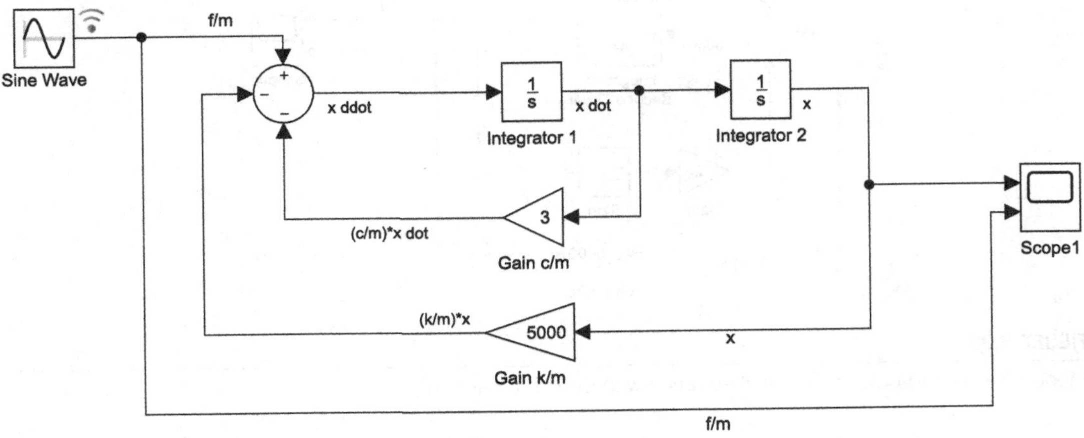

FIGURE 5.29

Model of a second-order differential equation, (5.62).

Simulating this equation would require two stages of differentiation. Although Simulink provides a differentiation process block, one should not simulate (5.62) in this way. Numerical differentiation is a noisy process that will lead to errors.

Example 5.5. In Example 5.4, we have used two integrator process blocks to determine the velocity and displacement from acceleration. Basically, we need as many stages of integration as the order of the ordinary differential equation. Because many ordinary differential equations arising from real world problems are second order, Simulink provides a second-order integrator, that is a block that combines two integrator blocks. For example, suppose we wish to simulate

$$\ddot{x} + F\text{sign}\dot{x} + kx = 0$$

Rearranging this equation in terms of $\ddot{x}$ gives

$$\ddot{x} = -F\text{sign}\dot{x} - kx$$

In this example, there is no external excitation, only the non-zero initial displacements and/or velocities cause the subsequent changes in x. Also, there is no $\dot{x}$ term. Energy dissipation is by a constant force F which opposes the motion and changes direction with the sign of $\dot{x}$. This form of energy dissipation is sometimes called Coulomb damping. The Simulink model for the system is shown in Simulink Fig. 5.30.

Note the ease with which we are able to model the non-linear Coulomb damping.

Example 5.6. In this example, we simulate the same system as simulated in Example 5.4. However, rather than a simulation based on the ODE that describes the system, we begin by converting (5.62) to an equivalent transfer function. To do this we take the Laplace Transform of (5.62). The Laplace transform is described in Section 10.12 of Chapter 10.

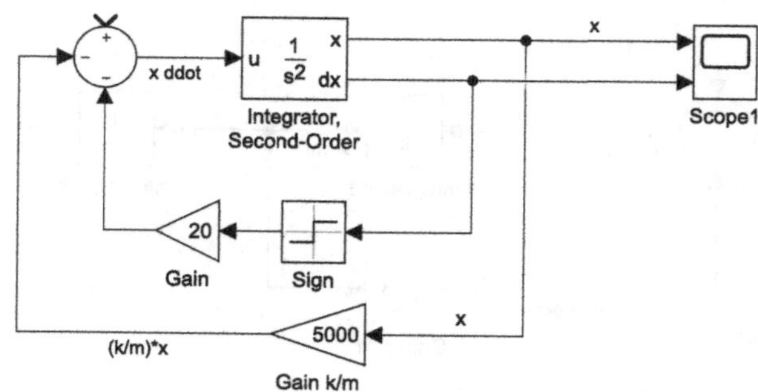

FIGURE 5.30

Model of a second-order differential equation with Coulomb damping.

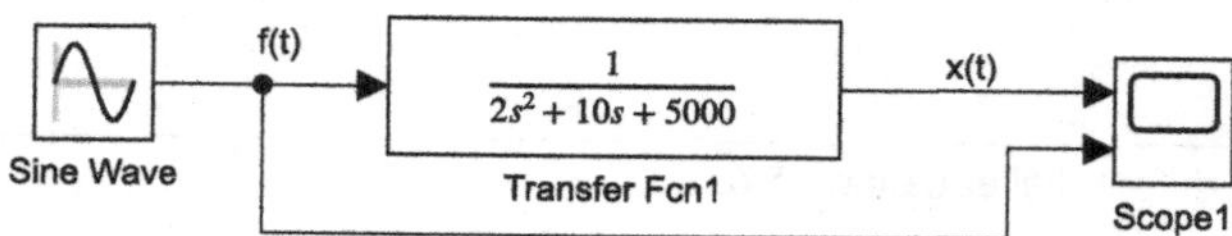

FIGURE 5.31

A second-order system modeled by a transfer function.

Denoting the Laplace transform of $x(t)$ by $X(s)$ then $\mathcal{L}(x(t)) = X(s)$ and $\mathcal{L}(\dot{x}(t)) = sX(s) + x(0)$ and $\mathcal{L}(\ddot{x}(t)) = s^2X(s) + sx(0) + v(0)$ and where $x(0)$ and $v(0)$ are the initial displacement and velocity, respectively. Applying this transform to (5.62) and assuming the initial conditions are zero gives

$$(ms^2 + cs + k)X(s) = F(s) \tag{5.64}$$

Rearranging, (5.64) becomes

$$\frac{X(s)}{F(s)} = \frac{1}{ms^2 + cs + k} \tag{5.65}$$

The expression on the right side of (5.65) is the transfer function for (5.62). Comparing Simulink Fig. 5.31 with Simulink Fig. 5.29, we see that the two integrators and two gain processes have been replaced by a Simulink transfer function block.

Example 5.7. We now give the Simulink model for Van der Pol's equation. Van der Pol's equation is

$$\ddot{x} - \mu(1 - x^2)\dot{x} + x = 0$$

The equation named after a Dutch electrical engineer who investigated the performance of an electrical circuit described by this equation. The equation has a non-linear damping term. To model it in

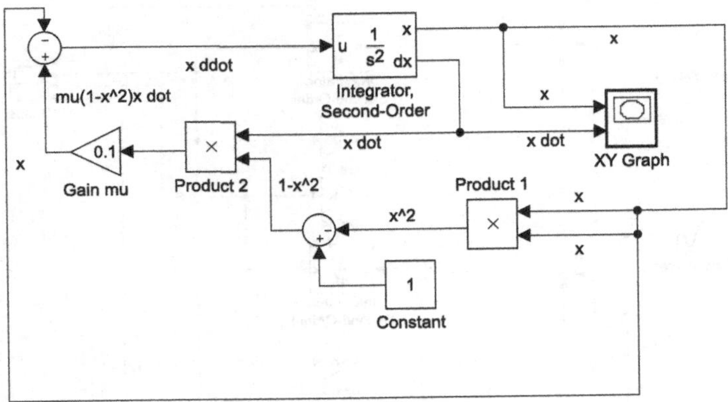

FIGURE 5.32

Model of Van der Pol's equation.

Simulink we express the equation as

$$\ddot{x} = \mu(1 - x^2)\dot{x} - x$$

The block diagram, Simulink Fig. 5.32 shows to stages of integration (using a double integration block). The x output is connected to both inputs of the product block to give the output signal x squared. In a summing block, this output is subtracted from unity. The output from the summing block is multiplied by x dot and then x is subtracted from it to generate x ddot; the input to the integrator. Note that instead of using a Scope block to view x or x dot against time, here we use an XY block and view x against x dot. If we start with a small initial condition, the oscillations grow and reach a *limit cycle* beyond which the vibration cannot grow. In contrast, if we start with a large initial condition the vibrations level decreases until a limit cycle is reached, below which the vibration level cannot fall. By examining the coefficient of $\dot{x}$ we can intuitively recognize this. When $x^2 < 1$ the $\dot{x}$ term is negative and the vibration level will increase. When $x^2 > 1$ the $\dot{x}$ term is positive and vibration levels will decrease.

Example 5.8. Consider the equations

$$m_{11}\ddot{x} + c_{11}\dot{x} + c_{12}\dot{y} + k_{11}x + k_{12}y = X(t) \tag{5.66}$$

$$m_{22}\ddot{y} + c_{21}\dot{x} + c_{22}\dot{y} + k_{21}x + k_{22}y = Y(t) \tag{5.67}$$

We will initially neglect the damping, ($\dot{x}$ and $\dot{y}$), terms. Thus, with some rearrangement (5.66) and (5.67) become

$$\ddot{x} = -(k_{11}/m_{11})x - (k_{12}/m_{11})y + X(t) \tag{5.68}$$

$$\ddot{y} = -(k_{21}/m_{22})x - (k_{22}/m_{22})y + Y(t) \tag{5.69}$$

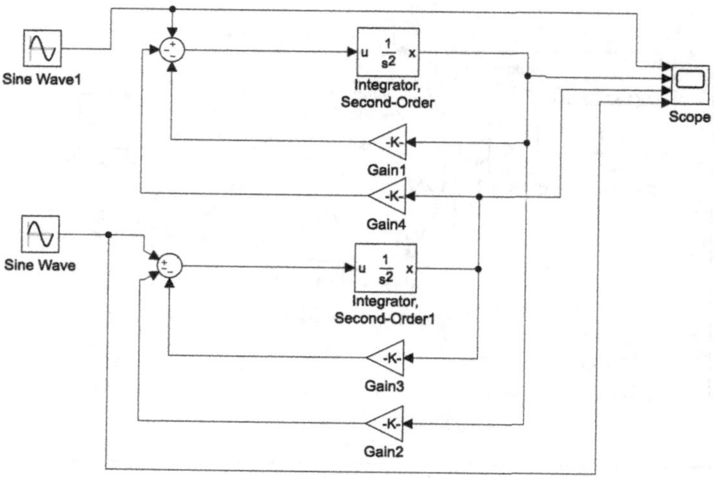

FIGURE 5.33

Model of a pair of simultaneous ordinary differential equations.

Fig. 5.33 illustrates the difficulty of simulating several simultaneous ODEs as separate ODE with interconnections. In this case we only have two equations and we have neglected $\dot{x}$ and $\dot{y}$ terms, but the flow diagram is already cluttered; many more ordinary differential equations in the system would make assembling a process diagram impossible.

However, if the two simultaneous equations, (5.66) and (5.67), are written in matrix notation we have

$$\mathbf{M}\ddot{\mathbf{q}} + \mathbf{C}\dot{\mathbf{q}} + \mathbf{K}\mathbf{q} = \mathbf{f}(t) \tag{5.70}$$

where $\mathbf{q}$ is a vector of system displacements, $[x\ y]^{\mathrm{T}}$, $\mathbf{M}$, $\mathbf{C}$, and $\mathbf{K}$ are the system mass, damping, and stiffness matrices and $\mathbf{f(t)}$ is a vector of input forces, $[X\ Y]^{\mathrm{T}}$. The easiest way of simulating this system in Simulink is to use a space-state representation. The classical state space model for a system is described by:

$$\begin{aligned} \dot{\mathbf{x}} &= \mathbf{A}\mathbf{x} + \mathbf{B}\mathbf{u} \\ \mathbf{y} &= \mathbf{C}\mathbf{x} + \mathbf{D}\mathbf{u} \end{aligned} \tag{5.71}$$

where $\mathbf{A}$ is called the state or system matrix, $\mathbf{B}$ is the input matrix, $\mathbf{C}$ is the output matrix and $\mathbf{D}$ is called the feed-through matrix. The vector $\mathbf{x}$ is the state vector, $\mathbf{u}$ is the input vector, and $\mathbf{y}$ is the output vector.

The question is, how do we rearrange (5.70) into a state space form? To obtain the state space representation we begin by multiplying (5.70) by $\mathbf{M}^{-1}$ to give

$$\ddot{\mathbf{q}} = -\mathbf{M}^{-1}\mathbf{C}\dot{\mathbf{q}} - \mathbf{M}^{-1}\mathbf{K}\mathbf{q} + \mathbf{M}^{-1}\mathbf{f}(t) \tag{5.72}$$

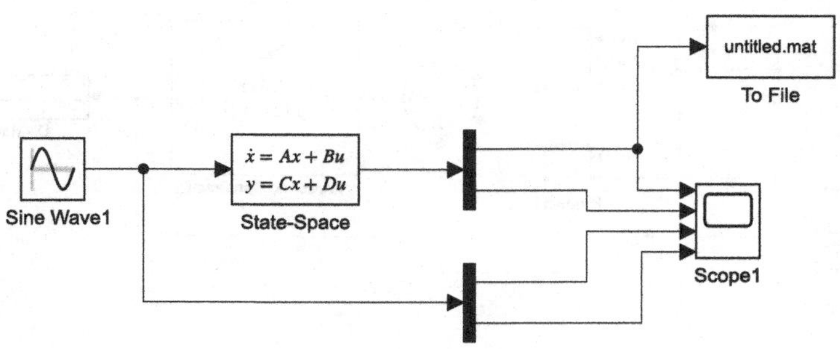

FIGURE 5.34

Two simultaneous ordinary differential equations modeled in state space form.

We now let $\mathbf{q} = \mathbf{x}_1$, $\dot{\mathbf{q}} = \mathbf{x}_2$, and $\mathbf{f}(t) = \mathbf{u}(t)$ so that (5.72) becomes

$$\dot{\mathbf{x}}_2 = -\mathbf{M}^{-1}\mathbf{C}\mathbf{x}_2 - \mathbf{M}^{-1}\mathbf{K}\mathbf{x}_1 + \mathbf{M}^{-1}\mathbf{u}(t) \tag{5.73}$$

Noting that

$$\dot{\mathbf{x}}_1 = \mathbf{x}_2 \tag{5.74}$$

we can combine (5.73) and (5.74) to give

$$\begin{bmatrix} \dot{\mathbf{x}}_1 \\ \dot{\mathbf{x}}_2 \end{bmatrix} = \begin{bmatrix} \mathbf{0} & \mathbf{I} \\ -\mathbf{M}^{-1}\mathbf{K} & -\mathbf{M}^{-1}\mathbf{C} \end{bmatrix} \begin{bmatrix} \mathbf{x}_1 \\ \mathbf{x}_2 \end{bmatrix} + \begin{bmatrix} \mathbf{0} \\ \mathbf{M}^{-1} \end{bmatrix} \mathbf{u}(t) \tag{5.75}$$

and

$$\mathbf{q} = [\mathbf{I} \quad \mathbf{0}] \begin{bmatrix} \mathbf{x}_1 \\ \mathbf{x}_2 \end{bmatrix} + \mathbf{0}\mathbf{u}(t) \tag{5.76}$$

Comparing (5.75) and (5.76) with (5.71) we see that

$$\mathbf{A} = \begin{bmatrix} \mathbf{0} & \mathbf{I} \\ -\mathbf{M}^{-1}\mathbf{K} & -\mathbf{M}^{-1}\mathbf{C} \end{bmatrix} \quad \mathbf{x} = \begin{bmatrix} \mathbf{q} \\ \dot{\mathbf{q}} \end{bmatrix} \quad \mathbf{B} = \begin{bmatrix} \mathbf{0} \\ \mathbf{M}^{-1} \end{bmatrix} \quad \mathbf{u}(t) = \begin{bmatrix} \mathbf{0} \\ \mathbf{I} \end{bmatrix} \mathbf{f}(t)$$

and

$$\mathbf{y} = \begin{bmatrix} \mathbf{q} \\ \mathbf{0} \end{bmatrix} \quad \mathbf{C} = [\mathbf{I} \quad \mathbf{0}] \quad \mathbf{D} = \mathbf{0}$$

Simulink has a state space process block. The matrices **A**, **B**, **C**, and **D** are specified and an input **u** must be provided. The output from the process, **y** can then be connected to a display. This is illustrated in Simulink Fig. 5.34.

The Simulink simulations shown in Examples 5.4 to 5.8 have one feature in common; they are all simulations of differential equations. It is possible to simulate algebraic equations, for example

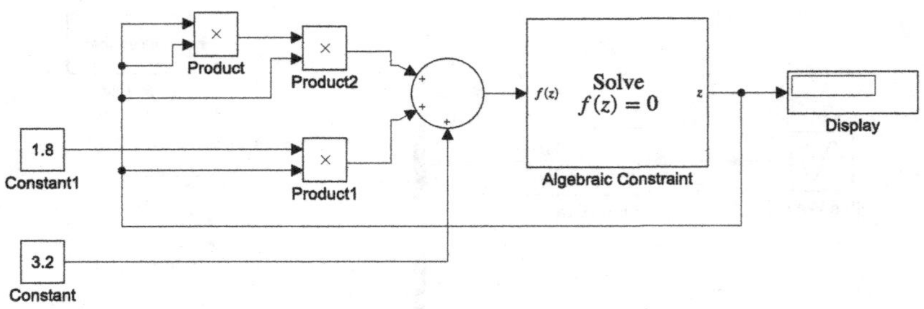

FIGURE 5.35

Model to determine the root of a cubic equation.

we could easily simulate $y = ax + b$ in Simulink. The danger when simulating algebraic equation is the accidental inclusion of an "algebraic loop". This is a loop exists in the Simulink block diagram with only direct feed-through blocks within the loop. A feed-through block output depends on its input and no integration is involved. Examples of a feed-through blocks include a gain block and math function blocks. It further illustrates an algebraic loop, consider a simulation that involves an input x to a summing block. If there are no other inputs to the summing block, the output from the block will also be x. However, suppose we take a connection from the output, pass it through a gain block, with a gain of 10, and connect it to the summing block. Then an input of $x = 1$ will produce a feedback of 10 into the summing junction which will give an output of $1 + 10 = 1$. This output is feedback and becomes $10 \times 11 = 110$ at the summing junction. Clearly, the output of the summing block rapidly tends a numeric overflow and Simulink stops. The following example solves an algebraic equation by simulation, but does contain an algebraic loop.

Example 5.9. Consider a model of a process to determine the smallest root of the cubic equation

$$x^3 + ax + b = 0$$

Note that there is no x^2 term in this equation. However, Simulink can find the lowest root of any equation.

The Simulink Fig. 5.35 shows the block diagram for the process. At the left side of the diagram we see two inputs. One input, the constant b, is connected directly to a summing block; the other input, the constant a, is multiplied by x and connected to a summing block. Also connected to the summing block is $x \times x^2 = x^3$. The output from the summing block, equal to $x^3 + ax + b$ is fed into the block labeled Solve $f(z) = 0$. This block iteratively determines a value of x to make the input to the block equal zero. At each iteration the current best estimate for x is fed back to allow an update value for $x^3 + ax + b$ to be determined. Once convergence has reached the tolerance level set in the Solve $f(z) = 0$ block, the value of x is passed to the display.

Simulink allows the user to mask a subsection of a block diagram. A mask hides the details and content of a selected subsystem so that it appears in the block diagram as a single block. (See Fig. 5.36.) If desired by the user, the masked block can have its own parameter dialog box and icon. Furthermore, it can be placed in the custom created block library.

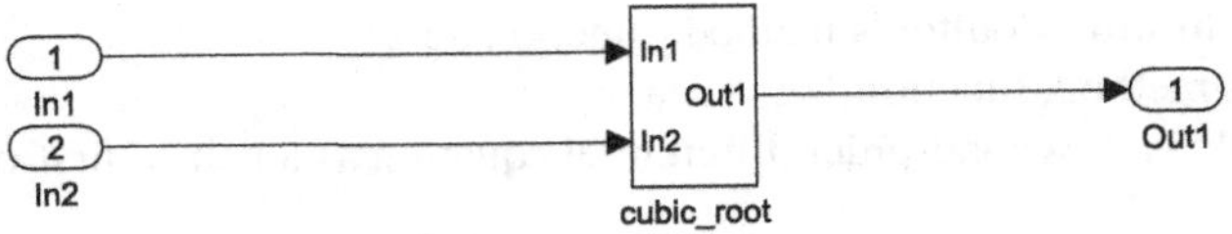

FIGURE 5.36

The Simulink model of Fig. 5.35 replaced by a single mask.

For readers with Simulink, an excellent example of a Simulink model can be found by from the MATLAB command window by entering

```
>> shower
```

This is a simulation of a temperature control system for a shower that uses fuzzy logic. There are four masked blocks. For example the hot water valve block itself contains seven lower level blocks.

5.18 SUMMARY

In this chapter, we have developed a range of MATLAB functions for solving differential equations and systems of differential equations which supplement those provided in MATLAB. We have demonstrated how these functions may be used to solve a wide variety of problems. One particular problem studied is that of a differential equation that is chaotic. We have also briefly introduced the MATLAB Simulink toolbox.

5.19 PROBLEMS

5.1. A radioactive material decays at a rate that is proportional to the amount that remains. The differential equation which models this process is

$$dy/dt = -ky \text{ where } y = y_0 \text{ when } t = t_0$$

Here y_0 represents the mass at the time t_0. Solve this equation for $t = 0$ to 10 given that $y_0 = 50$ and $k = 0.05$, using:
(a) the function `feuler`, with $h = 1, 0.1, 0.01$.
(b) the function `eulertp`, with $h = 1, 0.1$.
(c) the function `rkgen`, set for the classical method, with $h = 1$.
Compare your results with the exact solution, $y = 50\exp(-0.05t)$.

5.2. Solve $y' = 2xy$ with initial conditions $y_0 = 2$ when $x_0 = 0$ in the range $x = 0$ to 2. Use the classical, Merson, and Butcher variants of the Runge–Kutta method, all implemented in function `rkgen`, with step $h = 0.2$. Note that the exact solution is $y = 2\exp(x^2)$.

5.3. Repeat Problem 5.1 using the following predictor–corrector methods with $h = 2$ and for $t = 0$ to 50:

(a) Adams–Bashforth–Moulton's method, function `abm`.
(b) Hamming's method, function `fhamming`.

5.4. Express the following second-order differential equation as a pair of first-order equations:

$$xy'' - y' - 8x^3y^3 = 0$$

with the initial conditions $y = 1/2$ and $y' = -1/2$ at $x = 1$. Solve the pair of first-order equations using both `ode23` and `ode45` in the range 1 to 4. The exact solution is given by $y = 1/(1 + x^2)$.

5.5. Use function `fhermite` to solve:
(a) Problem 5.1 with $h = 1$.
(b) Problem 5.2 with $h = 0.2$.
(c) Problem 5.2 with $h = 0.02$.

5.6. Use the MATLAB function `rombergx` to solve the following problems. In each case use eight divisions.
(a) $y' = 3y/x$ with initial conditions $x = 1$, $y = 1$. Determine y when $x = 20$.
(b) $y' = 2xy$ with initial conditions $x = 0$, $y = 2$. Determine y when $x = 2$.

5.7. Consider the predator–prey problem described in Section 5.10.2. This problem may be extended to consider the effect of culling on the interacting populations by subtracting a term from both equations in (5.28) as follows:

$$dP/dt = K_1P - CPQ - S_1P$$
$$dQ/dt = K_2Q - DPQ - S_2Q$$

Here S_1 and S_2 are constants which provide the culling level for the populations. Use `ode45` to solve this problem with $K_1 = 2$, $K_2 = 10$, $C = 0.001$, and $D = 0.002$ and initial values of the population $P = 5000$ and $Q = 100$. Assuming that S_1 and S_2 are equal, experiment with values in the range 1 to 2. There is a wealth of experimental opportunity in this problem and the reader is encouraged to investigate different values of S_1 and S_2.

5.8. Solve the Lorenz equations given in Section 5.12.1 for $r = 1$, using `ode23`.

5.9. Use the Adams–Bashforth–Moulton method to solve $dy/dt = -5y$, with $y = 50$ when $t = 0$, in the range $t = 0$ to 6. Try step sizes, h, of 0.1, 0.2, 0.25, and 0.4. Plot the error against t for each case. What can you deduce from these results with regard to the stability of the method? The exact answer is $y = 50e^{-5t}$.

5.10. The following first-order differential equation represents the growth in a population in an environment that can support a maximum population of K:

$$dN/dt = rN(1 - N/K)$$

where $N(t)$ is the population at time t and r is a constant. Given $N = 100$ when $t = 0$, we use the MATLAB function `ode23`, solve this differential equation in the range 0 to 200, and plot a graph of N against time. Take $K = 10{,}000$ and $r = 0.1$.

5.11. The Leslie–Gower predator–prey problem takes the form

$$dN_1/dt = N_1(r_1 - cN_1 - b_1N_2)$$
$$dN_2/dt = N_2(r_2 - b_2(N_2/N_1))$$

where $N_1 = 15$ and $N_2 = 15$ at time $t = 0$. Use `ode45` to solve this equation given $r_1 = 1$, $r_2 = 0.3$, $c = 0.001$, $b_1 = 1.8$, and $b_2 = 0.5$. Plot N_1 and N_2 in the range $t = 0$ to 40.

5.12. By setting $u = dx/dt$, reduce the following second-order differential equation to two first-order differential equations.

$$\frac{d^2x}{dt^2} + k\left(\frac{1}{v_1} + \frac{1}{v_2}\right)\frac{dx}{dt} = 0$$

where $x = 0$ and $dx/dt = 10$ when $t = 0$. Use the MATLAB function `ode45` to solve this problem given that $v_1 = v_2 = 1$ and $k = 10$.

5.13. A model for a conflict between guerilla, g_2, and government forces, g_1, is given by the equations

$$dg_1/dt = -cg_2$$
$$dg_2/dt = -rg_2g_1$$

Given that the government forces number 2000 and the guerilla forces number 700 at time $t = 0$, use the function `ode45` to solve this system of equations, taking $c = 30$ and $r = 0.01$. You should solve the equations over the interval time interval 0 to 0.6. Plot a graph of the solution showing the changes in government and guerilla forces over time.

5.14. The following differential equation provides a simple model of a suspension system. The constant m gives the mass of moving parts, the constant k relates to stiffness of the suspension system and the constant c is a measure of the damping in the system. F is a constant force applied at $t = 0$.

$$m\frac{d^2x}{dt^2} + c\frac{dx}{dt} + kx = F$$

Given that $m = 1$, $k = 4$, $F = 1$ and both $x = 0$ and $dx/dt = 0$ when $t = 0$, use `ode23` to determine the response, $x(t)$ for $t = 0$ to 8 and plot a graph of x against t in each case. Assume the following values for c:

(a) $c = 0$ **(b)** $c = 0.3\surd(mk)$ **(c)** $c = \surd(mk)$

(d) $c = 2\surd(mk)$ **(e)** $c = 4\surd(mk)$

Comment on the nature of your solutions. The exact solutions are as follows:

For cases **(a)**, **(b)**, and **(c)** $x(t) = \frac{F}{k}\left[1 - \frac{1}{\sqrt{1-\zeta^2}}e^{-\zeta\omega_n t}\cos(\omega_d t - \phi)\right]$

For case **(d)** $x(t) = \frac{F}{k}\left[1 + (1 + \omega_n t)e^{-\omega_n t}\right]$

where

$$\omega_n = \sqrt{k/m}, \quad \zeta = c/(2\sqrt{mk}), \quad \omega_d = \omega_n\sqrt{1-\zeta^2}$$

$$\phi = \tan^{-1}(\zeta/\sqrt{1-\zeta^2})$$

For case **(e)**

$$x(t) = \frac{F}{k}\left[1 + \frac{1}{2q}\left(s_2e^{s_1t} - s_1e^{s_2t}\right)\right]$$

where

$$q = \omega_n \surd \zeta^1 -! \quad s_1 = -\zeta\omega_n + q \quad s_2 = -\zeta\omega_n - q$$

Plot these solutions and compare them with your numerical solutions.

5.15. Gilpin's system for modeling the behavior of three interacting species is given by the differential equations:

$$dx_1/dt = x_1 - 0.001x_1^2 - 0.001kx_1x_2 - 0.01x_1x_3$$
$$dx_2/dt = x_2 - 0.001kx_1x_2 - 0.001x_2^2 - 0.001x_2x_3$$
$$dx_3/dt = -x_3 + 0.005x_1x_3 + 0.0005x_2x_3$$

given $x_1 = 1000$, $x_2 = 300$, and $x_3 = 400$ at time $t = 0$. Taking $k = 0.5$, use `ode45` to solve this system of equations in the range $t = 0$ to $t = 50$ and plot the behavior of the population of the three species against time.

5.16. A problem which arises in planet formation is where a range of objects called planetesimals coagulate to form larger objects and this coagulation continues until a stable state is reached where a number planetary size objects have been created. To simulate this situation we assume that a minimum size object exists of mass m_1 and the masses of all other objects are integral multiples of the mass of this object. Thus there are n_k objects of mass m_k where $m_k = km_1$. Then the manner in which the number of objects of specific mass changes over time t is given by the *Coagulation Equation* as follows:

$$\frac{dn_k}{dt} = \frac{1}{2} \sum_{i+j=k} A_{ij}n_in_j - n_k \sum_{i=k+1}^{\max k} A_{ki}n_i$$

The values A_{ij} are the probabilities of collisions between the objects i and j. A simple interpretation of this equation is that number of bodies of mass n_k is increased by collisions between bodies of lesser mass but the decreased by collisions with larger bodies.

As an exercise write out the equations for this system for the case where there are only three different sizes of planetesimals and assign A_{ij} equal to $n_in_j/(1000(n_i + n_j))$. Note that the division by 1000 ensures that impacts are relatively rare which seems a plausible assumption in the vast volume of space considered. The initial values for the numbers of planetesimals n_1, n_2, n_3 are as taken as 200, 25, and 1, respectively.

Solve the resulting system using the MATLAB function `ode45` using a time interval of 2 units. Study the case where the values of the collision probabilities are calculated using the varying values of the number of planetesimals as time varies. Plot graphs of your results.

5.17. The following example studies the effect of life on a planetary environment. A relatively simple way of studying these effects is to consider the concept of daisy world. This envisages a world inhabited by only two life forms; white and black daisies. This situation can be modeled a pair of differential equations where the area covered by the black daisies a_b and the area covered by the white daisies a_w changes with time t as follows:

$$da_b/dt = a_b(x\beta_b - \gamma)$$
$$da_w/dt = a_w(x\beta_w - \gamma)$$

where $x = 1 - a_b - a_w$ represents the area not covered by either daisy assuming the total area of the planet is represented by unity. The value of γ gives the death rate for the daisies and β_b and β_w give the growth rate for the black and white daisies respectively. This is related to the energy they receive from the planetary Sun or the local temperature. Consequently, an empirical formula may be given for these values as follows:

$$\beta_b = 1 - 0.003265(295.5 - T_b)^2$$

and

$$\beta_w = 1 - 0.003265(295.5 - T_w)^2$$

where the values of T_b and T_w lie in the range 278 to 313 K, where K denotes degrees Kelvin. Outside this range growth is assumed zero. Taking $\gamma = 0.3$, $T_b = 295$ K, $T_w = 285$ K and initial values for $a_b = 0.2$, $a_w = 0.3$, solve the system of equations for $t = [0, 10]$ using the MATLAB function `ode45`. Plot graphs of the changes in a_b and a_w with respect to time. It should be noted that the extent of the areas covered by the black and white daisies will affect the overall temperature of the planet, since white and dark areas react differently in the way they absorb energy from the Sun.

5.18. In an article by Pearce and Shearer (2016), the authors describe how capillaries work in the circulatory system and how they can be modeled using a network of resistors and capacitors. In particular the manner in which the capacitor operates with an applied fluctuating current and consequently alternatively releasing and storing voltage through the resistor reflects the behavior of the circulatory system. The model of the corresponding basic electrical circuit behavior over time t, is well established and takes the form:

$$C\frac{dV}{dt} + \frac{V}{R} = I(t)$$

where V is the voltage, R the resistance, C the capacity, and I the current. In terms of blood flow this may be translated as:

$$C\frac{dP}{dt} + \frac{P}{R} = Q_0 \sin(\pi t / t_0)$$

where C is called the compliance of the aorta, which corresponds to the concept of capacity, R is the total resistance of the rest of the circulatory system and the right-hand side of the equation models the flow of the heart, t_0 is the initial time and Q_0 is the maximum flow rate. We can use this first-order differential equation to find the varying values of P.

Solve this equation using the MATLAB function for the solution of differential equations, `ode45`, taking $R = 10$, $C = 1.5$, $R = 40$, $Q_0 = 2.8$, and $t_0 = 0$.

CHAPTER

BOUNDARY VALUE PROBLEMS

6

Abstract

In this chapter algorithms for solving certain boundary value problems and problems with both boundary and initial values are given. The solution of a boundary value problem in one independent variable must satisfy specified conditions at two points, and the solution of a boundary value problem in two independent variables must satisfy specified conditions along a curve or set of lines enclosing a specified region.

6.1 CLASSIFICATION OF SECOND-ORDER DIFFERENTIAL EQUATIONS

In this chapter, we restrict the discussion to second-order differential equations in one or two independent variables and Fig. 6.1 shows how these equations may be classified. The general form of these equations for one and two independent variables is given by (6.1) and (6.2), respectively.

$$A(x)\frac{d^2z}{dx^2} + f\left(x, z, \frac{dz}{dx}\right) = 0 \tag{6.1}$$

$$A(x, y)\frac{\partial^2 z}{\partial x^2} + B(x, y)\frac{\partial^2 z}{\partial x\, \partial y} + C(x, y)\frac{\partial^2 z}{\partial y^2} + f\left(x, y, z, \frac{\partial z}{\partial x}, \frac{\partial z}{\partial y}\right) = 0 \tag{6.2}$$

These equations are linear in the second-order terms, but the terms

$$f\left(x, z, \frac{dz}{dx}\right) \text{ and } f\left(x, y, z, \frac{\partial z}{\partial x}, \frac{\partial z}{\partial y}\right)$$

may be linear or non-linear. In particular, (6.2) is classified as an elliptic, parabolic, or hyperbolic partial differential equation as follows:

If $B^2 - 4AC < 0$, the equation is elliptic.
If $B^2 - 4AC = 0$, the equation is parabolic.
If $B^2 - 4AC > 0$, the equation is hyperbolic.

Since the coefficients A, B, and C are, in general, functions of the independent variables, the classification of (6.2) may vary in different regions of the domain in which the problem is defined. We will commence with a study of (6.1).

Numerical Methods. https://doi.org/10.1016/B978-0-12-812256-3.00015-4

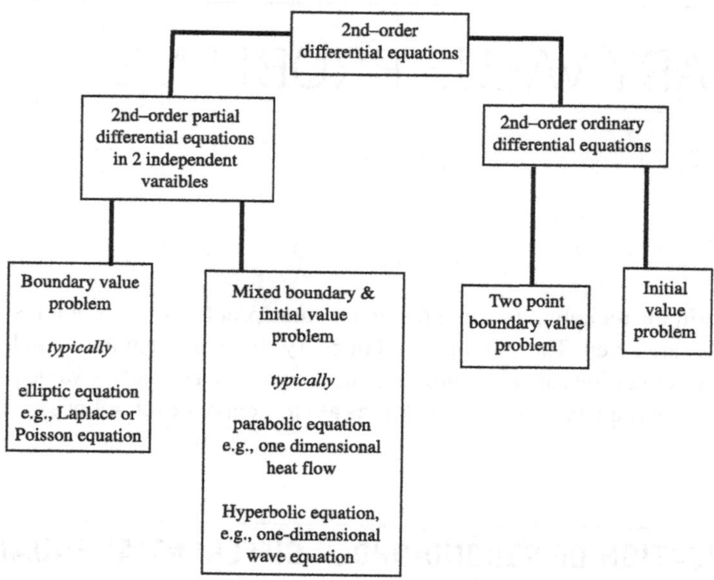

FIGURE 6.1

Second-order differential equations with one or two independent variables and their solutions.

6.2 THE SHOOTING METHOD

An initial value problem and a two-point boundary value problem derived from the same differential equation may have the same solution. For example, consider the differential equation

$$\frac{d^2y}{dx^2} + y = \cos 2x \tag{6.3}$$

Given the initial conditions that when $x = 0$, $y = 0$, and $dy/dx = 1$, the solution of (6.3) is

$$y = (\cos x - \cos 2x)/3 + \sin x$$

However, this solution also satisfies (6.3) with the two boundary conditions $x = 0$, $y = 0$, and $x = \pi/2$, $y = 4/3$.

This observation provides a useful method for solving two-point boundary value problems called "the shooting method". As an example, consider the equation

$$x^2\frac{d^2y}{dx^2} - 6y = 0 \tag{6.4}$$

with boundary conditions $y = 1$ when $x = 1$ and $y = 1$ when $x = 2$. We will treat this problem as an initial value problem where $y = 1$ when $x = 1$ and assume trial values for dy/dx when $x = 1$, denoted

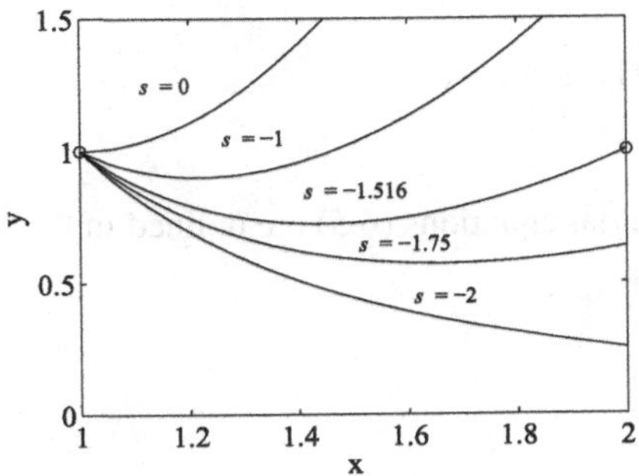

FIGURE 6.2

Solutions of $x^2(d^2y/dx^2) - 6y = 0$ with initial conditions $y = 1$ and $dy/dx = s$ when $x = 1$, for trial values of s.

by s. Fig. 6.2 shows the solution for various trial values of s. It is seen that when $s = -1.516$, the solution satisfies the required boundary condition that $y = 1$ when $x = 2$. The solution for (6.4) can be found by changing it into a pair of first-order differential equations by letting $z = dy/dx$ so that $d^2y/dx^2 = dz/dx$ and thus (6.4) is equivalent to

$$\begin{aligned} dy/dx &= z \\ dz/dx &= 6y/x^2 \end{aligned} \tag{6.5}$$

This pair of differential equations can be solved using any appropriate numerical method described in Chapter 5.

We must determine the slope dy/dx that gives the correct boundary condition. This could be achieved by trial and error, but this is tedious and in practice we can use interpolation. The script `e4s601.m` solves (6.5) for four trial slopes using the MATLAB function `ode45`. The vector `s` contains trial values of the slope dy/dx at $x = 1$. Vector `b` contains the corresponding values of y when $x = 2$, computed by `ode45`. From these values of y we can interpolate to determine the value of s required to give $y = 1$ when $x = 2$. The interpolation is carried out using the function `aitken` (described in Chapter 7). Finally, this interpolated value for the slope, `s0`, is used in `ode45` to determine the correct solution to (6.5).

```
% e4s601.m
f = @(x,y)[y(2); 6*y(1)/x^2];
option = odeset('RelTol',0.0005);
s = -1.25:-0.25:-2; s0 = [ ];
ncase = length(s); b = zeros(1,ncase);
for i = 1:ncase
    [x,y] = ode45(f,[1 2],[1 s(i)],option);
    [m,n] = size(y);
    b(1,i) = y(m,1);
end
```

```
s0 = aitken(b,s,1)
[x,y] = ode45(f,[1 2],[1 s0],option);
[x y(:,1)]
```

The right sides of the differential equations (6.5) are defined in the first line of this script. Running script e4s601 gives

```
s0 =
   -1.5161

ans =
    1.0000    1.0000
    1.0111    0.9836
    1.0221    0.9679
    1.0332    0.9529
    1.0442    0.9386
    1.0692    0.9084
    1.0942    0.8812
    1.1192
```

This output is very lengthy and part of it has therefore been deleted. The final stages are as follows:

```
              0.9293
    1.9442    0.9501
    1.9582    0.9622
    1.9721    0.9745
    1.9861    0.9871
    2.0000    1.0000
```

The interpolated value of the slope is -1.5161. The first column of ans gives the values of x and the second gives the corresponding values of y.

Whilst the shooting method is not particularly efficient, it does have the advantage of being able to solve non-linear boundary value problems. We now examine an alternative method for solving boundary problems: the finite difference method.

6.3 THE FINITE DIFFERENCE METHOD

In Chapter 4, it is shown how derivatives can be approximated by the use of finite differences. We can use the same approach for the solution of certain types of differential equations. The method effectively replaces the differential equation by a set of approximate difference equations. The central difference approximations for the first and second derivatives of z with respect to x are given by (6.6) and (6.7),

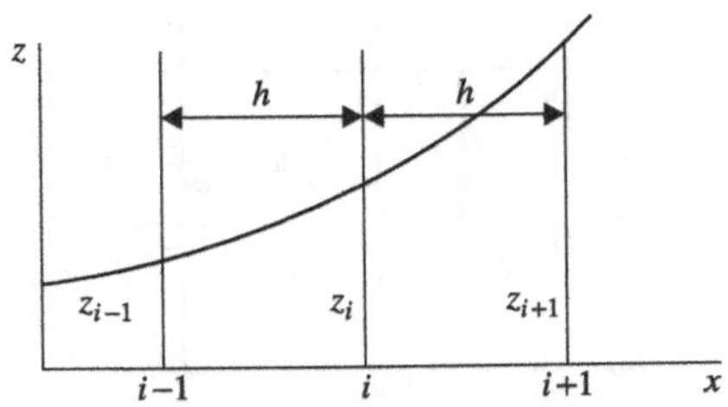

FIGURE 6.3

Equispaced nodal points.

which follow. In these and subsequent equations the operator D_x represents d/dx, $D_x^2 = d^2/dx^2$, etc. The subscript x is omitted where there is no danger of confusion. Thus at z_i,

$$Dz_i \approx (-z_{i-1} + z_{i+1})/(2h) \tag{6.6}$$

$$D^2 z_i \approx (z_{i-1} - 2z_i + z_{i+1})/h^2 \tag{6.7}$$

In (6.6) and (6.7), h is the distance between the nodal points (see Fig. 6.3) and these approximating formulae have errors of order h^2. Higher-order approximations can be generated which have errors of order h^4, but we do not require them. To achieve the same degree of accuracy we can make h smaller.

We can also determine the approximations for unevenly spaced nodal points. For example, it can be shown that (6.6) and (6.7) become

$$Dz_i \approx \frac{1}{h\beta(\beta+1)}\left\{-\beta^2 z_{i-1} - \left(1-\beta^2\right) z_i + z_{i+1}\right\} \tag{6.8}$$

$$D^2 z_i \approx \frac{2}{h^2\beta(\beta+1)}\{\beta z_{i-1} - (1+\beta)\, z_i + z_{i+1}\} \tag{6.9}$$

where $h = x_i - x_{i-1}$ and $\beta h = x_{i+1} - x_i$. Note that when $\beta = 1$, (6.8) and (6.9) simplify to (6.6) and (6.7), respectively. Approximation (6.8) has an error of order h^2, regardless of the value of β, and (6.9) has an error of order h for $\beta \neq 1$ and h^2 for $\beta = 1$.

Eqs. (6.6) through (6.9) are central difference approximations; that is the approximation for a derivative uses values of the function on either side of the point at which the derivative is to be determined. These are generally the most accurate approximations, but in some situations, for example at the left and right boundary nodes, it is necessary to use forward or backward difference approximations. For example, the forward difference approximation for Dz_i is

$$Dz_i \approx (-z_i + z_{i+1})/h \text{ with an error of order } h \tag{6.10}$$

The backward difference approximation for Dz_i is

$$Dz_i \approx (-z_{i-1} + z_i)/h \text{ with an error of order } h \tag{6.11}$$

To determine solutions for partial differential equations, we require the finite difference approximation for various partial derivatives in two or more variables. These approximations can be derived by

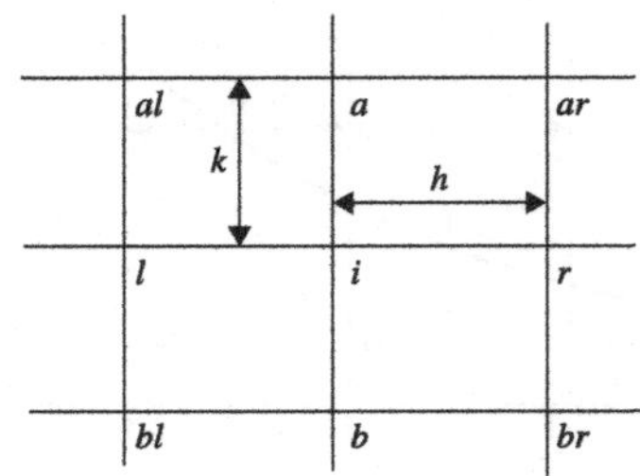

FIGURE 6.4

Grid mesh in rectangular coordinates.

combining some of the previous equations. For example, we can determine the finite difference approximation for $\partial^2 z/\partial x^2 + \partial^2 z/\partial y^2$ (that is, $\nabla^2 z$) from the approximation (6.7) or (6.9). To avoid double subscripts we will use the notation applied to the mesh shown in Fig. 6.4. Thus, from (6.7),

$$\begin{aligned}\nabla^2 z_i &\approx (z_l - 2z_i + z_r)/h^2 + (z_a - 2z_i + z_b)/k^2 \\ &\approx \left\{r^2 z_l + r^2 z_r + z_a + z_b - 2\left(1 + r^2\right) z_i\right\} / \left(r^2 h^2\right)\end{aligned} \tag{6.12}$$

where $r = k/h$. If $r = 1$, then (6.12) becomes

$$\nabla^2 z_i \approx (z_l + z_r + z_a + z_b - 4z_i)/h^2 \tag{6.13}$$

These central difference approximations for $\nabla^2 z_i$ have an error of $O(h^2)$.

The finite difference approximation for the second-order mixed derivative of z with respect to x and y, $\partial^2 z/\partial x \partial y$ or D_{xy}, is determined by applying (6.6) in the x direction to each term of (6.6) in the y direction thus:

$$\begin{aligned}\mathrm{D}_{xy} z_i &\approx \left[(z_r - z_l)_a/(2h) - (z_r - z_l)_b/(2h)\right]/(2k) \\ &\approx (z_{ar} - z_{al} - z_{br} + z_{bl})/(4hk)\end{aligned} \tag{6.14}$$

We can develop the finite difference approximations in other coordinate systems such as skew and polar coordinates, and we can have uneven spacing of the node points in any direction; see Salvadori and Baron (1961).

6.4 TWO-POINT BOUNDARY VALUE PROBLEMS

Before considering the application of finite difference methods to solve a differential equation, we first consider the nature of the solution. We begin by considering the following second-order inhomogeneous differential equation, that is a differential equation with a non-zero right side.

$$\left(1 + x^2\right)\frac{d^2 z}{dx^2} + x\frac{dz}{dx} - z = x^2 \tag{6.15}$$

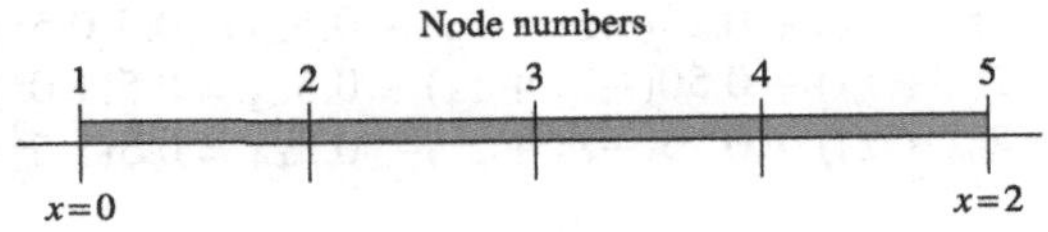

FIGURE 6.5

Node numbering used in the solution of (6.15).

subject to the boundary conditions $x = 0$, $z = 1$ and $x = 2$, $z = 2$. The solution of this equation is

$$z = -\frac{\sqrt{5}}{6}x + \frac{1}{3}\left(1 + x^2\right)^{1/2} + \frac{1}{3}\left(2 + x^2\right) \tag{6.16}$$

This is the only solution that satisfies both the equation and its boundary conditions. In contrast to this, consider the solution of the second-order homogeneous equation, that is a differential equation with a zero right side.

$$x\frac{d^2z}{dx^2} + \frac{dz}{dx} + \lambda x^{-1}z = 0 \tag{6.17}$$

subject to the conditions that $z = 0$ at $x = 1$ and $dz/dx = 0$ at $x = e$ (where $e = 2.7183...$). If λ is a given constant, this homogeneous equation has the trivial solution $z = 0$. However, if λ is an unknown, then we can determine values of λ to give non-trivial solutions for z. Eq. (6.17) is then a characteristic value or eigenvalue problem. Solving (6.17) gives an infinite number of solutions for λ and z as follows:

$$z_n = \sin\left\{\sqrt{\lambda_n}\log_e|x|\right\},\ \lambda_n = \{(2n+1)\pi/2\}^2 \text{ where } n = 0,\ 1,\ 2,\ ... \tag{6.18}$$

The values of λ that satisfy (6.18) are called characteristic values or eigenvalues, and the corresponding values of z are called characteristic functions or eigenfunctions. This particular type of boundary value problem is called a characteristic value or eigenvalue problem. It has arisen because both the differential equation and the specified boundary conditions are homogeneous.

The application of finite differences to the solution of boundary value problems is now illustrated by Examples 6.1 and 6.2.

Example 6.1. Determine an approximate solution for (6.15). We begin by multiplying (6.15) by $2h^2$ and writing d^2z/dx^2 as D^2z, etc., to give

$$2(1 + x^2)(h^2\mathrm{D}^2z) + xh\,(2h\mathrm{D}z) - 2h^2z = 2h^2x^2 \tag{6.19}$$

Using (6.6) and (6.7) we can replace (6.19) by

$$2(1 + x_i^2)(z_{i-1} - 2z_i + z_{i+1}) + x_ih(-z_{i-1} + z_{i+1}) - 2h^2z_i = 2h^2x_i^2 \tag{6.20}$$

Fig. 6.5 shows x divided into four segments ($h = 1/2$) with nodes numbered 1 to 5. Applying (6.20) to nodes 2, 3, and 4 gives

At node 2: $2(1+0.5^2)(z_1-2z_2+z_3)+0.25(-z_1+z_3)-0.5z_2=0.5(0.5^2)$
At node 3: $2(1+1.0^2)(z_2-2z_3+z_4)+0.50(-z_2+z_4)-0.5z_3=0.5(1.0^2)$
At node 4: $2(1+1.5^2)(z_3-2z_4+z_5)+0.75(-z_3+z_5)-0.5z_4=0.5(1.5^2)$

The problem boundary conditions are $x=0$, $z=1$ and $x=2$, $z=2$. Thus $z_1=1$ and $z_5=2$. Using these values, the preceding equations can be simplified and written in matrix form thus:

$$\begin{bmatrix} -44 & 22 & 0 \\ 28 & -68 & 36 \\ 0 & 46 & -108 \end{bmatrix} \begin{bmatrix} z_2 \\ z_3 \\ z_4 \end{bmatrix} = \begin{bmatrix} -17 \\ 4 \\ -107 \end{bmatrix}$$

This equation system can easily be solved using MATLAB as follows:

```
>> A = [-44 22 0;28 -68 36;0 46 -108];
>> b = [-17 4 -107].';
>> y = A\b

y =
     0.9357
     1.0987
     1.4587
```

Note that the rows in the previous matrix equation can always be scaled in order to make the coefficient matrix symmetrical. This is important in a large problem.

To increase the accuracy of the solution, we must increase the number of nodal points which consequently decrease h. However, formulating the finite difference approximation by hand for a large number of nodes is a tedious and error-prone process. The MATLAB function `twopoint` implements the process of solving the second-order boundary problem given by differential equation (6.21) together with appropriate boundary conditions.

$$C(x)\frac{d^2z}{dx^2}+D(x)\frac{dz}{dx}+E(x)z=F(x) \tag{6.21}$$

The user must supply a vector listing the values of nodal points chosen. These do not have to be equispaced. The user must also supply vectors listing the values of $C(x)$, $D(x)$, $E(x)$, and $F(x)$ for the nodal points. Finally, the user must provide the boundary conditions which can be expressed in terms of either z or dz/dx.

```
function y = twopoint(x,C,D,E,F,flag1,flag2,p1,p2)
% Solves 2nd order boundary value problem
% Example call: y = twopoint(x,C,D,E,F,flag1,flag2,p1,p2)
% x is a row vector of n+1 nodal points.
% C, D, E and F are row vectors
% specifying C(x), D(x), E(x) and F(x).
% If y is specified at node 1, flag1 must equal 1.
% If y' is specified at node 1, flag1 must equal 0.
```

```
% If y is specified at node n+1, flag2 must equal 1.
% If y' is specified at node n+1, flag2 must equal 0.
% p1 & p2 are boundary values (y or y') at nodes 1 and n+1.
n = length(x)-1;
h(2:n+1) = x(2:n+1)-x(1:n);
h(1) = h(2); h(n+2) = h(n+1);
r(1:n+1) = h(2:n+2)./h(1:n+1);
s = 1+r;
if flag1==1
    y(1) = p1;
else
    slope0 = p1;
end
if flag2==1
    y(n+1) = p2;
else
    slopen = p2;
end
W = zeros(n+1,n+1);
if flag1==1
    c0 = 3;
    W(2,2) = E(2)-2*C(2)/(h(2)^2*r(2));
    W(2,3) = 2*C(2)/(h(2)^2*r(2)*s(2))+D(2)/(h(2)*s(2));
    b(2) = F(2)-y(1)*(2*C(2)/(h(2)^2*s(2))-D(2)/(h(2)*s(2)));
else
    c0=2;
    W(1,1) = E(1)-2*C(1)/(h(1)^2*r(1));
    W(1,2) = 2*C(1)*(1+1/r(1))/(h(1)^2*s(1));
    b(1) = F(1)+slope0*(2*C(1)/h(1)-D(1));
end
if flag2==1
    c1 = n-1;
    W(n,n) = E(n)-2*C(n)/(h(n)^2*r(n));
    W(n,n-1) = 2*C(n)/(h(n)^2*s(n))-D(n)/(h(n)*s(n));
    b(n) = F(n)-y(n+1)*(2*C(n)/(h(n)^2*s(n))+D(n)/(h(n)*s(n)));
else
    c1 = n;
    W(n+1,n+1) = E(n+1)-2*C(n+1)/(h(n+1)^2*r(n+1));
    W(n+1,n) = 2*C(n+1)*(1+1/r(n+1))/(h(n+1)^2*s(n+1));
    b(n+1) = F(n+1)-slopen*(2*C(n+1)/h(n+1)+D(n+1));
end
for i = c0:c1
    W(i,i) = E(i)-2*C(i)/(h(i)^2*r(i));
    W(i,i-1) = 2*C(i)/(h(i)^2*s(i))-D(i)/(h(i)*s(i));
```

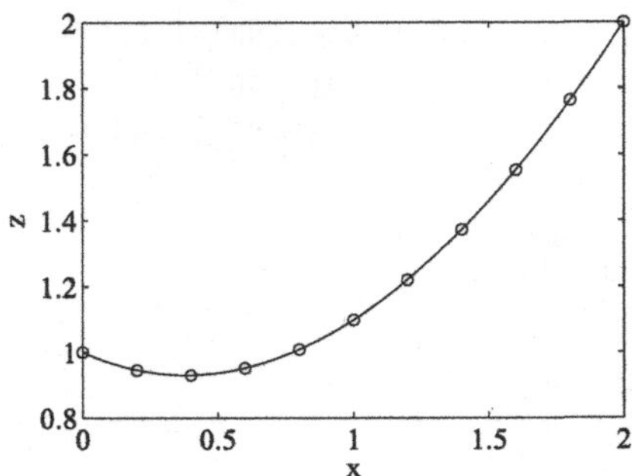

FIGURE 6.6

Finite difference solution of $(1+x^2)(d^2z/dx^2)+xdz/dx-z=x^2$. The o indicates the finite difference estimate; the continuous line is the exact solution.

```
    W(i,i+1) = 2*C(i)/(h(i)^2*r(i)*s(i))+D(i)/(h(i)*s(i));
    b(i) = F(i);
end
z = W(flag1+1:n+1-flag2,flag1+1:n+1-flag2)\b(flag1+1:n+1-flag2)';
if flag1==1 & flag2==1, y = [y(1); z; y(n+1)]; end
if flag1==1 & flag2==0, y = [y(1); z]; end
if flag1==0 & flag2==1, y = [z; y(n+1)]; end
if flag1==0 & flag2==0, y = z; end
```

We can use this function `twopoint` to solve (6.15) for nine nodes using script `e4s602`:

```
% e4s602.m
x = 0:.2:2;
C = 1+x.^2; D = x; E = -ones(1,11); F = x.^2;
flag1 = 1; p1 = 1; flag2 = 1; p2 = 2;
z = twopoint(x,C,D,E,F,flag1,flag2,p1,p2);
B = 1/3; A = -sqrt(5)*B/2;
xx = 0:.01:2;
zz = A*xx+B*sqrt(1+xx.^2)+B*(2+xx.^2);
plot(x,z,'o',xx,zz)
xlabel('x'); ylabel('z')
```

This script, `e4s602`, outputs the graph of Fig. 6.6.

The results from the finite difference analysis are very accurate. This is because the solution of the boundary problem, given by (6.16), is well approximated by a low-order polynomial.

Example 6.2. Determine the approximate solution of (6.17) subject to the boundary conditions that $z=0$ at $x=1$ and $dz/dx=0$ at $x=e$. The exact eigensolutions are given by $\lambda_n=\{(2n+1)\pi/2\}^2$ and $z_n(x)=\sin\{(2n+1)(\pi/2)\log_e|x|\}$, where $n=0,1,\ldots,\infty$. We will use the node-numbering scheme

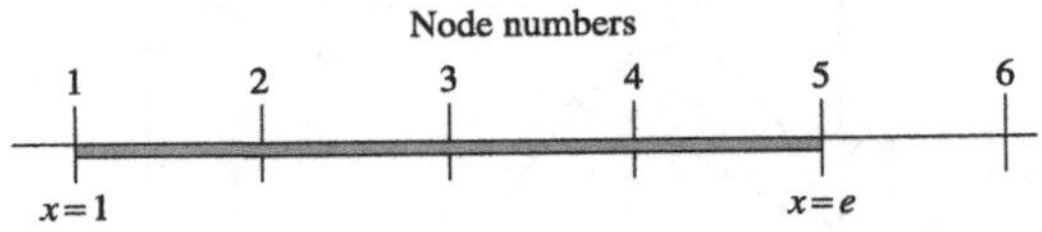

FIGURE 6.7

Node numbering used in the solution of (6.17).

shown in Fig. 6.7. To apply the boundary condition at $x = e$ we must consider the finite difference approximation for Dz at node 5 (that is, at $x = e$) and make $\mathrm{D}z_5 = 0$. Applying (6.6), we have

$$2h\mathrm{D}z_5 = -z_4 + z_6 = 0 \tag{6.22}$$

Note that, because we are using central difference approximations, we have been forced to introduce the fictitious node 6. (Alternatively, at the right boundary we could have chosen to use a backward difference approximation thereby avoiding the need for the ficitious node.)
However, from (6.22), $z_6 = z_4$. Multiplying (6.17) by $2h^2$ gives

$$2x(h^2\mathrm{D}^2z) + h\,(2h\mathrm{D}z) = -\lambda 2x^{-1}h^2z$$

Thus

$$2x_i\,(z_{i-1} - 2z_i + z_{i+1}) + h\,(-z_{i-1} + z_{i+1}) = -\lambda 2x_i^{-1}h^2z_i$$

Now $L = e - 1 = 1.7183$; thus $h = L/4 = 0.4296$. Applying (6.19) to nodes 2 through 5, we have

At #2: $2(1.4296)(z_1 - 2z_2 + z_3) + 0.4296(-z_1 + z_3) = -2\lambda(1.4296)^{-1}(0.4296)^2z_2$
At #3: $2(1.8591)(z_2 - 2z_3 + z_4) + 0.4296(-z_2 + z_4) = -2\lambda(1.8591)^{-1}(0.4296)^2z_3$
At #4: $2(2.2887)(z_3 - 2z_4 + z_5) + 0.4296(-z_3 + z_5) = -2\lambda(2.2887)^{-1}(0.4296)^2z_4$
At #5: $2(2.7183)(z_4 - 2z_5 + z_6) + 0.4296(-z_4 + z_6) = -2\lambda(2.7183)^{-1}(0.4296)^2z_5$

Letting $z_1 = 0$ and $z_6 = z_4$ leads to

$$\begin{bmatrix} -5.7184 & 3.2887 & 0 & 0 \\ 3.2887 & -7.4364 & 4.1478 & 0 \\ 0 & 4.1478 & -9.1548 & 5.0070 \\ 0 & 0 & 10.8731 & -10.8731 \end{bmatrix} \begin{bmatrix} z_2 \\ z_3 \\ z_4 \\ z_5 \end{bmatrix}$$
$$= \lambda \begin{bmatrix} -0.2582 & 0 & 0 & 0 \\ 0 & -0.1985 & 0 & 0 \\ 0 & 0 & -0.1613 & 0 \\ 0 & 0 & 0 & -0.1358 \end{bmatrix} \begin{bmatrix} z_2 \\ z_3 \\ z_4 \\ z_5 \end{bmatrix}$$

We can solve these equations using MATLAB as follows:

```
>> A = [-5.7184 3.2887 0 0;3.2887 -7.4364 4.1478 0;
0 4.1478 -9.1548 5.0070; 0 0 10.8731 -10.8731];
```

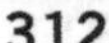

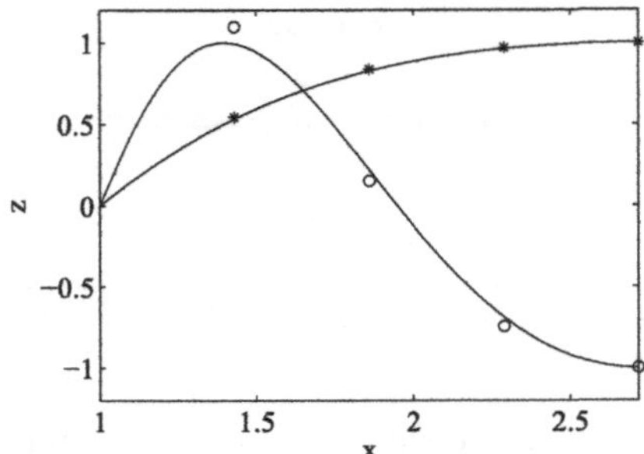

FIGURE 6.8

The finite difference estimates for the first and second eigenfunctions of $x(d^2z/dx^2) + dz/dx + \lambda z/x = 0$, denoted by a $*$ and a $\circ$ respectively: solid lines show the exact eigenfunctions $z_0(x)$ and $z_1(x)$.

```
>> B = diag([-0.2582 -0.1985 -0.1613 -0.1358]);
>> [u lambda] = eig(A,B)

u =
   -0.5424    1.0000   -0.4365    0.0169
   -0.8362    0.1389    1.0000   -0.1331
   -0.9686   -0.6793   -0.3173    0.5265
   -1.0000   -0.9112   -0.8839   -1.0000

lambda =
    2.5110         0         0         0
         0   20.3774         0         0
         0         0   51.3254         0
         0         0         0  122.2197
```

The exact values for the lowest four eigenvalues are 2.4674, 22.2066, 61.6850, and 120.9027. Because we are using only a small number of nodes, there are significant errors in computed eigenvalues, (`lambda`). The graph of Fig. 6.8 shows the first two eigenfunctions $z_0(x)$ and $z_1(x)$ and the estimates derived from the first and second columns of the preceding array `u`. Note that the values of `u` have been scaled to make those corresponding to the node z_5 either 1 or -1. The script `e4s603` evaluates and plots the exact eigenfunctions $z_0(x)$ and $z_1(x)$ and plots the scaled sample points which estimate these functions.

```
% e4s603.m
x = 1:.01:exp(1);
% compute eigenfunction values scaled to 1 or -1.
z0 = sin((1*pi/2)*log(abs(x)));
z1 = sin((3*pi/2)*log(abs(x)));
% plot eigenfuctions
plot(x,z0,x,z1), hold on
```

```
% Discrete approximations to eigenfunctions
% Scaled to 1 or -1.
u0 = [0.5424 0.8362 0.9686 1];
u1 = (1/0.9112)*[1 0.1389 -0.6793 -0.9112];
% determine x values for plotting
r = (exp(1)-1)/4;
xx = [1+r 1+2*r 1+3*r 1+4*r];
plot(xx,u0,'*',xx,u1,'o'), hold off
axis([1 exp(1) -1.2 1.2])
xlabel('x'), ylabel('z')
```

6.5 PARABOLIC PARTIAL DIFFERENTIAL EQUATIONS

The general second-order partial differential equation in terms of the independent variables x and y is given by (6.2). The equation is repeated here, except that y has been replaced by t.

$$A(x,t)\frac{\partial^2 z}{\partial x^2}+B(x,t)\frac{\partial^2 z}{\partial x\partial t}+C(x,t)\frac{\partial^2 z}{\partial t^2}+f\left(x,t,z,\frac{\partial z}{\partial x},\frac{\partial z}{\partial t}\right)=0 \tag{6.23}$$

This equation will be a parabolic equation if $B^2-4AC=0$. Parabolic equations are not defined in a closed domain, but propagate in an open domain. For example, the one-dimensional heat-flow equation, which describes heat flow assuming no energy generation, is

$$K\frac{\partial^2 u}{\partial x^2}=\frac{\partial u}{\partial t},\quad 0<x<L \text{ and } t>0 \tag{6.24}$$

where K is the thermal diffusivity and u is the temperature of the material. Comparing (6.24) with (6.23), we see that A, B, and C of (6.23) are K, 0, and 0, respectively, so that the term B^2-4AC is zero and the equation is parabolic.

To solve this equation, boundary conditions must be specified at $x=0$ and $x=L$ and initial conditions when $t=0$ must be given. To develop a finite difference solution, we divide the spatial domain into n sections, each of length h, so that $h=L/n$, and consider as many time steps as required, each time step of duration k. A finite difference approximation for (6.24) at node (i, j) is obtained by replacing $\partial^2u/\partial x^2$ by the central difference approximation (6.7) and $\partial u/\partial t$ by the forward difference approximation (6.10) to give

$$K\left(\frac{u_{i-1,j}-2u_{i,j}+u_{i+1,j}}{h^2}\right)=\left(\frac{-u_{i,j}+u_{i,j+1}}{k}\right) \tag{6.25}$$

or

$$u_{i,j+1}=u_{i,j}+\alpha(u_{i-1,j}-2u_{i,j}+u_{i+1,j}),\ i=0,\ 1,\ \ldots,\ n;\ \ j=0,\ 1,\ \ldots \tag{6.26}$$

In (6.26), $\alpha=Kk/h^2$. Node (i, j) is the point $x=ih$ and at time jk. Eq. (6.26) allows us to determine $u_{i,j+1}$, that is u at time $j+1$ from values of u at time j. Values of $u_{i,0}$ are provided by the initial conditions; values of $u_{0,j}$ and $u_{n,j}$ are obtained from the boundary conditions. This method of solution is called the explicit method.

In the numeric solution of parabolic partial differential equations, solution stability and convergence are important. It can be proved that when using the explicit method we must make $\alpha \le 0.5$ to ensure a steady decay of the entire solution. This requirement means that the grid separation in time must sometimes be very small, necessitating a very large number of time steps.

An alternative finite difference approximation for (6.24) is obtained by considering node $(i, j+1)$. We again approximate $\partial^2 u/\partial x^2$ by the central difference approximation (6.7), but we approximate $\partial u/\partial t$ by the backward difference approximation (6.11) to give

$$K\left(\frac{u_{i-1,j+1} - 2u_{i,j+1} + u_{i+1,j+1}}{h^2}\right) = \left(\frac{-u_{i,j} + u_{i,j+1}}{k}\right) \tag{6.27}$$

This equation is identical to (6.25) except that approximation is made at the $(j+1)$th time step instead of at the jth time step. Rearranging (6.27) with $\alpha = Kk/h^2$ gives

$$(1 + 2\alpha)u_{i,j+1} - \alpha(u_{i+1,j+1} + u_{i-1,j+1}) = u_{i,j} \tag{6.28}$$

where $i = 0, 1, ..., n$; $j = 0, 1, ...$ The three variables on the left side of this equation are unknown. However, if we have a grid of $n+1$ spatial points, then at time $j+1$ there are $n-1$ unknown nodal values and two known boundary values. We can assemble the set of $n-1$ equations of the form of (6.28) thus:

$$\begin{bmatrix} \gamma & -\alpha & 0 & \cdots & 0 \\ -\alpha & \gamma & -\alpha & \cdots & 0 \\ 0 & -\alpha & \gamma & \cdots & 0 \\ \vdots & \vdots & \vdots & & \vdots \\ 0 & 0 & 0 & \cdots & -\alpha \\ 0 & 0 & 0 & \cdots & \gamma \end{bmatrix} \begin{bmatrix} u_{1,j+1} \\ u_{2,j+1} \\ u_{3,j+1} \\ \vdots \\ u_{n-2,j+1} \\ u_{n-1,j+1} \end{bmatrix} = \begin{bmatrix} u_{1,j} + \alpha u_0 \\ u_{2,j} \\ u_{3,j} \\ \vdots \\ u_{n-2,j} \\ u_{n-1,j} + \alpha u_n \end{bmatrix}$$

where $\gamma = 1 + 2\alpha$. Note that u_0 and u_n are the known boundary conditions, assumed to be independent of time. By solving the preceding equation system, we determine $u_1, u_2, ..., u_{n-1}$ at time step $j+1$ from $u_1, u_2, ..., u_{n-1}$ at time step j. This approach is called the implicit method. Compared with the explicit method, each time step requires more computation; however, the method has the significant advantage that it is unconditionally stable. However, although stability does not place any restriction on α, h and k must be chosen to keep the discretization error small to maintain accuracy.

The following function `heat` implements an implicit finite difference solution for the parabolic differential equation (6.24).

```
function [u alpha] = heat(nx,hx,nt,ht,init,lowb,hib,K)
% Solves parabolic equ'n.
% e.g. heat flow equation.
% Example call: [u alpha] = heat(nx,hx,nt,ht,init,lowb,hib,K)
% nx, hx are number and size of x panels
% nt, ht are number and size of t panels
```

```
% init is a row vector of nx+1 initial values of the function.
% lowb & hib are boundaries at low and hi values of x.
% Note that lowb and hib are scalar values.
% K is a constant in the parabolic equation.
alpha = K*ht/hx^2;
A = zeros(nx-1,nx-1); u = zeros(nt+1,nx+1);
u(:,1) = lowb*ones(nt+1,1);
u(:,nx+1) = hib*ones(nt+1,1);
u(1,:) = init;
A(1,1) = 1+2*alpha; A(1,2) = -alpha;
for i = 2:nx-2
    A(i,i) = 1+2*alpha;
    A(i,i-1) = -alpha; A(i,i+1) = -alpha;
end
A(nx-1,nx-2) = -alpha; A(nx-1,nx-1) = 1+2*alpha;
b(1,1) = init(2)+init(1)*alpha;
for i = 2:nx-2, b(i,1) = init(i+1); end
b(nx-1,1) = init(nx)+init(nx+1)*alpha;
[L,U] = lu(A);
for j = 2:nt+1
    y = L\b; x = U\y;
    u(j,2:nx) = x'; b = x;
    b(1,1) = b(1,1)+lowb*alpha;
    b(nx-1,1) = b(nx-1,1)+hib*alpha;
end
```

We now use the function heat to study how the temperature distribution in a brick wall varies with time. The wall is 0.3 m thick and is initially at a uniform temperature of 100°C. For the brickwork, $K = 5 \times 10^{-7}$ m/s^2. If the temperature of both surfaces is suddenly lowered to 20°C and kept at this temperature, we wish to plot the subsequent variation of temperature through the wall at 440 s (7.33 min) intervals over a period of 22,000 s (366.67 min).

To study this problem we will use a mesh with 15 subdivisions of x and 50 subdivisions of t in e4s604 thus:

```
% e4s604.m
K = 5e-7; thick = 0.3; tfinal = 22000;
nx = 15; hx = thick/nx;
nt = 50; ht = tfinal/nt;
init = 100*ones(1,nx+1); lowb = 20; hib = 20;
[u al] = heat(nx,hx,nt,ht,init,lowb,hib,K);
alpha = al, surfl(u)
axis([0 nx+1 0 nt+1 0 120])
view([-217 30]), xlabel('x - node nos.')
ylabel('Time - node nos.'), zlabel('Temperature')
```

Running script e4s604 gives

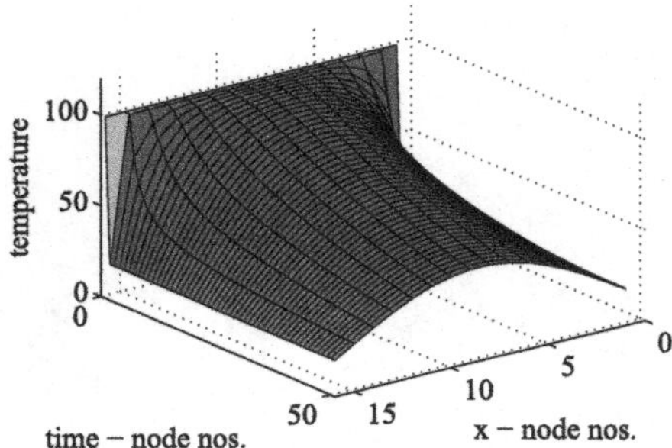

FIGURE 6.9

Plot shows how the distribution of temperature through a wall varies with time.

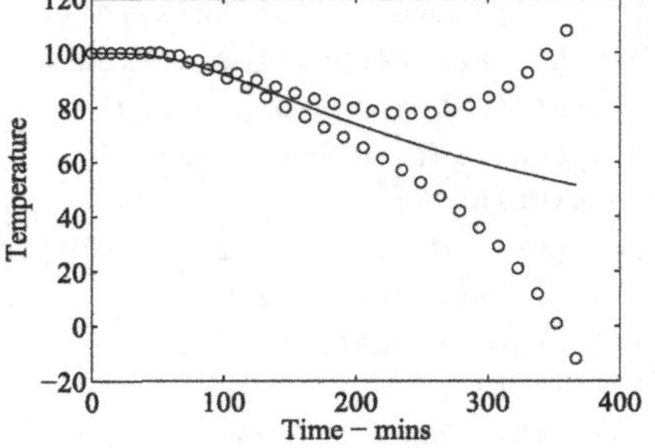

FIGURE 6.10

Variation in the temperature in the center of a wall. The steadily decaying solution denote by the solid line was generated using the implicit method of solution; the temperatures computed by the explicit method of solution are denoted by o. Note how the explicit method of solution gives temperatures that are oscillating and diverging with time.

```
alpha =
    0.5500
```

together with the plot shown in Fig. 6.9. The preceding plot shows how the temperature across the wall decreases with time. Fig. 6.10 shows the variation of temperature with time at the center of the wall, calculated by both the implicit method (using the MATLAB function `heat`) and the explicit method using the same mesh size. In the latter case a MATLAB function is not provided. It is seen that the solution determined using the explicit method becomes unstable with increasing time. We expect this because the mesh size has been chosen to make $\alpha = 0.55$.

6.6 HYPERBOLIC PARTIAL DIFFERENTIAL EQUATIONS

Consider the following equation:

$$c^2 \frac{\partial^2 u}{\partial x^2} = \frac{\partial^2 u}{\partial t^2}, \quad 0 < x < L \text{ and } t > 0 \tag{6.29}$$

This is the one-dimensional wave equation, and like the heat-flow problem of Section 6.5, its solution usually propagates in an open domain. Eq. (6.29) describes the wave in a taught string where c is the velocity of propagation of the waves in the string. Comparing (6.29) with (6.23), we see that $B^2 - 4AC = -4c^2(-1)$. Since c^2 is positive, $B^2 - 4AC > 0$ and the equation is hyperbolic. Eq. (6.29) is subject to boundary conditions at $x = 0$ and $x = L$ and also subject to initial conditions when $t = 0$.

We now develop equivalent finite difference approximations for these equations. Dividing L into n sections so that $h = L/n$ and consider time steps of duration k. Approximating (6.29) by central finite difference approximations based on (6.7) at node (i, j), we have

$$c^2\left(\frac{u_{i-1,j}-2u_{i,j}+u_{i+1,j}}{h^2}\right)=\left(\frac{u_{i,j-1}-2u_{i,j}+u_{i,j+1}}{k^2}\right)$$

or

$$(u_{i-1,j}-2u_{i,j}+u_{i+1,j})-(1/\alpha^2)(u_{i,j-1}-2u_{i,j}+u_{i,j+1})=0$$

where $\alpha^2=c^2k^2/h^2$, $i=0,1,...,n$, and $j=0,1,....$ Node (i, j) is the point $x=ih$ at time $t=jk$. Rearranging the preceding equation gives

$$u_{i,j+1}=\alpha^2(u_{i-1,j}+u_{i+1,j})+2(1-\alpha^2)u_{i,j}-u_{i,j-1} \tag{6.30}$$

When $j=0$, (6.30) becomes

$$u_{i,\,1}=\alpha^2(u_{i-1,\,0}+u_{i+1,\,0})+2(1-\alpha^2)u_{i,\,0}-u_{i,\,-1} \tag{6.31}$$

To solve a hyperbolic partial differential equation, initial values of $u(x)$ and $\partial u/\partial t$ must be specified. Let these values be U_i and V_i, respectively, where $i=0,1,...,n$. We can replace $\partial u/\partial t$ by its central finite difference approximation based on (6.6) thus:

$$V_i=(-u_{i,-1}+u_{i,1})/(2k)$$

Thus,

$$-u_{i,-1}=2kV_i-u_{i,1} \tag{6.32}$$

In (6.31) we replace $u_{i,0}$ by U_i and $u_{i,-1}$ by using (6.32) to give

$$u_{i,1}=\alpha^2(U_{i-1}+U_{i+1})+2(1-\alpha^2)U_i+2kV_i-u_{i,1}$$

so that

$$u_{i,1}=\alpha^2(U_{i-1}+U_{i+1})/2+(1-\alpha^2)U_i+kV_i \tag{6.33}$$

Eq. (6.33) is the starting equation and allows us to determine the values of u at time step $j=1$. Once we obtain these values, we can use (6.30) to provide an explicit method of solution. To ensure stability, the parameter α should be equal to or less than one. However, if α is less than one, the solution becomes less accurate.

The following function, `fwave`, implements an explicit finite difference solution for (6.29).

```
function [u alpha] = fwave(nx,hx,nt,ht,init,initslope,lowb,hib,c)
% Solves hyperbolic equ'n, e.g. wave equation.
% Example: [u alpha] = fwave(nx,hx,nt,ht,init,initslope,lowb,hib,c)
% nx, hx are number and size of x panels
% nt, ht are number and size of t panels
% init is a row vector of nx+1 initial values of the function.
% initslope is a row vector of nx+1 initial derivatives of
```

```
% the function.
% lowb is a column vector of nt+1 boundary values at the
% low value of x.
% hib is a column vector of nt+1 boundary values at hi value of x.
% c is a constant in the hyperbolic equation.
alpha = c*ht/hx;
u = zeros(nt+1,nx+1);
u(:,1) = lowb; u(:,nx+1) = hib; u(1,:) = init;
for i = 2:nx
    u(2,i) = alpha^2*(init(i+1)+init(i-1))/2+(1-alpha^2)*init(i) ...
    +ht*initslope(i);
end
for j = 2:nt
    for i = 2:nx
        u(j+1,i)=alpha^2*(u(j,i+1)+u(j,i-1))+(2-2*alpha^2)*u(j,i) ...
        -u(j-1,i);
    end
end
```

We now use the `fwave` function to examine the effect of displacing the boundary at one end of a taut string by 10 units in a positive direction for the time period $t = 0.1$ to $t = 4$ units.

```
% e4s605.m
T = 4; L = 1.6;
nx = 16; nt = 40; hx = L/nx; ht = T/nt;
c = 1; t = 0:nt;
hib = zeros(nt+1,1); lowb = zeros(nt+1,1);
lowb(2:5,1) = 10;
init = zeros(1,nx+1); initslope = zeros(1,nx+1);
[u al] = fwave(nx,hx,nt,ht,init,initslope,lowb,hib,c);
alpha = al, surfl(u)
axis([0 16 0 40 -10 10])
xlabel('Position along string')
ylabel('Time'), zlabel('Vertical displacement')
```

Running this script produces the following output, together with Fig. 6.11.

```
alpha =
     1
```

Fig. 6.11 shows that the disturbance at the boundary travels along the string. At the other boundary, it is reflected and becomes a negative disturbance. This process of reflection and reversal continues at each boundary. The disturbance travels at a velocity c and its shape does not change. Similarly, pressure fluctuations do not change as they travel along a speaking tube; if pressure fluctuations representing the sound “HELLO” enter the tube, the sound “HELLO” is detected at the other end. In practice, energy loss, which is not included in this model, would cause the amplitude of the disturbance to decay to zero over a period of time.

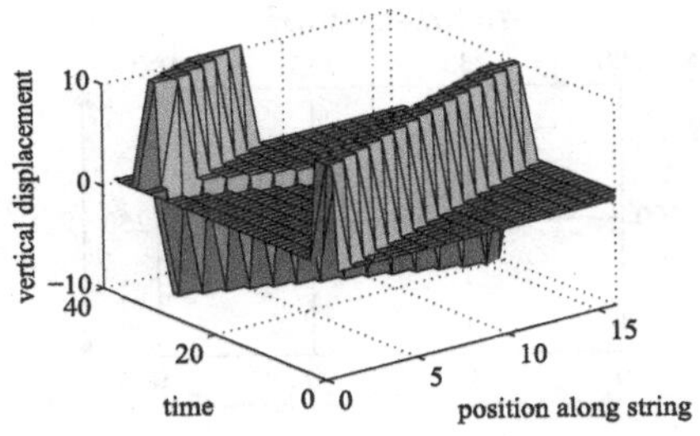

FIGURE 6.11

Solution of (6.29) subject to specific boundary and initial conditions.

6.7 ELLIPTIC PARTIAL DIFFERENTIAL EQUATIONS

The solution of a second-order elliptic partial differential equation is determined over a closed region, and the shape of the boundary and its condition at every point must be specified. Some important second-order elliptic partial differential equations, which arise naturally in the description of physical systems, are

$$\text{Laplace's equation: } \nabla^2 z = 0 \tag{6.34}$$

$$\text{Poisson's equation: } \nabla^2 z = F(x, y) \tag{6.35}$$

$$\text{Helmholtz's equation: } \nabla^2 z + G(x, y)z = F(x, y) \tag{6.36}$$

where $\nabla^2 z = \partial^2 z/\partial x^2 + \partial^2 z/\partial y^2$ and $z(x, y)$ is an unknown function. Note that the Laplace and Poisson equations are special cases of Helmholtz's equation. In general, these equations must satisfy boundary conditions that are specified in terms of either the function value or the derivative of the function which is normal to the boundary. Furthermore, a problem can have mixed boundary conditions. If we compare (6.34), (6.35), and (6.36) to the standard second-order partial differential equation in two variables, that is

$$A(x, y)\frac{\partial^2 z}{\partial x^2} + B(x, y)\frac{\partial^2 z}{\partial x \partial y} + C(x, y)\frac{\partial^2 z}{\partial y^2} + f\left(x, y, z, \frac{\partial z}{\partial x}, \frac{\partial z}{\partial y}\right) = 0$$

we see that in each case $A = C = 1$ and $B = 0$, so that $B^2 - 4AC < 0$, confirming that the equations are elliptic.

Laplace's equation is homogeneous, and if a problem has boundary conditions that are also homogeneous then the solution, $z = 0$, will be trivial. Similarly in (6.35), if $F(x, y) = 0$ and the problem boundary conditions are homogeneous, then $z = 0$. However, in (6.36) we can scale $G(x, y)$ by a factor λ, so that (6.36) becomes

$$\nabla^2 z + \lambda G(x, y)z = 0 \tag{6.37}$$

This is a characteristic or eigenvalue problem, and we can determine values of λ and corresponding non-trivial values of $z(x, y)$.

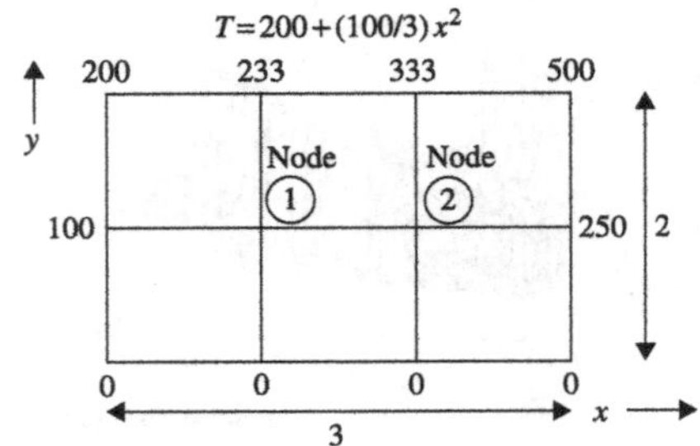

FIGURE 6.12

Temperature distribution around a plane section. Nodes 1 and 2 are shown.

The elliptic equations (6.34) through (6.37) can only be solved in a closed form for a limited number of situations. For most problems, it is necessary to use a numerical approximation. Finite difference methods are relatively simple to apply, particularly for rectangular regions. We will now use the finite difference approximation for $\nabla^2 z$, given by (6.12) or (6.13), to the solve some elliptic partial differential equations over a rectangular domain.

Example 6.3. *Laplace's equation.* Determine the distribution of temperature in a rectangular plane section, subject to a temperature distribution around its edges as follows:

$$x = 0, \quad T = 100y; \quad x = 3, \quad T = 250y; \quad y = 0, \quad T = 0; \quad \text{and} \quad y = 2, \quad T = 200 + (100/3)x^2$$

The section shape, the boundary temperature distribution and the two chosen nodes where we require the values of the temperature to be computed, are shown in Fig. 6.12.

The temperature distribution is described by Laplace's equation. Solving this equation by the finite difference method, we apply (6.13) to nodes 1 and 2 of the mesh shown in Fig. 6.12. This gives

$$(233.33 + T_2 + 0 + 100 - 4T_1)/h^2 = 0$$
$$(333.33 + 250 + 0 + T_1 - 4T_2)/h^2 = 0$$

where T_1 and T_2 are the unknown temperatures at nodes 1 and 2, respectively, and $h = 1$. Rearranging these equations gives

$$\begin{bmatrix} -4 & 1 \\ 1 & -4 \end{bmatrix} \begin{bmatrix} T_1 \\ T_2 \end{bmatrix} = \begin{bmatrix} -333.33 \\ -583.33 \end{bmatrix}$$

Solving this equation, we have $T_1 = 127.78$ and $T_2 = 177.78$.

If we require a more accurate solution of Laplace's equation, then we must use more nodes and the computation burden increases rapidly. The following MATLAB function `ellipgen` uses the finite difference approximation (6.12) to solve the general elliptic partial differential equations (6.34) through (6.37) for a rectangular domain only. The function is also limited to problems in which the boundary value is specified by values of the function $z(x, y)$, not its derivative. If the user calls the function with

10 arguments, the function solves (6.34) through (6.36); see Examples 6.4 and 6.5. Calling it with the first six arguments causes it to solve (6.37); see Example 6.6.

```
function [a,om] = ellipgen(nx,hx,ny,hy,G,F,bx0,bxn,by0,byn)
% Function either solves:
% nabla^2(z)+G(x,y)*z = F(x,y) over a rectangular region.
% Function call: [a,om]=ellipgen(nx,hx,ny,hy,G,F,bx0,bxn,by0,byn)
% hx, hy are panel sizes in x and y directions,
% nx, ny are number of panels in x and y directions.
% F and G are (nx+1,ny+1) arrays representing F(x,y), G(x,y).
% bx0 and bxn are row vectors of boundary conditions at x0 and xn.
% each beginning at y0. Each is (ny+1) elements.
% by0 and byn are row vectors of boundary conditions at y0 and yn.
% each beginning at x0. Each is (nx+1) elements.
% a is an (nx+1,ny+1) array of sol'ns, inc the boundary values.
% om has no interpretation in this case.
% or the function solves
% (nabla^2)z+lambda*G(x,y)*z = 0 over a rectangular region.
% Function call: [a,om]=ellipgen(nx,hx,ny,hy,G,F)
% hx, hy are panel sizes in x and y directions,
% nx, ny are number of panels in x and y directions.
% G are (ny+1,nx+1) arrays representing G(x,y).
% In this case F is a scalar and specifies the
% eigenvector to be returned in array a.
% Array a is an (ny+1,nx+1) array giving an eigenvector,
% including the boundary values.
% The vector om lists all the eigenvalues lambda.
nmax = (nx-1)*(ny-1); r = hy/hx;
a = zeros(ny+1,nx+1); p = zeros(ny+1,nx+1);
if nargin==6
    ncase = 0; mode = F;
end
if nargin==10
    test = 0;
    if F==zeros(nx+1,ny+1), test = 1; end
    if bx0==zeros(1,ny+1), test = test+1; end
    if bxn==zeros(1,ny+1), test = test+1; end
    if by0==zeros(1,nx+1), test = test+1; end
    if byn==zeros(1,nx+1), test = test+1; end
    if test==5
        disp('WARNING - problem has trivial solution, z = 0.')
        disp('To obtain eigensolution use 6 parameters only.')
        return
    end
    bx0 = bx0(1,ny+1:-1:1); bxn = bxn(1,ny+1:-1:1);
```

```
    a(1,:) = byn; a(ny+1,:) = by0;
    a(:,1) = bx0'; a(:,nx+1) = bxn'; ncase = 1;
end
for i = 2:ny
    for j = 2:nx
        nn = (i-2)*(nx-1)+(j-1);
        q(nn,1) = i; q(nn,2) = j; p(i,j) = nn;
    end
end
C = zeros(nmax,nmax); e = zeros(nmax,1); om = zeros(nmax,1);
if ncase==1, g = zeros(nmax,1); end
for i = 2:ny
    for j = 2:nx
        nn = p(i,j); C(nn,nn) = -(2+2*r^2); e(nn) = hy^2*G(j,i);
        if ncase==1, g(nn) = g(nn)+hy^2*F(j,i); end
        if p(i+1,j)~=0
            np = p(i+1,j); C(nn,np) = 1;
        else
            if ncase==1, g(nn) = g(nn)-by0(j); end
        end
        if p(i-1,j)~=0
            np = p(i-1,j); C(nn,np) = 1;
        else
            if ncase==1, g(nn) = g(nn)-byn(j); end
        end
            if p(i,j+1)~=0
                np = p(i,j+1); C(nn,np) = r^2;
            else
            if ncase==1, g(nn) = g(nn)-r^2*bxn(i); end
        end
        if p(i,j-1)~=0
            np = p(i,j-1); C(nn,np) = r^2;
        else
            if ncase==1, g(nn) = g(nn)-r^2*bx0(i); end
        end
    end
end
if ncase==1
    C = C+diag(e); z = C\g;
    for nn = 1:nmax
        i = q(nn,1); j = q(nn,2); a(i,j) = z(nn);
    end
else
    [u,lam] = eig(C,-diag(e));
```

```
    [om,k] = sort(diag(lam)); u = u(:,k);
    for nn = 1:nmax
        i = q(nn,1); j = q(nn,2);
        a(i,j) = u(nn,mode);
    end
end
```

We now give examples of the application of the `ellipgen` function.

Example 6.4. Use function `ellipgen` to solve Laplace's equation over a rectangular region subject to the boundary conditions shown in Fig. 6.12. The script `e4s606.m` calls the function to solve this problem using a 12 × 12 mesh. The example is the same as Example 6.3, but a finer mesh is used in the solution.

```
% e4s606.m
Lx = 3; Ly = 2;
nx = 12; ny = 12; hx = Lx/nx; hy = Ly/ny;
by0 = 0*[0:hx:Lx];
byn = 200+(100/3)*[0:hx:Lx].^2;
bx0 = 100*[0:hy:Ly];
bxn = 250*[0:hy:Ly];
F = zeros(nx+1,ny+1); G = F;
a = ellipgen(nx,hx,ny,hy,G,F,bx0,bxn,by0,byn);
aa = flipud(a);  colormap(gray)
surfl(aa)
xlabel('x direction')
ylabel('y direction')
zlabel('Temperature')
axis([0 12 0 12 0 500])
```

The output from this script is the surface plot shown in Fig. 6.13. The actual temperatures can be obtained from `aa`.

Example 6.5. *Poisson's equation*. Determine the deflection of a uniform square membrane, held at its edges and subject to a distributed load which can be approximated to a unit load at each node. This problem is described by Poisson's equation, (6.35), where $F(x, y)$ specifies the load on the membrane. We use the following script to determine the deflection of this membrane using the MATLAB function `ellipgen`.

```
% e4s607.m
Lx = 1; Ly = 1;
nx = 18; ny = 18; hx = Lx/nx; hy = Ly/ny;
by0 = zeros(1,nx+1); byn = zeros(1,nx+1);
bx0 = zeros(1,ny+1); bxn = zeros(1,ny+1);
F = -ones(nx+1,ny+1); G = zeros(nx+1,ny+1);
a = ellipgen(nx,hx,ny,hy,G,F,bx0,bxn,by0,byn);
```

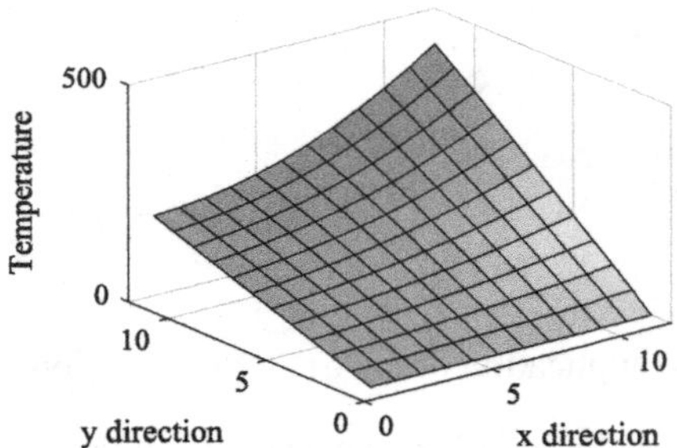

FIGURE 6.13

Finite difference estimate for the temperature distribution for the problem defined in Fig. 6.12.

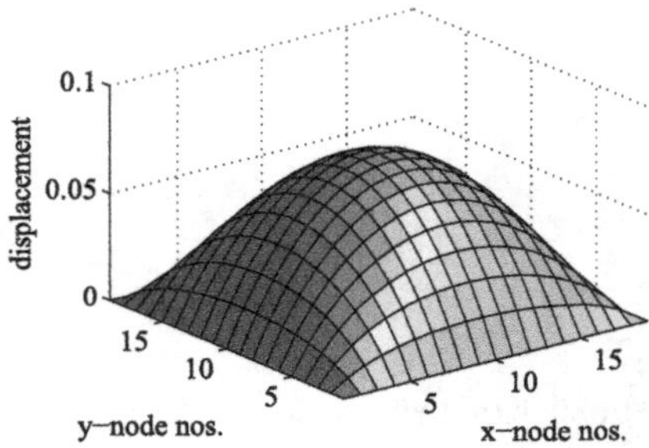

FIGURE 6.14

Deflection of a square membrane subject to a distributed load.

```
surfl(a)
axis([1 nx+1 1 ny+1 0 0.1])
xlabel('x-node nos.'), ylabel('y-node nos.')
zlabel('Displacement')
max_disp = max(max(a))
```

Running this script gives the output shown in Fig. 6.14 together with

```
max_disp =
    0.0735
```

This compares with the exact value of 0.0737.

Example 6.6. *Characteristic value problem.* Determine the natural frequencies and mode shapes of a freely vibrating square membrane held at its edges. This problem is described by the eigenvalue problem (6.37). The natural frequencies are related to the eigenvalues, and the mode shapes are the eigenvectors. The following MATLAB script, `e4s608`, determines the eigenvalues and vectors. It calls the function `ellipgen` and outputs a list of eigenvalues and provides Fig. 6.15, showing the second mode shape of the membrane.

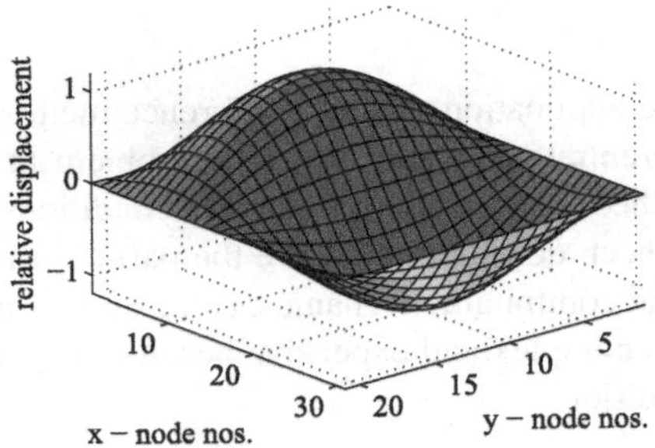

FIGURE 6.15

Finite difference approximation of the second mode of vibration of a uniform rectangular membrane.

Table 6.1 Finite difference (FD) approximations compared with exact eigenvalues for a uniform rectangular membrane

FD approximation	Exact	Error (%)
14.2318	14.2561	0.17
27.3312	27.4156	0.31
43.5373	43.8649	0.75
49.0041	49.3480	0.70
56.6367	57.0244	0.70

```
% e4s608.m
Lx = 1; Ly = 1.5;
nx = 20; ny = 30; hx = Lx/nx; hy = Ly/ny;
G = ones(nx+1,ny+1); mode = 2;
[a,om] = ellipgen(nx,hx,ny,hy,G,mode);
eigenvalues = om(1:5), surf(a)
view(140,30)
axis([1 nx+1 1 ny+1 -1.2 1.2])
xlabel('x - node nos.'), ylabel('y - node nos.')
zlabel('Relative displacement')
```

Running script `e4s608.m` gives

```
eigenvalues =
   14.2318
   27.3312
   43.5373
   49.0041
   56.6367
```

These eigenvalues compare with the exact values given in Table 6.1.

6.8 SUMMARY

In this chapter, we examined the application of finite difference methods to a broad range of second-order ordinary and partial differential equations. A major problem in the development of scripts is the difficulty of accounting for the wide variety of boundary conditions and boundary shapes that can occur. Software packages have been developed to solve the partial differential equations that arise in computational fluid dynamics and continuum mechanics, using either finite difference or finite element methods; however, they are both complex and expensive because they allow the user total freedom to define boundary shapes and conditions.

6.9 PROBLEMS

6.1. Classify the following second-order partial differential equations:

$$\frac{\partial^2 y}{\partial t^2} + a\frac{\partial^2 y}{\partial x \partial t} + \frac{1}{4}(a^2 - 4)\frac{\partial^2 y}{\partial x^2} = 0$$

$$\frac{\partial u}{\partial t} - \frac{\partial}{\partial x}\left(A(x,t)\frac{\partial u}{\partial x}\right) = 0$$

$$\frac{\partial^2 \varphi}{\partial x^2} = k\frac{\partial^2(\varphi^2)}{\partial y^2} \quad \text{where } k > 0$$

6.2. Use the shooting method to solve $y'' + y' - 6y = 0$, where the prime denotes differentiation with respect to x, given the boundary conditions $y(0) = 1$ and $y(1) = 2$. Note that an illustrative script for the shooting method is given in Section 6.2. Use trial slopes in the range $-3 : 0.5 : 2$. Compare your results with those you obtain using the finite difference method with 10 divisions. The finite difference method is implemented by function `twopoint`. Note that the exact solution is

$$y = 0.2657\exp(2x) + 0.7343\exp(-3x)$$

6.3. (a) Use the shooting method to solve $y'' - 62y' + 120y = 0$, where the prime denotes differentiation with respect to x, given the boundary conditions $y(0) = 0$ and $y(1) = 2$. Solve this equation by applying the shooting method, using trial slopes in the range $-0.5 : 0.1 : 0.5$. Note that the exact solution is

$$y = 1.751302152539304 \times 10^{-26}\{\exp(60x) - \exp(2x)\}$$

(b) By substituting $x = 1 - p$ in the original differential equation, show that $y'' + 62y' + 120y = 0$, where the prime denotes differentiation with respect to p. Note that the boundary conditions of this problem are $y(0) = 2$ and $y(1) = 0$. Solve this equation by applying the shooting method, using trial slopes in the range 0 to -150 with a step of -30 at $p = 0$. Note that a very good approximation to the solution is $y = 2\exp(-60p)$.

Compare the two answers you obtain for (a) and (b). Note that an illustrative script for the shooting method is given in Section 6.2. Also solve (a) and (b) using the finite difference method, implemented in `twopoint`. Use 10 divisions and repeat with 50 divisions. You should plot your answers and compare with a plot of the exact solution.

6.4. Solve the boundary value problem $xy'' + 2y' - xy = e^x$ given that $y(0) = 0.5$ and $y(2) = 3.694528$, using the finite difference method implemented by the function `twopoint`. Use 10 divisions in the finite difference solution and plot the results, together with the exact solution, $y = \exp(x)/2$.

6.5. Determine the finite difference equivalence of the characteristic value problem defined by $y'' + \lambda y = 0$, where $y(0) = 0$ and $y(2) = 0$. Use 20 divisions in the finite difference method. Then solve the finite difference equations using the MATLAB function `eig` to determine the lowest value of λ, that is the lowest eigenvalue.

6.6. Solve the parabolic equation (6.24) with $K = 1$, subject to the following boundary conditions: $u(0,t) = 0$, $u(1,t) = 10$, $u(x,0) = 0$ for all x except $x = 1$. When $x = 1$, $u(1,0) = 10$. Use the function `heat` to determine the solution for $t = 0$ to 0.5 in steps of 0.01 with 20 divisions of x. You should plot the solution for ease of visualization.

6.7. Solve the wave equation, (6.29) with $c = 1$, subject to the following boundary and initial conditions: $u(t,0) = u(t,1) = 0$, $u(0,x) = \sin(\pi x) + 2\sin(2\pi x)$, and $u_t(0,x) = 0$, where the subscript t denotes partial differentiation with respect to t. Use the function `fwave` to determine the solution for $t = 0$ to 4.5 in steps of 0.05, and use 20 divisions of x. Plot your results and compare with a plot of the exact solution, which is given by

$$u = \sin(\pi x)\cos(\pi t) + 2\sin(2\pi x)\cos(2\pi t).$$

6.8. Solve the equation

$$\nabla^2 V + 4\pi^2(x^2 + y^2)V = 4\pi\ \cos\{\pi(x^2 + y^2)\}$$

over the square region $0 \le x \le 0.5$ and $0 \le y \le 0.5$. The boundary conditions are

$$V(x,0) = \sin(\pi x^2),\quad V(x,0.5) = \sin\{\pi(x^2 + 0.25)\}$$

$$V(0,y) = \sin(\pi y^2),\quad V(0.5,y) = \sin\{\pi(y^2 + 0.25)\}$$

Use the function `ellipgen` to solve this equation with 15 divisions of x and y. Plot your results and compare with a plot of the exact solution, which is given by $V = \sin\{\pi(x^2 + y^2)\}$.

6.9. Solve the eigenvalue problem $\nabla^2 z + \lambda G(x,y)z = 0$ over a rectangular region bounded by $0 \le x \le 1$ and $0 \le y \le 1.5$, $z = 0$ at all boundaries. Use the function `ellipgen` with six divisions in x and nine divisions in y. The function $G(x,y)$ over this grid is given by the MATLAB statements `G = ones(10,7); G(4:7,3:5) = 3*ones(4,3);`. This represents a membrane with a central area thicker than its periphery. The eigenvalues are related to the natural frequencies of this membrane.

6.10. Solve Poisson's equation $\nabla^2\phi + 2 = 0$ over the region shown with $a = 1$ at the boundary where $\phi = 0$. You will have to assemble the finite difference equation by hand, applying (6.13) to each of the 10 nodes, and then use MATLAB to solve the resulting linear equation system.

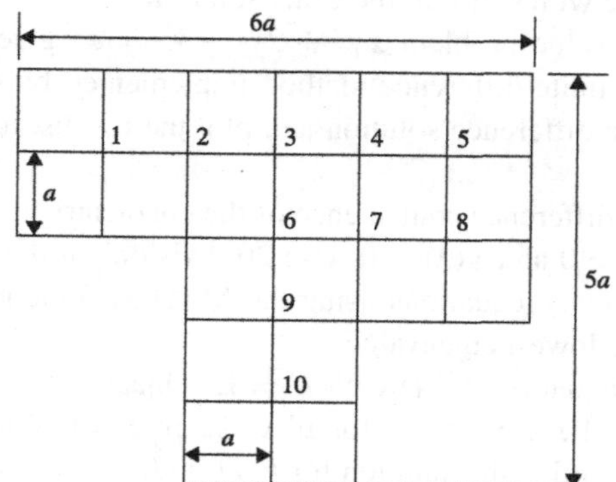

CHAPTER

ANALYZING DATA

7

Abstract

In this chapter we consider a variety of methods for interpolating and fitting functions to data; describe some of the MATLAB functions that are available for this purpose and develop some additional ones. Least squares methods are described together with regression analysis, the Kalman filter and principal component analysis.

7.1 INTRODUCTION

We fit functions to two general classes of data: data which is exact and data which is known to contain errors. When we fit a function to exact data, we fit to the points exactly. When we fit a function to data that is known to contain errors, we try to obtain the best fit to the trend of the data, using some criterion. The user must exercise skill in making a sensible choice of the function to fit the specific data.

We begin by examining polynomial interpolation which is an example of fitting to exact data.

7.2 INTERPOLATION USING POLYNOMIALS

Suppose y is some unknown function of x. Given a table of values of x and y, we may wish to obtain a value of y corresponding to a value of x that is not tabulated. Interpolation implies that the untabulated value of x is within the range of the tabulated data. If the untabulated x is outside this range, the process is called *extrapolation* and is often less accurate.

The simplest form of interpolation is linear interpolation. In this method only the pair of data points enclosing the required value are used. Thus if (x_0, y_0) and (x_1, y_1) are two adjacent data points in a tabulation, to obtain the value of y corresponding to an x where $x_0 < x < x_1$, we fit the straight line $y = ax + b$ to these points and evaluate y thus:

$$y = [y_0(x_1 - x) + y_1(x - x_0)]/(x_1 - x_0) \tag{7.1}$$

We may use the MATLAB function `interp1` for this purpose. For example, consider the function $y = x^{1.9}$, tabulated at $x = 1, 2, ..., 5$. If we require estimates of y for $x = 2.5$ and 3.8, we may use `interp1` setting the third parameter as `'linear'` to obtain linear interpolation as follows:

```
>> x = 1:5;
>> y = x.^1.9;
>> interp1(x,y,[2.5,3.8],'linear')
```

Numerical Methods. https://doi.org/10.1016/B978-0-12-812256-3.00016-6

```
ans =
    5.8979    12.7558
```

The exact answers are $y = 5.7028$ and $y = 12.6354$ corresponding to $x = 2.5$ and 3.8 respectively. For some applications the values obtained by linear interpolation may be sufficiently accurate.

Interpolation becomes more accurate when more of the tabulated data values are used because we can fit a higher-degree polynomial. A polynomial of degree n can be adjusted to pass through $n + 1$ data points. We do not need to know the coefficients of the polynomial explicitly, but they are used implicitly in the procedure to estimate y for a given value of x. For example, MATLAB allows cubic interpolation to be used by calling `interp1` with the third parameter set as `'pchip'`. The following example implements cubic interpolation using the same data as the previous example.

```
>> interp1(x,y,[2.5 3.8],'pchip')

ans =
    5.6938    12.6430
```

It can be seen that the cubic interpolation has given a much more accurate result.

An algorithm that provides an efficient method for fitting any degree polynomial to data is Aitken's algorithm. In this procedure a sequence of polynomial functions are fitted to the data. As the degree of the polynomial is increased, more of the data points are used and the accuracy of the interpolation improves.

Aitken's algorithm proceeds as follows. Suppose we have five pairs of data values labeled $1, 2, ..., 5$ and we wish to determine y^*, the value of y corresponding to a given x^*. Initially the algorithm determines straight lines (i.e., first-degree polynomials) that pass through data points 1 and 2, 1 and 3, 1 and 4, and 1 and 5, as shown in Fig. 7.1A. These four straight lines allow the procedure to determine four, probably poor, estimates for y^*.

Using $x_2, x_3, ..., x_5$ from the tabulated data and the four estimates of y^* determined from the first-degree polynomial, the algorithm repeats the procedure above using these new points but this now provides second-degree polynomials through the sets of data points $\{1, 2, 3\}$, $\{1, 2, 4\}$, and $\{1, 2, 5\}$, as shown in Fig. 7.1B. From these second-degree polynomials, the procedure determines three improved estimates for y^*.

Using x_3, x_4, x_5 from the tabulated data and the three new estimates for y^* obtained from the second-degree polynomials, the algorithm computes the third-degree polynomials passing through the sets of data points $\{1, 2, 3, 4\}$ and $\{1, 2, 3, 5\}$, as shown in Fig. 7.1C to allow the procedure to determine two further improved estimates for y^*. Finally a fourth-degree polynomial is computed that fits all the data. This fourth-degree polynomial provides the best estimate for y^*, as shown in Fig. 7.1D.

Aitken's algorithm has two advantages. First of all it is very efficient. Each new estimate for y^* requires only two multiplications and one division so that for $n + 1$ data points, the estimate using all the data requires $n(n + 1)$ multiplications and $n(n + 1)/2$ divisions. It is interesting to note that if we attempted to determine the coefficients of a polynomial passing through $n+1$ data points by assembling a set of $n + 1$ linear equations, then in addition to the computation required to assemble the equations, we would require $(n + 1)^3/2$ multiplications and divisions to solve them. The second advantage of Aitken's algorithm is that the process is adding an additional point at each stage. This allows the algorithm to fit higher-degree polynomials at each stage and, in general, to increase the accuracy of

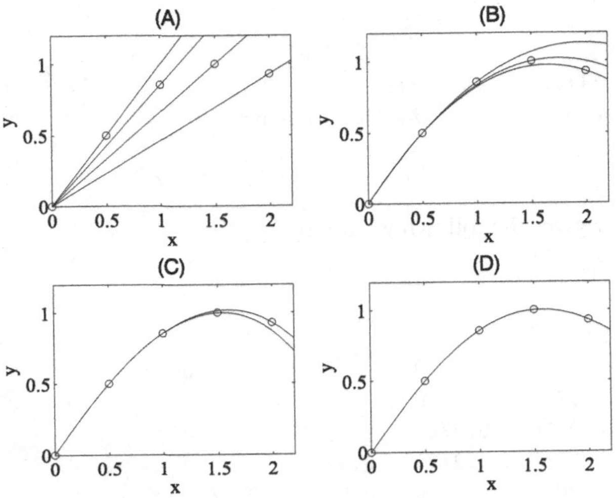

FIGURE 7.1

Increasing the degree of the polynomial to fit the given data. (A) 1st degree; (B) 2nd degree; (C) 3rd degree; (D) 4th degree.

the estimate. When there is no significant change in the estimate, the maximum obtainable accuracy has been reached and the process can be terminated, avoiding further and unnecessary computation. The MATLAB function `aitken` shown below implements Aitken's algorithm. The user must provide a set of data for the vectors `x` and `y`. The function then determines a value of `y` corresponding to `xval`. The function provides the best value and, if required, a table showing all the intermediate values of the Aitken algorithm can be obtained.

```
function [Q R] = aitken(x,y,xval)
% Aitken's method for interpolation.
% Example call: [Q R] = aitken(x,y,xval)
% x and y give the table of values. Parameter xval is
% the value of x at which interpolation is required.
% Q is interpolated value, R gives table of intermediate results.
n = length(x); P = zeros(n);
P(1,:) = y;
for j = 1:n-1
    for i = j+1:n
        P(j+1,i) = (P(j,i)*(xval-x(j))-P(j,j)*(xval-x(i)))/(x(i)-x(j));
    end
end
Q = P(n,n); R = [x.' P.'];
```

We now use this function to determine the reciprocal of 1.03 from a table of 10 equispaced values of x in the range 1 to 2 and $y = 1/x$. The script below calls the function `aitken` to solve this example:

```
% e4s701.m
x = 1:.2:2; y = 1./x;
[interpval table] = aitken(x,y,1.03);
fprintf('Interpolated value= %10.8f\n\n',interpval)
disp('Table = ')
disp(table)
```

Running script `e4s701.m` gives the following output:

```
interpolated value= 0.97095439

Table =
   1.0000    1.0000         0         0         0         0         0
   1.2000    0.8333    0.9750         0         0         0         0
   1.4000    0.7143    0.9786    0.9720         0         0         0
   1.6000    0.6250    0.9813    0.9723    0.9713         0         0
   1.8000    0.5556    0.9833    0.9726    0.9713    0.9710         0
   2.0000    0.5000    0.9850    0.9729    0.9714    0.9711    0.9710
```

Notice that the first column in this table contains the tabulated x values, the second column contains the tabulated y values and the remaining columns give successively higher-degree polynomial interpolants generated by Aitken's method. The zeros in this table are padding: The number of estimates in each column decreases as the estimates use more of the data. The exact value is $y = 0.970873786$; thus the Aitken interpolated value of $y = 0.97095439$ is correct to four decimal places. Linear interpolation gives 0.9750, a much poorer result with an error of approximately 0.2%.

Aitken's method provides an interpolated value of y for a given value of x by fitting a polynomial to the data but the coefficients of the polynomial are not determined explicitly. Conversely, we can fit a polynomial explicitly to the data, determine its coefficients, and then determine the required interpolated value by evaluating the polynomial. This approach may be less computationally efficient. The MATLAB function `polyfit(x,y,n)` fits a polynomial of degree n through the data given by x and y and returns the coefficients of descending powers of x. For an exact fit, n must equal $m - 1$ where m is the number of data points. The polynomial represented by the coefficients `p` can then be evaluated using the `polyval` function. For example, to determine the reciprocal of 1.03 from a table of 6 equispaced values of x in the range 1 to 2 and $y = 1/x$, we have

```
% e4s702.m
x = 1:.2:2; y = 1./x;
p = polyfit(x,y,5)
interpval = polyval(p,1.03);
fprintf('interpolated value = %10.8f\n',interpval)
```

Running script `e4s702.m` gives

```
p =
   -0.1033    0.9301   -3.4516    6.7584   -7.3618    4.2282

interpolated value = 0.97095439
```

Thus

$$y = -0.1033x^5 + 0.9301x^4 - 3.4516x^3 + 6.7584x^2 - 7.3618x + 4.2282$$

The interpolated value is identical to that given by Aitken's method, as indeed it must be (except for possible rounding errors in the computation) because there is only one polynomial that passes through all 6 data points and both methods have used it. We use the MATLAB function `polyfit` again in Section 7.7.

7.3 INTERPOLATION USING SPLINES

The spline is used to connect data points to each other using a curve which appears to the eye to be smooth, either for the purpose of visualization in design drawings or for interpolation. It has certain advantages over the use of a high-degree polynomial which has a tendency to oscillate between data values. This oscillation is illustrated in Fig. 7.4. Here an eight-degree polynomial if fitted to eight data points.

We begin with a historical example of ship design. Ships' hulls have always curved in a complex manner in two dimensions. Fig. 7.2 shows hull sections for a 74 gun British warship, circa 1813. The data points are taken from the original plans, and splines have been used to join the data points together smoothly. Each line shows a section of the ship; the innermost line is close to the stern, and the outermost line is near amidships. The graph gives a clear impression of the way the ship builder chose to reduce the ship's cross-section toward the stern.

Polynomials of varying degrees are used for splines, but here we only consider the cubic spline. The cubic spline is a series of cubic polynomials joining data points or "knots". Suppose we have n data points joined by $n-1$ polynomials. Each cubic polynomial has four unknown coefficients so that there are $4(n-1)$ coefficients to be determined. Obviously each polynomial must pass through the two data points it joins. This provides $2(n-1)$ equations that must be satisfied. In order that the polynomials join together smoothly, we require both continuity of slope (y') and curvature (y'') between adjacent polynomials at the $n-2$ internal data points. This gives $2(n-2)$ extra equations, making a total of $4n-6$ equations. With these equations we can determine an identical number of coefficients uniquely, and so to determine the $4n-4$ unknown coefficients, two further equations are required. The two remaining conditions can be chosen arbitrarily, but usually one of the following is used:

1. If the slope of the required curve is known at the outer ends, we can impose these two constraints. More often than not, these slopes are not known.
2. We can make the curvature at the outer ends zero, i.e., $y''_1 = y''_n = 0$. (These are called natural splines but have no particular advantage.)
3. We can make the curvature at x_1 and x_n equal to x_2 and x_{n-1}, respectively.
4. We can make the curvature at x_1 a linear extrapolation of the curvature at x_2 and x_3. Similarly, we make the curvature at x_n a linear extrapolation of the curvature at x_{n-1} and x_{n-2}.
5. We can make y''' continuous at x_2 and x_{n-1}. Since at any internal point, y, y', y'', and y''' are always made continuous, adding this condition is equivalent to using the same polynomial in the two outer panels. This is called the "not a knot" condition and is used in the MATLAB function `spline`.

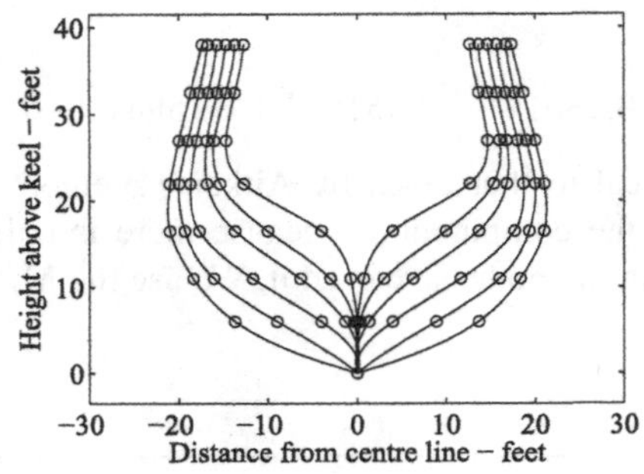

FIGURE 7.2

Use of splines to define cross-sections of a ship's hull.

Table 7.1 Data for spline fit

x	0	1	2	3	4
y	3	1	0	2	4

We now illustrate the two uses of the MATLAB function `spline` applied to the small set of data given in Table 7.1. Running the following script

```
% e4s703.m
x = 0:4; y = [3 1 0 2 4];
xval = 1.5; yval = spline(x,y,xval)
p = spline(x,y)
```

Running script `e4s703` gives

```
yval =
    0.1719

p =
     form: 'pp'
    breaks: [0 1 2 3 4]
     coefs: [4x4 double]
    pieces: 4
     order: 4
       dim: 1
```

where `yval` is the interpolated value. There may be some occasions when the user wishes to know the values of the coefficient of the polynomials. In this case the p-p form is required, where the abbreviation p-p means piecewise polynomial. The output `p` is a structure array that provides this information. In particular,

```
>> c = p.coefs

c =
    0.5417   -1.1250   -1.4167    3.0000
    0.5417    0.5000   -2.0417    1.0000
   -0.7083    2.1250    0.5833         0
   -0.7083   -0.0000    2.7083    2.0000
```

The parameter `c` gives all the required polynomial coefficient. In this example, these coefficients are combined with the powers of x as follows:

$$
\begin{aligned}
y &= c_{11}x^3 + c_{12}x^2 + c_{13}x + c_{14}, && 0 \le x \le 1 \\
y &= c_{21}(x-1)^3 + c_{22}(x-1)^2 + c_{23}(x-1) + c_{24}, && 1 \le x \le 2 \\
y &= c_{31}(x-2)^3 + c_{32}(x-2)^2 + c_{33}(x-2) + c_{34}, && 2 \le x \le 3 \\
y &= c_{41}(x-3)^3 + c_{42}(x-3)^2 + c_{43}(x-3) + c_{44}, && 3 \le x \le 4
\end{aligned}
$$

It is not necessary for the MATLAB user to know the details of how the p-p values are interpreted. MATLAB provides a function `ppval` which evaluates a composite polynomial provided its p-p values are known. If `x` and `y` are vectors of data, then `y1 = spline(x,y,x1)` is equivalent to the statements `p = spline(x,y); y2 = ppval(p,x1)`.

The script `e4s704.m` gives a plot of the spline fit to the data of Table 7.1.

```
% e4s704.m
x = 0:4; y = [3 1 0 2 4];
xx = 0:.1:4; yy = spline(x,y,xx);
plot(x,y,'o',xx,yy)
axis([0 4 -1 4])
xlabel('x'), ylabel('y')
```

Running script `e4s704.m` generates Fig. 7.3.

In Section 7.2 we show how polynomials are used in interpolation. However, their use is not always appropriate. When the data points are widely spaced, and when there are sudden changes in the y values, then polynomials can give very poor results. For example, the nine data points in Fig. 7.4 are taken from the function

$$y = 2\{1 + \tanh(2x)\} - x/10$$

This function changes abruptly, and if an eighth-degree polynomial is fitted to the data it oscillates and the path between data points bears no relationship to the true path. In contrast the spline fit is reasonably smooth and close to the true function.

The reader should note that the MATLAB function `interp1` can also be used to fit splines to data. The call `interp1(x,y,xi,'spline')` is identical to `spline(x,y,xi)`.

A special type of spline is the Bézier curve. This is a cubic function defined by four points. The two end points are used, together with two 'control' points. The slope of the curve at one end is a tangent to the line between that end point and one of the control points. Similarly, the slope at the other end

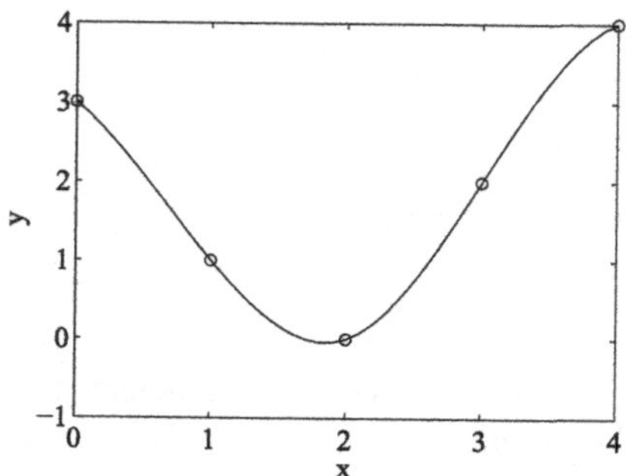

FIGURE 7.3

Spline fit to the data of Table 7.1 denoted by o.

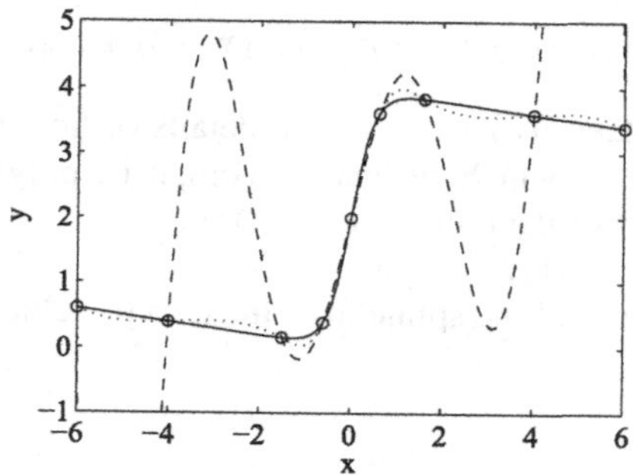

FIGURE 7.4

The *solid curve* shows the function $y = 2\{1 + \tanh(2x)\} - x/10$. The *dashed line* shows an eighth-degree polynomial fit; the *dotted line* shows a spline fit.

point is a tangent to the line between that end point and the other control point. In interactive computer graphics, the positions of the control points can be moved on the screen in order to adjust the slope of the curve at the end points.

7.4 MULTIPLE REGRESSION: LEAST SQUARES CRITERION

We now consider the problem of fitting a function to a relatively large amount of data that contains errors. It would not be sensible, nor computationally possible, to fit a very high-degree polynomial to a large amount of experimental data as may be done for interpolation. What is required is a function that smooths out fluctuations in the data due to errors and reveals any underlying trend. We therefore adjust the coefficients of a chosen function to provide a 'best fit' according to some criterion. For example, the criterion may be to minimize the maximum error, the sum of the modulus of the errors or the sum of the squares of the errors between the chosen function and the actual data points. The least squares method is the most widely used of these criteria. We now examine how this process is carried out.

Suppose we have an n component vector of observations $\mathbf{y}$ and p separate vectors of explanatory variables, $\mathbf{x}_1, \mathbf{x}_2, ..., \mathbf{x}_p$. The variables $\mathbf{x}_1, \mathbf{x}_2, ..., \mathbf{x}_p$ are generally independent vectors which are measured with negligible error or even controlled in an experiment and $\mathbf{y}$ is a single dependent variable, uncontrolled and containing random measurement errors. In some cases $\mathbf{x}_1$, $\mathbf{x}_2$, and so on, may be different functions of a single explanatory variable – in which case we call them *predictors* (see Sections 7.7 and 7.8). Our basic model consists of a *regression equation* and is said to be a regression of $\mathbf{y}$ upon the explanatory variables $\mathbf{x}_1, \mathbf{x}_2, ..., \mathbf{x}_p$. We assume a simple linear regression model, and so for observation x_i we have

$$y_i = \beta_0 + \beta_1 x_{1i} + \beta_2 x_{2i} + \ldots + \beta_p x_{pi} + \varepsilon_i,\ i = 1,\ 2,\ \ldots,\ n \tag{7.2}$$

where β_j $(j = 0, 1, 2..., p)$ are the unknown coefficients and ε_i are random errors. Initially we simply assume that these random errors are identically distributed with a zero mean and a common unknown variance, σ^2, and that they are independent of each other. This implies that

$$\beta_0 + \beta_1 x_{1i} + \beta_2 x_{2i} + ... + \beta_p x_{pi}$$

represents the mean value of y_i.

Our main task is to find an estimate b_j for each unknown β_j so that we can estimate the mean of y_i by fitting a function of the form

$$\hat{y}_j = b_0 + b_1 x_{1i} + b_2 x_{2i} + ... + b_p x_{pi},\ i = 1,\ 2,\ \ldots,\ n \tag{7.3}$$

The difference between the observed and the fitted value for the ith observation is

$$e_i = y_i - \hat{y}_i \tag{7.4}$$

The parameter e_i is called the *residual* and is an estimate of the corresponding random error ε_i. Our criterion for choosing the b_j estimates is that they should minimize the sum of the squares of the residuals, which is often called the *sum of squares of the errors* and is denoted by *SSE*. Thus

$$SSE = \sum_{i=1}^{n} e_i^2 = \sum_{i=1}^{n} (y_i - \hat{y}_i)^2 \tag{7.5}$$

To perform the necessary calculations efficiently, we rewrite the model in matrix form as follows. We define $\mathbf{b}$ as a $(p+1) \times 1$ vector such that

$$\mathbf{b} = [b_0 \quad b_1 \ldots b_p]^\mathrm{T}$$

Similarly we define $\mathbf{e}$, $\mathbf{y}$, $\mathbf{u}$, and $\mathbf{x}_j$ to be $n \times 1$ vectors thus:

$$\mathbf{e} = [e_1 \quad e_2 \ldots e_n]^\mathrm{T}$$

$$\mathbf{y} = [y_1 \quad y_2 \ldots y_n]^\mathrm{T}$$

$$\mathbf{u} = [1 \quad 1 \ldots 1]^\mathsf{T}$$

$$\mathbf{x}_j = [x_{j1} \quad x_{j2} \ldots x_{jn}]^\mathsf{T}, \quad j = 1, 2, \ldots, p$$

From these vectors the $n \times (p+1)$ matrix, $\mathbf{X}$, can be defined as follows:

$$\mathbf{X} = [\mathbf{u} \quad \mathbf{x}_1 \quad \mathbf{x}_2 \ldots \mathbf{x}_p]$$

This matrix corresponds to the coefficients of the equation system (7.3). From (7.4) the residuals may then be written as

$$\mathbf{e} = \mathbf{y} - \mathbf{Xb}$$

and the *SSE*, from (7.5), is then

$$SSE = \mathbf{e}^\mathsf{T}\mathbf{e} = (\mathbf{y} - \mathbf{Xb})^\mathsf{T}(\mathbf{y} - \mathbf{Xb}) \tag{7.6}$$

Differentiating the *SSE* with respect to $\mathbf{b}$, we get a vector of partial derivatives, as follows:

$$\frac{\partial}{\partial \mathbf{b}}(SSE) = -2\mathbf{X}^\mathsf{T}(\mathbf{y} - \mathbf{Xb})$$

Matrix differentiation is described in Appendix A. Equating this derivative to zero we have $\mathbf{Xb} = \mathbf{y}$ and this over-determined system of equations could be solved directly to obtain the coefficients $\mathbf{b}$ using the MATLAB operator \. However, in this instance it is more convenient to proceed as follows. Pre-multiplying $\mathbf{Xb} = \mathbf{y}$ by $\mathbf{X}^\mathsf{T}$, we obtain what are called the *normal equations* thus:

$$\mathbf{X}^\mathsf{T}\mathbf{Xb} = \mathbf{X}^\mathsf{T}\mathbf{y}$$

The formal solution of these equations is then

$$\mathbf{b} = (\mathbf{X}^\mathsf{T}\mathbf{X})^{-1}\mathbf{X}^\mathsf{T}\mathbf{y} = \mathbf{CX}^\mathsf{T}\mathbf{y} \tag{7.7}$$

where

$$\mathbf{C} = (\mathbf{X}^\mathsf{T}\mathbf{X})^{-1} \tag{7.8}$$

Note that $\mathbf{C}$ is a $(p+1) \times (p+1)$ square matrix.

Using the expression for $\mathbf{b}$ in (7.7), the vector of fitted values corresponding to $\mathbf{y}$ is

$$\hat{\mathbf{y}} = \mathbf{Xb} = \mathbf{XCX}^\mathsf{T}\mathbf{y}$$

So, defining $\mathbf{H} = \mathbf{XCX}^\mathsf{T}$, we can write

$$\hat{\mathbf{y}} = \mathbf{Hy} \tag{7.9}$$

The matrix $\mathbf{H}$, which converts $\mathbf{y}$ to $\hat{\mathbf{y}}$ ("y-hat") is called the *hat matrix* and plays an important role in the interpretation of the regression model. Among its important properties is that it is idempotent (see Appendix A).

From (7.7) it can be shown that minimum value of *SSE* is given by

$$SSE = \mathbf{y}^{\mathsf{T}}(\mathbf{I} - \mathbf{H})\mathbf{y} \tag{7.10}$$

where $\mathbf{I}$ is an $n \times n$ identity matrix. Our original data consisted of n sets of observations and we have introduced $p+1$ constraints into the system to estimate the parameters $\beta_0, \beta_1, ..., \beta_p$, so there are now $(n-p-1)$ degrees of freedom. Statistical theory shows that by dividing the minimum *SSE* from (7.10) by the number of degrees of freedom, we obtain an unbiased estimate of the unknown error variance, σ^2, which is

$$s^2 = \frac{SSE}{n-p-1} = \frac{\mathbf{y}^{\mathsf{T}}(\mathbf{I} - \mathbf{H})\mathbf{y}}{n-p-1} \tag{7.11}$$

On its own, the value of s obtained by fitting a single model is not very informative. However, we can also use $\mathbf{H}$ to evaluate the overall goodness of fit on an absolute scale and to examine how good each b_j is, as an estimate of the corresponding β_j.

The most widely used measure of the overall goodness of fit is the *coefficient of determination*, R^2, which is defined by

$$R^2 = \sum_{i=1}^{n} (\hat{y}_i - \bar{y})^2 \Big/ \sum_{i=1}^{n} (y_i - \bar{y})^2$$

where

$$\bar{y} = \frac{1}{n}\sum_{i=1}^{n} y_i$$

Thus, $\bar{y}$ is the mean of the observed $\mathbf{y}$ values. Using matrix notation, we can evaluate R^2 from the equivalent definition:

$$R^2 = \frac{\mathbf{y}^{\mathsf{T}}(\mathbf{H} - \mathbf{u}\mathbf{u}^{\mathsf{T}}/n)\mathbf{y}}{\mathbf{y}^{\mathsf{T}}(\mathbf{I} - \mathbf{u}\mathbf{u}^{\mathsf{T}}/n)\mathbf{y}} \tag{7.12}$$

The value of R^2 will lie between 0 and 1 and represents the proportion of the total observed variance of y that is accounted for by the explanatory variables. Thus a value close to 1 indicates that nearly all the observed variance is accounted for and we have a good fit. However, on its own this does not necessarily indicate that the model is satisfactory because the value of R^2 can always be increased by introducing more explanatory variables, even though their introduction can have other effects which are very undesirable. In the next section, we briefly consider some methods for deciding which explanatory variables should be included in the model.

7.5 DIAGNOSTICS FOR MODEL IMPROVEMENT

To see which variables might be removed in order to improve our original model, we now examine the b_j estimates in more detail to check whether they indicate that the corresponding β_j coefficients are non-zero. This would confirm that the corresponding x_j really do contribute to explaining y, or in the case of b_0, whether it is appropriate to include the constant term β_0 in the model.

If we make the assumption that the random errors, ε_i, are normally distributed, it can be shown that each of the b_j estimates behave as if they were observed values of normal random variables whose means are the β_j and whose covariance matrix is $s^2\mathbf{C}$ where $\mathbf{C}$ is defined by (7.8). Thus the covariance matrix is

$$s^2\mathbf{C} = s^2 \begin{bmatrix} c_{00} & c_{01} & c_{02} & \cdots & c_{0p} \\ & c_{11} & c_{12} & \cdots & c_{1p} \\ & & c_{22} & \cdots & c_{2p} \\ & & & & \vdots \\ & & & & c_{pp} \end{bmatrix}$$

Note that we numbered the rows and columns of $\mathbf{C}$ from zero to p. The matrix $\mathbf{C}$ is symmetric and so the sub-diagonal elements are not shown.

The variance of the distribution for b_j is s^2 times the corresponding diagonal element of $\mathbf{C}$, and the *standard error* (*SE*) of b_j is the square root of this variance thus:

$$SE(b_j) = s\sqrt{c_{jj}}$$

If β_j is really zero, the statistic

$$t = b_j / SE(b_j)$$

has a Student's t-distribution with $n - p - 1$ degrees of freedom. Thus a formal hypothesis test may be carried out to check whether it is reasonable to assume that β_j is zero and hence the predictor x_j does not make a significant contribution to explaining y. However, as an initial guide to whether x_j should be included in the regression model, it is usually sufficient to check whether the magnitude of the corresponding t-statistic is numerically greater than about 2. If $|t| > 2$, then x_j should be left in the model; otherwise, consideration should be given to removing it.

When there is more than one explanatory variable or predictor in the original regression, there is a possibility that two or more of these may be highly correlated with each other. This situation is called *multicollinearity*; when it occurs the columns in $\mathbf{X}$ corresponding to the correlated variables are almost linearly related, and this causes $\mathbf{X}^\mathsf{T}\mathbf{X}$ and its inverse $\mathbf{C}$ to be ill-conditioned. Although we shall be able to solve the normal equations for the given $\mathbf{b}$, as long as $\mathbf{X}^\mathsf{T}\mathbf{X}$ does not actually become singular, the solution is very sensitive to small changes in the data and there will be large off-diagonal elements in $\mathbf{C}$, indicating highly correlated b_j estimates. It is therefore worth calculating the *correlation matrix* which shows the correlation between y and each of the explanatory variables and the correlations between all pairs of explanatory variables. As in the case of the covariance matrix, we number the rows and columns of the correlation matrix from 0 to p. Thus the correlation matrix is

$$\begin{array}{c} \\ y \\ x_1 \\ x_2 \\ \\ \end{array} \begin{array}{c} \begin{array}{cccc} y & x_1 & x_2 & \end{array} \\ \begin{pmatrix} r_{00} & r_{01} & r_{02} & \cdots \\ r_{10} & r_{11} & r_{12} & \cdots \\ r_{20} & r_{21} & r_{22} & \cdots \\ \cdots & \cdots & \cdots & \cdots \end{pmatrix} \end{array}$$

where we define the typical element of the correlation matrix r_{ij} to be the correlation between x_i and x_j and r_{0j} to be the correlation between y and x_j.

In situations such as *polynomial regression*, which is considered in Section 7.7, there will always be high correlations between the predictors, but examination of the t-statistics may indicate that some of the predictors do not contribute significantly to explaining y. Those with the smallest $|t|$ and the smallest $|r_{0j}|$ are the most obvious candidates for discarding.

Another set of statistics which are useful in this context are the *variance inflation factors* (*VIFs*). To find the *VIF* for x_j, we regress x_j upon the other $p-1$ explanatory variables and calculate the coefficient of determination, R_j^2. The corresponding *VIF* is

$$VIF_j = \frac{1}{1 - R_j^2}$$

If x_j is almost entirely explained by the other variables, R_j^2 will be close to 1 and VIF_j will be large. A good working rule is to regard any x_j with $VIF_j > 10$ as a candidate for removal.

Note that when there are only two explanatory variables, they will always have equal variance inflation factors; if these are greater than 10, it is best to discard the variable which has the smallest correlation with y. When the model contains predictors that are different functions of some common explanatory variable, the corresponding *VIF* values may be very large, as in the case of polynomial regression described in Section 7.7. Such models would normally be considered for physical data when predictors of this type could be shown to have a causal relationship with y.

The following MATLAB function `mregg2` implements multiple regression; the diagnostics necessary for model improvement are also computed.

```
function [s_sqd R_sqd b SE t VIF Corr_mtrx residual] = mregg2(Xd,con)
% Multiple linear regression, using least squares.
% Example call:
%   [s_sqd R_sqd bt SEt tt VIFt Corr_mtrx residual] = mregg2(Xd,con)
% Fits data to y = b0 + b1*x1 + b2*x2 + ... bp*xp
% Xd is a data array. Each row of X is a set of data.
% Xd(1,:) = x1(:), Xd(2,:) = x2(:), ... Xd(p+1,:) = y(:).
% Xd has n columns corresponding to n data points and p+1 rows.
% If con = 0, no constant is used, if con ~= 0, constant term is used.
% Output arguments:
% s_sqd = Error variance, R_Sqd = R^2.
% b is the row of coefficients b0 (if con~=0), b1, b2, ... bp.
% SE is the row of standard error for the coeff b0 (if con~=0),
% b1, b2, ... bp.
% t is the row of the t statistic for the coeff b0 (if con~=0),
% b1, b2, ... bp.
% VIF is the row of the VIF for the coeff b0 (if con~=0), b1, b2, ... bp.
% Corr_mtrx is the correlation matrix
% residual is an arrray of 4 columns and n rows.
% For each row i, the residual
```

```
% array contains the value of y(i), the residual(i), the standardized
% residual(i) and the Cook distance(i) where i is the ith data value.
if con==0
    cst = 0;
else
    cst = 1;
end
[p1,n] = size(Xd);
p = p1-1;  pc = p+cst;
y = Xd(p1,:)';
if cst==1
    w = ones(n,1);
    X = [w Xd(1:p,:)'];
else
    X = Xd(1:p,:)';
end
C = inv(X'*X);  b = C*X'*y; b = b.';
H = X*C*X';  SSE = y'*(eye(n)-H)*y;
s_sqd = SSE/(n-pc); Cov = s_sqd*C;
Z = (1/n)*ones(n);
num = y'*(H-Z)*y; denom = y'*(eye(n)-Z)*y;
R_sqd = num/denom;
SE = sqrt(diag(Cov)); SE = SE.';
t = b./SE;
% Compute correlation matrix
V(:,1) = (eye(n)-Z)*y;
for j = 1:p
    V(:,j+1) = (eye(n)-Z)*X(:,j+cst);
end
SS = V'*V; D = zeros(p+1,p+1);
for j=1:p+1
    D(j,j) = 1/sqrt(SS(j,j));
end
Corr_mtrx = D*SS*D;
% Compute VIF
for j = 1+cst:pc
    ym = X(:,j);
    if cst==1
        Xm = X(:,[1 2:j-1,j+1:p+1]);
    else
        Xm = X(:,[1:j-1,j+1:p]);
    end
    Cm = inv(Xm'*Xm); Hm = Xm*Cm*Xm';
    num = ym'*(Hm-Z)*ym; denom = ym'*(eye(n)-Z)*ym;
```

```
    R_sqr(j-cst) = num/denom;
end
VIF = 1./(1-R_sqr); VIF = [0 VIF];
% Analysis of residuals
ee = zeros(length(y),1);   sr = zeros(length(y),1);
cd = zeros(length(y),1);
if nargout>7
    ee = (eye(n)-H)*y;
    s = sqrt(s_sqd);
    sr = ee./(s*sqrt(1-diag(H)));
    cd = (1/pc)*(1/s^2)*ee.^2.*(diag(H)./((1-diag(H)).^2));
    residual = [y ee sr cd];
end
```

To use `mregg2` it is necessary to provide the $(p+1) \times n$ data array, `Xd`, where p is the number of explanatory variables and n is the number of data sets. Rows 1 to p contain the values of the explanatory variables, x_1 to x_p, and row $p+1$ contains the corresponding value of y. If the parameter `con` is set to zero, then the constant term is removed from the regression model; otherwise, it is included. Examples of the use of the function `mregg2` are given in Sections 7.6 and 7.7.

The multiple regression model has wider applications. For example:

1. *Polynomial regression.* Here y is required to be a polynomial function of a single variable x, so that x_j in the general model is replaced by x^j. Thus we have

$$y_i = \beta_0 + \beta_1 x_i + \beta_2 x_i^2 + \ldots + \beta_p x_i^p + \varepsilon_i, \quad i = 1, 2, \ldots, n \tag{7.13}$$

   Although this is no longer linear in the explanatory variable x, it is still linear in the β_j coefficients and so the theory for the linear regression model still applies. We can use `mregg2` to carry out polynomial regression where the rows of data are $x, x^2, x^3, \ldots$ and the last row of data is y. (See Examples 7.3, 7.4, 7.5.)
2. *Multiple polynomial regression.* Suppose that we wished to fit data to the following regression model:

$$y_i = \beta_0 + \beta_1 x_1 + \beta_2 x_2 + \beta_3 x_1^2 + \beta_4 x_2^2 + \beta_5 x_1 x_2 + \epsilon_i, \; i = 1, 2, \ldots, n$$

   In this case the five predictors are x_1, x_2, x_1^2, x_2^2, and $x_1 x_2$. The predictors are still linear in the β_j coefficients. To use the function `mregg2` the six rows of the data array for must contain the values of x_1, x_2, x_1^2, x_2^2, and $x_1 x_2$ and y respectively.

7.6 ANALYSIS OF RESIDUALS

Besides considering the contributions made by each of the explanatory variables or predictors, it is important to consider how well the model fits at each data point and whether our assumptions about the error distribution are valid.

We recall from Section 7.9 that

$$\hat{\mathbf{y}} = \mathbf{H}\mathbf{y}$$

so we may write the residual vector as

$$\mathbf{e} = \mathbf{y} - \mathbf{H}\mathbf{y} = (\mathbf{I} - \mathbf{H})\,\mathbf{y}$$

It can be shown that the diagonal elements of $s^2(\mathbf{I} - \mathbf{H})$ represents the variances of the individual residuals, so the standard deviation of e_i is $s\sqrt{1-h_{ii}}$. Since the standard deviation varies from one data point to another, it is difficult to make a direct comparison between residuals at different points. However, if we *standardize* the residuals by dividing each by its standard deviation, we obtain statistics that are similar to the t-ratios which we used for analyzing $\mathbf{b}$. Thus the standardized residual, r_i is

$$r_i = \frac{e_i}{s\sqrt{1-h_{ii}}} \quad i = 1, 2, \ldots, n$$

To distinguish this from other kinds of standardized residuals, it is sometimes called the *Studentized residual*. If the assumptions behind our model are correct, the average value of the standardized residual should be close to zero but if any $|r_i|$ is larger than about 2 this may indicate one or more of the following:

1. that the point where this occurs is an *outlier*,
2. that the assumption about equal error variance at all points is incorrect,
3. there has been a fault in specifying the model.

In this context, an outlier is an observation that was not obtained under the same conditions as the others. It is not always easy to distinguish points that result from mistakes in observation from those that correspond to values of the explanatory variables that lie far away from those used for the other observations. Such points tend to have a large influence on the fitting process.

A useful statistic for measuring the influence of a particular point is the *Cook's distance*, d_i, which combines the size of the squared residual with the distance of a particular point from the mean value, called the *leverage*. This is the corresponding diagonal element of $\mathbf{H}$ determined by the values of the explanatory variables.

$$d_i = \left(\frac{1}{p+1}\right)\frac{e_i^2}{s^2}\left[\frac{h_{ii}}{(1-h_{ii})^2}\right] \quad i = 1, 2, \ldots\ n$$

Any point for which $d_i > 1$ will have a considerable effect on the regression and should be checked in detail. The point may have been correctly observed and provide information that is very useful in model building, but the modeler should be aware of its influence.

Example 7.1. Fit a regression model to the data given by `X0`, `X1`, and `Xd` in script `e4s705.m`. (To save space the data is given in the form required by the function `mregg2`.) The first row, second, and third rows of the matrices are the values of the explanatory variables x_1, x_2, and x_3, respectively, and the fourth row contains the corresponding values of y. The script e4s705.m implements this.

```
% e4s705.m
X0 = [1.00 1.00 1.00 1.00 1.00 2.00 2.00 2.00;
      2.00 2.00 4.00 4.00 6.00 2.00 2.00 4.00;
         0 1.00 0 1.00 2.00 0 1.00 0;
     -2.52 -2.71 -8.34 -8.40 -14.60 -0.62 -0.47 -6.49];
X1 = [2.00 3.00 3.00 3.00 3.00 3.00 3.00 3.00;
      6.00 2.00 2.00 2.00 4.00 6.00 6.00 6.00;
         0 0 1.00 2.00 1.00 0 1.00 2.00;
    -12.46 1.36 1.40 1.60 -4.64 -10.34 -10.43 -10.30];
Xd = [X0 X1];
[s_sqd R_sqd b SE t VIF Corr_mtrx res] = mregg2(Xd,1);

fprintf('Error variance = %7.4f      R_squared = %7.4f \n\n',s_sqd,R_sqd)
fprintf('                         Coeff      SE       t_ratio      VIF \n')
fprintf('Constant    :           %7.4f %7.4f %8.2f \n',b(1),SE(1),t(1))
fprintf('Coeff x1    :           %7.4f %7.4f %8.2f %8.2f\n',b(2),SE(2),t(2),VIF(2))
fprintf('Coeff x2    :           %7.4f %7.4f %8.2f %8.2f\n',b(3),SE(3),t(3),VIF(3))
fprintf('Coeff x3    :           %7.4f %7.4f %8.2f %8.2f\n\n',b(4),SE(4),t(4),VIF(4))
fprintf('Correlation matrix \n')
disp(Corr_mtrx)
fprintf('\n          y            Residual    St Residual    Cook dist\n')
for i = 1:length(Xd)
    fprintf('%12.4f %12.4f %12.4f %12.4f\n',res(i,1), ...
                                       res(i,2), res(i,3), res(i,4))
end
```

Running script e4s705.m gives

```
Error variance =   0.0147      R_squared =   0.9996

                  Coeff      SE      t_ratio      VIF
Constant    :    1.3484   0.1006     13.40
Coeff x1    :    2.0109   0.0358     56.10       1.03
Coeff x2    :   -2.9650   0.0179   -165.43       1.03
Coeff x3    :   -0.0001   0.0412     -0.00       1.04

Correlation matrix
    1.0000    0.2278   -0.9437   -0.0944
    0.2278    1.0000    0.1064    0.1459
   -0.9437    0.1064    1.0000    0.1459
   -0.0944    0.1459    0.1459    1.0000
```

y	Residual	St Residual	Cook dist
-2.5200	0.0508	0.4847	0.0196
-2.7100	-0.1390	-1.3223	0.1426
-8.3400	0.1609	1.5062	0.1617
-8.4000	0.1010	0.9274	0.0507
-14.6000	-0.1688	-1.9402	0.8832
-0.6200	-0.0601	-0.5458	0.0157
-0.4700	0.0901	0.8035	0.0270
-6.4900	0.0000	0.0002	0.0000
-12.4600	-0.0399	-0.3835	0.0130
1.3600	-0.0909	-0.8808	0.0730
1.4000	-0.0508	-0.4736	0.0154
1.6000	0.1493	1.5774	0.3958
-4.6400	-0.1607	-1.4227	0.0755
-10.3400	0.0692	0.6985	0.0602
-10.4300	-0.0206	-0.1930	0.0026
-10.3000	0.1095	1.1123	0.1589

The coefficient of x_3 is small, and, more importantly, the corresponding absolute value of the t-ratio is very small (it is in fact not zero, but -0.0032). This suggests that x_3 does not make a significant contribution to the model and can be removed.

If we change the last value of y to -8.3 (and showing the analysis of the residuals only), we have

y	Residual	St Residual	Cook dist
-2.5200	0.3499	0.7758	0.0503
-2.7100	-0.0756	-0.1670	0.0023
-8.3400	0.3095	0.6735	0.0323
-8.4000	0.0140	0.0300	0.0001
-14.6000	-0.6418	-1.7153	0.6903
-0.6200	0.1361	0.2876	0.0044
-0.4700	0.0507	0.1052	0.0005
-6.4900	0.0457	0.0941	0.0003
-12.4600	-0.1447	-0.3232	0.0093
1.3600	0.0024	0.0054	0.0000
1.4000	-0.1930	-0.4184	0.0120
1.6000	-0.2284	-0.5610	0.0501
-4.6400	-0.4534	-0.9331	0.0325
-10.3400	-0.1384	-0.3247	0.0130
-10.4300	-0.4638	-1.0077	0.0713
-8.3000	1.4308	3.3791	1.4664

For the observation $y = -8.3$, we see that the residual, the standard residual, and Cook's distance are all large, compared with the values for the rest of the data. Either we have a recording error in this particular observation or the data is correct and thus the model we are using fits this particular data point poorly.

Example 7.2. Using the same data as Example 7.1, fit a regression model using the explanatory variables x_1 and x_2 only. The data array is not shown in the script `e4s706.m`.

```
% e4s706.m
X0 ....
X1 ....
Xd = [X0 X1];
[s_sqd R_sqd b SE t VIF Corr_mtrx] = mregg2(Xd([1 2 4],:),1);
fprintf('Error variance = %7.4f      R_squared = %7.4f \n\n',s_sqd,R_sqd)
fprintf('                  Coeff      SE      t_ratio      VIF \n')
fprintf('Constant     :   %7.4f %7.4f  %8.2f \n',b(1),SE(1),t(1))
fprintf('Coeff x1     :   %7.4f %7.4f  %8.2f %8.2f\n',b(2),SE(2),t(2),VIF(2))
fprintf('Coeff x2     :   %7.4f %7.4f  %8.2f %8.2f\n\n',b(3),SE(3),t(3),VIF(3))
fprintf('Correlation matrix \n')
disp(Corr_mtrx)
```

Running script `e4s706.m` gives

```
Error variance =  0.0135      R_squared =  0.9996

                  Coeff     SE     t_ratio      VIF
Constant    :    1.3483  0.0960     14.04
Coeff x1    :    2.0109  0.0341     58.91      1.01
Coeff x2    :   -2.9650  0.0171   -173.71      1.01

Correlation matrix
    1.0000    0.2278   -0.9437
    0.2278    1.0000    0.1064
   -0.9437    0.1064    1.0000
```

This is a better model than that derived in Example 7.1 because the absolute values of the t-ratios are now all greater than 2. In fact, the original data was generated from the model

$$y = 1.5 + 2x_1 - 3x_2 + \text{random errors}$$

Thus x_3 was not linked to the model used to generate the data and variations in x_3 only appeared to influence y because of the random errors in the measurement of y.

Note that the standard errors (*SEs*) can used to construct confidence intervals for b_j. In this case, the true values of each β_j lie within the 95% confidence interval: that is, the interval within which we expect that β_j would lie with a probability of 0.95 if we did not know its true value. The precise width of the 95% confidence interval depends on the number of degrees of freedom (see Section 7.4), but to a reasonable approximation β_j lies between $b_j - 2SE(b_j)$ and $b_j + 2SE(b_j)$.

More comprehensive presentations of aspects of multiple regression, model improvement, and regression analysis can be found in Draper and Smith (1998), Walpole and Myers (1993), and Anderson et al. (1993).

7.7 POLYNOMIAL REGRESSION

The polynomial regression model, is given by (7.13) and is repeated here:

$$y_i = \beta_0 + \beta_1 x_i + \beta_2 x_i^2 + \ldots + \beta_p x_i^p + \varepsilon_i, \; i = 1, 2, \ldots, n$$

Although this is no longer linear in the explanatory variable x, it is still linear in the β_j coefficients and so the theory for the linear regression model still applies.

The processes of fitting, checking the b_j estimates and deciding whether some predictors can be discarded may be done in the same manner as in the general case described in Section 7.4. The diagnostics for model improvement and residual analysis also follow the general case described in Sections 7.5 and 7.6. It will inevitably be found that there are quite high correlations between the predictors, which are now all powers of the same explanatory variable. As discussed in Section 7.5, these high correlations between predictors can result in an ill-conditioned coefficient matrix $\mathbf{X}^\mathsf{T}\mathbf{X}$. For a large number of data points this matrix tends to the Hilbert matrix.

In Chapter 2, it is shown that the Hilbert matrix is very ill-conditioned. To illustrate the influence of this on the accuracy of computations, we note that the number of decimal places lost when working with an ill-conditioned matrix **A** is given approximately by the MATLAB expression `log10(cond(A))`. Thus, if we were fitting a fifth-degree polynomial, the number of decimal places that could be lost may be estimated by `log10(cond(hilb(5)))`. This equals 5.6782; that is five or six of the 16 significant digits that MATLAB uses are lost. One way to avoid this difficulty is to formulate the problem so that no system of linear equations has to be solved. An ingenious way of doing this is to use orthogonal polynomials. We will not describe this method here but refer the reader to Lindfield and Penny (1989). However, the worst effects of ill-conditioning can be avoided provided that p (which now represents the degree of the fitted polynomial) is kept reasonably small.

If the diagnostics which we have developed in Sections 7.4, 7.5, and 7.6 are required, then we can use the function `mregg2`, given in Section 7.5 for polynomial regression and in this case the data must be prepared as follows. The first row of the array `Xd` contains the values of x, the second row contains the values of x^2 and so on. The last row contains the corresponding values of y. In interpreting the output from `mregg2` note that all the *VIF* values are inevitably high because the powers of x are correlated with each other. The usual rule about discarding predictors with $VIF > 10$ should be ignored since we have good reason to suppose that a polynomial model is appropriate.

If the diagnostics are not required, the calculation of the b_j estimates can be done using the MATLAB function `polyfit`. This uses the method of least squares to fit a polynomial of specified degree to given data. The following examples illustrate some of these issues.

Example 7.3. Fit a cubic polynomial to the following data which has been generated from $y = 2 + 6x^2 - x^3$ with added random errors. The random errors have a normal distribution with a zero mean value and a standard deviation of 1. The script `e4s707.m` calls the MATLAB function `polyfit` to determine the coefficients of the cubic polynomial followed by `polyval` to evaluate it for plotting. It then calls the function `mregg2` to compute a regression model using x, x^2, and x^3 as the explanatory variables.

```
% e4s707.m
x = 0:.25:6;
y = [1.7660 2.4778 3.6898 6.3966 6.6490 10.0451 12.9240 15.9565 ...
     17.0079 21.1964 24.1129 25.5704 28.2580 32.1292 32.4935 34.0305 ...
     34.0880 32.9739 31.8154 30.6468 26.0501 23.4531 17.6940 9.4439 ...
     1.7344];
xx = 0:.02:6;
p = polyfit(x,y,3), yy = polyval(p,xx);
plot(x,y,'o',xx,yy)
axis([0 6 0 40]), xlabel('x'), ylabel('y')
Xd = [x; x.^2; x.^3; y];
[s_sqd R_sqd b SE t VIF Corr_mtrx] = mregg2(Xd,1);
fprintf('Error variance = %7.4f      R_squared = %7.4f \n\n',s_sqd,R_sqd)
fprintf('                    Coeff      SE       t_ratio      VIF \n')
fprintf('Constant    :      %7.4f %7.4f %8.2f \n',b(1),SE(1),t(1))
fprintf('Coeff x     :      %7.4f %7.4f %8.2f %8.2f\n',b(2),SE(2),t(2),VIF(2))
fprintf('Coeff x^2   :      %7.4f %7.4f %8.2f %8.2f\n',b(3),SE(3),t(3),VIF(3))
fprintf('Coeff x^3   :      %7.4f %7.4f %8.2f %8.2f\n\n',b(4),SE(4),t(4),VIF(4))
fprintf('Correlation matrix \n')
disp(Corr_mtrx)
```

Running script `e4s707.m` gives the following output, together with the graphs shown in Fig. 7.5. The polynomial coefficients from `polyfit` are in the order of descending powers of x. The diagnostic output from `mregg2` is self-explanatory.

```
p =
   -0.9855    5.8747    0.1828    2.2241

Error variance =  0.5191      R_squared =  0.9966

                    Coeff      SE      t_ratio      VIF
Constant    :      2.2241  0.4997      4.45
Coeff x     :      0.1828  0.7363      0.25      84.85
Coeff x^2   :      5.8747  0.2886     20.36     502.98
Coeff x^3   :     -0.9855  0.0316    -31.20     202.10

Correlation matrix
    1.0000    0.4917    0.2752    0.1103
    0.4917    1.0000    0.9659    0.9128
    0.2752    0.9659    1.0000    0.9858
    0.1103    0.9128    0.9858    1.0000
```

Note that the absolute value of the t-ratio for the explanatory variable x is less than 2, indicating that x should be removed (see Example 7.4).

The cubic polynomial that fits the data is

$$\hat{y} = 2.2241 + 0.1828x + 5.8747x^2 - 0.9855x^3$$

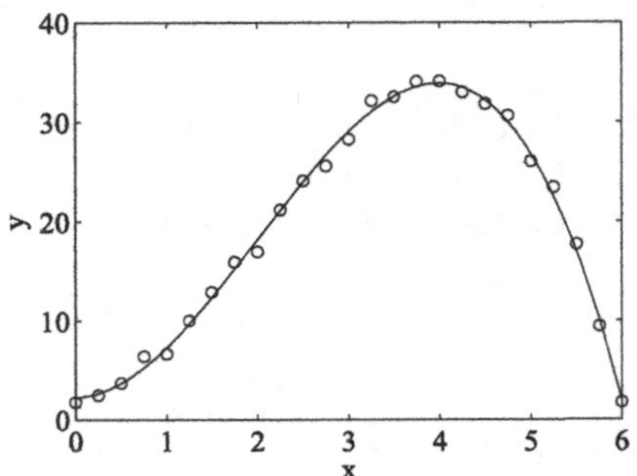

FIGURE 7.5

Fitting a cubic polynomial to data. Data points are denoted by o.

Example 7.4. Fit a cubic polynomial to data of Example 7.3 but use the explanatory variables x^2 and x^3 only. The script `e4s708.m` solves this problem:

```
% e4s708.m
x = 0:.25:6; y = 2+6*x.^2-x.^3;
y = y+randn(size(x)); Xd = [x.^2; x.^3; y];
[s_sqd R_sqd b SE t VIF Corr_mtrx] = mregg2(Xd,1);
fprintf('Error variance = %7.4f      R_squared = %7.4f \n\n',s_sqd,R_sqd)
fprintf('                          Coeff       SE       t_ratio       VIF \n')
fprintf('Constant    :       %7.4f %7.4f %8.2f \n',b(1),SE(1),t(1))
fprintf('Coeff x^2   :       %7.4f %7.4f %8.2f %8.2f\n',b(2),SE(2),t(2),VIF(2))
fprintf('Coeff x^3   :       %7.4f %7.4f %8.2f %8.2f\n\n',b(3),SE(3),t(3),VIF(3))
fprintf('Correlation matrix \n')
disp(Corr_mtrx)
```

Running script `e4s708.m` gives

```
Error variance =  0.4970     R_squared =  0.9965

                  Coeff      SE      t_ratio      VIF
Constant    :     2.3269  0.2741     8.49
Coeff x^2   :     5.9438  0.0750    79.21      35.52
Coeff x^3   :    -0.9926  0.0130   -76.61      35.52

Correlation matrix
    1.0000    0.2752    0.1103
    0.2752    1.0000    0.9858
    0.1103    0.9858    1.0000
```

Thus our improved model (compared with Example 7.3) is

$$\hat{y} = 2.32692.1793 + 5.9438x^2 - 0.99261.0084x^3$$

The true β_j are well within the 95% confidence limits given by this improved model. The error variance of 0.4970 is somewhat less than the unit variance of the random errors initially added to the data.

Example 7.5. Fit a third- and a fifth-degree polynomial to data generated from the function

$$y = \sin\{1/(x + 0.2)\} + 0.2x$$

contaminated with random noise, normally distributed with a standard deviation of 0.06 to simulate measurement errors as follows:

```
>> xs = [0:0.05:0.25 0.25:0.2:4.85];
>> us = sin(1./(xs+1))+0.2*xs+0.06*randn(size(xs));
>> save testdata1 xs us
```

The 30 data values are stored in the file `testdata1` so that it can be used in an example in Section 7.8.

The script `e4s709.m` loads the data, fits and plots the least squares polynomial.

```
% e4s709.m
load testdata
xx = 0:.05:5;
t1 = 'Error variance = %7.4f      R_squared = %7.4f \n\n';
t2 = '                       Coeff      SE      t_ratio        VIF \n';
t3 = 'Constant     :      %7.4f %7.4f %8.2f \n';
t4 = 'Coeff x      :      %7.4f %7.4f %8.2f %12.2f\n';
t4a = 'Coeff x     :      %7.4f %7.4f %8.2f \n';
t5 = 'Coeff x^2    :      %7.4f %7.4f %8.2f %12.2f\n';
t6 = 'Coeff x^3    :      %7.4f %7.4f %8.2f %12.2f\n';
t7 = 'Coeff x^4    :      %7.4f %7.4f %8.2f %12.2f\n';
t8 = 'Coeff x^5    :      %7.4f %7.4f %8.2f %12.2f\n';
t9 = 'Correlation matrix \n';
p = polyfit(xs,us,3), yy = polyval(p,xx);
Xd = [xs; xs.^2; xs.^3; us];
[s_sqd R_sqd b SE t VIF Corr_mtrx] = mregg2(Xd,1);
fprintf(t1,s_sqd,R_sqd), fprintf(t2)
fprintf(t3,b(1),SE(1),t(1))
fprintf(t4,b(2),SE(2),t(2),VIF(2))
fprintf(t5,b(3),SE(3),t(3),VIF(3))
fprintf([t6 '\n'],b(4),SE(4),t(4),VIF(4))
fprintf(t9), disp(Corr_mtrx)
[s_sqd R_sqd b SE t VIF Corr_mtrx] = mregg2(Xd,0);
fprintf(t1,s_sqd,R_sqd), fprintf(t2)
fprintf(t4a,b(1),SE(1),t(1))
```

```
fprintf(t5,b(2),SE(2),t(2),VIF(2))
fprintf([t6 '\n'],b(3),SE(3),t(3),VIF(3))
fprintf(t9), disp(Corr_mtrx)
plot(xs,us,'ko',xx,yy,'k'), hold on
axis([0 5 -2 2])
p = polyfit(xs,us,5), yy = polyval(p,xx);
Xd = [xs; xs.^2; xs.^3; xs.^4; xs.^5; us];
[s_sqd R_sqd b SE t VIF Corr_mtrx] = mregg2(Xd,1);
fprintf(t1,s_sqd,R_sqd), fprintf(t2)
fprintf(t3,b(1),SE(1),t(1))
fprintf(t4,b(2),SE(2),t(2),VIF(2))
fprintf(t5,b(3),SE(3),t(3),VIF(3))
fprintf(t6,b(4),SE(4),t(4),VIF(4))
fprintf(t7,b(5),SE(5),t(5),VIF(5))
fprintf([t8 '\n'],b(6),SE(6),t(6),VIF(6))
fprintf(t9)
disp(Corr_mtrx)
plot(xx,yy,'k--'), xlabel('x'), ylabel('y'), hold off
```

Fig. 7.6 shows the result of fitting a third- and a fifth-degree polynomial to the data and clearly displays the inadequacies of these polynomial approximations. The polynomials oscillate about the points and do not fit the data satisfactorily. In Section 7.8 we see that we can improve the fit by using different functions. The output from script `e4s709.m` is as follows:

```
p =
    0.0842   -0.6619    1.5324   -0.0448

Error variance =  0.0980      R_squared =  0.6215

                    Coeff      SE      t_ratio        VIF
Constant    :     -0.0448  0.1402      -0.32
Coeff x     :      1.5324  0.3248       4.72         79.98
Coeff x^2   :     -0.6619  0.1708      -3.87        478.23
Coeff x^3   :      0.0842  0.0239       3.52        193.93

Correlation matrix
    1.0000    0.5966    0.4950    0.4476
    0.5966    1.0000    0.9626    0.9049
    0.4950    0.9626    1.0000    0.9847
    0.4476    0.9049    0.9847    1.0000
```

The absolute value of the t-ratio for the constant terms is low, implying that it should not be included in the cubic model. The script also fits a cubic equation without a constant term to the data and gives the following output:

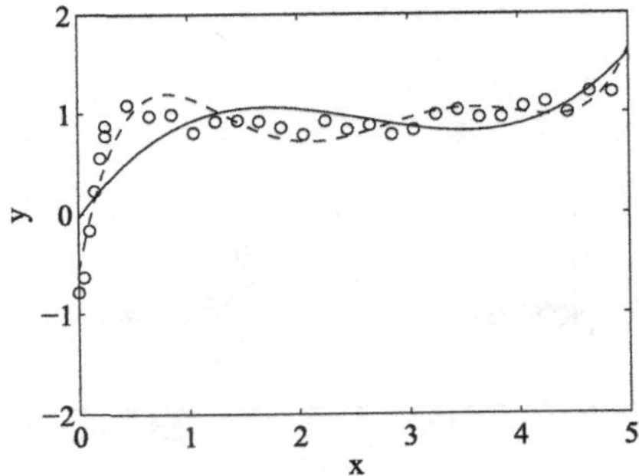

FIGURE 7.6

Fitting third- and fifth-degree polynomials (that is, a full and a dashed line, respectively) to a sequence of data. Data points are denoted by o.

```
Error variance =  0.0947       R_squared =  0.6200

                   Coeff       SE       t_ratio          VIF
Coeff x     :     1.4546   0.2116       6.87
Coeff x^2   :    -0.6285   0.1329      -4.73           35.13
Coeff x^3   :     0.0801   0.0199       4.02          299.50

Correlation matrix
    1.0000    0.5966    0.4950    0.4476
    0.5966    1.0000    0.9626    0.9049
    0.4950    0.9626    1.0000    0.9847
    0.4476    0.9049    0.9847    1.0000
```

This is a more robust model. Finally, script `e4s709.m` fits a fifth-degree polynomial to the data and gives the following, final, output:

```
p =
    0.0434   -0.5856    2.8998   -6.3340    5.7099   -0.5789

Error variance =  0.0341       R_squared =  0.8783

                   Coeff       SE       t_ratio          VIF
Constant    :    -0.5789   0.1122      -5.16
Coeff x     :     5.7099   0.6443       8.86          904.01
Coeff x^2   :    -6.3340   0.9052      -7.00        38560.71
Coeff x^3   :     2.8998   0.4918       5.90       234903.50
Coeff x^4   :    -0.5856   0.1137      -5.15       262672.06
Coeff x^5   :     0.0434   0.0094       4.62        38084.24
```

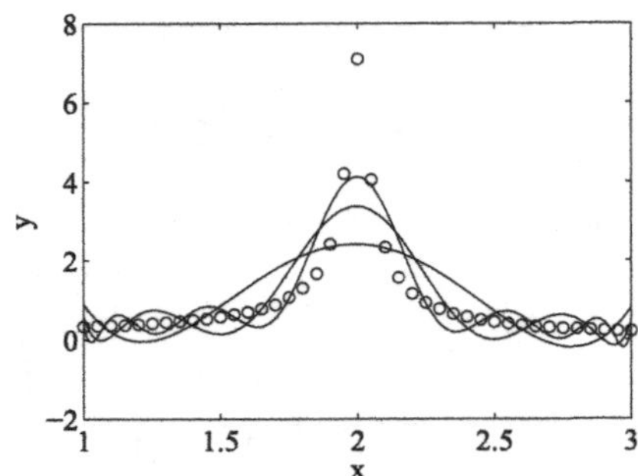

FIGURE 7.7

Polynomials of degree 4, 8, and 12 attempting to fit a sequence of data indicated by o in the graph.

```
Correlation matrix
    1.0000    0.5966    0.4950    0.4476    0.4172    0.3942
    0.5966    1.0000    0.9626    0.9049    0.8511    0.8041
    0.4950    0.9626    1.0000    0.9847    0.9555    0.9232
    0.4476    0.9049    0.9847    1.0000    0.9918    0.9742
    0.4172    0.8511    0.9555    0.9918    1.0000    0.9949
    0.3942    0.8041    0.9232    0.9742    0.9949    1.0000
```

Note the large values of *VIF*, caused by the regression being based on different functions of a single explanatory variable.

We now illustrate a difficulty that can sometime arise when trying to fit a polynomial function to data. To illustrate the problem we will begin by simulating experimental data based on the following relationship:

$$y = \frac{1}{\sqrt{0.02 + (4 - x^2)^2}} \tag{7.14}$$

Data values are generated by sampling this function from $x = 1$ to $x = 3$ in increments of 0.05, and small random errors are added to simulate measurement errors. The results of attempting to fit polynomials to these data values are shown in Fig. 7.7. The plot shows that as the degree of the polynomial is increased from 4 to 8, and finally to 12, the polynomial fits the data better in the sense that the total least squares error decreases, but the higher degree polynomials tend to oscillate between the data points. Thus even a twelfth-degree polynomial does not accurately represent the data, nor does it give us any insight into the underlying mathematical relationship between x and y. Furthermore, the twelfth-degree is polynomial is badly conditioned and MATLAB outputs a warning to the user concerning this ill-conditioning. We return to this problem is Section 7.10.

7.8 FITTING GENERAL FUNCTIONS TO DATA

We now consider a regression based on (7.2) with the separate explanatory variables x_j replaced by predictors which are various functions ϕ_j of a single explanatory variable x with values x_i:

$$y_i = \beta_0 + \beta_1\varphi_1(x_i) + \beta_2\varphi_2(x_i) + \ldots + \beta_p\varphi_p(x_i) + \varepsilon_i$$

The analysis presented in Section 7.4 extends directly to this regression model. Hence, we can use the MATLAB function `mregg2` to fit a set of any prescribed functions to data.

Consider again Example 7.5 of Section 7.7. We will fit the following function (or model) to the data:

$$\hat{y} = b_1 \sin\{1/(x + 0.2)\} + b_2 x$$

This function has been chosen because the data was originally generated from it with $b_1 = 1$ and $b_2 = 0.2$ with normally distributed random noise added. Note that there is no constant term in our model. The script `e4s710.m` calls the function `mregg2`. Note how the first row of the data matrix for `mregg2` contains the values of $\sin(1/(x + 0.2))$ and the second row contains the values of x.

```
% e4s710.m
load testdata
Xd = [sin(1./(xs+0.2)); xs; us];
[s_sqd R_sqd b SE t VIF Corr_mtrx] = mregg2(Xd,0);
fprintf('Error variance = %7.4f\n\n',s_sqd)
fprintf('                          Coeff      SE     t_ratio\n')
fprintf('sin(1/(x+0.2)):         %7.4f %7.4f %8.2f \n',b(1),SE(1),t(1))
fprintf('Coeff x         :       %7.4f %7.4f %8.2f \n\n',b(2),SE(2),t(2))
fprintf('Correlation matrix \n')
disp(Corr_mtrx)
xx = 0:.05:5; yy = b(1)*sin(1./(xx+0.2))+b(2)*xx;
plot(xs,us,'o',xx,yy,'k')
axis([0 5 -1.5 1.5]), xlabel('x'), ylabel('y')
```

Running this script gives

```
Error variance =  0.0044


                         Coeff      SE     t_ratio
sin(1/(x+0.2)):         0.9354  0.0257     36.46
Coeff x        :        0.2060  0.0053     38.55


Correlation matrix
    1.0000     0.7461     0.5966
    0.7461     1.0000    -0.0734
    0.5966    -0.0734     1.0000
```

and the graph of Fig. 7.8. The function that fits the data in a least squares sense is given by $\hat{y} = 0.9354 \sin\{1/(x + 0.2)\} + 0.2060x$. This is very close to the original function. Note also that

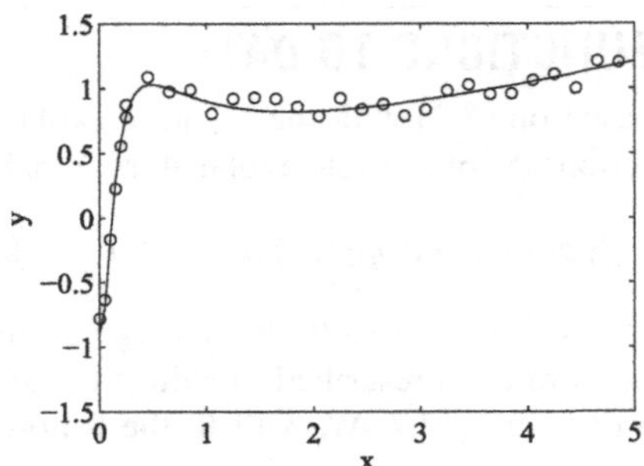

FIGURE 7.8

Data sampled from the function $y = \sin[1/(x + 0.2)] + 0.2x$. Data points denoted by o.

the error variance of 0.0044 compares quite well with 0.0036, the variance of the noise that was added to simulate measurement errors. If a constant term is included in the model, it has a very low absolute t-ratio, indicating that it should be removed.

7.9 NON-LINEAR LEAST SQUARES REGRESSION

We now consider the problem of fitting data to a function where the relationship between the unknown coefficients is non-linear. We still use the least squares criterion and there are several methods for fitting data to this type of model. Here we present a very simple iterative method based on Taylor's series.

Let $y = f(x, \mathbf{a})$ where f is a non-linear function of x and of the unknown coefficients $\mathbf{a}$. To determine these coefficients, let a trial set of coefficients by $\mathbf{a}^{(0)}$. Thus

$$y = f\left(x, \mathbf{a}^{(0)}\right)$$

This trial solution will not satisfy the requirement that the sum of the squares of the errors is a minimum. However, we can adjust the coefficients $\mathbf{a}^{(0)}$ in order to minimize the sum of the squares of the errors. Thus, letting the improved coefficients be $\mathbf{a}^{(1)}$, where

$$\mathbf{a}^{(1)} = \mathbf{a}^{(0)} + \Delta\mathbf{a}$$

we have

$$y = f\left(x, \mathbf{a}^{(1)}\right) = f\left(x, \mathbf{a}^{(0)} + \Delta\mathbf{a}\right)$$

Expanding the function as a Taylor series, and retaining only the first derivative terms in the Taylor series we have

$$y \approx f\left(x, \mathbf{a}^{(0)}\right) + \sum_{k=0}^{m} \Delta a_k \left[\partial f/\partial a_k\right]^{(0)}$$

Let $f_i^{(0)} = f\left(x_i, \mathbf{a}^{(0)}\right)$. The error between the function and y_i is given by

$$\varepsilon_i = y_i - f_i^{(0)} - \sum_{k=0}^{m} \Delta a_k \left[\partial f_i / \partial a_k\right]^{(0)}, \quad i = 1, 2, ...n$$

Thus the sum of the squares of the errors is

$$S = \sum_{i=0}^{n} \left\{ y_i - f_i^{(0)} - \sum_{k=0}^{m} \Delta a_k \left[\frac{\partial f_i}{\partial a_k}\right]^{(0)} \right\}^2$$

To determine the minimum of the sum of the squares of the errors we have

$$\frac{\partial S}{\partial (\Delta a_p)} = -2 \sum_{i=0}^{n} \left\{ y - f_i^{(0)} - \sum_{k=0}^{m} \Delta a_k \left[\frac{\partial f_i}{\partial a_k}\right]^{(0)} \right\} \left[\frac{\partial f_i}{\partial a_p}\right]^{(0)} = 0, \quad p = 0, 1, ...m$$

Rearranging we have

$$\sum_{i=0}^{n} \left(y - f_i^{(0)}\right) \left[\frac{\partial f_i}{\partial a_p}\right]^{(0)} = \sum_{k=0}^{m} \Delta a_k \left\{ \sum_{i=0}^{n} \left[\frac{\partial f_i}{\partial a_k}\right]^{(0)} \left[\frac{\partial f_i}{\partial a_p}\right]^{(0)} \right\}, \quad p = 0, 1, ...m$$

These equations can be expressed in matrix notation thus

$$\mathbf{K}(\Delta \mathbf{a}) = \mathbf{b}$$

where $\Delta \mathbf{a}$ has components Δa_p, $p = 0, 1, ..., m$, and

$$k_{pk} = \sum_{i=0}^{n} \left[\frac{\partial f_i}{\partial a_k}\right]^{(0)} \left[\frac{\partial f_i}{\partial a_p}\right]^{(0)} \quad b_p = \sum_{i=0}^{n} \left(y - f_i^{(0)}\right) \left[\frac{\partial f_i}{\partial a_p}\right]^{(0)}, \quad p, k = 0, 1, ...m$$

Solving for $\Delta \mathbf{a}$ we can then determine new values for the coefficients from

$$\mathbf{a}^{(1)} = \mathbf{a}^{(0)} + \Delta \mathbf{a}$$

Because we have discarded higher order terms in the Taylor series, $\mathbf{a}^{(1)}$ is not the exact solution, but it is a better solution than $\mathbf{a}^{(0)}$. We therefore iterate until the norm of $\Delta \mathbf{a}$ is less than a prescribed tolerance.

The function `nlls` listed below implements the above method for fitting a given non-linear function to a set of data.

```
function [a iter] = nlls(f,df,x,y,a0,err)
% Data given by vectors x and y are to be fitted to the function f(a)
% with an error of err. Function f(a) has n variables, a(1) ... a(n).
% a0 is a vector of n trial values for the unknown paramenters a.
% Function df is a column vector [df/da(1); df/da(2); .... df/da(n)].
iter = 0;   n = length(a0); a = a0;
```

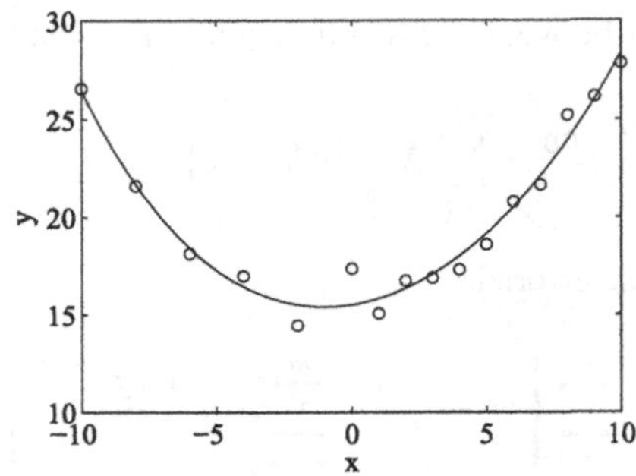

FIGURE 7.9

Fitting $y = a_1 e^{a_2 x} + a_3 e^{a_4 x}$ to data values indicated by o.

```
v = 10*err*ones(1,n);
while norm(v,2) > err
    p = feval(df,x,a);  q = y-feval(f,x,a);
    A = p*p'; b = q*p'; v = A\b';
    a = a + v';  iter = iter+1;
end
```

The script `e4s711.m` fits the function $y = a_1 e^{a_2 x} + a_3 e^{a_4 x}$ to 16 data points using the function `nlls`.

```
% e4s711.m
p = @(x,a) a(1)*exp(a(2)*x)+a(3)*exp(a(4)*x);
dp = @(x,a) [exp(a(2)*x); a(1)*x.*exp(a(2)*x);
             exp(a(4)*x); a(3)*x.*exp(a(4)*x)];
x = [-10:2:0  1:1:10]; xn = length(x);
xp = -10:0.05:10;
y = [26.56 21.60 18.14 17.00 14.46 17.38 15.07 16.76 ...
    16.90 17.32 18.61 20.79 21.65 25.22 26.16 27.84];
a = [7 -0.3 7 0.3];
[a iter] = nlls(p,dp,x,y,a,1e-5)
plot(x,y,'o',xp,p(xp,a))
xlabel('x'), ylabel('y')
```

Running script `e4s711.m` gives the following results:

```
a =
    5.4824   -0.1424   10.0343    0.0991


iter =
     7
```

together with Fig. 7.9.

7.10 TRANSFORMING DATA

We now consider an alternative approach to the problem of fitting data to functions where the relationship between the unknown coefficients is non-linear. This is to transform both the data and the function so that an equivalent function provides a linear relationship between y and the unknown coefficients. The only difficulty with this method is that no general rule can be given to provide a suitable transform; indeed such a transform may not even be possible. Consider the problem of fitting data to the following function:

$$\hat{y} = \frac{1}{\sqrt{a_0 + \left(a_1 - a_2 x^2\right)^2}} \tag{7.15}$$

Letting $\hat{Y} = 1/\hat{y}^2$ and $X = x^2$ we have

$$\begin{aligned} \hat{Y} &= 1/\hat{y}^2 = a_0 + (a_1 - a_2x^2)^2 = (a_0 + a_1^2) - 2a_1a_2x^2 + a_2^2x^4 \\ \hat{Y} &= a_0 + a_1^2 - 2a_1a_2X + a_2^2X^2 \\ \hat{Y} &= b_0 + b_1X + b_2X^2 \end{aligned} \tag{7.16}$$

Thus $\hat{Y}$ is a quadratic in X. If the data values are transformed by letting $Y_i = 1/y_i^2$ and $X_i = x_i^2$, then the process of fitting $Y = f(X)$ to these transformed data values will be a standard least squares polynomial fit giving b_0, b_1, and b_2. Hence estimates of a_0, a_1, and a_2 can be easily determined. Note, however, that the residual of the errors, $e_i = Y_i - \hat{Y}_i$, will not provide a good estimate of the measurement errors, $y_i - \hat{y}_i$ because of the transformations.

We illustrate the above process by considering a sequence of data values generated by (7.15) to which we have added normally distributed random errors having a zero mean value and a standard deviation of 1% to the dependent data, y_i. We can transform these data points using (7.15) and the script `e4s712.m` generates the required data, transforms the data points and fits a polynomial to them.

```
% e4s712.m
x = 1:.05:3; xx = 1:.005:3;
y = [0.3319   0.3454   0.3614   0.3710   0.3857   0.4030   0.4372 ...
     0.4605   0.4971   0.5232   0.5753   0.6363   0.6953   0.7782 ...
     0.8793   1.0678   1.3024   1.6688   2.4233   4.2046   7.0961 ...
     4.0581   2.3354   1.5663   1.1583   0.9278   0.7764   0.6480 ...
     0.5741   0.4994   0.4441   0.4005   0.3616   0.3286   0.3051 ...
     0.2841   0.2645   0.2407   0.2285   0.2104   0.2025];
Y = 1./y.^2; X = x.^2; XX = xx.^2;
p = polyfit(X,Y,2)
YY = polyval(p,XX);
for i = 1:length(xx)
    if YY(i)<0
        disp('Transformation fails with this data set');
        return
    end
end
```

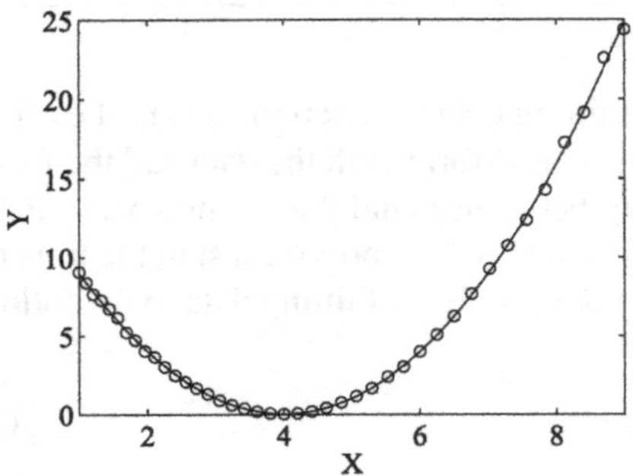

FIGURE 7.10

Fitting transformed data, denoted by "o" to a quadratic function.

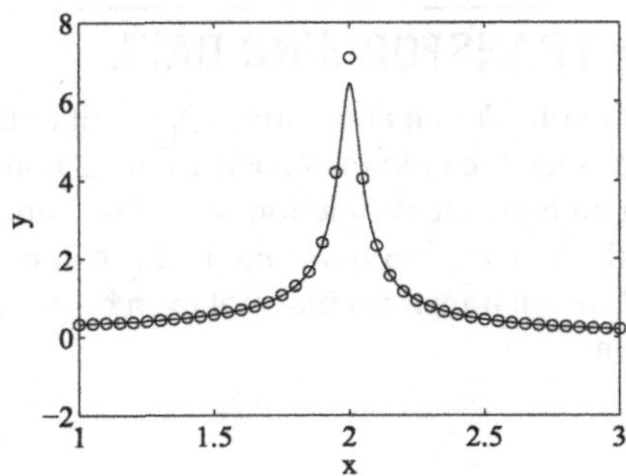

FIGURE 7.11

Fitting (7.15) to the given data denoted by o.

```
figure(1), plot(X,Y,'o',XX,YY)
axis([1 9 0 25]), xlabel('X'), ylabel('Y')
yy = 1./sqrt(YY);
figure(2), plot(x,y,'o',xx,yy)
axis([1 3 -2 8]), xlabel('x'), ylabel('y')
```

Running script `e4s712.m` gives the following results:

```
p =
    0.9944   -7.9638   15.9688
```

together with the plots shown in Fig. 7.7, Fig. 7.10, and Fig. 7.11. From the output of the script we see that the relationship between X and $\hat{Y}$ is

$$\hat{Y} = 0.9944X^2 - 7.9638X + 15.9688$$

We can deduce the values of the unknown coefficients by comparing the above equation with (7.15) to give

$$a_2^2 = 0.9944, \text{ hence } a_2 = \pm 0.9972. \text{ (Take the positive value as in (7.14).)}$$
$$-2a_1a_2 = 7.9748, \text{ hence } a_1 = 3.9931.$$
$$\text{Also } a_1^2 + a_0 = 15.9688, \text{ hence } a_0 = 0.0149.$$

We use these values in the original function (7.15) and fit it to the given data. This is shown in Fig. 7.11. This function provides a much better fit than the polynomials, shown earlier. However, even this fit does not pass through the peak value. This is caused by the sensitivity of the process to small random errors in the data. If the random errors are removed, the fit is exact. If the script `e4s712.m` is rerun with the random errors, the fit may be worse, and if the size of the random errors is increased, the process may fail. This is because in the region of $x = 2$, the value of y is essentially only dependent on a_0. This has a small value which may vary in sign.

Table 7.2 Transformations from non-linear to linear equations

Original equ'n	Substitution	Transformed equ'n
$y = ax^b$	$Y = \log_e(y),\ X = \log_e(x),\ A = \log_e(a)$	$Y = A + bX$
$y = ae^{bx}$	$Y = \log_e(y),\ A = \log_e(a)$	$Y = A + bx$
$y = axe^{bx}$	$Y = \log_e(y/x),\ A = \log_e(a)$	$Y = A + bx$
$y = 1/(a + bx)$	$Y = 1/y$	$Y = a + bx$
$y = 1/(a + bx)^2$	$Y = 1/\sqrt{y}$	$Y = a + bx$
$y = x/(a + bx)$	$Y = 1/y,\ X = 1/x$	$Y = aX + b$
$y = ax/(b + x)$	$Y = 1/y,\ X = 1/x,\ A = 1/a,\ B = b/a$	$Y = A + BX$

Table 7.2 lists some functions with a non-linear relationship between $\hat{y}$ and the coefficients of the function and the corresponding transformations that linearize these relationships so they have the form $\hat{Y} = BX + C$.

The script e4s713.m implements the first two relationships shown in Table 7.2. It also determines the sum of the squares of the errors for the two original relationships and plots a graph of the data and the two fitted functions.

```
% e4s713.m
x = 0.2:0.2:4;
y = 2*exp(0.5*x).*(1+0.2*rand(size(x)));
X = log(x);  Y = log(y);
% Case 1: Fit y = a*x^b
v = polyfit(X,Y,1);
A1 = v(2); b1 = v(1); a1 = exp(A1);
e1 = y-a1*x.^b1; s1 = e1*e1';
fprintf('\n y = %8.4f*x^(%8.4f): SSE = %8.4f',a1,b1,s1)
% case 2: Fit y = a*exp(b*x)
v = polyfit(x,Y,1);
A2 = v(2); b2 = v(1); a2 = exp(A2);
e2 = y-a2*exp(b2*x); s2 = e2*e2';
fprintf('\n y = %8.4f*exp(%8.4f*x): SSE = %8.4f \n',a2,b2,s2)
% Plotting
n = length(x);
r = x(n)-x(1); inc = r/100;
xp = [x(1):inc:x(n)];
yp1 = a1*xp.^b1;  yp2 = a2*exp(b2*xp);
plot(x,y,'ko',xp,yp1,'k:',xp,yp2,'k')
xlabel('x'), ylabel('f(x)')
```

Running script e4s713.m gives

```
 y =   4.5129*x^(  0.6736): SSE =  78.3290
 y =   2.2129*exp(  0.5021*x): SSE =   2.0649
```

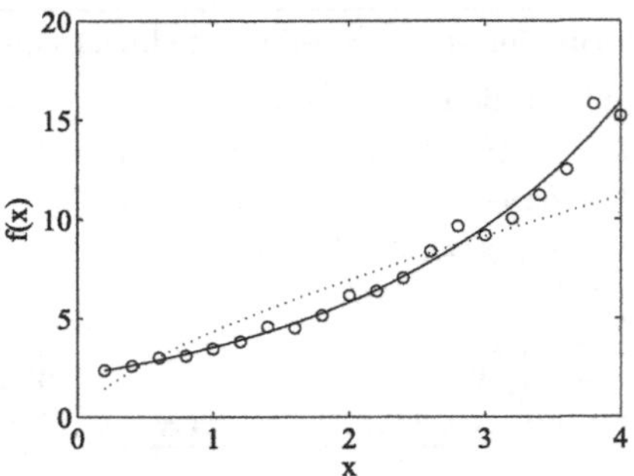

FIGURE 7.12

The graph shows the original data, denoted by o, and the fits obtained from $y = be^{ax}$ shown by the full line and $y = ax^b$ shown by the dotted line.

Fig. 7.12 and the above output confirms that the best fit is, as one would expect, the exponential function. Clearly this is a warning to the user to use discrimination in selecting the function to fit. This MATLAB function could be adapted to fit data to a wide range of mathematical functions.

7.11 THE KALMAN FILTER

The Kalman filter is an algorithm developed by Rudolf Kalman and described in Kalman (1960). Richard Bucy also contributed to this work and the algorithm is often called the Kalman–Bucy Filter, see Kalman and Bucy (1961).

The Kalman algorithm is used to analyze systems which are affected by statistical noise and other inaccuracies. The method aims to produce estimates of the key variables which describe the system as accurately as possible, thus separating the underlying signal from the noise. There are many applications of the Kalman filter, particular in navigation systems, and the everyday use of the Kalman filters in mobile phones. The method was used in the Apollo missions navigation computer and other navigation system; there are also important applications in robotics and computer vision and other fields. The Kalman filter analyzes the system variables over time and at each time step. An estimate of the system is made based on the current and previous states of the dynamic system and is used recursively over the whole time period. The system is assumed to be linear and the noise in the system is assumed to be Gaussian. We can use the normal distribution with appropriate parameters to estimate the noise.

The method calculates optimally the *Kalman gain matrix* to minimize the mean square error using the covariance matrix. The term filter indicates that the method is attempting to filter out the best results from noisy data. Kalman acknowledges the early work of Wiener on this type of problem and Wiener's characterization of it as a solution of the Wiener–Hopf integral equation (Wiener, 1949).

Before describing the Kalman filter it is useful to remind the reader of the meaning of the term expectation denoted by E. This may be formally defined for a single discrete random variable with discrete outcomes x_i with probability distribution p_i as:

$$E(x) = \sum_{i=1}^{k} x_i p_i$$

Note that

$$\sum_{i=1}^{k} p_i = 1$$

Consequently, $E(x)$ is the weighted average. For the continuous case the summation may be replaced by the integral for the continuous set of values and a continuous probability distribution. This can be extended from single variables to vectors directly.

A key measure of the accuracy of the results obtained from the Kalman filter for a given system where multiple observations are involved is the mean squared error function (MSE), we can write:

$$\mathbf{P}_k = E(\mathbf{e}_k \mathbf{e}_k^{\mathsf{T}})$$

where $\mathbf{e}_k$ represents the error in the system and $\mathbf{P}_k$ is called the error covariance matrix and E stands for the expectation of the value $\mathbf{e}_k \mathbf{e}_k^{\mathsf{T}}$. The error vectors $\mathbf{e}_k$ provide the differences between the correct and estimated values for the system and $\mathbf{P}_k$ is the covariance matrix at time interval k. The errors which are the differences between the correct and estimated values are given by:

$$\mathbf{e}_k = \mathbf{x}_k - \hat{\mathbf{x}}_k$$

Consequently:

$$\mathbf{P}_k = E\left((\mathbf{x}_k - \hat{\mathbf{x}}_k)(\mathbf{x}_k - \hat{\mathbf{x}}_k)^{\mathsf{T}}\right)$$

The expectation function has a range of useful properties including linearity, but in general expectations are not multiplicable. This description of expectation is useful since these properties are used in the derivation of the Kalman algorithm.

The nature of multiplicity for the expectation function is illustrated by the result:

$$\mathbf{C}(\mathbf{x}, \mathbf{y}) = E(\mathbf{x}\mathbf{y}^{\mathsf{T}}) - E(\mathbf{x})E(\mathbf{y}^{\mathsf{T}})$$

Here we introduce the concept of the covariance matrix denoted by $\mathbf{C}(\mathbf{x}, \mathbf{y})$ and this provides the covariance of the $\mathbf{x}$ and $\mathbf{y}$ variables. In this case the covariance matrix gives a measure of the extent to which expectation values are not multiplicable where

$$E(\mathbf{x}\mathbf{y}^{\mathsf{T}}) = \mathbf{C}(\mathbf{x}, \mathbf{y}) + E(\mathbf{x})E(\mathbf{y}^{\mathsf{T}})$$

here the covariance matrix supplies a measure of how the variables $\mathbf{x}$ and $\mathbf{y}$ interact. However if $\mathbf{x}$ and $\mathbf{y}$ are un-correlated, for example independent variables then the expectations are multiplicable.

To further clarify this concept of covariance, which will play a major role in our discussion of the Kalman filter, we note that the term variance refers to the square of the standard deviation and arises in many probability distributions, for example the normal distribution. It is often written $N(\mu, \sigma^2)$ where σ^2 is the variance, here the variance indicates the breadth of the spread of the distribution values, μ gives the mean position of the variable of the normal curve for example if $\mu = 0$ the curve is symmetric about 0. Extending the concept of variance to more than one variable we can ask do changes or errors in one variable affect the other. If not, then we can write the covariance matrix, **C**, for the two variables as

$$\mathbf{C} = \begin{bmatrix} \sigma_x^2 & 0 \\ 0 & \sigma_y^2 \end{bmatrix}$$

If, however, the errors in one variable affect the other then we must introduce cross terms reflecting this in the covariance matrix, **C**, thus:

$$\mathbf{C} = \begin{bmatrix} \sigma_x^2 & \sigma_{xy}^2 \\ \sigma_{xy}^2 & \sigma_y^2 \end{bmatrix}$$

This description clearly applies to two variables but is easily extended to many variables, just by extending the size of the matrix, a ten-by-ten matrix provides the information for ten variables and their various interactions. Having described these useful preliminaries we now return to the description of the Kalman filter. The description we provide is for the discrete form of the Kalman filter. There is a continuous form described in, Kalman and Bucy (1961) which is formulated as a system of first order differential equations. We can formally represent the equations which govern the behavior of the system used in the Kalman filter by the equation:

$$\mathbf{x}_{k|k-1} = \mathbf{F}_k \mathbf{x}_{k-1|k-1} + \mathbf{B}_k \mathbf{u}_k + \mathbf{w}_k \quad k = 0,\ 1,\ 2, \ldots \tag{7.17}$$

Here the notation $x_{k|k-1}$ means the estimate of x_k before the kth measurement has been taken into account. Here we are estimating the new n component state vector of the system $\mathbf{x}_k$ from the previous state vector $\mathbf{x}_{k-1}$ and the p component control vector $\mathbf{u}_k$. The $n \times n$ component matrix $\mathbf{F}_k$ and the $n \times p$ matrix $\mathbf{B}_k$ provide the model which arises out of the analysis of the specific dynamic system being studied. This may, for example, be based on the dynamics or the electrical behavior of the system over time and the specific form is dependent on the particular application. The matrix $\mathbf{B}_k$ providing the control system values for the control vector $\mathbf{u}_k$ such as the velocity and acceleration, and the matrix $\mathbf{F}_k$ describing how the model transits from one state to another. The n component vector $\mathbf{w}_k$ provides the noise in the system, this is drawn from a normal distribution with mean zero and covariance $\mathbf{Q}_k$ hence,

$$\mathbf{w}_k = N(0, \mathbf{Q}_k)$$

To complete the model we also make a measurement of the observed values at time interval k, denoted by the s component vector $\mathbf{z}_k$ of the key parameters of the model which we wish to observe and these constitute the required output from the model. This is given by:

$$\mathbf{z}_k = \mathbf{H}_k \mathbf{x}_k + \mathbf{v}_k \tag{7.18}$$

where the $s \times n$ matrix $\mathbf{H}_k$ is the model of the observed system and the s component vector $\mathbf{v}_k$ the noise in the system drawn from the normal distribution with mean zero and covariance $\mathbf{R}_k$. The derivation of specific models or determining $\mathbf{H}_k$, $\mathbf{F}_k$, and $\mathbf{B}_k$ is based on the dynamical features of the specific problem, for example, the motion of a rocket, the laws of dynamics and specific guidance features are required to be formulated in the construction of the model. However, the determination of the noise factors which are given by the covariance matrices $\mathbf{Q}_k$ and $\mathbf{R}_k$, and the vectors $\mathbf{v}_k$ and $\mathbf{w}_k$, present significant difficulties in formulating and must be estimated from the data we have for the particular system. Least squares routines are available for example in MATLAB for such calculations.

We now describe how the Kalman filter is implemented. Essentially there are two phases in the application of the Kalman filter:

1. The predict stage, which uses (7.17) without the error term to predict the state of the system:

$$\mathbf{x}_{k|k-1} = \mathbf{F}_k \mathbf{x}_{k-1|k-1} + \mathbf{B}_k \mathbf{u}_k \tag{7.19}$$

Together with $\mathbf{x}_{k|k-1}$ we have an estimate of the current state covariance matrix given by

$$\mathbf{P}_{k|k-1} = \mathbf{F}_k \mathbf{P}_{k-1|k-1} \mathbf{F}_k^{\mathsf{T}} + \mathbf{Q}_k \tag{7.20}$$

This provides the error estimation, which can directly be derived from the definition of the covariance matrix and the use of the properties of the expectation function. This is calculated from the square of the difference between the unknown true value and the estimate of this value as we have described above.

2. The update stage, where each of the elements of the model is updated and an important feature related to the quality of the model called the Kalman gain is calculated. This is the stage at which the bulk of the computation is performed and provides the practical implementation of the algorithm. We can calculate a measurement of the residual: $\mathbf{y}_k$ as

$$\mathbf{y}_k = \mathbf{z}_k - \mathbf{H}_k \mathbf{x}_{k|k-1} \tag{7.21}$$

which we can see from (7.18) is the difference between the observation and the state variable $\mathbf{x}_{k|k-1}$. We then calculate the innovation covariance matrix thus:

$$\mathbf{S}_k = \mathbf{H}_k \mathbf{P}_{k|k-1} \mathbf{H}_k^{\mathsf{T}} + \mathbf{R}_k \tag{7.22}$$

The value of this provides us with a measure of the reliability of the residual based on the noise $\mathbf{R}_k$ and error matrix $\mathbf{P}_{k|k-1}$.
Calculation of the Kalman gain $\mathbf{K}_k$:

$$\mathbf{K}_k = \mathbf{P}_{k|k-1} \mathbf{H}_k^{\mathsf{T}} \mathbf{S}_k^{-1} \tag{7.23}$$

This is a key feature of the computational process, the expression given by (7.23) gives the optimal gain for the feedback system and consequently the optimal gain filter. $\mathbf{K}_k$ is computed to provide the minimum of the error covariance matrix. Note that we can obtain the expression for the matrix $\mathbf{S}_k$ from (7.22). Now use the optimum Kalman gain to update the state variable estimates:

$$\mathbf{x}_{k|k} = \mathbf{x}_{k|k-1} + \mathbf{K}_k \mathbf{y}_k \tag{7.24}$$

Update the estimate of the system covariance matrix:

$$\mathbf{P}_{k|k} = (\mathbf{I} - \mathbf{K}_k\mathbf{H}_k)\mathbf{P}_{k|k-1} \tag{7.25}$$

The $\mathbf{P}_{k|k}$ matrix should be symmetric and should be positive definite at each stage of the process. However, this equation is only true for the optimal value of $\mathbf{K}_k$.

Each of the steps can be derived formally but we will provide only the derivation of the expression for the Kalman gain matrix since this provides the key element in ensuring the efficiency of the algorithm.

The important features of this derivation are that we can write a new estimate of the data by combining the previous estimate with the measurement data residual from (7.21). This is in fact what (7.24) implements, and uses the Kalman gain matrix. However, the important feature of this process is that we wish to select the matrix **K** for optimum results. Thus, we wish to select **K** to minimize the sum of squares of the errors for the process, i.e. the MSE. Now we can write the covariance matrix **P** in terms of the error differences, which in turn will be dependent on **K** the Kalman gain and in particular the trace terms of the covariance matrix will give the sum of squares of the errors terms. Thus the problem resolves into finding the minimum of the trace of the covariance matrix **P** with respect to the **K** matrix. This is achieved by differentiating the trace of the **P** matrix, which is the diagonal elements, with respect to the matrix **K** and equating to zero. This then gives the required expression for the matrix **K**. We can now derive this key result for the optimal Kalman filter matrix formally: We now state the definition of the covariance matrix **P** given earlier:

$$\mathbf{P}_k = E(\mathbf{e}_k\mathbf{e}_k^{\mathsf{T}}) = E(\mathbf{x}_k - \hat{\mathbf{x}}_k)(\mathbf{x}_k - \hat{\mathbf{x}}_k)^{\mathsf{T}} \tag{7.26}$$

However, we wish to select the estimate of using the Kalman gain (7.24) and substituting from (7.21) leads to:

$$\mathbf{x}_{k|k} = \mathbf{x}_{k|k-1} + \mathbf{K}_k\mathbf{y}_k = \mathbf{x}_{k|k-1} + \mathbf{K}_k(\mathbf{z}_k - \mathbf{H}_k\mathbf{x}_{k|k-1}) \tag{7.27}$$

Now substituting for the difference between the **x** values which equals the current error gives:

$$\mathbf{e}_k = \mathbf{x}_{k|k} - \mathbf{x}_{k|k-1} = \mathbf{K}_k(\mathbf{z}_k - \mathbf{H}_k\mathbf{x}_{k|k-1}) \tag{7.28}$$

In addition, we note that from Eq. (7.18) we can substitute for $\mathbf{z}_k$ thus giving:

$$\mathbf{e}_k = \mathbf{K}_k(\mathbf{H}_k\mathbf{x}_k + \mathbf{v}_k - \mathbf{H}_k\mathbf{x}_{k|k-1}) \tag{7.29}$$

We can now substitute this expression for the error into the expression for the covariance matrix **P**, given by Eq. (7.26). Then we obtain an expression for the covariance matrix in terms of the Kalman gain matrix **K** as follows after considerable algebraic manipulation:

$$\mathbf{P}_k = (\mathbf{I} - \mathbf{K}_k\mathbf{H}_k)E(\mathbf{x}_k - \hat{\mathbf{x}}_k)(\mathbf{x}_k - \hat{\mathbf{x}}_k)^{\mathsf{T}}(\mathbf{I} - \mathbf{K}_k\mathbf{H}_k) + \mathbf{K}_kE(\mathbf{v}_k\mathbf{v}_k^{\mathsf{T}})\mathbf{K}_k \tag{7.30}$$

Since the expression $E(\mathbf{x}_k - \hat{\mathbf{x}}_k)(\mathbf{x}_k - \hat{\mathbf{x}}_k)^{\mathsf{T}}$ represents the previous covariance matrix we have an expression for the new covariance matrix as

$$\mathbf{P}_{k|k} = (\mathbf{I} - \mathbf{K}_k\mathbf{H}_k)\mathbf{P}_{k|k-1}(\mathbf{I} - \mathbf{K}_k\mathbf{H}_k) + \mathbf{K}_kE(\mathbf{v}_k\mathbf{v}_k^{\mathsf{T}})\mathbf{K}_k \tag{7.31}$$

Setting $\mathbf{R} = E(\mathbf{v}_k\mathbf{v}_k^{\mathsf{T}})$ we obtain:

$$\mathbf{P}_{k|k} = (\mathbf{I} - \mathbf{K}_k\mathbf{H}_k)\mathbf{P}_{k|k-1}(\mathbf{I} - \mathbf{K}_k\mathbf{H}_k) + \mathbf{K}_k\mathbf{R}\mathbf{K}_k \tag{7.32}$$

Finally, we wish to minimize trace of this equation with respect to the Kalman matrix since the diagonal elements give the sum of squares of the errors, thus we differentiate the trace of the covariance matrix and equate it to zero, giving:

$$\frac{\partial \mathrm{tr}(\mathbf{P}_{k|k})}{\partial \mathbf{K}_k} = -2(\mathbf{H}_k\mathbf{P}_{k|k-1})^{\mathsf{T}} + 2\mathbf{K}_k(\mathbf{H}_k\mathbf{P}_{k|k-1}\mathbf{H}_k^{\mathsf{T}} + \mathbf{R}) = \mathbf{0} \tag{7.33}$$

where tr denotes the trace of a matrix. See Appendix A for an explanation of matrix differentiation. Solving (7.33) gives the required expression for the Kalman gain:

$$\mathbf{K}_k = (\mathbf{H}_k\mathbf{P}_{k|k-1})^{\mathsf{T}}(\mathbf{H}_k\mathbf{P}_{k|k-1}\mathbf{H}_k^{\mathsf{T}} + \mathbf{R})^{-1} \tag{7.34}$$

Using (7.22) this can be simplified to:

$$\mathbf{K}_k = (\mathbf{H}_k\mathbf{P}_{k|k-1})^{\mathsf{T}}\mathbf{S}_k^{-1} \tag{7.35}$$

This will be a minimum if the second derivative of the trace is positive definite, which implies that the covariance matrix should be positive definite.

The set of Eqs. (7.19) to (7.25) are the steps in the process which define the Kalman filter algorithm. Although the implementation of the algorithm would appear relatively straightforward there are significant problems in its practical implementation particularly in assessing the estimation of noise in a practical real world situation. The estimation of initial values for the covariance matrices may also present problems.

We now consider a specific application of the Kalman filter: the motion of a projectile launched with a specific velocity and at an angle α. The equations of motion are given in (7.36) and (7.37).

$$v = v_0 + at$$

and

$$s = v_0t + at^2/2$$

These equations give the velocity v and the distance traveled by an object traveling linearly with acceleration a and with initial velocity v_0 at time t. We can use these equations to study the flight of a projectile projected at an angle by identifying the components of the velocity in the x and y co-ordinate directions and replacing the general acceleration by gravity, g. This leads to the well known results for the velocity of the object and the distance traveled at time t, in component terms:

$$\begin{aligned} x &= v_0\cos(\alpha)t \\ y &= v_0\sin(\alpha)t - gt^2/2 \\ v_y &= v_0\sin(\alpha)t - gt \end{aligned} \tag{7.36}$$

Expressing these equations in discrete form for the time interval Δt and using the vector $\mathbf{x}_k = [x_k^1, x_k^2, x_k^3]$ to represent the component state variables $[x, y, v_y]$, each element denoted by a superscript at time interval k this gives the equations in discrete form as:

$$\begin{aligned} x_{k+1}^1 &= x_k^1 + v_0 \cos(\alpha)\Delta t \\ x_{k+1}^2 &= x_k^2 + v_0 \sin(\alpha)\Delta t - g(\Delta t)^2/2 - gt\Delta t \\ x_{k+1}^3 &= x_k^4 + v_0 \sin(\alpha)\Delta t - g\Delta t \end{aligned} \tag{7.37}$$

This enables us to track the behavior of the projectile over time t. Now we reformulate these expressions in the form of (7.17):

$$\mathbf{x}_k = \mathbf{F}_k x_{k-1} + \mathbf{B}_k \mathbf{u}_k + \mathbf{w}_k$$

This requires that we define the matrices $\mathbf{F}_k$, $\mathbf{B}_k$, and $\mathbf{w}_k$. The first two matrices arise from the basic dynamics of the system, the matrix $\mathbf{B}_k$ whose elements relate to factors such as the imposed force in this case gravity and the noise matrix can be generated from a suitable normal distribution. Rewriting the dynamics equations in matrix form implies:

$$\mathbf{F}_k = \begin{bmatrix} 1 & 0 & 0 \\ 0 & 1 & 0 \\ 0 & 0 & 1 \end{bmatrix} \tag{7.38}$$

Note that in this example $\mathbf{F}_k$ is a unit matrix.

$$\mathbf{B}_k = \begin{bmatrix} 0 & 0 & 0 & 0 & \Delta t & 0 & 0 & 0 \\ 0 & -(\Delta t)^2/2 & 0 & -t\Delta t & 0 & \Delta t & 0 & 0 \\ 0 & 0 & 0 & 0 & -\Delta t & 0 & 0 & 1 \end{bmatrix} \tag{7.39}$$

$$\mathbf{U}_k = \begin{bmatrix} 0 \\ g \\ 0 \\ g \\ v_0 \cos(\alpha) \\ v_0 \sin(\alpha) \\ v_0 \cos(\alpha) \\ v_0 \sin(\alpha) \end{bmatrix} \tag{7.40}$$

In this example we can generate a noise vector in an arbitrary way using MATLAB as: `w_k=0.1*randn(4,1)`. We can also generate a noise vector for the observation shown in equation using MATLAB for example as `v_k=0.1*randn(4,1)`. Now using this information the problem has been defined and we can use the Kalman algorithm to solve this specific problem, which is to examine how effective the Kalman filter is in extracting the signal from the noise and reflecting the true underlying nature of the physical dynamics of the process. It is important to emphasize that initial values have to be set for the covariance matrices and the noise vectors, this has been done arbitrarily in this case but will need careful attention in a real world application.

We have defined the key matrices $\mathbf{F}_k$, $\mathbf{B}_k$, and $\mathbf{U}_k$ for this problem, the equations are given in (7.36) and the matrices are given by (7.38), (7.39), and (7.40). These are used in the Kalman filter script e4s714.m thus:

```
% e4s714.m
% Kalman filter for tracking a projectile
clear all
% Example projectile at angle
% Set up matrices F, B, U and constants
dt = 0.01; alpha = pi/3; v0 = 10; g = 9.81;
x = [0  0  0]'; timev = dt; n = 3;
% time of flight, maximum height, maximum horizontal distance
tf = 2*v0*sin(alpha)/g
ymax = (v0*sin(alpha))^2/(2*g)
xmax = tf*v0*cos(alpha)
F = eye(3); timeaxis = [ ];
xh = [0 0 0]';
H = eye(3);
% The calculation for the velocity and position in the falling object example
xdset = [ ]; timev = dt
while timev<=tf
    B = [0  0 0  0  dt 0 0 0;
         1  -0.5*dt^2  0 -dt*timev  0  dt  0  0;
         0  0 0 -dt 0 0 0 dt];
    U = [0  g  0 g  v0*cos(alpha) v0*sin(alpha) v0*cos(alpha) v0*sin(alpha)]';
    timev = timev+dt;
    xd = F*x+B*U;
    xdset = [xdset,xd];
    timeaxis = [timeaxis,timev];
    x = xd;
end
% The state covariance matrices are assigned initial approximations
P = 0.02^2*eye(3);
Q = 0.04^2*eye(3);
R = 0.40^2*eye(3);
plotxset = [ ]; plotzset = [ ]; xhset = [ ];
timeaxis = [ ]; timev = dt;
while timev <=tf
    B = [0  0 0  0  dt 0 0 0;
         0  -0.5*dt^2  0 -dt*timev  0  dt  0  0;
         0  0 0 -dt 0 0 0 dt;];
    U = [0  g  0 g  v0*cos(alpha)  v0*sin(alpha) v0*cos(alpha) v0*sin(alpha)]';
    timev = timev+dt;
    timeaxis = [timeaxis, timev];
    % The noise matrices w and v are sampled from the normal distribution
```

```
    w = 0.04*randn(3,1);
    v = 0.1*randn(3,1);
    % Kalman filter the current state:
    x = F*x+B*U+w;
    % calculate observed/measured values which include noise
    z = H*x+v;
    plotzset = [plotzset z];
    % estimate state
    xh = F*xh+B*U;
    % Estimate covariance matrix P
    P = F*P*F'+Q;
    % Kalman filter the update stage: Caculate the residual
    y = z-H*xh;
    % Calculate the innovation covariance
    S = H*P*H'+R;
    % The optimum Kalman gain is, uses inverse of S using matlab /:
    K = P*H'/S;
    % calculate the estimate of the current state
    xh = xh+K*y;
    xhset = [xhset, xh];
    % recalculate error convergence matrix
    P = (eye(n)-K*H')*P;
    x = xh;
end
plot(timeaxis,xhset(2,:),'k-')
hold on
plot(timeaxis,plotzset(2,:),'k*')
plot(timeaxis,xdset(2,:),'k--')
hold off
xlabel('Time of flight')
ylabel('Height')
```

Running script e4s714.m gives the results shown in Fig. 7.13 and Fig. 7.14 and the following data:

```
tf =
    1.7656

ymax =
    3.8226

xmax =
    8.8280

timev =
    0.0100
```

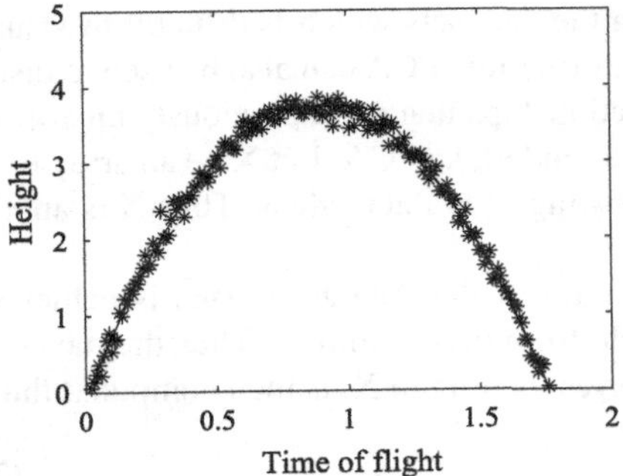

FIGURE 7.13

Changes in the height of a projectile over the time of fight. Graph shows the path of the projectile without noise as the dashed line, the observed values including noise as asterisks and the path generated by the Kalman filter as the continuous line.

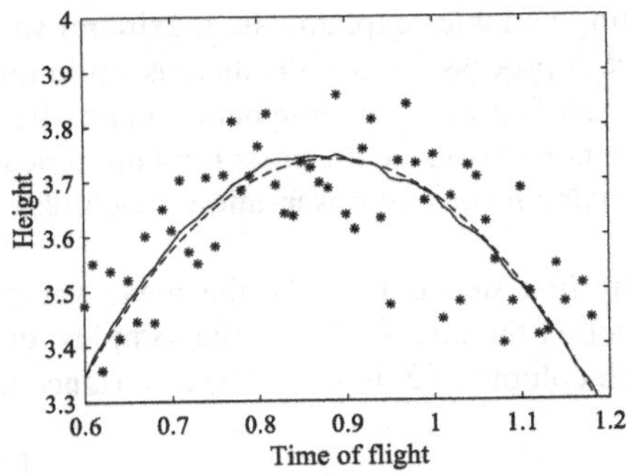

FIGURE 7.14

Considerably expanded graph of the height of projectile over time of flight showing observations subject to noise by asterisks, the dashed line of the flight of the projectile subject only to laws of dynamics and the output of the Kalman filter after processing the noisy data.

Before completing our discussion of the Kalman filter we mention briefly two important considerations. Firstly, the description of the Kalman filter we have provided here is applied to a linear dynamic model. If the model is non-linear a modified version of the Kalman filter is used called the extended Kalman filter (EKF). The EKF employs an iterative linearization of the original non-linear model. Secondly, an alternative method for the solution of noisy problems is to use a least squares method or one of its modifications such as the recursive least squares method (RLS). Many studies have been performed comparing the EKF and RLS methods for specific applications and the results may be viewed as problem dependent. An interesting study performed on a theoretical basis is given in Leondes (1970) which provides comparison of the Kalman filter method with least squares methods and Baysian approaches to the problem. The papers edited by Leondes, include a wide range of applications of the Kalman filter including some details of the guidance of the Apollo mission vehicle.

7.12 PRINCIPAL COMPONENT ANALYSIS

Principal component analysis (PCA) was developed at the beginning of the 20th century by Karl Pearson and subsequently has been rediscovered several times. For example, Hotelling independently rediscovered the method in the 1930s. Other, closely related methods include the discrete Karhunen–Loève transform (KLT) used in signal processing and proper orthogonal decomposition (POD) applied in mechanical engineering. Singular value decomposition (SVD) is a similar process and is often used in the computation of PCA. PCA is used to analyze and visualize large data sets and also used in data reduction. In this brief introduction to PCA we will concentrate on data reduction.

PCA is a dimensionality reduction technique used to transform high-dimensional data sets into data sets with fewer dimensions consequently clarifying the structure of complex data sets. The set of resulting variables explains the maximum variance within the data sets which is difficult to visualize. The PCA process ensures the there is a minimum loss of information. PCA can also be used to discover important features of a large data set and often reveals relationships that were previously unsuspected.

We now introduce the basis for a numerical procedure to undertake PCA. Let $\mathbf{X}$ be an array of data, comprising n rows of observations, each observation consisting of m data values. Thus $\mathbf{X}$ is an $n \times m$ array.

The first step is to make the mean value of each column of the data zero. This is achieved by subtracting the mean value of the samples for each variable from these samples. Thus the mean value of each column of $\mathbf{X}$ is zero. The covariance matrix, $\mathbf{C}$, corresponding to $\mathbf{X}$, is then computed thus:

$$\mathbf{C} = \tfrac{1}{n-1}\mathbf{X}^\mathsf{T}\mathbf{X} \tag{7.41}$$

The covariance matrix will be an $m \times m$ symmetrical matrix.

If we wish, we can also standardize the data by dividing each element of a column of $\mathbf{X}$ by the standard deviation of the column. Thus every column of data has a standard deviation of one. This process creates the correlation matrix, which we also denote by $\mathbf{C}$.

When scales of variables are similar, the covariance matrix is always preferred to the correlation matrix, as the correlation matrix will lose information when standardizing the variance. However, when the variance of the variables are very different, PCA applied to un-normalized variables could lead to large loadings for variables with high variance. In turn, this will lead to dependence of a principal component on the variable with high variance which is misleading.

We can now perform eigenvalue decomposition to determine the eigenvalues and eigenvectors of $\mathbf{C}$ so that

$$\mathbf{C} = \mathbf{V}\mathbf{D}\mathbf{V}^\mathsf{T} \tag{7.42}$$

This is a restatement of (2.68). Here $\mathbf{D}$ is a diagonal matrix of eigenvalues and $\mathbf{V}$ is a matrix of the corresponding eigenvectors.

Alternatively, we can obtain the same result using the more efficient and robust singular value decomposition. Let

$$\mathbf{X} = \mathbf{U}\mathbf{S}\mathbf{V}^\mathsf{T} \tag{7.43}$$

This is a restatement of (2.23). Here $\mathbf{S}$ is a diagonal matrix of singular values, $\mathbf{U}$ is an orthogonal matrix such that $\mathbf{U}^\mathsf{T}\mathbf{U} = \mathbf{I}$. Now

$$\mathbf{C} = \tfrac{1}{n-1}\mathbf{X}^\mathsf{T}\mathbf{X} = \tfrac{1}{n-1}\left(\mathbf{V}\mathbf{S}^\mathsf{T}\mathbf{U}^\mathsf{T}\right)\left(\mathbf{U}\mathbf{S}\mathbf{V}^\mathsf{T}\right) = \tfrac{1}{n-1}\mathbf{V}\mathbf{S}^\mathsf{T}\mathbf{S}\mathbf{V}^\mathsf{T} \equiv \mathbf{V}\mathbf{D}\mathbf{V}^\mathsf{T} \tag{7.44}$$

Note that because $\mathbf{C}$ is a square, $m \times m$ matrix, so also are $\mathbf{U}$, $\mathbf{V}$, and $\mathbf{S}$. Clearly $\mathbf{V}$ is identical in both transforms. Let $\mathbf{S}_0$ be the $m \times m$ diagonal array of the non-zero singular values of $\mathbf{S}$. Then

$$\tfrac{1}{n-1}\mathbf{S}^\mathsf{T}\mathbf{S} = \tfrac{1}{n-1}\mathbf{S}_0^2 = \mathbf{D}$$

As will become apparent, we are more concerned with the ranking, and relative values of $\mathbf{D}$ or $\mathbf{S}_0^2$ rather than their actual values.

The matrix $\mathbf{U}$ is a set of orthogonal linearly uncorrelated vectors called the principal components. The first principal component is the one corresponding to the largest eigenvalue, the second principal component corresponds to the second largest eigenvalue and so on. The first principal component has the largest possible variance (that is, it accounts for as much of the variability in the data as possible). Each element of a principal component corresponds to a variable. Since the eigenvalues are a measure of amplitude, and hence the relative importance of each principal component, it is helpful to express the eigenvalues as a percentage of the total; this shows the relative importance of the corresponding principal component. Providing the cumulative sum of the eigenvalues is an alternative way of providing the same information.

The array score, $\mathbf{S}_c$, is given by $\mathbf{S}_c = \mathbf{XU}$. The array $\mathbf{S}_c$ is the data generated by transforming the original data array into the space of the principal components. The values of the corresponding eigenvalues are the variance of the columns of $\mathbf{S}_c$.

We can regenerate the original data analyzed the PCA by using (7.43). However, if we wish to use only the p most significant principal components, then, using the subscript red to indicate a reduced model, we have

$$\mathbf{X}_{red} = \mathbf{U}_{red}\mathbf{S}_{red}\mathbf{V}_{red}^{\mathsf{T}}$$

where $\mathbf{U}_{red}$ is an $n \times p$ matrix, $\mathbf{S}_{red}$ is a $p \times p$ diagonal matrix and $\mathbf{V}_{red}$ is an $m \times p$ matrix, comprising columns 1 to p of $\mathbf{U}$ and $\mathbf{V}$ respectively and $\mathbf{S}_{red}$ comprising only the first p rows and columns of $\mathbf{S}$. The resultant matrix, $\mathbf{X}_{red}$, is, of course, still an $m \times n$ matrix.

If we carry out this process using the full set of principal components, then the data is regenerated and is identical to the original data. However, if we choose to use only the p most significant principal components, where $p < m$, then the regenerated data will not be exactly identical to the original data.

We can determine a matrix $\mathbf{W}$, where

$$\mathbf{W}_{red} = \mathbf{U}_{red}\mathbf{S}_{red}$$

Then the communality $\mathbf{c}_o$ consists of the diagonal elements of $\mathbf{W}_{red}\mathbf{W}_{red}^{\mathsf{T}}$. Communality is the total amount of variance an original variable shares with all the other variables in the analysis. When $p = m$ the communality is unity (or 100%) for every variable, meaning that the full set of principal components explain 100% of each variable, as they must. When $p < m$ the p principal components used in the analysis explain less than 100% of each variable.

The function `fprcomal` carries out a principal component analysis of the data given in the array `Z`. Options allow the principal components to be determined for the covariance matrix or the correlation matrix. The number of principal components used to determine the communality and to regenerate the data can be chosen. The function outputs a range of data, including the eigenvalues and principal components.

```
function [aver, stan, C,eigenvalues,percent_eig,cuml_eig,princ_comp, score, ...
                comm, Xrec] = fprcomal(Z,flag,p)
% Z = [1 2 3 4;
%      2 3 4 5] for example (variable x samples)
% flag == 1 ensures mean of data X is zero
% flag == 2 ensures mean of data X is zero and the variance is unity.
% p is number of principal components to be included in the analysis
% aver is average of each variable
% stan is the standard dev for each variable
```

```
% C is covariance or correlation matrix of X
% eigenvalues, percent_eig are the actual and percentage of X.
% cuml_eig are the cumulative sum of the eigenvalues of X.
% princ_comp are the proncipal components of X
% score is X transformed into the space of the principal components.
% comm is the communality.
% Xrec is the reconstructed value of X using p coordinates
Z = Z'; Z1 = Z; % Z and z1 are (samples x variables)
if flag == 1, disp('Using covariance matrix'), end
if flag == 2, disp('Using correlation matrix'), end
[n,m] = size(Z); aver = mean(Z); stan = std(Z);
if flag == 1 | flag == 2; Z = Z-aver; end
if flag == 2;  Z = Z./stan;  end
C = (1/(n-1))*Z'*Z;  % Coherence or correlation matrix
X1 = (1/sqrt(n-1))*Z';  % X1 (variable x samples)
[u,s,v] = svd(X1); %
sr = s(1:m,1:m); % singular values
eigenvalue = diag(sr.^2); eigenvalues = eigenvalue';
percent_eig = 100*eigenvalues./sum(eigenvalues);
cuml_eig = (100*cumsum(eigenvalue)/sum(eigenvalues))';
princ_comp = u;
score = Z*u;  %Z1*u;
vp = 1:p;
X1rec = u(:,vp)*s(vp,vp)*v(:,vp)'; ZrecT = X1rec'*(sqrt(n-1));
Xrec = ZrecT'; XrecT = Xrec';
if flag == 2,  ZrecT = ZrecT.*stan; end
if flag == 1 | flag == 2, XrecT = ZrecT+aver; end, Xrec = XrecT;
W = u(:,vp)*s(vp,vp);
comm = 100*diag(W*W')';
if flag == 1, comm = comm./(stan.^2); end
```

Example 7.6. Undertake a principal component analysis on the follow data comprising 8 samples of 4 variables. Rows 1 to 4 of samples of variables 1 to 4.

```
X =[1.0    2.0    3.0    4.0    5.0    6.0    7.0    8.0;
      0    1.0    1.6    2.0    2.3    2.6    2.8    3.0;
    1.5    4.7    7.0    9.0   10.8   12.4   13.9   15.4;
    1.9    1.0    1.1    2.6    0.3    1.9    2.2    1.2];
```

The script e4s715.m is as follows

```
% e4s715.m
[aver, stan, C,eigenvalues,percent_eig,cuml_eig,princ_comp, score,...
                comm, Xrec]  = fprcomal(X,2,4)
figure(1)
subplot(2,2,1), plot(Xrec(:,1),Xrec(:,2),'k*--', X(1,:),X(2,:),'ko-')
```

```
xlabel('x1'), ylabel('x2')
subplot(2,2,2), plot(Xrec(:,1),Xrec(:,3),'k*--', X(1,:),X(3,:),'ko-')
xlabel('x1'), ylabel('x3')
subplot(2,2,3), plot(Xrec(:,1),Xrec(:,4),'k*--', X(1,:),X(4,:),'ko-')
xlabel('x1'), ylabel('x4')
subplot(2,2,4), plot(Xrec(:,2),Xrec(:,3),'k*--', X(2,:),X(3,:),'ko-')
xlabel('x2'), ylabel('x3')
```

Running e4s715.m with p set to 4 gives the following output:

```
Using correlation matrix

aver =
    4.5000    1.9125    9.3375    1.5250

stan =
    2.4495    1.0134    4.7449    0.7517

C =
    1.0000    0.9582    0.9913    0.0466
    0.9582    1.0000    0.9874    0.0033
    0.9913    0.9874    1.0000    0.0237
    0.0466    0.0033    0.0237    1.0000

eigenvalues =
    2.9590    1.0001    0.0409    0.0001

percent_eig =
   73.9740   25.0014    1.0234    0.0013

cuml_eig =
   73.9740   98.9754   99.9987  100.0000

princ_comp =
   -0.5757    0.0105    0.6829    0.4495
   -0.5746   -0.0346   -0.7284    0.3714
   -0.5813   -0.0135    0.0449   -0.8124
   -0.0217    0.9993   -0.0318   -0.0029

score =
    2.8564    0.5711    0.3089   -0.0029
    1.6883   -0.6643   -0.0628    0.0027
    0.8284   -0.5541   -0.1977    0.0120
    0.0782    1.4249   -0.2509   -0.0060
   -0.4811   -1.6437   -0.0735   -0.0119
```

```
   -1.1284    0.4728   -0.0628    0.0015
   -1.6692    0.8648    0.0737    0.0004
   -2.1726   -0.4714    0.2652    0.0042

comm =

  100.0000  100.0000  100.0000  100.0000

Xrec =
    1.0000   -0.0000    1.5000    1.9000
    2.0000    1.0000    4.7000    1.0000
    3.0000    1.6000    7.0000    1.1000
    4.0000    2.0000    9.0000    2.6000
    5.0000    2.3000   10.8000    0.3000
    6.0000    2.6000   12.4000    1.9000
    7.0000    2.8000   13.9000    2.2000
    8.0000    3.0000   15.4000    1.2000
```

From this output, we can deduce that the first and second principal components provide 98% of the variance in the data. The first principal component has high values corresponding to first three variables. The second principal component has a high value corresponding to the fourth variable. Fig. 7.16 shows that the original data can be well represented using only first two principal components.

If `p` is set to 1 the output from `e4s715.m` is unchanged except for the following part of the output:

```
comm =

   98.0790   97.7077   99.9701    0.1392

Xrec =

    0.4718    0.2491    1.4596    1.4784
    2.1191    0.9294    4.6813    1.4975
    3.3318    1.4301    7.0529    1.5115
    4.3897    1.8669    9.1217    1.5237
    5.1784    2.1926   10.6643    1.5328
    6.0913    2.5696   12.4496    1.5434
    6.8540    2.8846   13.9412    1.5522
    7.5639    3.1777   15.3295    1.5604
```

Note that the first principal component gives a high communality for all except the fourth variable. This is further illustrated in Fig. 7.15 where the correspondence between the original and regenerated data for the fourth variable is very poor.

The data for this problem was generated as follows. Letting the four variables be x_1, x_2, x_3, and x_4, then we have set $x_2 = \log_2(x_1)$ and $x_3 = x_1 + 2x_2 + 0.5\sqrt{x_1}$. Thus, x_1, x_2, and x_3 are closely related. However, we have set $x_4 = 0.5\sqrt{x_1}$ together with a normally distributed random number with a standard deviation of unity. Thus x_4 is only weakly related to the other variables.

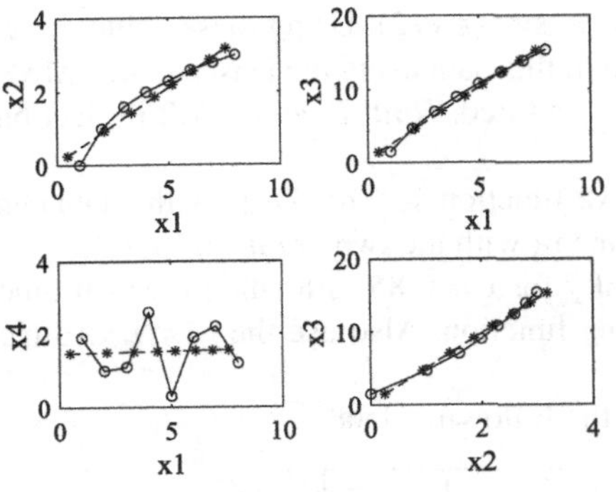

FIGURE 7.15

Relationship between selected variables. Dashed line is generated using only the first principal component.

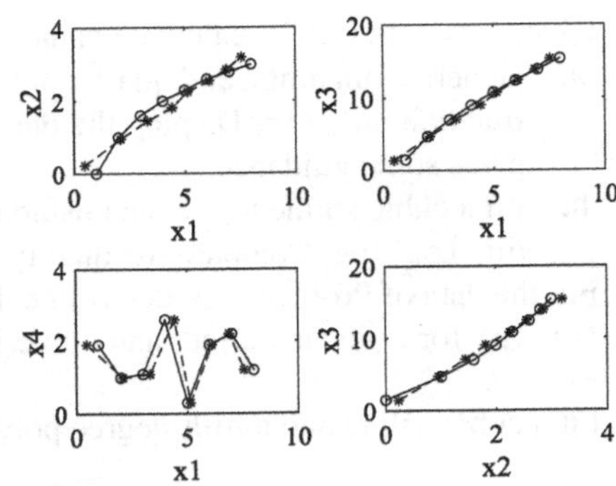

FIGURE 7.16

Relationship between selected variables. Dashed line is generated using only the first and second principal component.

7.13 SUMMARY

Methods have been described for fitting functions to data for the purposes of interpolation. These have included Aitken's method and spline fits. Regression analysis discussed least squares approximations to experimental data using polynomial and more general functions. We provide an explanation of the widely used Kalman filter and conclude the chapter with an introduction to principal component analysis.

Note: For readers who a have particular interest in apply MATLAB to data analysis using curve fitting and statistical methods, MATLAB provides a Curve Fitting Toolbox, a Statistical and Machine Learning Toolbox and a Control System Toolbox.

7.14 PROBLEMS

7.1. The following tabulation gives values of the complete elliptic integral

$$E(\alpha) = \int_0^{\pi/2} \sqrt{(1 - \sin^2 \alpha \sin^2 \theta)}\, d\theta$$

α	0°	5°	10°	15°	20°	25°	30°
$E(\alpha)$	1.57079	1.56780	1.55888	1.54415	1.52379	1.49811	1.46746

Determine $E(\alpha)$ for $\alpha = 2°$, 13°, and 27° using the MATLAB function `aitken`.

7.2. Generate a table of values of $f(x) = x^{1.4} - \sqrt{x} + 1/x - 100$ for $x = 20:2:30$. Find the value of x corresponding to $f(x) = 0$ using the MATLAB function `aitken`. This is an example of inverse interpolation since we are finding the value of x corresponding to a given value of $f(x)$. In particular, this gives an approximation to the root of the equation $f(x) = 0$. Compare your solution with that of Problem 3.2 in Chapter 3.

7.3. Given $x = -1 : 0.2 : 1$, calculate values of y from $y = \sin^2(\pi x/2)$. Using these values of x:

a. Generate quadratic and quartic polynomials to fit this data using the least squares MATLAB function `polyfit`. Display the data and the curve fitted. *Hint*: Example 7.2 in Section 7.6 gives some guidance.

b. Fit a cubic spline to the data using the MATLAB function `spline`. Display the data and the fitted `spline`. Compare the quality of this spline fit with the two graphs from (a).

7.4. For the data of Problem 7.3, determine the values of y for $x = 0.85$ using the MATLAB function `interp1` for a linear, spline, and cubic interpolating function. Also use the MATLAB function `aitken`.

7.5. Fit a cubic spline and a fifth-degree polynomial to the following data.

x	-2	0	2	3	4	5
y	4	0	-4	-30	-40	-50

Plot the data points, the spline, and the polynomial on the same graph. Which curve appears to give the more realistic representation of any underlying function from which the data might have been taken?

7.6. For the data given by the vectors $x = 0 : 0.25 : 3$ and

$$y = [6.3806\ 7.1338\ 9.1662\ 11.5545\ 15.6414\ 22.7371\ 32.0696\ ...\\ 47.0756\ 73.1596\ 111.4684\ 175.9895\ 278.5550\ 446.4441]$$

fit the following functions:

a. $f(x) = a + be^x + ce^{2x}$ using the MATLAB function `mregg2`.

b. $f(x) = a + b/(1 + x) + c/(1 + x)^2$ using the MATLAB function `mregg2`.

c. $f(x) = a + bx + cx^2 + dx^3$ using the MATLAB functions `polyfit` or `mregg2`.

You should plot the three trial functions and the data. How well do these functions fit the data? The data values were in fact generated from $f(x) = 3 + 2e^x + e^{2x}$ with a small amount of random noise added.

7.7. The following values of x and corresponding values of y_u and y_l define an airfoil section:

$$x = [0\ 0.005\ 0.0075\ 0.0125\ 0.025\ 0.05\ 0.1\ 0.2\ 0.3\ 0.4\ ...\\ 0.5\ 0.6\ 0.7\ 0.8\ 0.9\ 1]$$

$$y_u = [0\ 0.0102\ 0.0134\ 0.0170\ 0.0250\ 0.0376\ 0.0563\ 0.0812\ ...\\ 0.0962\ 0.1035\ 0.1033\ 0.0950\ 0.0802\ 0.0597\ 0.0340\ 0]$$

$$y_l = [0\ -0.0052\ -0.0064\ -0.0063\ -0.0064\ -0.0060\ -0.0045\ ...\\ -0.0016\ 0.0010\ 0.0036\ 0.0070\ 0.0121\ 0.0170\ 0.0199\ 0.0178\ 0]$$

The (x, y_u) coordinates define the upper surface and (x, y_l) coordinates define the lower surface. Use the MATLAB function `spline` to fit separate splines to the upper and lower surfaces and plot the results as a single figure.

7.8. Consider the approximation

$$\prod_{p<P}\left(1 + \frac{1}{p}\right) \approx C_1 + C_2 \log_e P$$

where the product is taken of all the prime numbers p less than a prime number P. Write a script to generate these products from the list of prime numbers provided and fit the function $C_1 + C_2 \log_e P$ to the points given by the primes P and the corresponding values of the products using the MATLAB function `polyfit`. Generate a list of prime numbers using the MATLAB function `primes(103)`.

7.9. The gamma function may be approximated by a fifth-degree polynomial as

$$\Gamma(x+1) = a_0 + a_1 x + a_2 x^2 + a_3 x^3 + a_4 x^4 + a_5 x^5$$

Use the MATLAB function `gamma` to generate values of $\Gamma(x+1)$ for $x = 0 : 0.1 : 1$. Then, using the MATLAB function `polyfit`, fit a fifth-degree polynomial to this data. Compare your answers with the approximation for the Γ function given by Abramowitz and Stegun (1965), which gives $a_0 = 1$, $a_1 = -0.5748666$, $a_2 = 0.9512363$, $a_3 = -0.6998588$, $a_4 = 0.4245549$, $a_5 = -0.1010678$. These coefficients give an accuracy for the Γ function in the range $0 \le x \le 1$ of less than or equal to 5×10^{-5}.

7.10. Generate a table of values of z from the function

$$z(x, y) = 0.5\left(x^4 - 16x^2 + 5x\right) + 0.5\left(y^4 - 16y^2 + 5y\right)$$

in the range $x = -4 : 0.2 : 4$ and $y = -4 : 0.2 : 4$. Use this data and the MATLAB function `interp2` to interpolate a value for z at $x = y = -2.9035$. Use both linear and cubic interpolation and check your answer by direct substitution in the function. This point gives the global minimum of this function.

7.11. The difference between the mean Sun and the real Sun is called the equation of time. Thus the value of the equation of time

$$E = (\text{mean Sun time} - \text{real Sun time})$$

The following table gives E in minutes at 20 equispaced intervals during the year, beginning January 1.

$$E = [-3.5\ -10.5\ -14.0\ -14.25\ -9.0\ -4.0\ 1.0\ 3.5\ 3.0 \ldots$$
$$-0.25\ -3.5\ -6.25\ -5.5\ -1.75\ 4.0\ 10.5\ 15.0\ 16.25\ 12.75\ 6.5]$$

Plot a graph of the data values E against time of year. Then use the function `interpft` to interpolate 300 points and plot E over a period of one year. (Use the MATLAB `help` function to obtain information on the MATLAB `interpft`.) Finally, use the command `[x,y]=ginput(4)` to read from the graph the values of the two minimum and two maximum values of E. At what times do these maxima and minima occur?

7.12. The cost of producing an electronic component varied over a four year period as follows:

Year	0	1	2	3
Cost	$30.2	$25.8	$22.2	$20.2

Assuming the equation relating production cost and time is (a) a cubic and (b) a quadratic polynomial, estimate the cost of production in year 6. A small error was discovered in the data.

The cost of production in year 2 should have been \$22.5 and in year 3, \$20.5. Recompute the estimated production cost in year 6 using a cubic and a quadratic equation as before. What conclusions can you draw from the results?

7.13. From the table of the Gamma function given, use inverse interpolation to find the value of x in the range $x = 2$ to $x = 3$ which makes $\Gamma(x) = 1.3$. Use the MATLAB function `interp1` with the cubic option selected; also use the function `aiken`.

x	2	2.2	2.4	2.6	2.8	3
$\Gamma(x)$	1.0000	1.1018	1.2422	1.4296	1.6765	2.0000

7.14. The following table gives the value of the integral

$$I = \int_0^{\pi/2} \frac{d\varphi}{\sqrt{1 - \sin^2\alpha \sin^2\varphi}}$$

for various values of α. (This integral is the complete elliptical integral of the first kind.)

α	0°	5°	10°	15°	20°	25°
I	1.57080	1.57379	1.58284	1.59814	1.62003	1.64900

Using polynomial interpolation, find I when $\alpha = 2°$. Then use inverse interpolation to find the value of α such that $I = 1.58$. In both cases, use the MATLAB function `interp1` with the cubic option selected; also use the function `aiken`.

7.15. It is required to find a formula to calculate the number of nodes in one corner of a cube. If n is the number of equally spaced nodes on an edge of the cube and f_n is the number of nodes on three half-faces, including nodes on the face diagonals, then the following table shows values of f_n for given values of n.

n	1	2	3	4
f_n	1	4	10	20

By fitting a cubic function to this data (using `polyfit`), find a general formula for the relationship between f_n and n and verify that when $n = 5$, $f_n = 35$.

7.16. It is required to fit a regression model of the form $z = f(x, y)$ to the following data:

x	0.5	1.0	1.0	2.0	2.5	2.0	3.0	3.5	4.0
y	2.0	4.0	5.0	2.0	4.0	5.0	2.0	4.0	5.0
z	−0.19	−0.32	−1.00	3.71	4.49	2.48	6.31	7.71	8.51

a. Use the function `mregg2` to generate a model of the form $z = a + bx + cy$ and also $z = a + bx + cy + dxy$. Which model do you consider the best fit to the data? It is important to consider the differences in the error variance.

b. By analysis of the residuals (particularly the Cook distance), decide whether any data point could be considered to be an *outlier*?

7.17. One of the data values given in Problem 7.16 has been found to be in error. The value of z corresponding to $x = 4$, $y = 5$ should have been recorded as 9.51, not 8.51, a common human error. Use the function `mregg2` to generate models of the form $z = a + bx + cy$ and also $z = a + bx + cy + dxy$. Again, assess the quality of the models.

7.18. Using following the table, obtain the pressure across a shock wave when the upstream Mach Number is 4.4 by:

a. linear interpolation

b. Aitken's method

c. spline interpolation

Mach number	1.00	2.00	3.00	4.00	5.00
p_2/p_1	1.00	4.50	10.33	18.50	29.00

CHAPTER

ANALYZING DATA USING DISCRETE TRANSFORMS

8

Abstract

In this chapter we consider a variety of transformations for the analysis of data. A description of Fourier analysis for discrete data is provided along with data analysis using the Hilbert and Walsh transforms, concluding with a short introduction to wavelet analysis.

8.1 INTRODUCTION

In this chapter, we restrict the analysis to problems in one dimension. In one dimension, the transforms are frequently but not exclusively in terms of the time domain. Thus, we will frequently use the variable t, which clearly can represent time but might be applied to other domains.

The discrete Fourier transform (DFT) is an example of fitting a set of orthogonal harmonic functions exactly to data. If the data has an underlying harmonic content, the DFT will reveal it and help the engineer or scientist to identify the causes of the harmonic content of the data. Similarly the discrete Walsh transform fits a set of orthogonal square wave functions exactly to the data and can reveal underlying patterns in waveforms predominantly composed of abrupt steps.

The Fourier transform should be applied to stationary data, that is data in which its harmonic content does no change with time. The Hilbert transform allow us to analyze the harmonic content of this non-stationary data and examine how the harmonic content of the data changes with time. Non-stationary data can be analyzed using wavelet transforms, either by discrete or continuous wavelets. In the last 25 years a great amount of research has been conducted on wavelet analysis, both for one-dimensional time series analysis but also for two-dimensional image analysis. Here we give a brief introduction to wavelet analysis applied to time series analysis.

An important 'rule of thumb' is that the best results are obtained when the transform or wavelet shape is close to the underlying nature of the data. This explains the wide spread use of Fourier analysis, since naturally occurring data is frequently generated from harmonic processes.

8.2 FOURIER ANALYSIS OF DISCRETE DATA

Fourier analysis, in its various forms, is an important tool for the scientist or engineer engaged in the interpretation of data where a knowledge of the frequencies present in the data or function may give some insight into the mechanism that has generated it. For continuous periodic functions, the frequency content is determined from the coefficients of the terms in the well-known Fourier series; for non-periodic functions, it is determined from the Fourier integral transform. The frequency content

Numerical Methods. https://doi.org/10.1016/B978-0-12-812256-3.00017-8

of a sequence of data can be determined by Fourier analysis, in this case from the *discrete* Fourier transform (DFT). The harmonic functions fit the data exactly and allows the user to determine the harmonic content of the data.

The data can come from many sources. For example, the radial forces acting at discrete points around a cylinder constitute a sequence of data which must be periodic. The most frequently occurring form of data is a time series in which the value of some quantity is given at equal intervals of time – for example, data sampled from a signal from a transducer – and for this reason the analysis which follows is developed in terms of the independent variable t which represents time. Note that in this chapter, the term signal means a function of time, and thus signal and function are often used interchangeably. It must be stressed, however, that the DFT can be applied to any data regardless of the domain from which it originates. Determining the DFT for a sequence of data points is straightforward, although the computation is tedious.

We begin by defining a periodic function. A function $y(t)$ is periodic if it has the property that for any value of time t, $y(t) = y(t+T)$ where T is the time period, typically measured in seconds. The reciprocal of the period is equal to the frequency, denoted by f and it is measured in cycles/second. In the SI system of units, 1 Hertz (Hz) is defined as 1 cycle/second. If we are concerned with a periodic function $z(x)$, where x is a spatial variable, then for any value of x, $z(x) = z(x+X)$ where X is the spatial period or wavelength, typically measured in meters. The frequency $f = 1/X$ is then measured in cycles/meter.

The Fourier series for a periodic function $F(t)$ is given by

$$F(t) = a_0/2 + \sum_{n=1}^{\infty} \{a_n \cos(nt) + b_n \sin(nt)\}$$

where

$$a_0 = (1/\pi) \int_{-\pi}^{\pi} f(t)dt$$

$$a_n = (1/\pi) \int_{-\pi}^{\pi} f(t)\cos(nt)dt$$

$$b_n = (1/\pi) \int_{-\pi}^{\pi} f(t)\sin(nt)dt$$

We now examine how to fit a finite set of trigonometric functions to n data points (t_r, y_r) where $r = 0, 1, 2, ..., n-1$. We assume the data points are equispaced and the number of data points, n, is even. Data values may be complex but in most practical situations they are real. The data points are numbered as shown in Fig. 8.1. The point following the $(n-1)$th is assumed to equal the value of the zero point. Thus the DFT assumes the data is periodic with a period T equal to the range of the data.

Let the relationship between y_r and t_r be given by a finite set of sine and cosine functions thus:

$$y_r = \frac{1}{n}\left[A_0 + \sum_{k=1}^{m-1} \{A_k \cos(2\pi k t_r/T) + B_k \sin(2\pi k t_r/T)\} + A_m \cos(2\pi m t_r/T)\right] \tag{8.1}$$

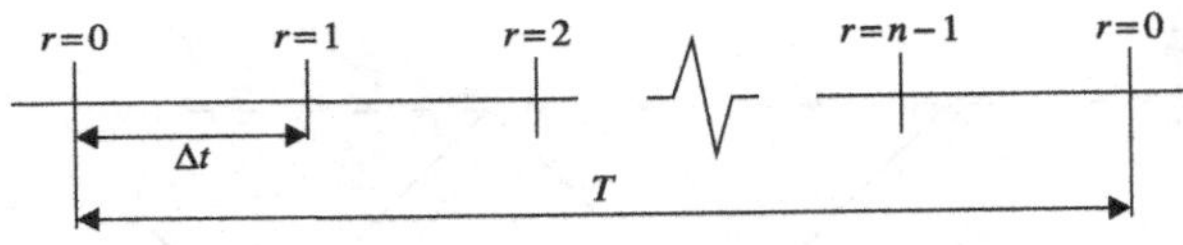

FIGURE 8.1

Numbering scheme for data points.

where $r = 0, 1, 2, ..., n-1$, $m = n/2$ and the range of the data is 0 to T as shown in Fig. 8.1. The n coefficients A_0, A_m, A_k, and B_k (where $k = 1, 2, \ldots, m-1$) must be determined. Since we have n data values and n unknown coefficients, (8.1) can be made to fit the data exactly. The factor $1/n$ in (8.1) is omitted by some authors, and omitting it has the effect of reducing the size of the coefficients A_0, A_m, A_k, and B_k by the factor n. The reason for choosing $m+1$ coefficients multiplied by a cosine function, including $\cos 0$ which equals one and multiplies A_0, and $m-1$ coefficients multiplied by a sine function in (8.1) will become apparent.

Each sine or cosine term of (8.1) represents k complete cycles in the range of the data T. Thus, the period of each sine term is T/k, where $k = 1, 2, \ldots, (m-1)$, and the period of each cosine term is T/k, where $k = 1, 2, \ldots, m$. The corresponding frequencies are given by k/T. Thus the frequencies present in (8.1) are $1/T, 2/T, \ldots, m/T$. Letting Δf be the frequency increment between components and f_{max} be the maximum frequency, then

$$\Delta f = 1/T \tag{8.2}$$

and

$$f_{max} = m\Delta f = (n/2)\Delta f = n/(2T) \tag{8.3}$$

The data values t_r are equally spaced in the range T and may be expressed as

$$t_r = rT/n, \quad r = 0, 1, 2, ..., n-1 \tag{8.4}$$

Letting Δt be the sampling interval (see Fig. 8.1), then

$$\Delta t = T/n \tag{8.5}$$

Let T_0 be the period corresponding to f_{max}, the maximum frequency in (8.1). Then, from (8.3),

$$f_{max} = 1/T_0 = n/(2T)$$

Thus $T = T_0 n/2$. Substituting this relationship in (8.5), we have $\Delta t = T_0/2$. This tells us that the maximum frequency component in the DFT contains two samples of data per cycle. The maximum frequency, f_{max}, is called the Nyquist frequency, and the corresponding sampling rate is called the Nyquist sampling rate. In order to take at least two samples per cycle of the highest frequency, f_{max}, the sampling rate must be at equal to $2f_{max}$.

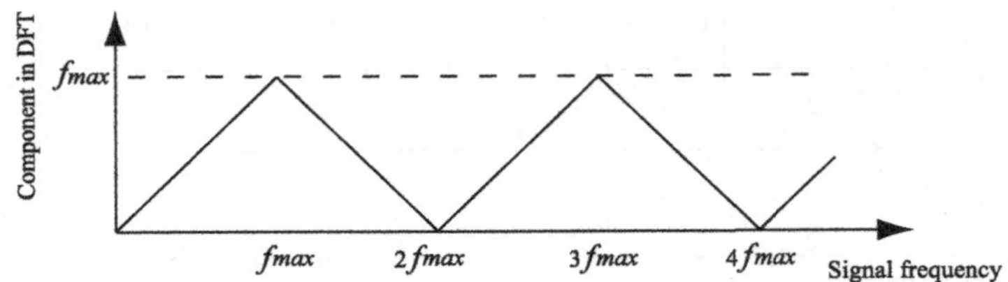

FIGURE 8.2

Graph shows the relationship between a signal frequency and its component in the DFT. Thus, for example, a signal frequency of twice Nyquist frequency, $2f_{max}$, will give a component of zero frequency in the DFT.

A harmonic with a frequency that is exactly equal to the Nyquist frequency cannot be properly detected because at this frequency the DFT has a cosine term but no corresponding sine term. This result has an important implication when data is sampled from a continuously varying function or signal. It implies that there must be *more than two* data samples per cycle at the highest frequency *present in the function or signal.* If there are frequencies in the signal higher than the Nyquist frequency then, because of the periodic nature of the DFT itself, they appear as frequency components in the DFT at a lower frequency. This phenomenon is called "aliasing". For example, if data is sampled at 0.005 second intervals, that is, 200 samples per second, then the Nyquist frequency, f_{max}, is 100 Hz. A frequency of 125 Hz in this signal would appear as a frequency component at 75 Hz. A frequency of 225 Hz would appear as 25 Hz. The relationship between frequencies in a signal and the frequency components in the DFT is shown in Fig. 8.2. Frequency aliasing should be avoided because it makes it difficult or impossible to relate the frequency components in the DFT to their physical causes.

We now return to the task of determining the n coefficients A_0, A_m, A_k, and B_k in (8.1). Replacing $t_r = rT/n$ in (8.1), we obtain

$$y_r = \frac{1}{n}\left[A_0 + \sum_{k=1}^{m-1}\{A_k\cos(2\pi kr/n) + B_k\sin(2\pi kr/n)\} + A_m\cos(\pi r)\right] \tag{8.6}$$

where $r = 0, 1, 2, ..., n-1$. It was previously noted that the coefficients B_0 and B_m are absent from (8.1). It is now clear that had we introduced these coefficients they would be multiplied by $\sin(0)$ and $\sin(\pi r)$, both of which are zero.

In (8.6) the n unknown coefficients are real. However, (8.6) can be expressed more concisely in terms of complex exponentials with complex coefficients. Using the fact that

$$\cos(2\pi kr/n) = \{\exp(\imath 2\pi kr/n) + \exp(-\imath 2\pi kr/n)\}/2$$

$$\sin(2\pi kr/n) = \{\exp(\imath 2\pi kr/n) - \exp(-\imath 2\pi kr/n)\}/2\imath$$

and

$$\exp\{\imath 2\pi(n-k)r/n\} = \exp(-\imath 2\pi kr/n)$$

where $k = 1, 2, ..., m - 1$ and $\imath = \sqrt{-1}$, then it can be shown that (8.6) reduces to

$$y_r = \frac{1}{n}\sum_{k=0}^{n-1} Y_k \exp(\imath 2\pi kr/n),\ r = 0,\ 1\,2, \ldots,\ n-1 \tag{8.7}$$

In (8.7),

$$Y_0 = A_0 \text{ and } Y_m = A_m, \text{ where } m = n/2$$

$$Y_k = (A_k - \imath B_k)/2 \text{ and } Y_{n-k} = (A_k + \imath B_k)/2, \text{ for } k = 1, 2, ..., m-1$$

Note that if y_r is real, then A_k and B_k are also real, so that Y_{n-k} is the complex conjugate of Y_k, for $k = 1, 2, ..., (n/2 - 1)$. To find the values of the unknown complex coefficients of (8.7) we make use of the following orthogonal property of exponential functions sampled at n equispaced points:

$$\sum_{r=0}^{n-1} \exp(\imath 2\pi rj/n)\exp(\imath 2\pi rj/n) = \begin{cases} 0 & \text{if } |j-k| \neq 0,\ n,\ 2n \\ n & \text{if } |j-k| = 0,\ n,\ 2n \end{cases} \tag{8.8}$$

Multiplying (8.7) by $\exp(-\imath 2\pi rj/n)$ and summing over the n values of r, and then using (8.8), an expression for the unknown coefficients can be found thus:

$$Y_k = \sum_{r=0}^{n-1} y_r \exp(\imath 2\pi kr/n) \quad k = 0,\ 1,\ 2, \ldots,\ n-1 \tag{8.9}$$

If we let $W_n = \exp(-\imath 2\pi/n)$, where W_n is a complex constant, then (8.9) becomes

$$Y_k = \sum_{r=0}^{n-1} y_r W_n^{kr} \quad k = 0,\ 1,\ 2, \ldots,\ n-1 \tag{8.10}$$

Note that in (8.10) W_n is raised to the power of the product of k and r. Alternatively, we can write (8.10) in matrix notation, giving

$$\mathbf{Y} = \mathbf{W}\mathbf{y} \tag{8.11}$$

where y is the vector of samples, W_n^{kr} is the element in the $(k+1)$th row and $(r+1)$th column of $\mathbf{W}$, since k and r both start at zero. Note that $\mathbf{W}$ is an $n \times n$ array of complex coefficients. Thus, $\mathbf{Y}$ is a *vector* of the complex Fourier coefficients and in this instance we are departing from our usual convention that emboldened upper case letters represent arrays.

We can obtain the coefficients Y_k from the equispaced data (t_r, y_r) by using (8.9), or (8.10), or (8.11). These equations are alternative statements of the DFT. Furthermore, by replacing k in (8.9) by $k + np$, where p is any integer, it can be shown that $Y_{k+np} = Y_k$. Thus the DFT is periodic over the range n. The inverse of the DFT is called the inverse discrete Fourier transform (IDFT) and is implemented by (8.7). By replacing r in (8.7) by $r + np$, where p is an integer, it can be shown that the IDFT is also periodic over the range n. Both y_r and Y_k may be complex, although, as previously stated, the samples y_r are usually real. These transforms constitute a pair: If the data values are transformed by the DFT to determine the coefficients Y_k, then they can be recovered in their entirety by means of the IDFT.

To evaluate the coefficients of the DFT it would appear convenient to use (8.11). Although using these equations is satisfactory for small sequences of data, calculating the DFT for n real data points requires $2n^2$ multiplications. Thus, to transform a sequence of 4096 data points would require approximately 33 million multiplications. In 1965 this situation was dramatically changed with the publication of the fast Fourier transform (FFT) algorithm (Cooley and Tukey, 1965). The FFT algorithm is extremely efficient, and approximately $2n \log_2 n$ multiplications are required to compute the FFT for real data. With this development, together with the increase in computation speed due to the improvements in computing hardware that have occurred in the past fifty years, it is now possible to compute the FFT for a relatively large number of data points on a personal computer.

Many refinements have been made to the basic FFT algorithm since it was first formulated and several variants have been developed. Here one of the simplest forms of the algorithm is outlined.

To develop the basic FFT algorithm one further restriction must be placed on the data. In addition to the data being equispaced, the number of data points must be an integer power of 2. This allows a sequence of data to be successively subdivided. For example, 16 data points can be divided into two sequences of 8, four sequences of 4 and finally eight sequences of only 2 data points. A crucial relationship on which the FFT algorithm is based is now developed from (8.9) as follows. Let y_r be the sequence of n data points for which we require the DFT. We can subdivide y_r into two sequences of $n/2$ data points u_r and v_r thus:

$$\left.\begin{aligned} u_r &= y_{2r} \\ v_r &= y_{2r+1} \end{aligned}\right\} \tag{8.12}$$

Note that alternate points in the original data sequence are placed in different subsets. We now determine the DFTs of the data sets u_r and v_r from (8.9), with n replaced by $n/2$, thus:

$$\left.\begin{aligned} U_k &= \sum_{r=0}^{n/2-1} u_r \exp\{-\imath 2\pi kr/(n/2)\} \\ V_k &= \sum_{r=0}^{n/2-1} v_r \exp\{-\imath 2\pi kr/(n/2)\} \end{aligned}\right\} k = 0,\ 1,\ 2, \ldots,\ n/2 - 1 \tag{8.13}$$

The DFT, Y_k, for the original data sequence y_r is given by using (8.9) as follows:

$$\begin{aligned} Y_k &= \sum_{r=0}^{n-1} y_r \exp(-\imath 2\pi kr/n) \\ &= \sum_{r=0}^{n/2-1} y_{2r} \exp\{-\imath 2\pi k 2r/n\} + \sum_{r=0}^{n/2-1} y_{2r+1} \exp\{-\imath 2\pi k(2r+1)/n\} \end{aligned}$$

where $k = 0, 1, 2, \ldots, n$. Substituting for y_{2r} and y_{2r+1} from (8.12), we have

$$Y_k = \sum_{r=0}^{n/2-1} u_r \exp\{-\imath 2\pi kr/(n/2)\} + \exp(-\imath 2\pi k/n) \sum_{r=0}^{n/2-1} v_r \exp\{-\imath 2\pi kr/(n/2)\}$$

Comparing this equation with (8.13), we see that

$$Y_k = U_k + \exp(-\imath 2\pi k/n)V_k = U_k + (W_n^k)V_k \tag{8.14}$$

where $W_n^k = \exp(-\imath 2\pi k/n)$ and $k = 0, 1, 2, ..., n/2 - 1$.

Eq. (8.14) provides only half of the required DFT. However, using the fact that U_k and V_k are periodic in k, it can be proved that

$$Y_{k+n/2} = U_k - \exp(-\imath 2\pi k/n)V_k = U_k - (W_n^k)V_k \tag{8.15}$$

We can use (8.14) and (8.15) to determine efficiently the DFT of the original data from the DFTs of subsets composed of alternate points of the original data. Of course, we can determine the DFTs of each subset of data by further subdividing these subsets until the final division leaves subsets consisting of a single data point. For a sequence of data comprising a single data point, we see from (8.9) with $n = 1$ that the DFT is equal to the value of the single data point. This is essentially how the FFT algorithm works.

In the preceding discussion we started from a sequence of data and continuously subdivided it (with alternate points in different subsets) until the subdivisions produced single data points. What we require is a method of starting from single data points and ordering them in such a way that successively combining the DFTs of the subsets ultimately forms the required DFT of the original data. This can be achieved by the "bit reversed algorithm", and we illustrate it and the subsequent stages of the FFT by assuming a sequence of eight data points, y_0 to y_7. To determine the correct order for combining the data we express the subscript denoting the original position of each data point as a binary number and reverse the order of the binary digits (or bits). This reversed-order binary number determines the position of each data point in the reordered sequence and is shown for eight data points in Fig. 8.3. The diagram also shows the stages of the FFT algorithm which repeatedly uses (8.14) and (8.15) thus:

Stage 1. Determine $\mathbf{Y}_{04}$ from Y_0 and Y_4, determine $\mathbf{Y}_{26}$ from Y_2 and Y_6, determine $\mathbf{Y}_{15}$ from Y_1 and Y_5, determine $\mathbf{Y}_{37}$ from Y_3 and Y_7.
Stage 2. Determine $\mathbf{Y}_{0246}$ from $\mathbf{Y}_{04}$ and $\mathbf{Y}_{26}$, determine $\mathbf{Y}_{1357}$ from $\mathbf{Y}_{15}$ and $\mathbf{Y}_{37}$.
Stage 3. Determine $\mathbf{Y}_{01234567}$ from $\mathbf{Y}_{0246}$ and $\mathbf{Y}_{1357}$.

Note that there are three stages in the above process. For an n point DFT, using the FFT algorithm, the number of stages equals $\log_2 n$. In this small example, $n = 8$ and hence $\log_2 8 = \log_2 2^3 = 3$. Thus the process requires three stages.

MATLAB provides both the function `fft` to determine the DFT of a sequence of data values using the FFT algorithm, and the function `ifft` to determine the IDFT using a slight modification of the FFT algorithm. Thus to determine the DFT of the data in `y` we use the `fft` function as the script `e4s801.m` illustrates:

```
% e4s801.m
v = 0:15;
y = [2.8 -0.77 -2.2 -3.1 -4.9 -3.2 4.83 -2.5 3.2 ...
    -3.6 -1.1 1.2 -3.2 3.3 -3.4 4.9];
s = sum(y), Y = fft(y);
[v' Y.']
```

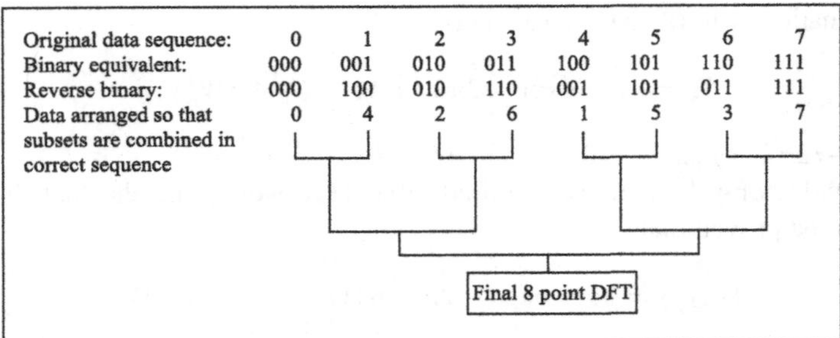

FIGURE 8.3

Stages in the FFT algorithm.

Running script e4s801.m gives the following results:

```
s =
   -7.7400

ans =
   0.0000 + 0.0000i  -7.7400 + 0.0000i
   1.0000 + 0.0000i   3.2959 + 8.3851i
   2.0000 + 0.0000i  13.9798 +10.9313i
   3.0000 + 0.0000i   8.0796 - 6.6525i
   4.0000 + 0.0000i  -0.2300 + 4.7700i
   5.0000 + 0.0000i   4.3150 + 6.8308i
   6.0000 + 0.0000i  14.2202 + 1.4713i
   7.0000 + 0.0000i -17.2905 +15.0684i
   8.0000 + 0.0000i  -0.2000 + 0.0000i
   9.0000 + 0.0000i -17.2905 -15.0684i
  10.0000 + 0.0000i  14.2202 - 1.4713i
  11.0000 + 0.0000i   4.3150 - 6.8308i
  12.0000 + 0.0000i  -0.2300 - 4.7700i
  13.0000 + 0.0000i   8.0796 + 6.6525i
  14.0000 + 0.0000i  13.9798 -10.9313i
  15.0000 + 0.0000i   3.2959 - 8.3851i
```

We have already noted that for real data, Y_{n-k} is the complex conjugate of Y_k, for $k = 1, 2, ..., (n/2 - 1)$. The above results illustrate this relationship and in this case $Y_{15}, Y_{14}, ..., Y_9$ are the complex conjugates of $Y_1, Y_2, ..., Y_7$, respectively, and provide no extra information. It is also seen that Y_0 is equal to the sum of the original data values y_r.

We now give examples of the use of the `fft` function to examine the frequency content of data sequences sampled from continuous functions. Note that in each of the examples the DFT index is related to frequency using (8.2) through (8.5).

Example 8.1. Determine the DFT of a sequence of 64 equispaced data points, sampled at intervals of 0.05 s from the function $y = 0.5 + 2\sin(2\pi f_1 t) + \cos(2\pi f_2 t)$, where $f_1 = 3.125$ Hz and $f_2 = 6.25$ Hz. The script `e4s802.m` calls the `fft` function and displays the resulting DFT in various ways:

```
% e4s802.m
clf
nt = 64; dt = 0.05; T = dt*nt
df = 1/T, fmax = (nt/2)*df
t = 0:dt:(nt-1)*dt;
y = 0.5+2*sin(2*pi*3.125*t)+cos(2*pi*6.25*t);
f = 0:df:(nt-1)*df;   Y = fft(y);
figure(1)
subplot(121), bar(real(Y),'r')
axis([0 63 -100 100])
xlabel('Index k'), ylabel('real(DFT)')
subplot(122), bar(imag(Y),'r')
axis([0 63 -100 100])
xlabel('index k'), ylabel('imag(DFT)')
fss = 0:df:(nt/2-1)*df;
Yss = zeros(1,nt/2); Yss(1:nt/2) = (2/nt)*Y(1:nt/2);
figure(2)
subplot(221), bar(fss,real(Yss),'r')
axis([0 10 -3 3])
xlabel('Frequency (Hz)'), ylabel('real(DFT)')
subplot(222), bar(fss,imag(Yss),'r')
axis([0 10 -3 3])
xlabel('Frequency (Hz)'), ylabel('imag(DFT)')
subplot(223), bar(fss,abs(Yss),'r')
axis([0 10 -3 3])
xlabel('Frequency (Hz)'), ylabel('abs(DFT)')
```

Running script `e4s802.m` gives

```
T =
    3.2000

df =
    0.3125

fmax =
    10
```

together with Fig. 8.4 and Fig. 8.5. Note that in script `e4s802.m` we have used the `bar` rather than the `plot` statement to emphasize the discrete nature of the DFT. Fig. 8.4 shows the amplitudes of the 64 real and imaginary components of the DFT plotted against the index number k. Note that components 63 to

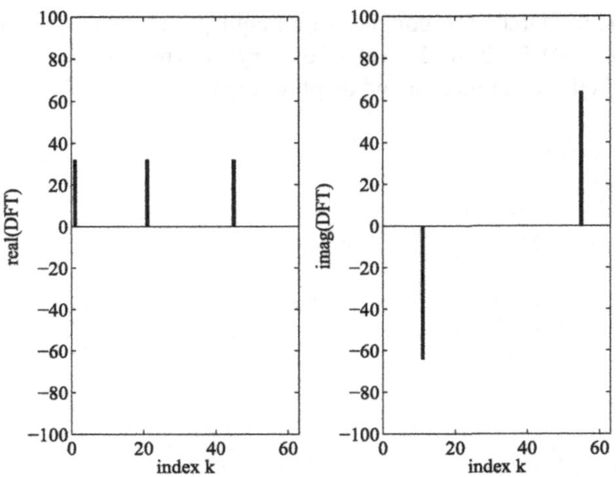

FIGURE 8.4

Plots of the real and imaginary part of the DFT.

33 are the complex conjugates of components 1 to 31. Whilst these plots display the DFT, the amplitude and frequency of the harmonic components in the original signal cannot easily be recognized. To achieve this the DFT must be scaled and displayed as shown in Fig. 8.4. In the real part of the DFT there are components at $k = 0, 20$, and 44, each with an amplitude of 32, and in the imaginary part of the DFT there are components at $k = 10$ and 54, with amplitudes of 64 and -64 respectively. Since they contain no extra information, we ignore the components above $k = 32$ (that is, the component $k = 44$ and 54) and consider only the components in the range $k = 0, 1, ..., 31$; in this case specifically $k = 0$, 10, and 20. We can convert the DFT index number to frequency by multiplying by Δf ($= 0.3125$ Hz) to give components at 0 Hz, 3.125 Hz, and 6.25 Hz, respectively. We now scale the DFT in the range $k = 1, 2, ..., 31$ by dividing it by $(n/2)$, in this case by 32. The plots of the 31 scaled DFT components (most of which are zero) corresponding to frequencies in the range 0 to 9.6875 Hz are shown in Fig. 8.5. We now see that the real component at 6.25 Hz has an amplitude of 1 and the imaginary component at 3.125 Hz has an amplitude of -2. These components correspond, respectively, to the cosine component and the negative of the sine component in the original signal from which the data was sampled. If we only wish to know the amplitude of the frequency components, then we can display the absolute values of the scaled DFT. The component at $f = 0$ Hz is equal to *twice* the mean value of the data, in this case we have $2 \times 0.5 = 1$. These plots are called frequency spectra or periodograms. If sampling is over an integer number of cycles of the harmonics present in the signal, the amplitude of the components in the scaled DFT equal the amplitude of the corresponding harmonics, as shown in this example. If the sampling is not over an integer number of cycles of any harmonic present in the signal, then the component in the DFT closest to the frequency of the harmonic is reduced in amplitude and spread into other frequencies. This phenomenon is called "spectral leakage", or "leakage" and is further discussed in Problem 8.4.

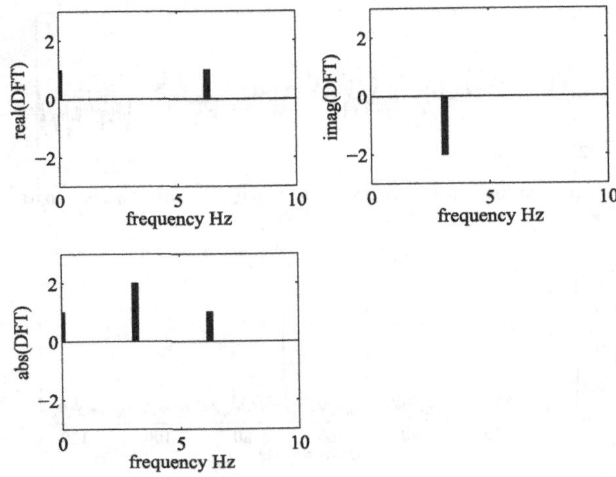

FIGURE 8.5

Frequency spectra.

Example 8.2. Determine the spectrum of a sequence of 512 data points sampled over a period of 2 seconds from the function

$$y = 0.2\cos(2\pi f_1 t) + 0.35\sin(2\pi f_2 t) + 0.3\sin(2\pi f_3 t) + \text{random noise}$$

where $f_1 = 20$ Hz, $f_2 = 50$ Hz, and $f_3 = 70$ Hz. The random noise is normally distributed with a standard deviation of 0.5 and a mean of zero. Script `e4s803.m` plots the time series and the DFT scaled by the factor $n/2$.

```
% e4s803.m
clf
f1 = 20; f2 = 50; f3 = 70;
nt = 512; T = 2; dt = T/nt
t_final = (nt-1)*dt; df = 1/T
fmax = (nt/2)*df
t = 0:dt:t_final;
dt_plt = dt/25;
t_plt = 0:dt_plt:t_final;
y_plt = 0.2*cos(2*pi*f1*t_plt)+0.35*sin(2*pi*f2*t_plt) ...
                                +0.3*sin(2*pi*f3*t_plt);

y_plt = y_plt+0.5*randn(size(y_plt));
y = y_plt(1:25:(nt-1)*25+1); f = 0:df:(nt/2-1)*df;
figure(1)
subplot(211), plot(t_plt,y_plt)
axis([0 0.04 -3 3])
```

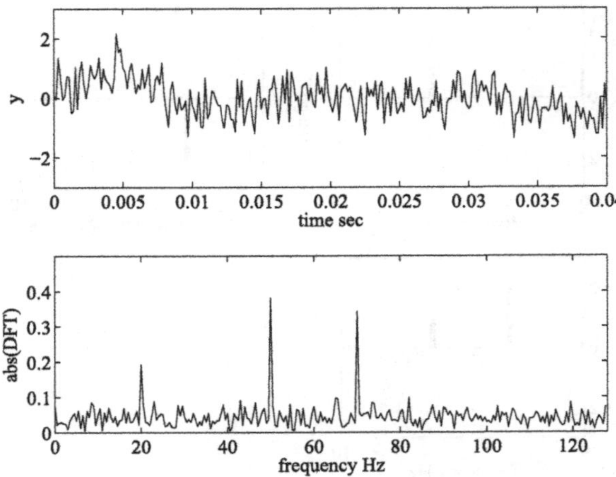

FIGURE 8.6

The top graph shows the data in the time domain and the bottom graph shows the corresponding frequency spectrum. Note frequency components at 20, 50, and 70 Hz.

```
xlabel('Time (sec)'), ylabel('y')
yf = fft(y);
yp(1:nt/2) = (2/nt)*yf(1:nt/2);
subplot(212), plot(f,abs(yp))
axis([0 fmax 0 0.5])
xlabel('Frequency (Hz)'), ylabel('abs(DFT)');
```

Running the script `e4s803.m` gives

```
dt =
    0.0039

df =
    0.5000

fmax =
   128
```

together with Fig. 8.6. The lower plot of Fig. 8.6 shows that random noise in the signal does not prevent the frequency components 20, 50, and 70 Hz from revealing themselves in the spectrum. These components are not obviously visible in the original time series data shown in the upper plot of Fig. 8.6.

We now repeat the discrete Fourier analysis for this example, but increase the sampling period T to 2.5 and reduce the number of samples to 256. Running the script then gives `dt = 0.0098`, `df = 0.4000` and `fmax = 51.2000` together with the plots shown in Fig. 8.7. The lower plot shows frequency compo-

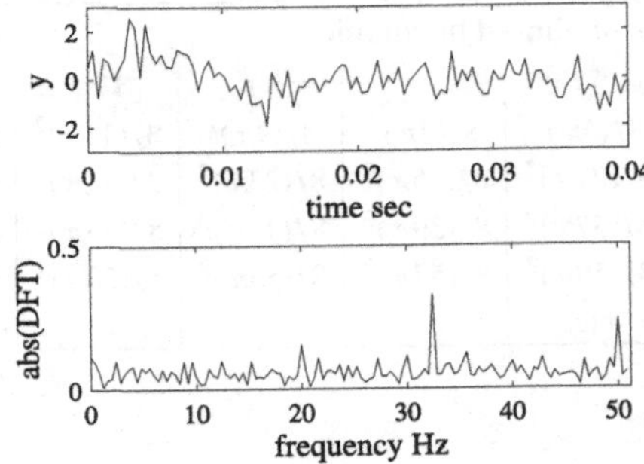

FIGURE 8.7

The top graph shows the data in the time domain and the bottom graph shows the corresponding frequency spectrum. Note that due to aliasing, frequency components are at 20, approximately 32.4 and 50 Hz.

nents of 20, 32.4 and 50 Hz in the spectrum. Since we know the spectral composition of the input data we know that this is not correct. The 20 and 50 Hz components are genuine but due to the fact that the maximum frequency that can be detected is only 51.2 Hz, the 70 Hz component is aliased and appears at 32.4 Hz. Note that this aliased frequency arises because $70 - 51.2 = 18.8$ and $51.2 - 18.8 = 32.4$. Clearly, if we did not know the spectral composition of the input data *a priori* then we would be mislead by this spectrum.

Example 8.3. Determine the spectrum of a triangular wave of amplitude ± 1 and period 1 second, sampled at intervals of 1/32 second over one cycle. The script `e4s804.m` outputs the DFT scaled by the factor $n/2$.

```
% e4s804.m
nt = 32; T = 1, dt = T/nt
t = 0:dt:(nt-1)*dt;
df = 1/T, fmax = nt/(2*T)
f = 0:df:df*(nt/2-1);
y = 0.125*[8 7 6 5 4 3 2 1 0 -1 -2 -3 -4 -5 -6 -7 -8 ...
    -7 -6 -5 -4 -3 -2 -1 0 1 2 3 4 5 6 7];
Yss = zeros(1,nt/2); Y = fft(y);
Yss(1:nt/2) = (2/nt)*Y(1:nt/2);
[f' abs(Yss)']
```

Running script `e4s804.m` gives

```
T =

     1


dt =

    0.0313
```

Table 8.1 Coefficients of aliased harmonics

f	$3f$	$5f$	$7f$	$9f$	$11f$	$13f$	$15f$
$8/\pi^2$	$8/(3\pi)^2$	$8/(5\pi)^2$	$8/(7\pi)^2$	$8/(9\pi)^2$	$8/(11\pi)^2$	$8/(13\pi)^2$	$8/(15\pi)^2$
$8/(31\pi)^2$	$8/(29\pi)^2$	$8/(27\pi)^2$	$8/(25\pi)^2$	$8/(23\pi)^2$	$8/(21\pi)^2$	$8/(19\pi)^2$	$8/(17\pi)^2$
$8/(33\pi)^2$	$8/(35\pi)^2$	$8/(37\pi)^2$	$8/(39\pi)^2$	$8/(41\pi)^2$	$8/(43\pi)^2$	$8/(45\pi)^2$	$8/(47\pi)^2$
$8/(63\pi)^2$	$8/(61\pi)^2$	$8/(59\pi)^2$	$8/(57\pi)^2$	$8/(55\pi)^2$	$8/(53\pi)^2$	$8/(51\pi)^2$	$8/(49\pi)^2$
$8/(65\pi)^2$	$8/(67\pi)^2$	etc.					

```
df =

     1

fmax =

    16

ans =

         0         0
    1.0000    0.8132
    2.0000         0
    3.0000    0.0927
    4.0000         0
    5.0000    0.0352
    6.0000         0
    7.0000    0.0194
    8.0000         0
    9.0000    0.0131
   10.0000         0
   11.0000    0.0100
   12.0000         0
   13.0000    0.0085
   14.0000         0
   15.0000    0.0079
```

The Fourier series for the triangular wave of this example is

$$f(t) = \frac{8}{\pi^2}\left(\cos(2\pi t) + \frac{1}{3^2}\cos(6\pi t) + \frac{1}{5^2}\cos(10\pi t) + \frac{1}{7^2}\cos(14\pi t) + \ldots\right)$$

The first eight frequency components in the scaled DFT at frequencies 1, 3, 5 Hz, and so on, are not equal to values we expect; that it $8/\pi^2$, $8/(3\pi)^2$, $8/(5\pi)^2$ (i.e., 0.8106, 0.0901, 0.0324), and so on, because of the effect of aliasing. A triangular wave contains an infinite number of harmonics and because of aliasing these appear as components in the DFT as shown in Table 8.1. Thus the size of the 3 Hz component in the DFT is $(8/\pi^2)(1/3^2 + 1/29^2 + 1/35^2 + 1/61^2 + \ldots)$. By summing a large number of terms down the columns of Table 8.1, the terms in the DFT are obtained. The DFT as shown above is correct, and the inverse DFT recovers the original data. However, when using it to provide information about the contribution of frequency components in the original data, the DFT must interpreted with care.

Example 8.4. Determine the DFT of a sequence of 128 data points sampled from a signal at intervals of 0.0625 seconds. The signal has a constant amplitude of 1 unit which, after 10 samples, is switched to zero.

The script `e4s805` determines the DFT for the data.

```
% e4s805.m
clf
nt = 128; nb = 10;
y = [ones(1,nb) zeros(1,nt-nb)];
dt = 0.0625; T = dt*nt
df = 1/T, fmax = (nt/2)*df;
f = 0:df:(nt/2-1)*df;
yf = fft(y);
yp = (2/nt)*yf(1:nt/2);
figure(1), bar(f,abs(yp),'w')
axis([0 fmax 0 0.2])
xlabel('Frequency (Hz)'), ylabel('abs(DFT)')
```

Running script `e4s805.m` gives

```
T =
     8

df =
     0.1250
```

together with the graphical output shown in Fig. 8.8. The plot shows that the frequency spectrum is continuous and the largest components are clustered near the zero frequency. This is in contrast to the spectrum of Examples 8.1 and 8.2 which show sharp peaks due to the presence of periodic components in the original data. Note that because the original signal is a step and not periodic, the amplitude of its DFT is dependent on the sampling period.

In this section we have provided examples of how the DFT (computed using the FFT) can be used to study the distribution of frequency components in data. There are other applications of the DFT. It is sometimes used for interpolation, as is any procedure that fits mathematical functions to a sequence of data. MATLAB provides the function `interpft` to allow interpolation using the DFT.

The advances in computer hardware have extended the range of problems to which the DFT can usefully be applied, and this in turn has encouraged the development of new and powerful variants of

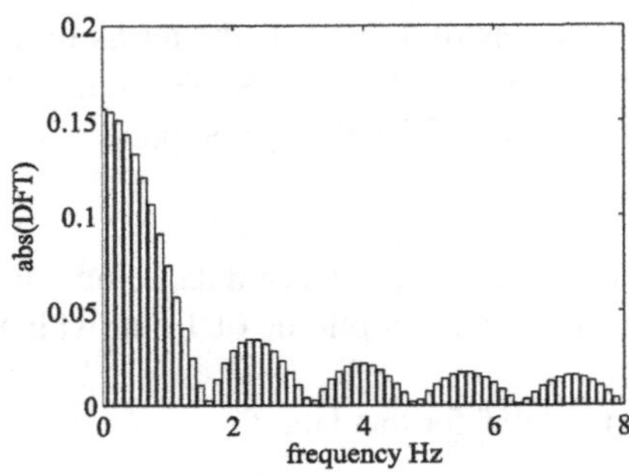

FIGURE 8.8

Spectrum of a sequence of data.

the FFT algorithm. A detailed description of the FFT algorithm is given by Brigham (1974), and a straightforward introduction which emphasizes the practical problems in using this type of analysis is given by Ramirez (1985).

8.3 THE HILBERT TRANSFORM

Data is sometimes sampled from signals of short duration that may change substantially over their duration. For example, musical instruments and earthquakes do not produce infinite duration harmonics and cannot be analyzed successfully using the FFT. One approach is to use the short-time Fourier transform (STFT). This is a basic form of time–frequency analysis which has limitations and which we do not describe. The Hilbert transform, and its extension, the Hilbert–Huang transform (HHT) can be utilized to solve this type of problem and a description of the transform and its application is given.

The Hilbert transform of a real valued continuous function or signal $x(t)$ is defined as

$$\mathcal{H}\{x(t)\} = \tilde{x}(t) = \frac{1}{\pi}\int_{-\infty}^{\infty} \frac{x(\tau)}{(t-\tau)}\,d\tau \tag{8.16}$$

Here $\mathcal{H}\{\bullet\}$ indicates the Hilbert transform, and $\tilde{x}(t)$ is the Hilbert transform of $x(t)$. It can be seen that the integrand of (8.16) has a singularity at $t = \tau$ and care must be exercised in evaluating this integral. In contrast to the Fourier transform, the Hilbert transform maps a function in one domain into a different function in the same domain. For example, consider the Hilbert transform of $\cos(2\pi f t)$, then

$$\mathcal{H}\{\cos(2\pi f t)\} = \frac{1}{\pi}\int_{-\infty}^{\infty} \frac{\cos(2\pi f \tau)}{(t-\tau)}\,d\tau \tag{8.17}$$

Letting $p = t - \tau$ then $dp = -d\tau$ and when $\tau = \infty$, $p = -\infty$, and when $\tau = -\infty$, $p = \infty$. Thus we have

$$\mathcal{H}\{\cos(2\pi f t)\} = \frac{1}{\pi}\int_{\infty}^{-\infty} \frac{\cos(2\pi f (t-p))}{p}\,(-dp) \tag{8.18}$$

Noting that

$$\cos(2\pi f(t-p)) = \cos(2\pi ft - 2\pi fp) = \cos(2\pi ft)\cos(2\pi fp) + \sin(2\pi ft)\sin(2\pi fp)$$

$$\mathcal{H}\{\cos(2\pi ft)\} = \frac{1}{\pi}\left[\cos(2\pi ft)\int_{-\infty}^{\infty}\frac{\cos(2\pi fp)}{p}\,dp + \sin(2\pi ft)\int_{-\infty}^{\infty}\frac{\sin(2\pi fp)}{p}\,dp\right] \tag{8.19}$$

Now $\frac{\cos(2\pi fp)}{p}$ is an odd function, that is $\frac{\cos(-2\pi fp)}{p} = -\frac{\cos(2\pi fp)}{p}$, and consequently, over the range $\pm\infty$, it integrates to zero. It can be shown that the integral of $\frac{\sin(2\pi fp)}{p}$ over the range $\pm\infty$ is π. Thus

$$\mathcal{H}\{\cos(2\pi ft)\} = \sin(2\pi ft) \tag{8.20}$$

Thus, the Hilbert transform has shifted the phase of the cosine function by $-90°$. In fact, we will show that the effect of the Hilbert transform is to rotate *all* the frequency components of the signal by $-90°$.

A more efficient method of determining the Hilbert transform, and at the same time illustrating that all frequency components in the data are rotated by $-90°$, is to take the Fourier transform of (8.16).

The convolution integral is defined by

$$(f \otimes g)(t) = \int_{-\infty}^{\infty} f(\tau)g(t-\tau)\,d\tau = \int_{-\infty}^{\infty} f(t-\tau)g(\tau)\,d\tau \tag{8.21}$$

Note that the symbol $\otimes$ indicates that $f(t)$ is convolved with $g(t)$. Comparing (8.16) with (8.21) we see that the Hilbert transform is a convolution integral of $x(t)$ and $(\pi t)^{-1}$. Thus we may write

$$\tilde{x}(t) = x(t) \otimes (\pi t)^{-1} \tag{8.22}$$

Now it can be proved that a convolution integral in one domain becomes a product in the corresponding Fourier domain. Thus

$$\mathcal{F}\{\tilde{x}(t)\} = \mathcal{F}\left\{x(t) \otimes (\pi t)^{-1}\right\} = \mathcal{F}\{x(t)\}\,\mathcal{F}\left\{(\pi t)^{-1}\right\} \tag{8.23}$$

Now $\mathcal{F}\left\{(\pi t)^{-1}\right\} = -\jmath\,\mathrm{sign}(f)$. Writing $\mathcal{F}\{x(t)\} = X(f)$, we have

$$\mathcal{F}\{\tilde{x}(t)\} = -\jmath X(f)\mathrm{sign}(f) \tag{8.24}$$

Taking the inverse transform of this equation gives the following expression for $\tilde{x}(t)$:

$$\tilde{x}(t) = \mathcal{F}^{-1}\{-\jmath X(f)\mathrm{sign}(f)\} \tag{8.25}$$

Thus the simplest way of describing the Hilbert transform is that it gives all the positive frequency components in the data, a phase shift of $-\jmath$ (that is $-\pi/2$ or $-90°$) and all negative frequency components, a phase shift of $+\jmath$ or $\pi/2$. Then data is generated in the domain of $x(t)$ from these phase shifted frequency components using the inverse Fourier transform. We have already shown that $\mathcal{H}\{\cos(2\pi ft)\} = \sin(2\pi ft)$. Knowing that all positive frequency components in the data have a phase shift of $-\pi/2$, it can be deduced that $\mathcal{H}\{\sin(2\pi ft)\} = -\cos(2\pi ft)$.

From the original data, $x(t)$, and the Hilbert transform of the data, $\tilde{x}(t)$ we can generate the so-called "analytic" signal thus

$$\bar{\tilde{x}}(t) = x(t) + \jmath\tilde{x}(t) \tag{8.26}$$

The analytical signal (8.26) can be expressed in the form

$$\bar{\tilde{x}}(t) = |\bar{\tilde{x}}(t)| \exp(\jmath\theta(t)) \tag{8.27}$$

where $|\bar{\tilde{x}}(t)| = \sqrt{x(t)^2 + \tilde{x}(t)^2}$. The term $|\bar{\tilde{x}}(t)|$ is the amplitude of the signal or function at t. The instantaneous phase at t is

$$\theta(t) = \tan^{-1}\left(\frac{\tilde{x}(t)}{x(t)}\right) \tag{8.28}$$

We have seen that if $x(t) = \cos(2\pi f t)$ then $\tilde{x}(t) = \sin(2\pi f t)$. Thus $\bar{\tilde{x}}(t) = \cos(2\pi f t) + \jmath\sin(2\pi f t) = \exp(\jmath 2\pi f t)$. We can consider (8.27), and hence (8.26), as generalizations of this relationship for any function.

The instantaneous frequency of the data can be obtained from the derivative of (8.28) and hence

$$\omega(t) = \frac{d\theta(t)}{dt} \tag{8.29}$$

For example, if $x(t) = \cos(\alpha t^2)$ then $\tilde{x}(t) = \sin(\alpha t^2)$ and so

$$\theta(t) = \tan^{-1}(\sin(\alpha t^2)/\cos(\alpha t^2)) = \alpha t^2 \tag{8.30}$$

Thus

$$\omega(t) = \frac{d(\alpha t^2)}{dt} = 2\alpha t \tag{8.31}$$

Hence, the instantaneous frequency at time t is $2\alpha t$.

So far we have defined the Hilbert transform for continuous functions or signals. We can apply the discrete Hilbert transform to discrete (sampled) data. In this case, (8.25) becomes

$$\mathcal{H}\{\mathbf{x}\} = \tilde{\mathbf{x}} = \mathcal{F}^{-1}\{\mathbf{X}^m\} \tag{8.32}$$

where $\mathcal{H}\{\bullet\}$ is the discrete Hilbert transform (DHT), $\mathcal{F}^{-1}\{\bullet\}$ is the inverse DFT, and $\mathbf{X}^m$ is a vector of modified values of $\mathbf{X}$ and is defined by

$$X_i^m = -\jmath X_i \text{sign}(f_i) \quad i = 0,\ 1,\ \ldots,\ n-1 \tag{8.33}$$

where X_i, for $i = 0, 1, \ldots, n-1$, is the DFT of x_j $(j = 0, 1, \ldots, n-1)$ and f_i is the positive or negative frequency component at i.

The amplitude and phase angle of the analytical signal are given by (8.34) and (8.35) respectively, as follows:

$$|\bar{\tilde{x}}_i| = \sqrt{x_i^2 + \tilde{x}_i^2} \quad i = 0,\ 1,\ \ldots,\ n-1 \tag{8.34}$$

and

$$\theta_i = \tan^{-1}(\tilde{x}_i/x_i) \quad i = 0,\ 1, ...,\ n-1 \tag{8.35}$$

The rate of change of phase is the instantaneous frequency. Thus

$$\omega_i = \frac{d\theta_i}{dt} \tag{8.36}$$

If we take the DFT of (8.26), we have

$$\bar{\bar{\mathbf{X}}} = \mathbf{X} + \jmath\tilde{\mathbf{X}}$$

Let $\mathbf{X}(f^+)$ denote the vector of values selected from $\mathbf{X}$ with positive frequency components and zero elsewhere, and let $\mathbf{X}(f^-)$ denote the vector selected from $\mathbf{X}$ with negative frequency components and zero elsewhere. Thus we have

$$\bar{\bar{\mathbf{X}}} = \left[\mathbf{X}(f^+) + \mathbf{X}(f^-)\right] + \jmath\left[-\jmath\mathbf{X}(f^+) + \jmath\mathbf{X}(f^-)\right] = 2\mathbf{X}(f^+)$$

The negative frequency components of $\mathbf{X}$ cancel, while the positive frequency components double. This occurs because the Hilbert transform provides a 90° phase shift at all negative frequencies, and a −90° phase shift at all positive frequencies. The discrete analytic signal is determined directly for the DFT using the following algorithm, due to Marple (1999).

$$\bar{\bar{X}}_i = \begin{cases} X_i, & \text{if } i = 0 \\ 2X_i, & \text{if } 1 \le i \le n/2 - 1 \\ X_i, & \text{if } i = n/2 \\ 0, & \text{if } n/2 + 1 \le i \le n - 1 \end{cases} \tag{8.37}$$

and hence $\bar{\bar{\mathbf{x}}} = \mathcal{F}^{-1}(\bar{\bar{\mathbf{X}}})$.

The MATLAB function `fhilb` generates the analytical data corresponding to the input data.

```
function xdbar = fhilb(x)
% xdbar is the analytic signal for x
% x must be row a vector. dbar is also row a vector
n = length(x);
X = fft(x);
sft = [1 2*ones(1,n/2-1) 1 zeros(1,n/2-1)];
xdbar = ifft(sft.*X);
```

Script `e4s806.m` illustrates that the function `fhilb.m` generates the analytical data, thus:

```
% e4s806.m
clear all
x = [1 3 5 3 -4 -3 0 3];
y = fhilb(x);
xdbar = y.'
```

Running script e4s806.m gives

```
xdbar =
   1.0000 - 0.6213i
   3.0000 - 2.0000i
   5.0000 + 0.6213i
   3.0000 + 5.5355i
  -4.0000 + 3.6213i
  -3.0000 - 2.0000i
   0.0000 - 3.6213i
   3.0000 - 1.5355i
```

Here we see that the real part of xdbar is the original data, the imaginary part is the Hilbert transform.

Consider the function $y = (0.5 + 0.6e^{-0.10t})\cos(7t + 6\log(\cosh((t-10)/2)))$. Clearly the amplitude and frequency of this function varies with time. The MATLAB script e4s807 generates a vector of data from this function over 24 seconds at intervals of 0.01 s and uses the Hilbert transform to determine the variation of frequency and amplitude with time. Running e4s807 generates Figs. 8.9, 8.10, 8.11, and 8.12.

```
% e4s807.m
close all, clear all
dt = 0.01;
t = dt:dt:24; n = length(t);
omega0 = 7+3*tanh((t-10)/2);
theta = 7*t+6*log(cosh((t-10)/2));
A0 = 0.5+0.6*exp(-0.10*t);
y = A0.*cos(theta);
ya = fhilb(y);
ye = abs(ya);
figure(1), plot(t,y,'k',t,ye,'k--')
xlabel('Time, s'), ylabel('y')
axis([0 24 -1.5 1.5])
figure(2), plot3(t(1:end-15),real(ya(1:end-15)),imag(ya(1:end-15)),'k')
xlabel('Time, s'), ylabel('real(ya)'), zlabel('imag(ya)')
axis([0 24 -1.5 1.5 -1.5 1.5])
grid
d_theta = zeros(size(t));
thetas = angle(ya);
theta = unwrap(thetas);
d_theta(1) = (theta(2)-theta(1))/dt;
d_theta(n) = (theta(n-1)-theta(n))/dt;
for k = 2:n-1
   d_theta(k) = (theta(k+1)-theta(k-1))/(2*dt);
end
```

```
figure(3), plot(t,omega0/2/pi,'k--',t,d_theta/2/pi,'k')
xlabel('Time, s'), ylabel('Frequency, Hz')
axis([0 24 0 2.5])
nt = 2048; T = t(nt);
df = 1/T; f = df*(0:nt/2-1);
yf = fft(y(1:nt));
yp = (2/nt)*yf(1:nt/2);
figure(4)
plot(f, abs(yp),'k')
axis([0 3 0 0.4])
ylabel('y')
xlabel('Frequency, Hz')
```

Fig. 8.9 shows a plot of y against time, t. The dashed line shows the envelope of the function, derived from the amplitude of the analytic signal. Fig. 8.10 is a three-dimensional plot and shows that when the real and imaginary parts of the analytic signal are plotted against time, a spiral is generated. Fig. 8.11 shows the variation of frequency with time. The dashed line represents the exact frequency and the full line is an estimate of the frequency, computed from the derivative of the phase of the analytic signal. Note that except at the ends of the sample period the agreement is very good. Finally, Fig. 8.12 shows the scaled Fourier transform of the data in the range zero to 20.48 s. The periodogram shows the frequency spread between about 0.6 and 1.6 Hz. However, the Fourier transform tells us nothing about how the frequency and amplitude varies with time.

We have seen how, using the Hilbert transform we can examine a single harmonic component in a set of data as it changes in amplitude and frequency with time. This basic analysis does not allow us to examine multiple harmonic components in a data set. In order to do this we must decompose the data into its *intrinsic mode functions* (IMF). An IMF is defined as a function that satisfies the following requirements:

1. In the whole data set, the number of extrema and the number of zero-crossings must either be equal or differ by, at most, by one.
2. At any point, the mean value of the envelope defined by the local maxima and the envelope defined by the local minima are zero.

This definition guarantees the Hilbert transform of an IMF is well-behaved.

The IMF represents a simple oscillatory mode and is equivalent in some respects to the simple harmonic function, but it is much more general: instead of constant amplitude and frequency in a simple harmonic function, an IMF can have variable amplitude and frequency along the time axis. Intrinsic modes comprise only one harmonic component and the Hilbert transform can be applied to each intrinsic mode separately. This definition guarantees a well-behaved Hilbert transform of the IMF.

To decompose the data into its intrinsic modes, we use *Empirical Mode Decomposition* (EMD). The decomposition was first presented by Huang et al. (1998) and for that reason, EMD followed by the application of the Hilbert transform to each intrinsic mode is often called the Hilbert–Huang transform.

In deconstructing signals or time series data into its various intrinsic modes, EMD can be compared with other analysis methods such as the Fourier transform and the Wavelet transform. Using the EMD method, any complicated data set can be decomposed into a finite and often small number of components. These components form a complete and nearly orthogonal basis for the original signal.

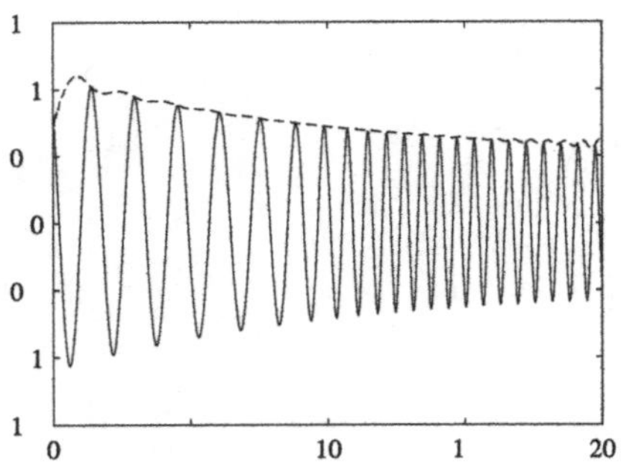

FIGURE 8.9

Plot of data y against time, in seconds. The dashed line is the envelope derived from the absolute value of the analytic data.

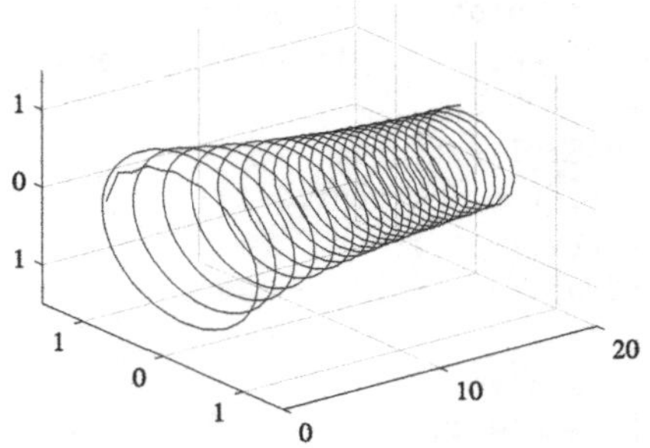

FIGURE 8.10

A three-dimensional plot of the real and imaginary parts of the analytic data against time, in seconds, showing an exponentially decaying spiral.

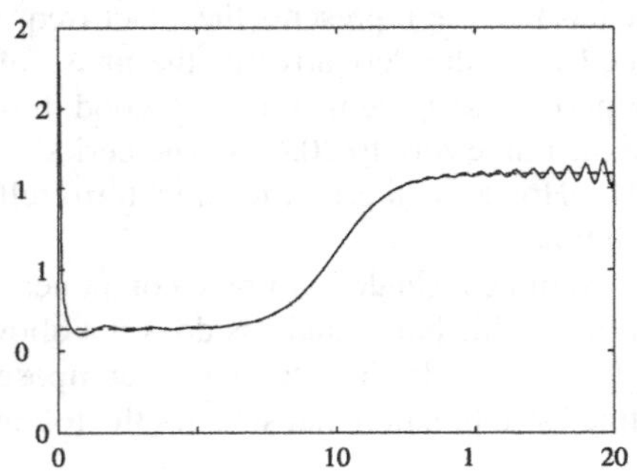

FIGURE 8.11

Plot of frequency, in Hz, derived from the Hilbert transform, against time, in seconds. The dashed line is the exact frequency.

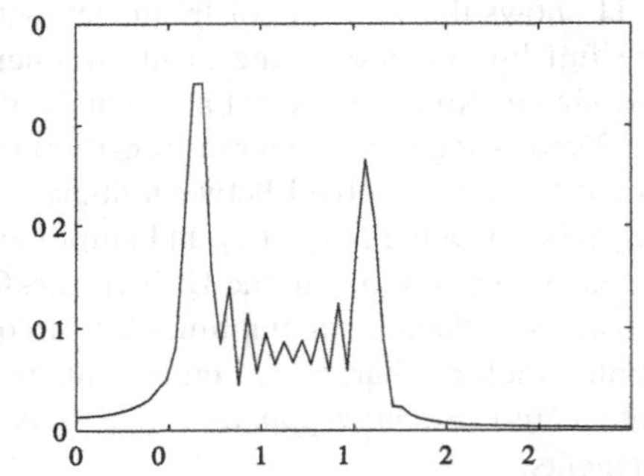

FIGURE 8.12

Fourier transform of the data, showing a spectrum between 0.5 Hz and 1.5 Hz, but the transform gives no information about the variation of frequency with time.

The first IMF usually carries the highest frequency component and it can be rejected to remove high-frequency components such as example, random noise. The EMD based smoothing algorithms have been widely used in seismic data processing.

The EMD algorithm is adaptive and highly efficient, see Huang et al. (1998). Since the decomposition is based on the local characteristic time scale of the data, it can be applied to both non-linear and non-stationary processes.

The procedure of extracting an IMF is called sifting. The sifting process connects all the local maxima by a cubic spline line which is the upper envelope. This procedure is then repeated for the local minima to produce the lower envelope. The upper and lower envelopes should cover all the data between them. The difference between the original data and the mean of these two envelopes is the first estimate of the IMF. The process is repeated iteratively to obtain a satisfactory estimate of the first IMF. A stopping criterion determines the number of sifting steps to produce an estimate of the IMF and currently researchers have developed four criteria. Once a stopping criterion is selected, the first IMF, can be obtained. Overall, this IMF should contain the finest scale or the shortest period component of the signal. Once the first IMF is obtained it is subtracted from the original data and the IMF is determined from the remainder. This IMF is the second IMF of the original data. Then, using the same process the third, fourth IMFs, etc., can be determined. The process finally stops when the residue is a monotonic function from which no more IMFs can be extracted.

Thus, a decomposition of the data into empirical modes is achieved. The components of the EMD are usually physically meaningful, for the characteristic scales are defined by the physical data.

Having obtained the intrinsic mode function components, the instantaneous frequency and amplitude can be computed using the Hilbert transform applied to each IMF.

The significance of each IMF is determined by its correlation coefficient. This is defined by

$$\gamma_p = \frac{\sum_{i=1}^{n} C_{pi} x_i}{\sqrt{\sum_{i=1}^{n} C_{pi}^2 \times \sum_{i=1}^{n} x_i^2}} \quad p = 1,\ 2,\ \ldots,\ m \tag{8.38}$$

or, equivalently,

$$\gamma_p = \frac{\mathbf{C}_p \mathbf{x}'}{\sqrt{\mathbf{C}_p \mathbf{C}_p' \mathbf{x}\mathbf{x}'}} \quad p = 1,\ 2,\ \ldots,\ m \tag{8.39}$$

where C_{pi} is the amplitude of the pth intrinsic mode for the ith data value, x_i.

In summary, Hilbert spectral analysis (HSA) is a method for examining each of the IMF's instantaneous frequencies as functions of time. The final result is a frequency–time distribution of signal amplitude, called as the Hilbert spectrum, designated by $H(t, \omega)$, which permits the identification of localized features.

Example 8.5. Consider data sampled from the function

$$x = 4\cos(22t + 0.5t^2) + 3e^{-0.1t}\cos(14t - t^2/10), \quad t < 15 \tag{8.40}$$

$$= 4\cos(22t + 0.5t^2) + 3e^{-0.1t}\cos(14t - t^2/10) + (t - 15)\sin(18t) \quad t \geq 15 \tag{8.41}$$

The function is sampled in the time range 0.01 to 20.48 s at intervals of 0.01 s, i.e. 2048 samples. The frequency resolution is $df = 1/T = 1/20.48 = 0.0488$ Hz. The maximum recommended frequency is approximately $1/(4 \times 0.01) = 25$ Hz. Applying the EMD algorithm we find that the signal is decomposed into 10 IMFs (the last being the monotonic residual function). Computing the correlation coefficients for the 10 IMFs using (8.38) gives 0.8989, 0.4688, 0.0657, 0.0146, 0.0097, 0.0054, 0.0061, 0.0040, 0.0062, and, for the monotonic residual, 0.0057. Clearly, in this example, only the first two or possibly three intrinsic modes are significant. This is also illustrated by Fig. 8.13 that shows the original data and the first 5 intrinsic modes. It is seen that the first two intrinsic modes are the most significant. This can be further demonstrated by regenerating the original data, by combining only the first two intrinsic modes, see Fig. 8.14. This recreates the original data almost exactly. We now take the discrete Hilbert transform of each intrinsic mode. Since only one frequency component is present in each IMF the discrete Hilbert transform will allow us to compute the amplitude and frequency of this component. By applying the discrete Hilbert transform to all the IMFs we can determine all the

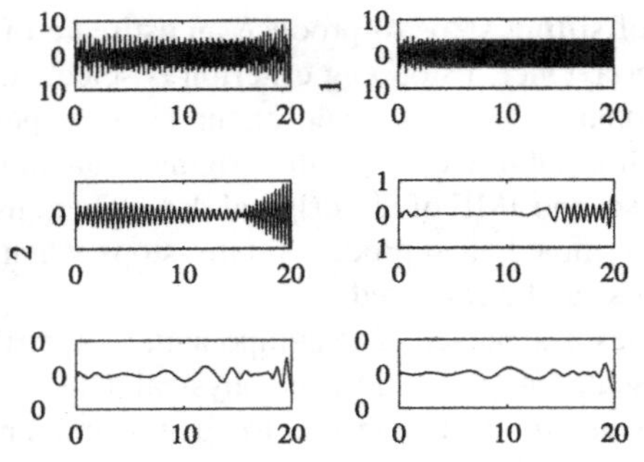

FIGURE 8.13

Original data is shown in the first plot in the left column. The remaining plots are of the first 5 intrinsic mode functions derived from it.

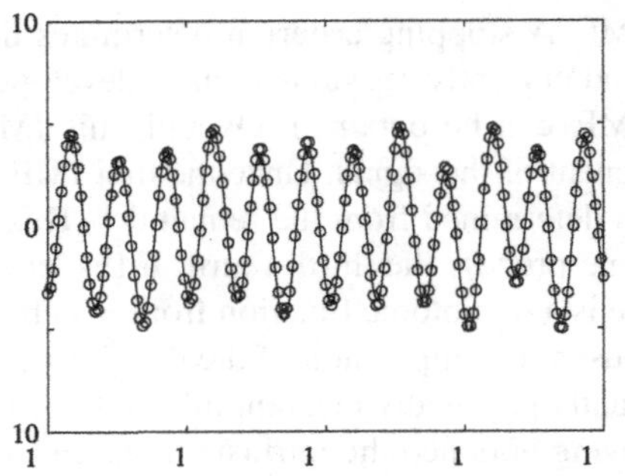

FIGURE 8.14

Plot of the original data over the interval from $t = 14.5$ s to 16.5 s and data points reconstructed from the first and second intrinsic mode functions indicated by o. Note the very close agreement.

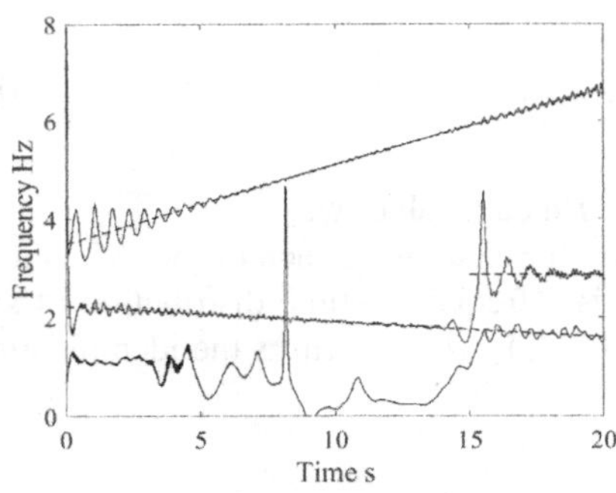

FIGURE 8.15

Plot showing the variation with time of the frequency components. The full lines are data from the intrinsic mode functions. The dashed lines are the actual frequency components.

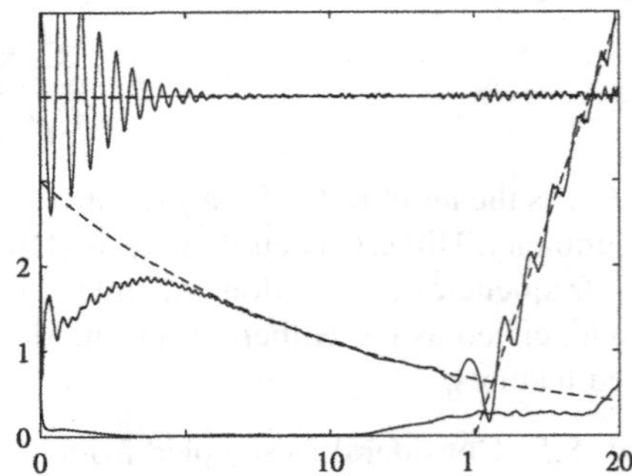

FIGURE 8.16

Plot showing the variation with time of the amplitude of the frequency components. The full lines are data from the intrinsic mode functions The dashed lines are the actual amplitudes of the frequency components.

frequency components and their amplitudes, see Figs. 8.15 and 8.16. These two figures only show the data from the first three intrinsic modes. In the frequency plot, Fig. 8.15, the first IMF is seen to track the highest frequency component in the data and the second IMF is seen to tracks the lower frequency component, but at approximately 15 s it jumps to track the frequency component that begins at 15 s. The third IMF has a very low amplitude and frequency below $t = 15$ s but, at $t = 15$ s, its frequency increases and it tracks the lowest frequency component in the data that only exists above 15 s. A similar pattern can be discerned in the amplitude plot, Fig. 8.16. Note how the second IMF fails to give the correct amplitude of the lower frequency below about 5 s, but then gives and accurate estimate of the

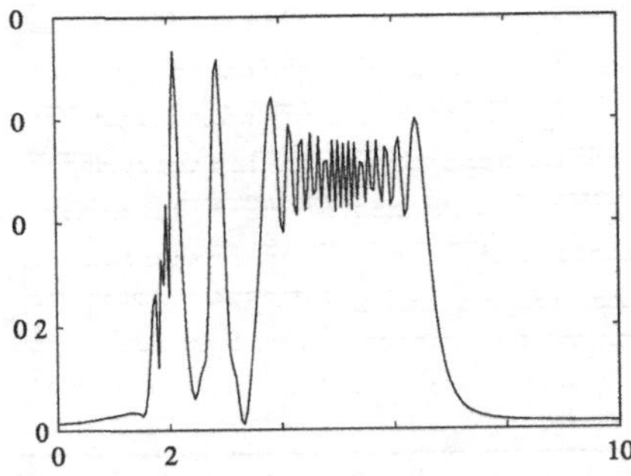

FIGURE 8.17

Fourier transform of the data of Example 8.5. Note that this plot gives no information about the variation of frequency with time.

amplitude of the lower frequency component, until at $t = 15$ s it switches to give the rapidly increasing amplitude of the frequency component that begins at $t = 15$ s.

Fig. 8.17 shows the amplitude of Fourier transform. The frequency components in the data are, approximately, a component varying between 1.6 Hz and 2.2 Hz, a component varying between 3.5 Hz and 6.8 Hz, and a component of 2.8 Hz. These can be identified in the Fourier transform, but not that the spectrum is changing with time.

The Hilbert and the Hilbert–Huang transforms have been widely applied to the analysis of the behavior of non-stationary and non-linear systems, for example, Feldman (2011).

8.4 THE WALSH TRANSFORMS

The Walsh function is an orthogonal function and can be used as the basis for a continuous or discrete transform.

Considering first the Walsh function: this function forms an ordered set of rectangular waveforms that can take only two values, +1 and −1.

There are at least four methods of generating Walsh functions. Here we consider only the Hadamard matrix approach. The 2×2 Hadamard matrix is

$$\mathbf{H}_2 = \begin{bmatrix} 1 & 1 \\ 1 & -1 \end{bmatrix}$$

Higher-order matrices can be obtained using the recursive relationship

$$\mathbf{H}_{2^k} = \begin{bmatrix} \mathbf{H}_{2^{k-1}} & \mathbf{H}_{2^{k-1}} \\ \mathbf{H}_{2^{k-1}} & -\mathbf{H}_{2^{k-1}} \end{bmatrix} \tag{8.42}$$

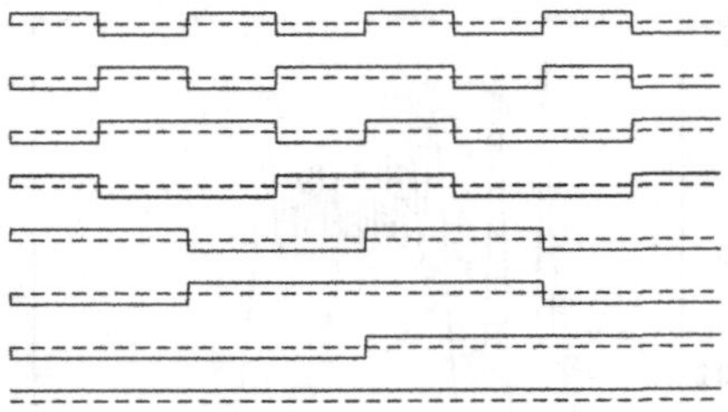

FIGURE 8.18

Walsh functions in the range $t = 0$ to 1, in ascending sequency order from WAL(0, t), with no zero crossings to WAL(7, t) with seven zero crossings.

Letting $k = 2$ we generate the following matrix.

$$\mathbf{H}_4 = \begin{bmatrix} 1 & 1 & 1 & 1 \\ 1 & -1 & 1 & -1 \\ 1 & 1 & -1 & -1 \\ 1 & -1 & -1 & 1 \end{bmatrix} \begin{matrix} row\ 0 \\ row\ 1 \\ row\ 2 \\ row\ 3 \end{matrix}$$

The rows of **H** are the values of the Walsh function, but the order is not the required sequency order. In this ordering, the functions are referenced in ascending order of zero crossings in the function in the range $0 < t < 1$. To convert **H** to the sequency order, the row number (beginning at zero) must be converted to binary, then the binary code converted to Gray code, then the order of the binary digits in the Gray code is reversed, and finally these binary digits are converted to decimal (that is they are treated as binary numbers, not Gray code). The definition of Gray code is provided by Weisstein (2017). The following shows the application of this procedure to the 4×4 Hadamard matrix.

row	0	1	2	3
binary	00	01	10	11
Gray	00	01	11	10
order reversed	00	10	11	01
decimal	0	2	3	1

The decimal value is the new position of the row. Hence

$$\mathbf{W} = \begin{bmatrix} 1 & 1 & 1 & 1 \\ 1 & 1 & -1 & -1 \\ 1 & -1 & -1 & 1 \\ 1 & -1 & 1 & -1 \end{bmatrix} \begin{matrix} row\ 0 \\ row\ 2 \\ row\ 3 \\ row\ 1 \end{matrix} \tag{8.43}$$

The first 8 Walsh functions are shown in Fig. 8.18. It should be noted that the Walsh functions can be logically ordered (and listed) in more than one way.

In Fig. 8.18 the functions are in *sequency* order. In this ordering, the functions are referenced in ascending order of zero crossings in the function in the range $0 < t < 1$ and for time signals, sequency is defined in terms of zero crossings per second or zps. This is similar to the ordering of Fourier components in increasing harmonic number (that is half the number of zero crossings). Another ordering is the natural or the Paley order. The functions are then called Paley functions, so that, for example, the 15th Walsh function and 8th Paley function are identical. Here we only consider sequency ordering.

There does not appear to be a standard notation for the Walsh function. Here we follow Beauchamp (1975) and denote the nth Walsh function at time t by $\text{WAL}(n,t)$ where $n=0,\ 1,\ 2,\ldots$.

It has been previously stated that the Walsh functions form an orthogonal set. Thus

$$\int_0^T \text{WAL}(i,t)\text{WAL}(j,t)\,dt = \begin{cases} 0 & i \neq j \\ T & i = j \end{cases} \tag{8.44}$$

or, for a set of n point discrete Walsh functions,

$$\sum_{k=0}^{n-1} \text{WAL}(i,k)\text{WAL}(j,k) = \begin{cases} 0 & i \neq j \\ n & i = j \end{cases} \tag{8.45}$$

It is easy to verify that continuous or discrete Walsh functions are orthogonal.

The discrete Walsh transform (DWT) is defined by

$$X_k = \sum_{j=0}^{n-1} x_j \text{WAL}(k,j), \quad k = 0,\ 1,\ 2,\ldots,n-1 \tag{8.46}$$

Defining $\mathbf{W}_{ij} = \text{WAL}(i,\ j)$ then

$$\mathbf{X} = \mathbf{W}\mathbf{x} \tag{8.47}$$

where $\mathbf{X}$ is the discrete Walsh transform of $\mathbf{x}$ and $\mathbf{W}$ is an $n \times n$ symmetrical matrix so that $\mathbf{W}^\mathsf{T} = \mathbf{W}$. Since $\mathbf{W}$ is also an orthogonal matrix $\mathbf{W}^\mathsf{T} = n\mathbf{W}^{-1}$.

From (8.47), we can obtain

$$\mathbf{x} = \mathbf{W}^{-1}\mathbf{X} = (1/n)\mathbf{W}^\mathsf{T}\mathbf{X} \tag{8.48}$$

and thus the inverse discrete Walsh transform is

$$\mathbf{x} = (1/n)\mathbf{W}^\mathsf{T}\mathbf{X} \tag{8.49}$$

The position of the factor n is somewhat arbitrary because it may appear in the transpose or the inverse transpose.

A Walsh transform of n real values has n real values in its transform. In contrast a Fourier transform of n real values has 2 real values and $n/2-1$ pairs of complex conjugate values, as explained in Section 8.2. Thus an n point Fourier transform has n discrete values, just as an n point Walsh transform produces n discrete values. We also note that like the discrete Fourier transform, the DWT satisfies Parseval's Theorem, that is

$$\sum_{i=0}^{n-1} x_i^2 = (1/n)\sum_{k=0}^{n-1} X_k^2 \tag{8.50}$$

A fast Walsh transform algorithm has been developed in a similar fashion to the fast Fourier transform. Here we provide the less efficient function dwht which computes the DWT using (8.47), or the inverse DWT based on (8.49)and as follows:

```
function X  = dwht(x,flag)
% x is an n point row vector.
% flag = 1 for DWT, flag ~= 1 for inverse DWT.
% X is row vector of DWT or IDWT
n = length(x);
W = fwalsh(n);
if flag == 1
    % Walsh transform
    X = (W*x')';
else
    % Inverse walsh transform
    X = (1/n)*(W*x')';
end
end
%-------------------------
function W  = fwalsh(n)
m = log2(n);
H = hadamard(n);
w = zeros(1,n);
for n = 0:n-1
     nb = dec2bin(n,m);
     ng = bin2gray(nb);
     gr = fliplr(ng);
     w(n+1) = bin2dec(gr);
end
W = H(w+1,:);
end
%-----------------------------------------
function g = bin2gray(b)
g(1) = b(1);
for k = 2:length(b);
    x = xor(str2num(b(k-1)), str2num(b(k)));
    g(k) = num2str(x);
end
end
```

The Walsh functions can be classified in terms of even CAL(k,t) and odd SAL(k,t) waveform symmetry. The names CAL and SAL are those used by Beauchamp (1975) and the functions are defined thus

$$\begin{aligned} \text{CAL}(k,t) &= \text{WAL}(2k,t), & k &= 0,\ 1, \ldots \\ \text{SAL}(k,t) &= \text{WAL}(2k-1,t), & k &= 1,\ 2, \ldots \end{aligned} \tag{8.51}$$

Thus, Fig. 8.18, by showing WAL$(0,t)$ to WAL$(7,t)$, also shows CAL$(0,t)$ to CAL$(3,t)$ and SAL$(1,t)$ to SAL$(4,t)$. The CAL and SAL functions are analogous to the cosine and sine functions of Fourier analysis. However, although the sine and cosine functions can be combined into a single, complex, function $\exp(j2\pi kt)$, it is not possible to combine CAL(k,t) and SAL(k,t) in such a simple and revealing relationship.

The power spectrum of the DWT can be determined as follows. In the DWT, consider the coefficients of WAL, CAL, and SAL functions, w_i, c_i, and s_i respectively. Then

$$\begin{aligned} P_0 &= c_0^2 = w_0^2 \\ P_k &= c_k^2 + s_k^2 = w_{2k}^2 + w_{2k-1}^2 \text{ where } 1 \leq k \leq n/2 - 1 \end{aligned} \tag{8.52}$$

The MATLAB function `walshps` extracts the CAL and SAL functions and also the power spectrum from the DWT, as follows

```
function [Cal,Sal,Pow] = walshps(Wal)
% Computes CAL. SAL and Power from WAL
% Wal is a row vector of Walsh values
% Cal, Sal and Pow are row vectors of CAL, SAL and Power values
% Index of WAL is from 0 to n-1. Computer Wal(i), i = 1 to n
% Index of CAL is from 0 to n/2-1. Computer Cal(i), i = 1 to n/2
% Index of SAL is from 1 to n/2. Computer Sal(i), i = 2 to n/2+1.
% Index i of Wal, Sal and Cal must be i-1 for index of actual function
n = length(Wal);
Cal(1) = Wal(1); Pow(1) = Wal(1)^2; Sal(1) = 0;
for i = 2:n/2
    Cal(i) = Wal(2*i-1);
    Sal(i) = Wal(2*i-2);
    Pow(i) = Cal(i)^2+Sal(i)^2;
end
Sal(n/2+1) = Wal(n); Pow(n/2+1) = Wal(n)^2;
```

The relationship between the WAL, CAL, and SAL indices, and sequency are similar to the relationship between the discrete Fourier index and frequency. In Section 8.2 we show that the frequency increment df is equal to $1/T$ where T is the time over which data is sampled. Since there are two zero crossings per cycle of a periodic waveform, the sequency is equal to twice the frequency. Hence $ds = 2df = 2/T$.

Consider an n point Walsh transform. The transform outputs the amplitude of n WAL functions denoted by WAL(k,t), where $k = 0$ to $n - 1$. We cannot easily relate the WAL index to sequency because a single sequency relates to pairs of WAL function indices. However, separating the WAL function amplitudes into the amplitudes of the CAL and SAL functions gives $n/2$ CAL functions denoted by CAL(k,t), where $k = 0$ to $n/2 - 1$ and $n/2$ SAL functions denoted by SAL(k,t), where $k = 1$ to $n/2$. The amplitude of the sequency power is related to the CAL and SAL functions as shown in (8.52) and is denoted by P_k, where $k = 0$ to $n/2 - 1$. Note that in the function `walshps`, the indices of the functions WAL, CAL, SAL, and P are increased by 1 since by definition the functions WAL, CAL, and P start with an index of 0, which is not allowed in MATLAB. Although the function SAL starts with an index of 1, this index is increased by 1 for consistency with WAL, CAL, and P. Care must be taken when using these functions to ensure that allowance is made for this adjustment.

The following script e4s808 calls the functions dwht and walshps to determine the coefficients of the Walsh, SAL, CAL, and power sequency spectrum of the input data. The input data is based on the weighted sum of WAL(63) and WAL(88) together with a random number selected from a normal or Gaussian distribution with a mean of zero and a standard deviation 2. The details of this can be seen in the script e4s808.m.

```
%e4s808.m
clear all, close all
clf
% WAL(63)=SAL(32) and WAL(88)=CAL(44)
P = [1 1 -1 -1]; P1 = [P P P P];
W63 = [P1 P1 P1 P1 P1 P1 P1 P1];
S = ones(1,1); L = ones(1,2);
R1 = [-S L -S S -L S -S L -S S -L L -L S -S L -S S -L S -S L -S];
W88 = [R1 R1 R1 R1];
y = 3*W63+4*W88+2*randn(size(W88));
n = length(y);
T = 4; dt = T/n; ds = 2/T;
t_rng = dt:dt:T;
dt_plt = dt/25;
t_plt = dt_plt:dt_plt:T;
s_cal = [0:(n/2-1)]*ds; s_pwr = s_cal;
s_sal = [0:(n/2)]*ds;
smax = n/T;
X  = dwht(y,1);
[CC,CS,CP] = walshps(X);
figure(1), subplot(2,1,1), plot(t_rng,y,'k')
xlabel('time s'), ylabel('x(t)')
axis([0 T -15 15])
subplot(2,1,2), bar(s_pwr,CP,'k')
xlabel('Sequency z/s'), ylabel('Power')
axis([0 smax 0 3e5])
figure(2), subplot(2,1,1), bar(s_cal, CC,'k')
xlabel('Sequency z/s'), ylabel('Coef of CAL')
axis([0 smax -600 600])
subplot(2,1,2), bar(s_sal, CS,'k')
xlabel('Sequency z/s'), ylabel('Coef of SAL')
axis([0 smax -600 600])
```

Fig. 8.19 is generated by e4s808 and shows the data and the power spectrum plot. Since the initial data is generated from two Walsh functions, we are able, as might be expected, to identify these components in the power spectrum. Fig. 8.20, also generated by e4s808 shows the DWT separated out

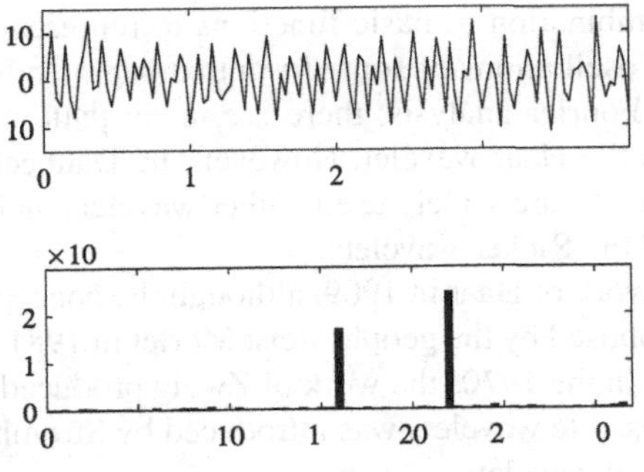

FIGURE 8.19

Upper figure shows plot of time series. Lower figure shows power sequency spectrum of the time series.

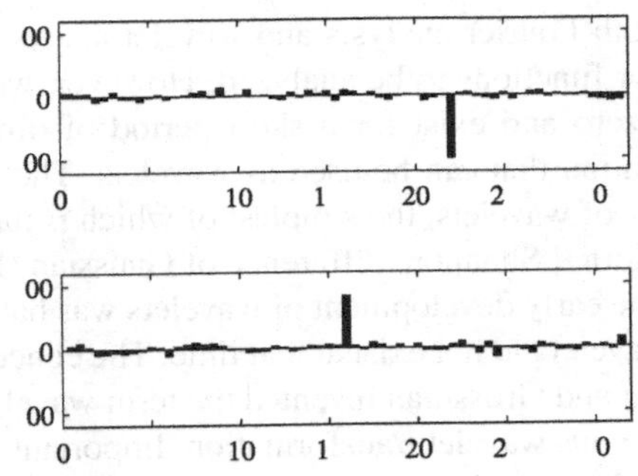

FIGURE 8.20

Plots show the coefficients of CAL and SAL sequency spectrum for the time series shown in Fig. 8.19.

into the SAL and CAL components. It is seen that the SAL sequency spectrum has a significant component of 16 z/s and the CAL sequency spectrum has a significant component of 22 z/s. The random component in the data causes the small levels across the sequency spectra.

8.5 INTRODUCTION TO WAVELET ANALYSIS

When a signal arises from a non-linear or non-stationary system (meaning a system in which the properties of the frequency components vary with time), the Fourier transform is not well suited as a basis for analysis. This is because the Fourier transform and its variants such as the DFT are based on the assumption that the spectrum of frequencies in the function or data to be analyzed remain constant over the period of the analysis. If these properties are changing, then one option is to take a series of DFT of the data over short time periods. For example, suppose we have 10 s of a sampled function or signal. Then, if the frequency content is changing, it would seem better to take a sequence of ten transforms, each of 1 s duration, rather than a single transform over the 10 s. However, Gabor (1946) noted a limitation in the use of Fourier analysis. By reference to quantum mechanics, an analog of the Heisenberg uncertainty principle may be developed, showing that resolution in time and frequency cannot be arbitrarily small because

$$\Delta t \Delta f \geq 1/(4\pi)$$

where Δf is the frequency bandwidth and Δt is sampling period or window.

From this inequality we can see that a good frequency resolution (that is a small frequency bandwidth) may only be obtained if the sampling time period of the analysis is sufficiently long. In contrast, if the analysis is over a short sampling time, then the time location is better, but the frequency bandwidth will be greater and the frequency resolution poorer. This is a significant limitation on Fourier analysis and it is this limitation that wavelet analysis can overcome. Wavelets have been used in many fields where the signals are non-stationary; for example, in the analysis of earthquake, medical, atmospheric, and oceanographic data.

Both Fourier analysis and wavelet analysis use a combination of basic functions to represent the data or functions to be analyzed. However, wavelets are oscillations of some form that begin and end with zero and exist for a short period of time. Unlike Fourier analysis, there are many patterns of oscillation that can be used as wavelets. The simplest is the Haar wavelet. However, the Daubechies family of wavelets, the simplest of which *is* the Haar wavelet, are widely used. Other wavelets include the Morlet, Shannon, difference of Gaussian (DOG), and the Ricker wavelets.

The early development of wavelets was based on the work of Haar in 1909, although the concept of the wavelet did not exist at that time. The concept was proposed by the geophysicist Morlet in 1981 and Morlet and Grossman invented the term wavelet in 1984. In the 1970s the work of Zweig produced the continuous wavelet transformation. Important work on discrete wavelets was introduced by Stromberg in 1983 and by Daubechies (1988) in relation to orthogonal wavelets.

A wavelet $\psi(t)$ is chosen to make

$$\int_{-\infty}^{\infty} |\psi(t)| dt < \infty \text{ and } \int_{-\infty}^{\infty} |\psi(t)|^2 dt < \infty \tag{8.53}$$

This ensures that the wavelet can be formulated so that:

$$\int_{-\infty}^{\infty} \psi(t) dt = 0 \text{ and } \int_{-\infty}^{\infty} |\psi(t)|^2 dt = 1 \tag{8.54}$$

A wavelet $\psi_{a,b}(t)$ can be expressed in terms of its *mother* wavelet $\psi(t)$ as follows.

$$\psi_{a,b}(t) = \frac{1}{\sqrt{|a|}} \psi\left(\frac{t-b}{a}\right) \quad a \neq 0 \tag{8.55}$$

where t is the variable representing time, b is a time shift and a is a dilation or scale factor. Thus for a given scaling a the wavelet can be translated by b. The mother wavelet $\psi(t)$ is the basic definition of a particular wavelet and is the function with $a = 1$ and $b = 0$. When $a \neq 1$ and $b \neq 0$ the wavelet is often referred to as a *daughter* wavelet.

The wavelet transform of $x(t)$ is defined by

$$W_{a,b} = \int_{-\infty}^{\infty} x(t) \frac{1}{\sqrt{|a|}} \psi^*\left(\frac{t-b}{a}\right) dt \tag{8.56}$$

where a and b are real and if $\psi(t)$ is complex, then $\psi^*(t)$ is its complex conjugate. Many wavelets are real functions, so that $\psi^*(t) = \psi(t)$. Note that the resulting wavelet transform is a function of two variables. The factor $(1/\sqrt{|a|})$ is a normalizing factor to ensure the energy is constant for all a and b. Thus

$$\int_{-\infty}^{\infty} |\psi_{a,b}(x)|^2 dt = \int_{-\infty}^{\infty} |\psi(x)|^2 dt \tag{8.57}$$

Wavelet transforms can take one of two basic forms, either discrete or continuous wavelet analysis, respectively. When applied to sampled data, both the discrete and the continuous wavelets must be sampled. The terms discrete or continuous refer to the dilation and shift parameters a and b. If these parameters are allowed to take any values then the transform is a continuous wavelet transform (CWT) whereas if these parameters are restricted to a particular set of discrete values, then the transform is a discrete wavelet transform (DWT). Note that abbreviation DWT is also used to denote the discrete Walsh transform. The discrete Walsh transform is not a wavelet transform.

Discrete transforms are based on orthogonal wavelets such as those of Haar or Daubechies and the transforms are non-redundant. Scales and shifts are imposed and consequently very fast algorithms for the transform have been developed. Discrete wavelets are wide, but not exclusively, used in image analysis and compression. Continuous transforms, such as those based on the Morlet or DOG wavelets are widely used for data analysis because the scales and shifts can be freely chosen, but the transforms are redundant and computationally expensive.

Initially we consider the discrete wavelet transform (DWT).

8.6 DISCRETE WAVELET TRANSFORMS

Consider (8.55). If we replace a by 2^{-j} and b by $2^{-j}k$, where $j = 0, 1, 2, \ldots$ and $k = 0, 1, 2, \ldots, 2^j - 1$ then (8.55) becomes

$$\psi_{j,k}(t) = \frac{1}{2^{-j/2}}\psi\left(\frac{t - 2^{-j}k}{2^{-j}}\right) = 2^{j/2}\psi(2^j t - k)$$

Hence (8.56) becomes

$$W_{j,k} = \int_{-\infty}^{\infty} x(t)2^{j/2}\psi^*(2^j t - k)\,dt$$

This is the discrete wavelet transform of $x(t)$. If we assume that n equispaced samples, x_i, of the signal $x(t)$ are taken of the signal over the period of interest, then

$$W_{j,k} = \sum_{i=0}^{n-1} x_i 2^{j/2}\psi^*(2^j t_i - k)$$

8.6.1 HAAR WAVELET

The Haar wavelet is the simplest example of a Discrete Wavelet. Haar wavelets are defined by

$$\psi(0,t) = 1 \quad \text{for } 0 < t \le 1$$

$$\psi(1,t) = \begin{cases} 1 & \text{for } 0 \le t < \frac{1}{2} \\ -1 & \text{for } \frac{1}{2} \le t \le 1 \end{cases}$$

$$\psi(2,t) = \begin{cases} \sqrt{2} & \text{for } 0 \le t < \frac{1}{4} \\ -\sqrt{2} & \text{for } \frac{1}{4} \le t < \frac{1}{2} \\ 0 & \text{for } \frac{1}{2} \le t \le 1 \end{cases}$$

$$\psi(3,t) = \begin{cases} 0 & \text{for } 0 \le t < \frac{1}{2} \\ \sqrt{2} & \text{for } \frac{1}{2} \le t < \frac{3}{4} \\ -\sqrt{2} & \text{for } \frac{3}{4} \le t \le 1 \end{cases}$$

and, in general, for $2^p + n \ge 1$,

$$\psi(2^p+n,t) = \begin{cases} \sqrt{2^p} & \text{for } n/2^p \le t < (n+\frac{1}{2})/2^p \\ -\sqrt{2^p} & \text{for } (n+\frac{1}{2})/2^p \le t < (n+1)/2^p \\ 0 & \text{elsewhere} \end{cases} \tag{8.58}$$

for $p = 0, 1, 2, ...,$ and $n = 0, 1, 2, 4, ..., 2^p - 1$ and $0 \le t \le 1$. Fig. 8.22 shows the first 8 Haar wavelets.

An alternative notation for the Haar wavelet may be given in terms of $\psi_{j,k}(t)$ thus

$$\begin{aligned} &\psi(0,t) \quad \text{is a constant and is described as Level } -1 \\ &\psi(1,t) = \psi_{0,0}(t) \equiv \text{Level } 0 \\ &\psi(2,t) = \psi_{1,0}(t) \equiv \text{Level } 1 \\ &\psi(3,t) = \psi_{1,1}(t) \equiv \text{Level } 1 \\ &\psi(4,t) = \psi_{2,0}(t) \equiv \text{Level } 2 \\ &\psi(5,t) = \psi_{2,1}(t) \equiv \text{Level } 2 \\ &\psi(6,t) = \psi_{2,2}(t) \equiv \text{Level } 2 \\ &\psi(7,t) = \psi_{2,3}(t) \equiv \text{Level } 2 \\ &\text{etc.} \end{aligned} \tag{8.59}$$

In addition to the values of the indices j and k, the Levels, as shown in Fig. 8.21, are given. The constant term is designated Level -1 so that then the remaining levels are then equal to the index j.

The Haar wavelets form a complete set of orthogonal functions and it can be shown that

$$\int_0^T \psi(i,t)\psi(j,t)\,dt = \begin{cases} 0 & \text{for } i \ne j \\ T & \text{for } i = j \end{cases} \tag{8.60}$$

or, for a set of n point discrete Haar wavelets,

$$\sum_{k=0}^{n-1} \psi(i,k)\psi(j,k) = \begin{cases} 0 & i \ne j \\ n & i = j \end{cases} \tag{8.61}$$

The discrete Haar wavelet transform, X_k of x_j is defined by

$$X_k = \sum_{j=0}^{n-1} x_j \psi(k,j), \quad k = 0, 1, ..., n-1 \tag{8.62}$$

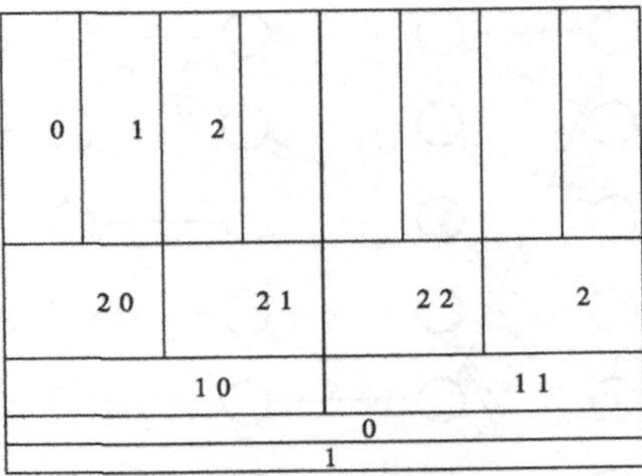

FIGURE 8.21

Diagram showing the partitioning of the time–frequency plane in the DWT.

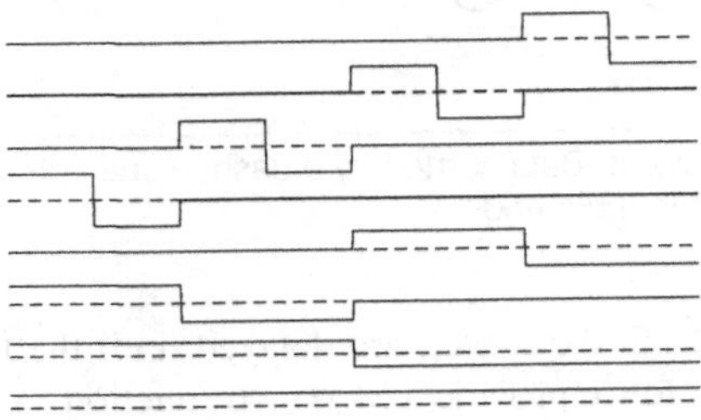

FIGURE 8.22

The Haar wavelets in ascending order from $\psi(0, t)$ to $\psi(7, t)$ over the range $0 < t < 1$.

In matrix notation this becomes

$$\mathbf{X} = \mathbf{H}\mathbf{x} \tag{8.63}$$

In (8.63), $\mathbf{H}$ is an $n \times n$ orthogonal matrix, so that $n\mathbf{H}^{-1} = \mathbf{H}^{\mathrm{T}}$. Thus the inverse transform can be derived from

$$\mathbf{x} = (1/n)\mathbf{H}^{\mathrm{T}}\mathbf{X} \tag{8.64}$$

The Haar wavelet also satisfies Parseval's equation as follows

$$\sum_{i=0}^{n-1} x_i^2 = (1/n) \sum_{k=0}^{n-1} X_k^2 \tag{8.65}$$

Like the Fourier and Walsh transforms, a fast algorithm has been developed for the Haar transform and the flow diagram for this algorithm is shown in Fig. 8.23. In fact, the fast Haar transform is the fastest of all discrete transforms, requiring only $2(n-1)$ additions or subtractions. All the

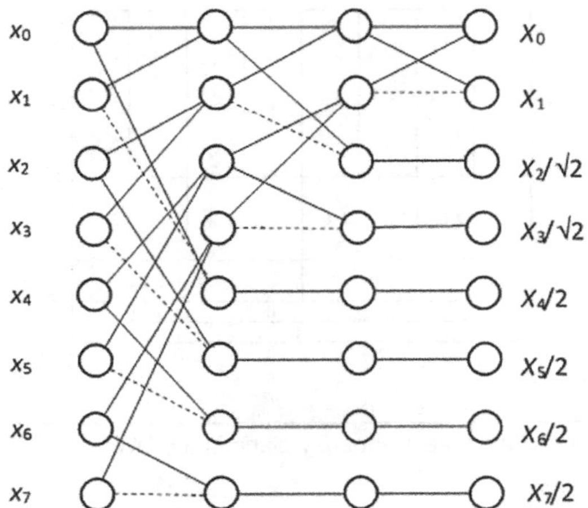

FIGURE 8.23

Flow diagram for the fast Haar transform. Data carried by a dashed line entering a node is negated and added to the data carried by the full line entering a node.

multiplications of the factors $\sqrt{2}$, 2, $2\sqrt{2}$, etc., are delayed until the transform is complete, so only n multiplications are required. Here we provide a MATLAB function `fasthaar` that determines the discrete Haar wavelet transform, provided the number of data values is a power of 2. The function is based on the fast transform algorithm but it is not optimized for maximum efficiency.

```
% z is the Haar transform of x
% x and z can be row or column vectors
function z = fasthaar(x)
n = length(x);
m = log2(n);
for k = 1:m
    p = 2^(m-k);
    y = x;
    for j = 1:p
        j2 = 2*j;
        x(j)   = y(j2-1)+y(j2);
        x(p+j) = y(j2-1)-y(j2);
    end
end
z = zeros(size(x));
z(1:2) = x(1:2);
cs = 2;
for k = 1:m-1
    for j = 1:2^k
```

```
        cs = cs+1;
        z(cs) = x(cs)*sqrt(2)^k;
    end
end
```

Fig. 8.24 shows a signal $x(t)$ decomposed into a constant value, Level −1, and 6 levels of Haar wavelets, a total of 63 individual wavelets. For example, the plot of Level 2 comprises 4 individual wavelets. When the separate levels are added together, as shown in Fig. 8.25, the original signal $x(t)$ is reconstructed. Thus the Haar wavelet transform, and other discrete wavelets, can decompose a sequence of data and then, from the wavelet levels, the data can be reconstructed.

Because the wavelet transform is a function of both the indices j and k, it is helpful to plot a contour plot showing the wavelet amplitude as a function of both time or shift and wavelet scale or level. This is often called a scalogram. The MATLAB function `waveletmap` has been developed to generate a scalogram from the Haar transform of a set of data.

```
function Cg = waveletmap(t_rng,X,option,cl,f1,f2)
% t_rng is a row vector of the time range,
% X is a row vector of Haar transform coefficients
% in sequency order
% option == 1 plots the log(magnitude) of the w'let coefficient
% option == 2 plots the magnitude of the wavelet coefficient
% option == 3 plots the magnitude^2 of the wavelet coefficient
% cl is the number of contours required in the contour plot
% f1 is the figure number given to the contour plot
% f2 is the figure number given to the surface plot
n = length(X);
Cg = zeros(log2(n)+1,n);
Cg(1,:) = X(1)*ones(1,n);
Cg(2,:) = X(2)*ones(1,n);
for m = 3:log2(n)+1
    for k = 1:n/2^(log2(n)+2-m)
        p = 2^(log2(n)+2-m);
        Cg(m,1+p*(k-1):p*k) = X(2^(m-2)+k)*ones(1,p);
    end
end
figure(f1)
if option == 1
    contourf(t_rng, -1:log2(n)-1, log10(abs(Cg)),cl)
    title(['Log(Magnitude) Scalogram. ' num2str(cl) ' Contours'])
elseif option == 2
    contourf(t_rng, -1:log2(n)-1, (abs(Cg)),cl)
    title(['Magnitude Scalogram. ' num2str(cl) ' Contours'])
else
    contourf(t_rng, -1:log2(n)-1, Cg.^2,cl)
    title(['Magnitude Squared Scalogram. ' num2str(cl) ' Contours'])
```

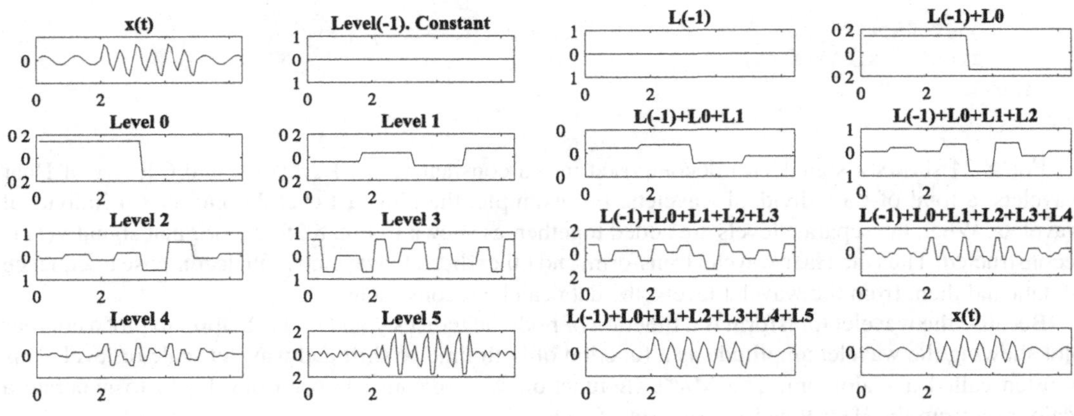

FIGURE 8.24

Decomposition of $x(t)$ into a constant term and 6 levels of Haar wavelets.

FIGURE 8.25

Reconstruction of $x(t)$ from its Haar wavelet components. Adding the constant term (Level -1) and all the Haar wavelets from Level 0 to Level 5 together provides and exact reconstruction $x(t)$.

```
end
colormap gray, colorbar
shading interp
ylabel('Level'), xlabel('Time')
figure(f2)
if option == 1
    surfl(t_rng, -1:log2(n)-1, log10(abs(Cg))), zlabel('log(Amplitude)')
elseif option == 2
    surfl(t_rng, -1:log2(n)-1, (abs(Cg))), zlabel('Amplitude')
else
    surfl(t_rng, -1:log2(n)-1, Cg.^2), zlabel('Amplitude^2')
end
ylabel('Level'), xlabel('time')
shading interp
end
```

To illustrate the use of the Haar wavelet we examine three cases. In each case 2048 samples of the signal are taken over the period $T = 25.6$ s. Level 0 comprises a single Haar wavelet and thus its period is 25.6 s and the corresponding frequency of a harmonic wave is $1/25.6 = 0.0390625$ Hz. At Level 1 there are two Haar wavelets in 25.6 s and hence the period is halved to $25.6/2 = 12.8$ s and the frequency doubles to 0.07812 Hz. Thus at Level L the period is $T/2^L$ and the frequency is $2^L/T$. Since the Haar wavelet require a minimum two data samples to define it, the minimum period is $2T/N$ and thus the maximum frequency is $N/(2T) = 2048/(2 \times 25.6) = 40$ Hz. From this we can deduce that the maximum Level is $\log_2(N) - 1 = 10$. Table 8.2 gives the periods and equivalent frequencies for this analysis.

Table 8.2 Period and frequency at the DWT levels

Level	Period, s	Frequency, Hz
−1	Constant value	
0	25.60	0.0390625
1	12.80	0.0781250
2	6.40	0.15625
3	3.20	0.3125
4	1.60	0.6250
5	0.80	1.25
6	0.40	2.5
7	0.20	5.00
8	0.10	10.00
9	0.05	20.00
10	0.025	40.00

Example 8.6. A signal defined by (8.66) comprises a square wave of period 0.8 s, together with two further square wave components of periods 0.1 s and 3.2 s that only exist from 3 to 11.2 s and from 12.8 to 22.4 s respectively. Letting sqr(T_p, t) denote a unit amplitude square wave with a period of T_p the signal is defined by (8.66) thus,

$$\begin{aligned}
x_1(t) &= 20\,\mathrm{sqr}(0.1, t) \text{ where } 3 <= t <= 11.2\\
x_2(t) &= 4\,\mathrm{sqr}(3.2, t) \text{ where } 12.8 <= t <= 22.4\\
x_3(t) &= 10\,\mathrm{sqr}(0.80, t)\\
x(t) &= x_1(t) + x_2(t) + x_3(t)
\end{aligned} \tag{8.66}$$

Fig. 8.26 shows the scalogram for this signal. The three square wave components in the signal are clearly seen. Of course, since the signal comprises three square waves, a Haar wavelet is an ideal base for this transform. The scalogram clearly shows components at levels 3, 5, and 8 and the components in the signal begin and end as expected. In fact, the periods of the components in the signal exactly coincide with the periods given by the levels 3, 5, and 8 and the components in the signal begin and end at the boundary between two wavelets.

Example 8.7. Eq. (8.67) defines a signal that comprises a sine wave of frequency 1.25 Hz, together with two further sine wave components of frequencies of 10 Hz and 0.3125 Hz that only exist 3 to 11.2 s and 12.8 to 22.4 s respectively. Normally distributed random noise with a standard deviation of 2 is also added.

$$\begin{aligned}
x_1(t) &= 20\sin(2\pi 10t) \text{ where } 3 <= t <= 11.2\\
x_2(t) &= 4\sin(2\pi 0.3125t) \text{ where } 12.8 <= t <= 22.4\\
x_3(t) &= 10\sin(2\pi 1.25t)\\
x(t) &= x_1(t) + x_2(t) + x_3(t) + \text{random noise}
\end{aligned} \tag{8.67}$$

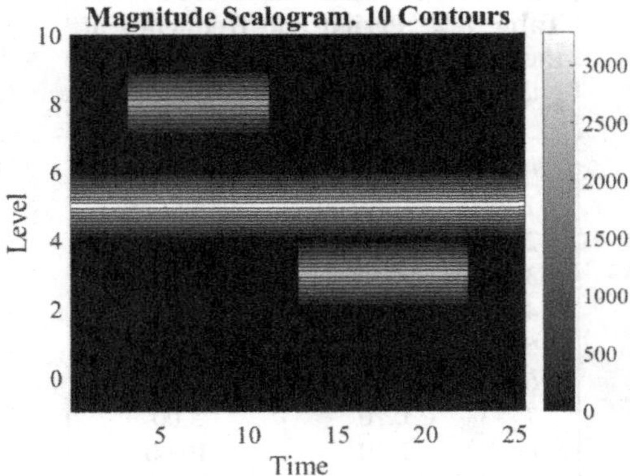

FIGURE 8.26

Contour plot of DWT of signal defined by (8.66), a composition of square waves. Responses can be clearly seen at levels 5, 3, and 8.

The script `e4s809.m` implements this equation and calls the `fasthaar` and `waveletmap` to generate the scalogram for this problem.

```
% e4s809.m
clear all, close all
n = 2048; T = 25.6;
dt = T/n;  nn = 1:n;
t_rng = dt:dt:T;
% Sine waves plus noise
k = 0;
for t = t_rng
k = k+1;
    x1(k) = 20*sin(2*pi*10*t)*(t>=3. & t<=11.2);
    x2(k) = 4*sin(2*pi*0.3125*t)*(t>=12.8 & t<=22.4);
    x3(k) = 10*sin(2*pi*1.25*t);
end
x = x1+x2+x3 + 2*randn(size(x1));
X = fasthaar(x);
Cg = waveletmap(t_rng,X,1,30,1,2);
```

In Fig. 8.27 the three sine wave components in the signal can be seen at levels 3, 5, and 8. The frequencies of the components exactly coincide with the level frequencies and the components begin and end at the boundary between two wavelets. However, there is a lot of background noise, in contrast to the uncluttered response in Fig. 8.26.

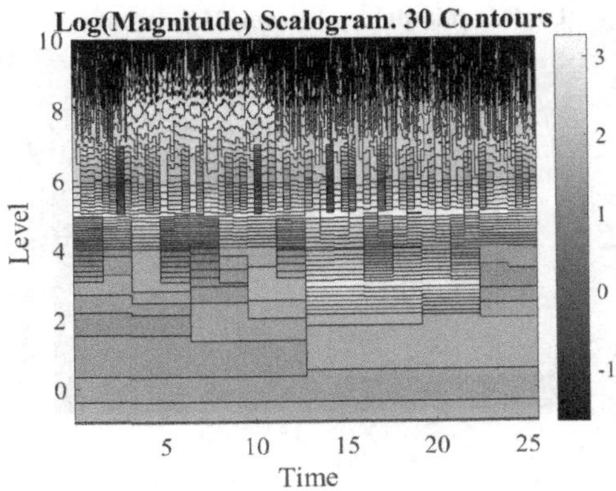

FIGURE 8.27

Contour plot of DWT of signal defined by (8.67), a composition of sine waves. Responses can be observed at levels 5, 3, and 8.

Example 8.8. A signal defined by (8.68) comprises five bursts of energy. Each burst occurs at different instants of time, at a different frequency and at different exponential decay rates. The signal mean value is zero.

$$\begin{aligned}
x_1(t) &= 8\exp(-0.10(t-3.2))\sin(2\pi 1.25(t-3.2)) \text{ where } t >= 3.2\\
x_2(t) &= 14\exp(-0.18(t-6.4))\sin(2\pi 10(t-6.4)) \text{ where } t >= 6.4\\
x_3(t) &= 9\exp(-0.07(t-11.2))\sin(2\pi 5(t-11.2)) \text{ where } t >= 11.2\\
x_4(t) &= 20\exp(-0.07(t-17.6))\sin(2\pi 20(t-17.6)) \text{ where } t >= 17.6\\
x_5(t) &= 7\exp(-0.30(t-19.2))\sin(2\pi 0.3125(t-19.2)) \text{ where } t >= 19.2\\
x(t) &= x_1(t)+x_2(t)+x_3(t)+x_4(t)+x_5(t)\\
x(t) &= x(t) - \text{mean}\,(x(t))
\end{aligned} \tag{8.68}$$

Fig. 8.28 shows the scalogram for this signal. The bursts of energy can be seen at level 5 beginning at $t = 3.2$ s, at level 8 beginning at $t = 6.4$ s, at level 7 beginning at $t = 11.2$ s, at level 9 beginning at $t = 17.6$ s and at level 3 beginning at $t = 19.2$ s. However, whilst these bursts of energy can be recognized in the scalogram, there are other areas in the scalogram could give the user misleading results. For example, there is a spurious high level of wavelet for $T < 3$ s from Level 2 to 10.

8.6.2 DAUBECHIES WAVELETS

The Haar wavelet is of limited effectiveness in a DWT. It obviously works well for signals comprising predominantly square waves but, as we have seen, does not give clear information when processing more general harmonic waves. Daubechies invented what are called compactly supported orthonormal wavelets and their shape has made discrete wavelet analysis attractive and practicable. The Daubechies wavelets cannot be expressed in a closed form. Essentially, a *scaling function* is derived by iterating an equation with $2N$ coefficients, and from this scaling function the wavelet function can be derived,

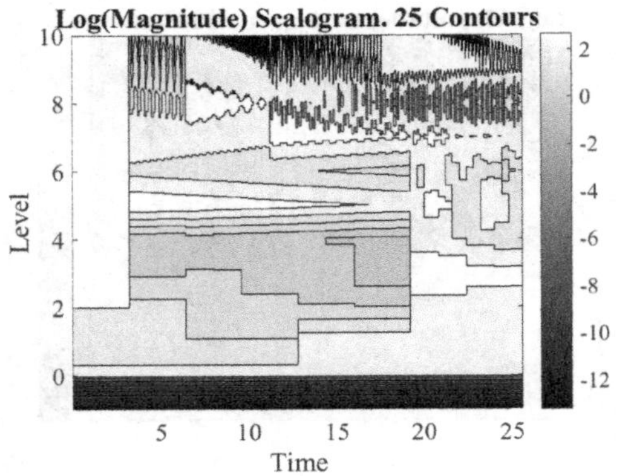

FIGURE 8.28

Contour plot of DWT of signal comprising bursts of exponentially decaying components, (8.68). Response at levels 5 (at $t = 3.2$), 8 (at $t = 6.4$), 7 (at $t = 11.2$), 9 (at $t = 17.6$), and 3 (at $t = 19.2$) can be observed.

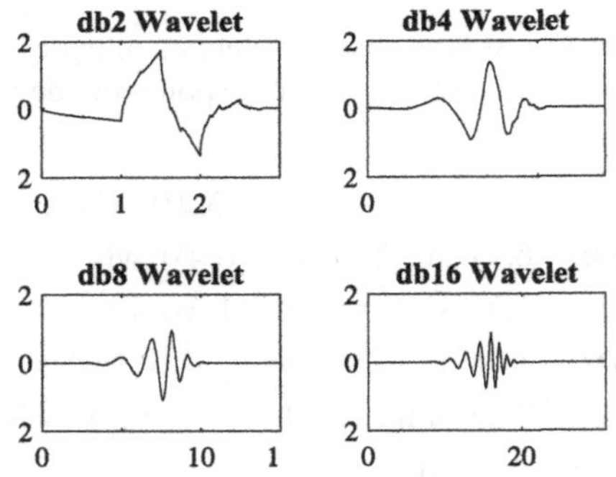

FIGURE 8.29

Plots of the wavelets db2, db4, db8, and db16.

again with $2N$ coefficients. The coefficients have been determined and tabulated for $N = 1$ to $N = 45$. The names of the Daubechies family wavelets are written dbN, where N is the order of the wavelet. The order is the number of vanishing moments for the particular Daubechies wavelet. A wavelet has N vanishing moments if

$$\int_{-\infty}^{\infty} t^k \psi(t)\, dt = 0 \text{ for } 0 \le k < N$$

The Daubechies wavelets are orthogonal, biorthogonal, but they are not symmetrical. The compact support, that is the range over which they are non-zero is $[0,\ 2N - 1]$. Daubechies wavelet are widely used in the DWT but also in the CWT. Fig. 8.29 shows the Daubechies wavelets db2, 4, 8, and 16.

The Haar wavelet is, in fact, a member of the Daubechies family. The Haar wavelet is the 1st-order Daubechies wavelet; that is db1. Since $N = 1$ the compact support is over the range [0, 1], and the only moment equation satisfied is

$$\int_{-\infty}^{\infty} \psi(t)\,dt = 0$$

8.7 CONTINUOUS WAVELET TRANSFORMS

A wavelet is defined in (8.55) and is repeated here

$$\psi_{a,b}(t) = \frac{1}{\sqrt{|a|}}\psi\left(\frac{t-b}{a}\right) \quad a \neq 0$$

where t is the variable representing time, b is a time shift and a is a dilation or scale factor. Thus for a given scaling a the wavelet can be translated by b.

The wavelet transform of $x(t)$ is defined by

$$W_{a,b} = \int_{-\infty}^{\infty} x(t)\frac{1}{\sqrt{|a|}}\psi^*\left(\frac{t-b}{a}\right) dt$$

where $\psi(t)$ is the analyzing (continuous) wavelet. In contrast to the discrete wavelet transform, the continuous wavelet transform permits the shift, a, and the dilation, b, to be infinitely variable. Thus, "continuous" refers to the variation of a and b, rather than the transform itself. When analyzing sampled or discrete data, both the continuous and discrete transforms are discrete. In the continuous wavelet the number of scales between each doubling of frequency, or octave, is referred to as the number of *voices per octave*.

To develop a continuous wavelet transform a specific wavelet function must be chosen. Here we consider a Morlet wavelet, a commonly used wavelet function. There is some flexibility in its definition and here we give the version that is used by the functions in the MATLAB Wavelet toolbox.

$$\psi(t) = e^{j5t}e^{-t^2/2} \tag{8.69}$$

Other definitions of this wavelet use a marginally different factor in the complex exponential term and also scale the function by dividing by $\pi^{1/4}$. This wavelet is complex and can be separated into its real and imaginary components thus:

$$\begin{aligned} \text{real}(\psi(t)) &= \cos(5t)e^{-t^2/2} \\ \text{imag}(\psi(t)) &= \sin(5t)e^{-t^2/2} \end{aligned} \tag{8.70}$$

This mother wavelet may be shifted by b and dilated by a to give

$$\begin{aligned} \text{real}(\psi_{a,b}(t)) &= (1/\sqrt{a})\cos(5(t-b)/a)e^{-((t-b)/a)^2/2} \\ \text{imag}(\psi_{a,b}(t)) &= (1/\sqrt{a})\sin(5(t-b)/a)e^{-((t-b)/a)^2/2} \end{aligned} \tag{8.71}$$

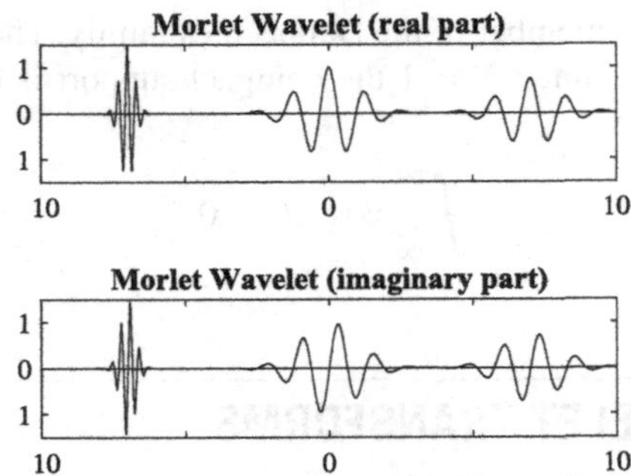

FIGURE 8.30

Plots of the real and imaginary parts of the Morlet wavelet. The mother wavelet in the middle of the plot with $a = 1$ and $b = 0$, that is, the wavelet is neither dilated nor shifted. The wavelet at the right of the plot is the wavelet shifted by $b = 7$ but it is not dilated. The wavelet at the left of the plot is both shifted by $b = -7$ and dilated by $a = 1/4$.

The frequency of this wavelet appears to be $5/a$ rad/s or $5/(2\pi a) = 0.7958/a$ Hz. However, a sine or cosine wave of this frequency only fits the Morlet wave over a short time span when t tends to b. An approximate frequency fit over a wider range is when frequency is equal to $0.8125/a$. This is called the center frequency. The wavelet is not orthogonal, nor is it bi-orthogonal.

Three Morlet wavelets are shown in Fig. 8.30. In the center is the mother wavelet, neither dilated nor shifted. The wavelet at the right of the plot is shifted by 7 time units but not dilated. The wavelet at the left is shifted by −7 time units and also dilated by 1/4.

From Fig. 8.30 we notice the wavelet develops and then dies out roughly in the window given approximately by the interval [−4 4]. This an important requirement of wavelet functions that their nature is transient.

Beside the Morlet wavelet, there are many other wavelet functions which are used and here we provide a description of the Ricker or Mexican Hat wavelet. This wavelet is very similar to the difference of Gaussian (DOG) wavelet function. The Ricker wavelet is a real function and the mother wavelet is defined by

$$\psi(t) = \frac{2}{\pi^{1/4}\sqrt{3}}(1 - t^2)\exp\left(-\frac{t^2}{2}\right)$$

This mother wavelet can be dilated and shifted thus

$$\psi_{a,b}(t) = \frac{2}{\pi^{1/4}\sqrt{3a}}\left(1 - \frac{(t-b)^2}{a^2}\right)\exp\left(-\frac{(t-b)^2}{2a^2}\right)$$

Fig. 8.31 shows the Ricker wavelet.

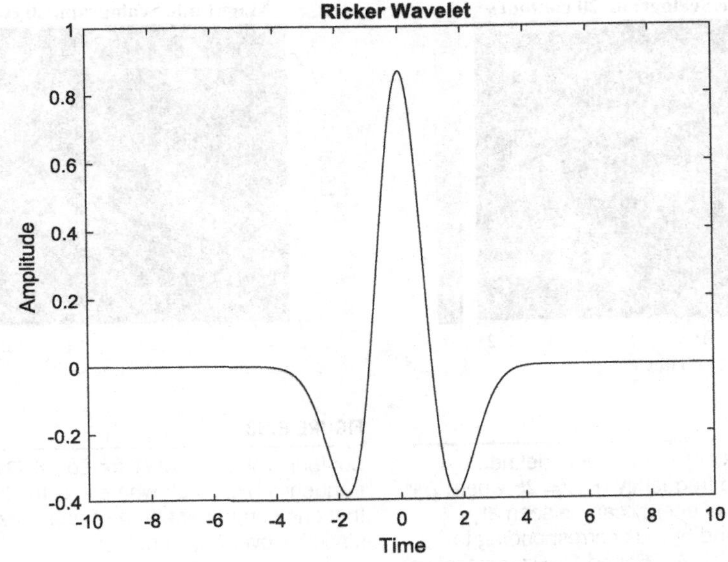

FIGURE 8.31

Ricker wavelet.

We do not provide MATLAB scripts or functions to generate a CWT scalogram, but merely provide examples. In the following CWT scalograms, the complex Morlet wavelet is used with 10 voices/octave. The frequencies are related to the levels by $f = 2^L$ where f is the frequency (Hz) and L is the scalogram level.

We now consider two examples. In each case there are 2048 sample points, over a time period of 25.6 s.

Example 8.9. This problem is similar to **Example 8.8** except that the frequencies of the bursts of energy have been changes to 1, 16, 4, 32, and 0.25 Hz, corresponding to Levels 0, 4, 2, 5, and −2 respectively. These values have been chosen to make it easy to locate on scalogram the energy bursts.

$$
\begin{aligned}
&x_1(t) = 24\exp(-0.10(t-3.2))\sin(2\pi 1(t-3.2)) \text{ where } t >= 3.2 \\
&x_2(t) = 14\exp(-0.18(t-6.4))\sin(2\pi 16(t-6.4)) \text{ where } t >= 6.4 \\
&x_3(t) = 9\exp(-0.07(t-11.2))\sin(2\pi 4(t-11.2)) \text{ where } t >= 11.2 \\
&x_4(t) = 20\exp(-0.07(t-17.6))\sin(2\pi 32(t-17.6)) \text{ where } t >= 17.6 \\
&x_5(t) = 40\exp(-0.30(t-19.2))\sin(2\pi 0.25(t-19.2)) \text{ where } t >= 19.2 \\
&x(t) = x_1(t) + x_2(t) + x_3(t) + x_4(t) + x_5(t) \\
&x(t) = x(t) - \text{mean}\,(x(t))
\end{aligned}
\tag{8.72}
$$

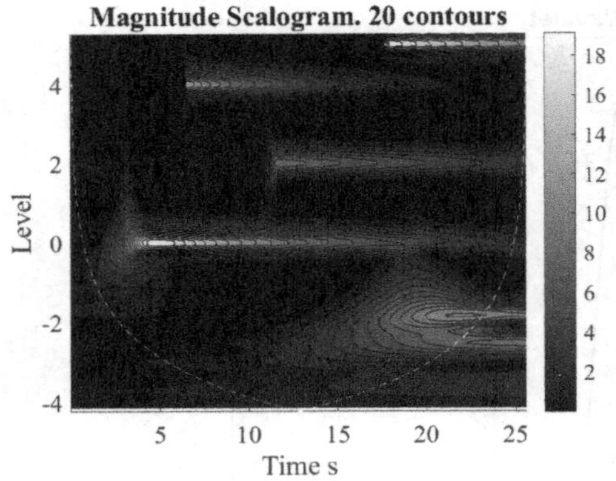

FIGURE 8.32

Contour plot of the CWT of the signal defined by (8.72). Note that the frequency (Hz) $= 2^L$ where L is the level. The burst of energy can be seen at levels -2, 0, 2, 4, and 5, thus corresponding to frequencies of 0.25, 1, 4, 16, and 32 Hz, respectively.

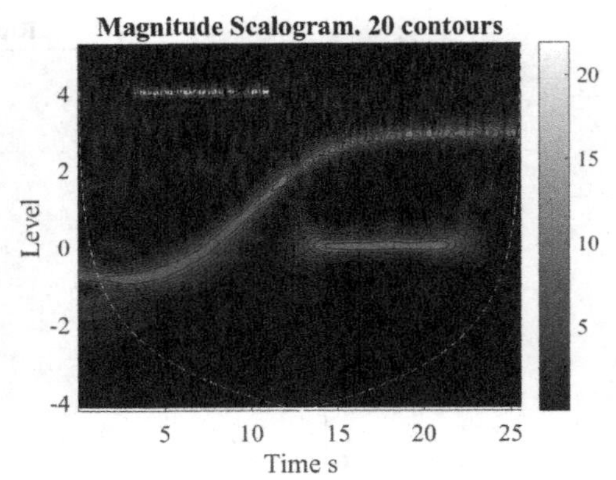

FIGURE 8.33

Contour plot of the CWT for Eq. (8.73). Note that the frequency (Hz) $= 2^L$ where L is the level. It is seen that one component of the signal clearly increases smoothly over the sampling time.

Fig. 8.32 shows the scalogram for this problem. The energy bursts can easily be located in time and level, and hence frequency. Also shown on this figure, and also on Fig. 8.33 is a dashed line which is the *cone of influence*. Outside this line is the region of the wavelet scalogram in which edge effects become important and the scalogram is less accurate.

Example 8.10. Eq. (8.73) defines a signal that comprises a sine wave that varies in frequency according to $f(t) = (25.5 + 22.5\tanh((t - 12.8)/4)))/(2\pi)$. Hence, when $t = 0$, $f(0) = (25.5 - 22.5)/(2\pi) = 0.48$ Hz. When $t = 25.6$, $f(25.6) = (25.5 + 22.5)/(2\pi) = 7.64$ Hz. These two frequencies equate to Levels -1.07 and 2.93 respectively. Additionally the signal has two further sine wave components of frequencies of 16 Hz and 1 Hz that only exist in the intervals 3 to 11.2 s and 12.8 to 22.4 s respectively. Also added is normally distributed random noise with a standard deviation of 4.

$$\begin{aligned}
x_1(t) &= 20\sin(2\pi 16t) \text{ where } 3 <= t <= 11.2 \\
x_2(t) &= 10\sin(2\pi t) \text{ where } 12.8 <= t <= 22.4 \\
x_3(t) &= 10\sin\left[25.5t + 90\log(\cosh(t/4 - 3.2))\right]; \\
x &= x1 + x2 + x3 + \text{random noise}
\end{aligned} \tag{8.73}$$

Fig. 8.33 clearly shows that one component of $x(t)$ varies smoothly from a lower to a higher frequency.

Wavelet analysis has found wide applications in many fields, including geophysics, medicine, and engineering; see for example Yan et al. (2014), Unser and Aldroubi (1996), Addison (2005), and Kumar and Foufoula-Georgiou (1997).

8.8 SUMMARY

For periodic data, we have examined the use of fast Fourier transform and the Walsh transform. For non-stationary systems, we have used the Hilbert transform and both discrete and continuous wavelet transforms.

Note: For readers who a have particular interest in applying MATLAB to data analysis using discrete transforms, MATLAB provides a Signal Processing Toolbox and a Wavelet Toolbox.

8.9 PROBLEMS

8.1. Determine the real and imaginary parts of the DFT, using the MATLAB function `fft`, for the following periodic data where the 32 data points are sampled at intervals of 0.1 second. Examine the amplitude and frequency of its components. What conclusions can you draw from these results?

$$y = [2\ -0.404\ 0.2346\ 2.6687\ -1.4142\ -1.0973\ 0.8478\ -2.37\ 0$$
$$2.37\ -0.8478\ 1.0973\ 1.4142\ -2.6687\ -0.2346\ 0.404\ -2$$
$$1.8182\ 1.7654\ -1.2545\ 1.4142\ -0.3169\ -2.8478\ 0.9558$$
$$0\ -0.9558\ 2.8478\ 0.3169\ -1.4142\ 1.2545\ -1.7654\ -1.8182]$$

8.2. Determine the DFT of $y = 32\sin^5(2\pi ft)$ where $f = 30$ Hz. Use 512 points sampled over 1 second. From the imaginary part of the DFT estimate the coefficients a_0, a_1, a_2 in the relationship

$$32\sin^5(2\pi ft) = a_0 \sin[2\pi ft] + a_1 \sin[2\pi(3f)t] + a_2 \sin[2\pi(5f)t]$$

Repeat the process for $y = 32\sin^6(2\pi ft)$ where $f = 30$ Hz. Use 512 points sampled over 1 second. From the real part of the DFT, estimate the coefficients b_0, b_1, b_2, b_3 in the relationship

$$32\sin^6(2\pi ft) = b_0 + b_1 \cos[2\pi(2f)t] + b_2 \cos[2\pi(4f)t] + b_3 \cos[2\pi(6f)t]$$

8.3. Determine the DFT of a set of 512 data points sampled over a 1 second period from

$$y = \sin(2\pi f_1 t) + 2\sin(2\pi f_2 t)$$

where $f_1 = 30$ Hz and $f_2 = 400$ Hz. Explain why there is a large component in the spectrum at 112 Hz.

8.4. Determine the DFT of a set of 256 data points sampled over 1 second from $y(t) = \sin(2\pi ft)$ for $f = 25$, 30.27, and 35.49 Hz. Plot the absolute value of the DFT against frequency for all three values of f in the same figure. It will be noted that even though the amplitude of the sine function from which the samples are taken is the same in each case, the frequency components corresponding to f have different amplitudes. This is because, in the case of the 30.27 Hz and 35.49 Hz wave the sampling is not over an integer number of periods of y. This phenomenon is known as "spectral leakage" and part of the pure sine wave seems to have leaked into adjacent frequencies. Its effect may be reduced by applying a "window" to the data. The

Hanning window is $w(t) = 0.5\{1 - \cos(2\pi t/T)\}$ where T is the sampling period. Multiply $y(t)$ by $w(t)$ and determine the DFT of the resulting data. Plot the absolute value of this DFT against frequency for all three values of f in the same figure. It will be seen that the amplitude variation of the frequency components corresponding to f and the smearing into other frequencies has been reduced significantly.

8.5. The following 32 data points are sampled over a period of 0.0625 second.

$$y = [0\ 0.9094\ 0.4251\ -0.6030\ -0.6567\ 0.2247\ 0.6840\ 0.1217 \ldots$$
$$-0.5462\ -0.3626\ 0.3120\ 0.4655\ -0.0575\ -0.4373\ -0.1537 \ldots$$
$$0.3137\ 0.2822\ -0.1446\ -0.3164\ -0.0204\ 0.2694\ 0.1439\ -0.1702 \ldots$$
$$-0.2065\ 0.0536\ 0.2071\ 0.0496\ -0.1594\ -0.1182\ 0.0853 \ldots$$
$$0.1441\ -0.0078]$$

(a) Determine the DFT and estimate the frequency of the most significant component present in the data. What is the frequency increment in the DFT?

(b) To the end of the existing data, add an additional 480 zero values, thus increasing the number of data points to 512. This process is called "zero padding" and is used to improve the frequency resolution in the DFT. Determine the DFT of the new data set and estimate the frequency of the most significant component. What is the frequency increment in the DFT?

8.6. The following function is sampled at intervals of 0.01 second over a period of 10.24 seconds.

$$y(t) = 3\sin(2\pi 4t) \text{ for } t < 3 \quad (8.74)$$
$$4\sin(2\pi 7t) \text{ for } 3 \leq t \leq 6 \quad (8.75)$$
$$6\sin(2\pi 5t) \text{ for } t > 6 \quad (8.76)$$

Use `fhilb` to determine the Hilbert transform for the data. Then, using (8.34), (8.35), and (8.36), write some appropriate MATLAB code to determine the instantaneous frequencies. Plot these frequencies against time.

8.7. Determine the 32 point Walsh transform, and the SAL, CAL, and power spectrum for the following data, generated from the MATLAB statements given. The data is sampled over a period of 8 seconds. For each case determine the predominant sequency of the data. Is the data based on an even (CAL) or odd (SAL) function?

a. Generate the data from

```
p = [1 1 -1 -1];
x = [ p p p p p p p p];
```

b. Generate the data from

```
p = [1 1 -1 -1];
x = [1 -1 -1 p p p p p p p 1];
```

8.8. Consider the function

$$y(t) = 0.2\cos(2\pi f1t) + 0.35\sin(2\pi f2t) + 0.3\sin(2\pi f3t)$$

where $f_1 = 20$ Hz, $f_2 = 50$ Hz, and $f_3 = 70$ Hz. 512 equispaced samples of the function are taken over 2 seconds. Determine the Walsh power spectrum of the data. The function is closely

related to the function of Example 8.2. How does the Walsh power spectrum compare with the Fourier power spectrum, shown in Fig. 8.6?
Repeat the above analysis for the function:

$$y(t) = 0.2\mathrm{sign}(\cos(2\pi f1t)) + 0.35\mathrm{sign}(\sin(2\pi f2t)) + 0.3\mathrm{sign}(\sin(2\pi f3t))$$

and again compare the Walsh and Fourier power spectra.

8.9. Determine the 64 point Haar transform for the following sets of data, using the function `fasthaar`. Then, from the Haar transform data, generate a wavelet map, using the function `waveletmap`. The data is sample over a period of 5.12 seconds.

a. Generate the data from

```
p = [1 1 1 1 -1 -1 -1 -1];
x = [p p p p p p p p];
x = x+0.5*ones(size(x))+0.01*randn(size(x));
```

b. Generate the data from

```
p = [1 1 1 1 -1 -1 -1 -1];
p1 = [1 1 -1 -1];
x = [p p p1 p1 p1 p1 p p p p];
```

c. Generate the data from

```
p = [1 1 1 1 1 -1 -1 -1 -1 -1];
x = [p p p p p p p p];
```

For each case, what components do you see in the data?

8.10. MATLAB has available test data sets including data collected for the sunspot activity. This may be obtained using the MATLAB statement

```
load sunspot.dat
```

The set of data `sunspot(:,1)` gives the year of the observed sunspot activity and `sunspot(:,2)`, the Wolfer number which indicates the level of sunspot activity for that year. Let `wolfer = sunspot(:,2)`. Generate a simple plot of the wolfer number against the year. To further analyze this data take the fast Fourier transform of the variable `wolfer`. Scaling this data will help in its interpretation. Do this by using the transformations `Power=abs(Y(1:N/2)).^2` and `freq=(1:N/2)/N` where `N` is the length of the vector `Y`. Plot `freq` against `power`.

CHAPTER

OPTIMIZATION METHODS 9

Abstract

The purpose of this chapter is to bring together a selection of algorithms for optimizing linear and non-linear functions which have applications in science and engineering. We describe methods to solve constrained linear optimization problems and both constrained and unconstrained non-linear optimization problems.

9.1 INTRODUCTION

The major types of optimization problems considered in this chapter are:

1. the solution of linear programming problems by both exterior and interior point methods;
2. the optimization of single-variable non-linear functions;
3. the solution of multi-variable non-linear optimization problems using conjugate gradient methods, simulated annealing, and evolutionary algorithms;
4. solution of systems of linear equations using conjugate gradient methods;
5. the solution of constrained non-linear optimization problems using the Sequential Unconstrained Minimization Technique (SUMT)

It is not our intention to describe fully the theoretical basis for these methods but to give some indication of the ideas that lie behind them. We begin with a discussion of linear programming problems.

9.2 LINEAR PROGRAMMING PROBLEMS

Linear programming (LP) is normally considered to be an operational research (OR) method and has a very wide range of applications. In this context the word *programming* has nothing to do with computer programming. An early application of LP, in World War 2, was to military problems and programming was a military term. The simplex method for solving linear programming problems was published by Dantzig in 1947.

A detailed description of the problem and associated theory is beyond the scope of this text but this information can be obtained from Dantzig (1963) and Sultan (1993). The problem may be expressed in standard form as

$$\begin{aligned} &\text{Minimize } f = \mathbf{c}^\mathsf{T}\mathbf{x} \\ &\text{subject to } \mathbf{A}\mathbf{x} = \mathbf{b} \\ &\text{and } \mathbf{x} \geq \mathbf{0} \end{aligned} \tag{9.1}$$

Numerical Methods. https://doi.org/10.1016/B978-0-12-812256-3.00018-X

where $\mathbf{x}$ is the column vector of n components that we wish to determine. The function to be minimized is called the *objective function* and the remaining equations are the *constraint equations*. Note that each element of $\mathbf{x}$ is constrained to be greater than or equal to zero. This is a common requirement in this type of optimization because most practical optimization problems will require non-negative values for $\mathbf{x}$. For example, if each element of $\mathbf{x}$ is the numbers of workers with a particular skill set employed by an organization, the number of workers in any group cannot be negative. The given constants of the system are provided by an m component column vector $\mathbf{b}$, an $m \times n$ matrix $\mathbf{A}$, and an n component column vector $\mathbf{c}$. Clearly all the constraints and the function we wish to minimize are linear in form.

The optimization problem is sometimes of the form

$$\begin{aligned} \text{Minimize } & f = \mathbf{c}^\mathsf{T}\mathbf{x} \\ \text{subject to } & \mathbf{A}\mathbf{x} \geq \mathbf{b} \\ \text{and } & \mathbf{x} \geq \mathbf{0} \end{aligned} \tag{9.2}$$

If the problem, called the *primal* problem, is posed in the form of (9.2), then it has a *dual* problem of the form

$$\begin{aligned} \text{Maximize } & g = \mathbf{b}^\mathsf{T}\mathbf{y} \\ \text{subject to } & \mathbf{A}^\mathsf{T}\mathbf{y} \leq \mathbf{c} \\ \text{and } & \mathbf{y} \geq \mathbf{0} \end{aligned} \tag{9.3}$$

The problems described by (9.2) and (9.3) are a pair in the sense that whichever is the primal problem, the other problem is its dual. It is also important to note that a minimization problem can be converted to a maximization problem by multiplying the objective function by -1 and vice versa. Also any constraint equation can be converted from a greater or equal relationship to a less than or equal relationship by multiplying the equation by -1 and vice versa. The problems defined by (9.2) can be converted to the standard form (9.1) by adding extra positive-valued variables, called *surplus variables*, to the problem and subtracting them from the equations of constraint. Similarly, problem (9.3) can be converted to the standard form by adding extra positive variables, called *slack variables*, to the problem and adding them to the equations of constraint.

The importance of this type of problem lies in the fact that it corresponds to the general aim of optimizing the use of scarce resources to meet a specific objective. Some practical examples where linear programming has been applied are:

1. the hospital diet problem, requiring food costs to be minimized while dietary constraints are satisfied;
2. the problem of minimizing cutting pattern loss;
3. the problem of optimizing profit subject to constraints on the availability of specified materials;
4. the problem of optimizing the routing of telephone call.

An important numerical algorithm for solving this problem is called the simplex method; see Dantzig (1963). This was applied to wartime problems of troop and material distribution. However, here we consider more recent developments which have provided new algorithms which are theoretically better. These are based on the work of Karmarkar (1984) who produced an algorithm which differed greatly in principle from the simplex method of Dantzig. While the theoretical complexity of Dantzig's method is exponential in the number of variables of the problem, some versions of Karmarkar's algorithm have a complexity which is of the order of the cube of the number of variables. It has been reported that for some problems this leads to substantial saving of computational effort. Here we describe an algorithm due to Barnes (1986) that provides an elegant modification of Karmarkar's algorithm but preserves its fundamental principles.

We do not describe the theoretical details of these complex algorithms but it is useful to compare, in broad terms, the nature of the Karmarkar and Dantzig algorithms. The application of the simplex method of Dantzig is best illustrated by considering a simple linear programming problem as follows. In a factory producing electronic components, let x_1 be the number of batches of resistors and x_2 the number of batches of capacitors produced. Each batch of resistors manufactured gains 7 units of profit and each batch of capacitors gains 13 units of profit. Each is manufactured in a two-stage process. Stage 1 is limited to 18 units of time per week and stage 2 is limited to 54 units of time per week. A batch of resistors requires 1 unit of time in stage 1 and 5 units of time in stage 2. A batch of capacitors requires 3 units of time in the first stage and 6 in the second. The aim of the manufacturer is to maximize profitability while meeting the time constraints; this leads to the following linear programming problem.

$$\text{Maximize } z = 7x_1 + 13x_2 \text{ (where } z \text{ is the profit)}$$

$$\text{subject to}$$

$$x_1 + 3x_2 \leq 18 \quad \text{(stage 1 process)}$$

$$5x_1 + 6x_2 \leq 54 \quad \text{(stage 2 process)}$$

$$\text{and } \ x_1, x_2 \geq 0$$

To see how the simplex algorithm works we give a geometric interpretation of this problem in Fig. 9.1. In this figure the region lying under the shaded lines and confined by the x_1- and x_2-axes represents the feasible region. This is the region in which all possible solutions to the problem lie. Clearly there is an infinity of such points. Fortunately, it can be shown that the only true candidates for the optimum solution are the points that lie at the vertices of the feasible region. In fact, we can find this optimum using simple geometric principles. The objective function is represented in Fig. 9.1 by the dashed line of constant slope and variable intercept proportional to the value of the objective function. If we move this line parallel to itself until it just leaves the feasible region, it leaves at the vertex which gives the maximum value of the objective function. Clearly, beyond this point the values of x_1 and x_2 no longer satisfy the constraints. For this problem the optimum solution is given by $x_1 = 6$, $x_2 = 4$ so that the profit $z = 94$.

Although this provides a solution for this simple two-variable problem, linear programming problems often involve thousands or hundreds of thousands of variables. For practical problems a well-specified numerical algorithm is required. This is provided by Dantzig's simplex algorithm. We do not describe this in detail here but the general principle of its operation is to generate a sequence of points which correspond mathematically to the vertices of the multidimensional feasible region. The algorithm proceeds from one vertex to another, each time improving the value of the objective function, until the optimum is found. These points are all on the surface of the feasible region and for larger problems there may be a huge number of them.

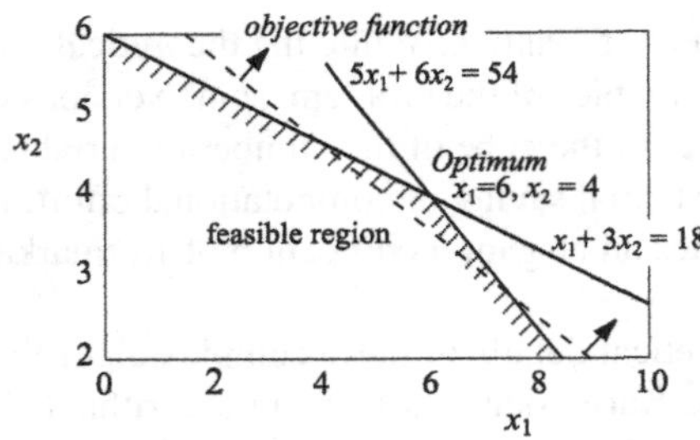

FIGURE 9.1

Graphical representation of an optimization problem. The *dashed line* represents the objective function and the *solid lines* represent the constraints.

The algorithm proposed by Karmarkar deals with the linear programming problem in a different way. The algorithm was developed at AT&T to solve very large linear programming problems concerned with routing telephone calls in the Pacific Basin. This algorithm transforms the problem to a more convenient form and then searches through the interior of the feasible region using a good direction of search toward its surface. Because this type of algorithm uses interior points, it is often described as an *interior point* method. Since its discovery, many improvements and modifications have been made to this algorithm and here we describe a form which, although conceptually complex, leads to a remarkably simple and elegant linear programming algorithm. This formulation was given by Barnes (1986).

The Barnes algorithm may be applied to any linear programming problem once it is converted to the form of (9.1). However, one important initial modification is required to ensure the algorithm starts at an interior point $\mathbf{x}^0 > \mathbf{0}$. This modification is achieved by introducing an additional column, that is a new last column, to the **A** matrix, the elements of which are the **b** vector minus the sum of the columns of the **A** matrix. We associate an additional variable with this additional column and, to avoid a superfluous variable in the solution, we introduce an extra element in the vector **c**. We make the value of this element very large to ensure that the new variable is driven to zero when the optimum is reached. Now we find that $\mathbf{x}^0 = [1\ 1\ 1 \ldots 1]^\mathsf{T}$ satisfies this set of constraints and clearly $\mathbf{x}^0 \geq \mathbf{0}$. We now describe the Barnes algorithm:

- Step 0: Assuming n variables in the original problem,

$$\text{Set } a\,(i, n+1) = b\,(i) - \sum_j a\,(i, j) \quad \text{and} \quad c\,(n+1) = 10{,}000.$$

$$\mathbf{x}^0 = [1\ 1\ 1 \ldots 1], \qquad k = 0.$$

- Step 1: Set $\mathbf{D}^k = \text{diag}(\mathbf{x}^k)$ and compute an improved point using the equation

$$\mathbf{x}^{k+1} = \mathbf{x}^k - \frac{s(\mathbf{D}^k)^2(\mathbf{c} - \mathbf{A}^\mathsf{T}\boldsymbol{\lambda}^k)}{\text{norm}(\mathbf{D}^k(\mathbf{c} - \mathbf{A}^\mathsf{T}\boldsymbol{\lambda}^k))}$$

where the vector λ^k is given by

$$\lambda^k = (\mathbf{A}(\mathbf{D}^k)^2\mathbf{A}^\top)^{-1}\mathbf{A}(\mathbf{D}^k)^2\mathbf{c}$$

The step s is chosen such that

$$s = \min\left\{\frac{\text{norm}\left((\mathbf{D}^k)\left(\mathbf{c}-\mathbf{A}^\top\lambda^k\right)\right)}{x_j^k\left(c_j-\mathbf{A}_j^\top\lambda^k\right)}\right\} - \alpha$$

where $\mathbf{A}_j$ is the jth column of the matrix $\mathbf{A}$ and α is a small preset constant value. Here the minimum is taken for the values

$$\left(c_j - \mathbf{A}_j^\top\lambda^k\right) > 0 \text{ only.}$$

Note also that λ^k provides an approximation for the solution of the dual problem (see for example, Problems 9.1 and 9.2).

- Step 2: Stop if the *primal* and *dual* values of the objective functions are approximately equal. Else set $k = k + 1$ and repeat from step 1.

Note that in step 2 we use an important result in linear programming. This is that if a solution exists, every primal problem (that is, the original problem) and its corresponding dual problem have equal optimal values of their objective functions. There are several other termination criteria that could be used and Barnes suggested a more complex but more reliable one.

The algorithm provides an iterative improvement starting from the initial point $\mathbf{x}^0$ by taking the maximum step which ensures that $\mathbf{x}^k > \mathbf{0}$ in the normalized direction given by $(\mathbf{D}^k)^2(\mathbf{c} - \mathbf{A}^\top\lambda^k)$. It is this direction which is the crucial element of the algorithm. This direction is a projection of the objective function coefficients into the constraint space. For a proof that this direction reduces the objective function, while ensuring the constraints are satisfied, the reader is referred to Barnes (1986).

The reader should be warned that this algorithm is deceptively simple. In fact the computation of the direction is very difficult for large problems. This is because the algorithm requires the solution of an extremely ill-conditioned equation system. Many alternatives have been suggested for finding the direction of search, including the use of a conjugate gradient method that is discussed in Section 9.6. The MATLAB function `barnes` provided here solves the ill-conditioned equation system in a direct manner using the MATLAB \ operator. The function `barnes` is easily modified to use the conjugate gradient solver given in Section 9.6.

```
function [xsol,basic,objective] = barnes(A,b,c,tol)
% Barnes' method for solving a linear programming problem.
% to minimize c'x subject to Ax = b.Assumes problem is non-degenerate.
% Example call: [xsol,basic]=barnes(A,b,c,tol)
% A is the matrix of coefficients of the constraints.
% b is the right side column vector and c is the row vector of
% cost coefficients. xsol is the solution vector, basic is the
% list of basic variables.
x2 = [ ];  x = [ ];
```

```
[m n] = size(A);
% Set up initial problem
aplus1 = b-sum(A(1:m,:)')';
cplus1 = 1000000;
A = [A aplus1]; c = [c cplus1]; B = [ ];
n = n+1;
x0 = ones(1,n)'; x = x0;
alpha = .0001; lambda = zeros(1,m)';
iter = 0;
% Main step
while abs(c*x-lambda'*b)>tol
    x2 = x.*x;
    D = diag(x); D2 = diag(x2); AD2 = A*D2;
    lambda = (AD2*A')\(AD2*c');
    dualres = c'-A'*lambda;
    normres = norm(D*dualres);
    for i = 1:n
        if dualres(i)>0
            ratio(i) = normres/(x(i)*(c(i)-A(:,i)'*lambda));
        else
            ratio(i)=inf;
        end
    end
    R = min(ratio)-alpha;
    x1 = x-R*D2*dualres/normres;
    x = x1;
    basiscount = 0;
    B = [ ]; basic = [ ];
    cb = [ ];
    for k = 1:n
        if x(k)>tol
            basiscount = basiscount+1;
            basic = [basic k];
        end
    end
    % Only used if problem non-degenerate
    if basiscount==m
        for k = basic
            B = [B A(:,k)];   cb = [cb c(k)];
        end
        primalsol = b'/B';
        xsol = primalsol;
        break
    end
```

```
    iter = iter+1;
end
objective = c*x;
```

We now solve the linear programming problem

$$\text{Maximize } z = 2x_1 + x_2 + 4x_3$$
$$\text{subject to}$$
$$x_1 + x_2 + x_3 \leq 7$$
$$x_1 + 2x_2 + 3x_3 \leq 12$$
$$x_1, x_2, x_3 \geq 0$$

The requirement that $x_1, x_2, x_3 \geq 0$ are called the non-negativity constraints. This linear programming problem can be easily transformed to the standard form by adding new positive-valued variables, called *slack variables*, to the left sides of the inequalities and changing the signs of the coefficients in the objective function so that it is converted to a minimization problem subject to equality constraints thus:

$$\text{Minimize } p = -z = -(2x_1 + x_2 + 4x_3)$$
$$\text{subject to}$$
$$x_1 + x_2 + x_3 + x_4 = 7$$
$$x_1 + 2x_2 + 3x_3 + x_5 = 12$$
$$x_1, x_2, x_3, x_4, x_5 \geq 0$$

The variables x_4 and x_5 are the slack variables and they represent the difference between the available resources and the resources used. Note that if the constraints were of the form greater than or equal to zero, then we would subtract the *surplus variables* to produce equality. Thus, we have

$$\mathbf{c} = \begin{bmatrix} -2 & -1 & -4 & 0 & 0 \end{bmatrix}$$

Note that **c** now has 5 elements. We use the script `e4s901.m` to solve this problem.

```
% e4s901.m
c = [-2 -1 -4 0 0];
A = [1 1 1 1 0;1 2 3 0 1 ]; b = [7 12]';
[xsol,ind,object] = barnes(A,b,c,0.00005);
fprintf('objective = %8.4f', object)
i = 1;
fprintf('\nSolution is:');
for j = ind
    fprintf('\nx(%1.0f) =%8.4f',j,xsol(i))
    i = i+1;
end;
fprintf('\nAll other variables are zero\n')
```

Running script `e4s901.m` provides the result

```
objective = -19.0000
Solution is:
x(1) =   4.5000
x(3) =   2.5000
All other variables are zero
```

Since the original problem was to maximize the objective function, its value is 19. This solution illustrates an important theorem of linear programming. The number of non-zero primal variables is at most equal to the number of independent constraints (excluding non-negativity constraints). In this problem there are only two main constraints. Thus, there are only two non-zero variables, x_1 and x_3. The slack variables x_4 and x_5 are zero and so is x_2.

The MATLAB function `lsqnonneg`, discussed in Section 2.12, provides a method for finding a solution to an equation system in which all components of the solution are non-negative. This corresponds to a basic feasible solution for the system but it is generally non-optimal for a specific objective function.

Having examined the process for solving linear optimization problems we now consider methods which are used to solve non-linear optimization problems.

9.3 OPTIMIZING SINGLE-VARIABLE FUNCTIONS

We sometimes need to determine the maximum or minimum value of a one-variable non-linear function. Throughout this discussion we assume we are seeking the minimum value of the function. If we require the maximum value, then we merely have to change the sign of the original function.

The most obvious way of determining the minimum of a function is to differentiate it and find the value of the independent variable that makes this derivative zero. However, there are situations in which it is not practical to find the derivative directly; see for example (9.6). A method is now described which provides an approximation to the minimum to any required accuracy.

Consider a function $y = f(x)$ and let us assume that in the range $[x_a\ x_b]$ there is a single minimum, as shown in Fig. 9.2. Two additional points, x_1 and x_2, are chosen arbitrarily so that the range is divided into three intervals. Assuming that $x_a < x_1 < x_2 < x_b$, then:

If $f(x_1) < f(x_2)$ then the minimum value must lie in the range $[x_a\ x_2]$.
If $f(x_1) > f(x_2)$ then the minimum value must lie in the range $[x_1\ x_b]$.

Either of these ranges must provide a smaller interval than $[x_a\ x_b]$ in which the minimum lies. This interval reduction can be repeated continuously in successively smaller ranges until an acceptably small interval is found for the minimum.

It might be assumed that the most efficient procedure is to select x_1 and x_2 so that the range $[x_a\ x_b]$ is subdivided into three equal intervals. In fact, this is not so and for a more efficient procedure we take

$$x_1 = x_a + r(1 - g), \quad x_2 = x_a + rg$$

where $r = x_b - x_a$ and

$$g = \frac{1}{2}\left(-1 + \sqrt{5}\right) \approx 0.61803$$

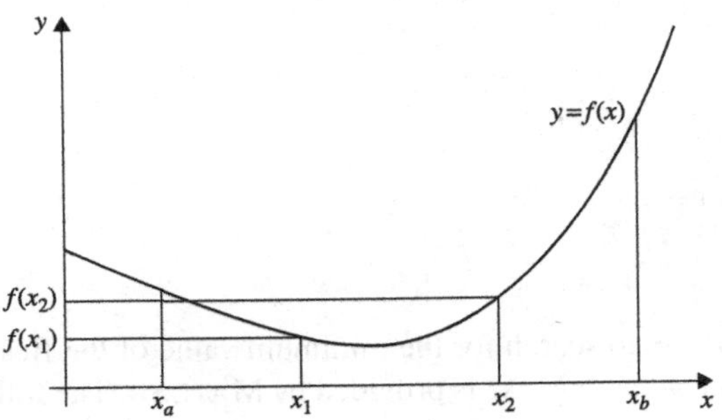

FIGURE 9.2

Graph of a function with a minimum in the range $[x_a\ x_b]$.

The quantity g is called the golden ratio. This quantity has many interesting properties. For example, it is one of the roots of the equation

$$x^2 + x + 1 = 0$$

This *golden ratio* is also related to the famous Fibonacci series. This series is 1, 1, 2, 3, 5, 8, 13, ... and it is generated from:

$$N_{k+1} = N_k + N_{k-1}, \quad k = 2, 3, 4, \ldots$$

where $N_2 = N_1 = 1$ and N_k is the kth term in the series. As k tends to infinity the ratio N_k/N_{k+1} tends to the golden ratio, g.

This search algorithm using the golden ratio is implemented in MATLAB thus:

```
function [f,a,iter] = golden(func,p,tol)
% Golden search for finding min of one variable non-linear function.
% Example call: [f,a] = golden(func,p,tol)
% func is the name of the user defined non-linear function.
% p is a 2 element vector giving the search range.
% tol is the tolerance. a is the optimum values of the function.
% f is the minimum of the function. Iter is the number of iterations
if p(1)<p(2)
    a = p(1);  b = p(2);
else
    a = p(2);  b = p(1);
end
g = (-1+sqrt(5))/2;
r = b-a;  iter = 0;
while r>tol
    x = [a+(1-g)*r a+g*r];
    y = feval(func,x);
    if y(1)<y(2)
```

```
        b = x(2);
    else
        a = x(1);
    end
    r = b-a; iter = iter+1;
end
f = feval(func,a);
```

We can use the function `golden` to search for the minimum value of the Bessel function of the second kind of order 2. The function `bessely(2,x)` is provided by MATLAB. The following command provides the output shown:

```
>> format long
>> [f,x,iter] = golden(@(x) bessely(2,x),[4 10],0.000001)

f =
  -0.279275263440711

x =
   8.350724427010965

iter =
    33
```

Note that if we had divided the search interval into three equal sections, rather than using the golden ratio, then 39 iterations would have been required.

The search algorithm has been developed assuming that there is only one minimum value of the function in the search range. If there are several minima in the search range, then the procedure locates one, but the one located is not necessarily the global minimum in the range. For example, a Bessel function of the second kind of order 2 has 3 minima in the range 4 to 25, as shown in Fig. 9.3. If we use the function `golden` and search in the ranges 4 to 24, 4 to 25, and 4 to 26, we obtain the results given in Table 9.1. In this table we see that a different minimum has been found for each of the search ranges, even though all three minima lie within each of the search ranges used. Ideally we require the search to determine the global minimum of the function, something that the function `golden` has failed to accomplish in two out of three tests. Obtaining a *global* solution is a major problem in minimization.

In this particular example we can verify the accuracy of these solutions by calculus. The derivative of the Bessel function of the second kind of order n is given by

$$\frac{d}{dx}\{Y_n(x)\} = \frac{1}{2}\{Y_{n-1}(x) - Y_{n+1}(x)\}$$

where $Y_n(x)$ is the Bessel function of the second kind of order n. (Sometimes the notation $N_n(x)$ is used instead of $Y_n(x)$.)

The minimum (or maximum) value of a function occurs when the derivative of the function is zero. Hence, with $n=2$ we have a minimum (or maximum) occurring when

$$Y_1(x) - Y_3(x) = 0$$

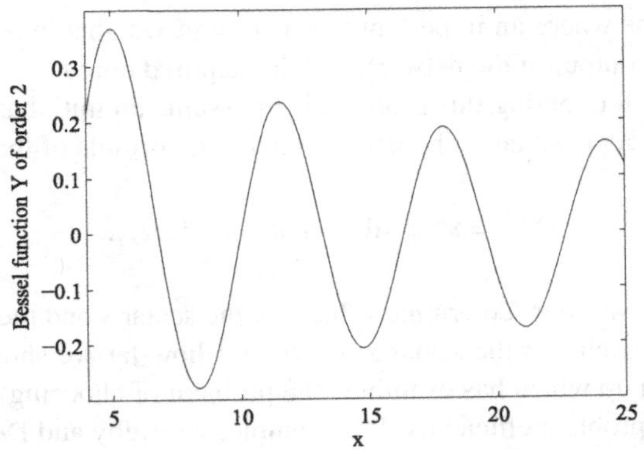

FIGURE 9.3

A plot of the Bessel function of the second kind showing three minima.

Table 9.1 Effect of different search ranges

Search range	Value of $f(x)$ at min	Value of x at min
4 to 24	−0.20844576503764	14.76085144779431
4 to 25	−0.17404548213116	21.09284729991696
4 to 26	−0.27927526323841	8.35068549680869

We cannot escape from the need to use a numerical method because the only way to find the roots of this equation is to use a numerical procedure such as that implemented by the MATLAB function `fzero`. Thus, using `fzero` to determine a root near 8 we have

```
>> format long
>> fzero(@(x) bessely(1,x)-bessely(3,x),8)

ans =
    8.350724701413078
```

We can also use `fzero` to find roots at 14.76090930620768 and 21.09289450441274. These results are in good agreement with the minima found by the function `golden`.

9.4 THE CONJUGATE GRADIENT METHOD

Here we confine ourselves to solving the problem

$$\text{Minimize} \quad f(\mathbf{x}) \text{ for all } \mathbf{x} \in \mathrm{R}^n$$

where $f(\mathbf{x})$ is a non-linear function of $\mathbf{x}$ and $\mathbf{x}$ is an n component column vector. This is called a non-linear unconstrained optimization problem. These problems arise in many applications – for example, in neural network problems where an important aim is to find weights in a network which minimize the difference between the output of the network and the required output.

The standard approach for solving this problem is to assume an initial approximation $\mathbf{x}^0$ and then to proceed to an improved approximation by using an iterative formula of the form

$$\mathbf{x}^{k+1} = \mathbf{x}^k + s\mathbf{d}^k \text{ for } k = 0, 1, 2, \ldots \tag{9.4}$$

Clearly, to use this formula we must determine values for the scalar s and the vector $\mathbf{d}^k$. The vector $\mathbf{d}^k$ represents a direction of search and the scalar s determines how far we should step in this direction. A vast literature has grown up which has examined the problem of choosing the best direction and the best step size to solve this problem efficiently. For example, see Adby and Dempster (1974). A simple choice for a direction of search is to take $\mathbf{d}^k$ as the negative gradient vector at the point $\mathbf{x}^k$. For a sufficiently small step value this can be shown to guarantee a reduction in the function value. This leads to an algorithm of the form

$$\mathbf{x}^{k+1} = \mathbf{x}^k - s\nabla f\left(\mathbf{x}^k\right) \text{ for } k = 0, 1, 2, \ldots \tag{9.5}$$

where $\nabla f(\mathbf{x}) = (\partial f/\partial x_1, \partial f/\partial x_2, \ldots, \partial f/\partial x_n)$ and s is a small constant value. This is called the steepest descent algorithm. The minimum is reached when the gradient is zero, as in the ordinary calculus approach. We also assume that there exists only one local minimum which we wish to find in the range considered. The problem with this method is that although it reduces the function value, the step may be very small and therefore the algorithm is very slow. An alternative approach is to choose the step which gives the maximum reduction in the function value in the current direction. This may be described formally as

$$\text{For each } k \text{ find the value of } s \text{ that minimizes } f(\mathbf{x}^k - s\nabla f(\mathbf{x}^k)) \tag{9.6}$$

This procedure is known as a line-search. The reader will note that this is also a minimization problem. However, since $\mathbf{x}^k$ is known, it is a *one-variable* minimization problem in the step size s. Although it is a difficult problem, numerical procedures are available to solve it, one of which is the search method given in Section 9.3. Eqs. (9.5) and (9.6) provide a workable algorithm but it is still slow. One reason for this poor performance lies in our choice of direction $-\nabla f(\mathbf{x}^k)$.

Consider the function we wish to minimize in (9.6). Clearly the value of s which minimizes $f(\mathbf{x}^k - s\nabla f(\mathbf{x}^k))$ is such that the derivative of $f(\mathbf{x}^k - s\nabla f(\mathbf{x}^k))$ with respect to s is zero. Now, differentiating $f(\mathbf{x}^k - s\nabla f(\mathbf{x}^k))$ with respect to s gives

$$\frac{df\left(\mathbf{x}^k - s\nabla f\left(\mathbf{x}^k\right)\right)}{ds} = -\left(\nabla f\left(\mathbf{x}^{k+1}\right)\right)^{\mathsf{T}} \nabla f\left(\mathbf{x}^k\right) = 0 \tag{9.7}$$

This shows that the successive directions of search are orthogonal. This is not the best way of getting from our original approximation to the optimum value since the changes in direction are so large.

The conjugate gradient method provides a significant improvement in performance by taking a combination of the previous direction and the new direction to approach the optimum more directly. It uses the same step size choice procedure given by (9.6), so we must now consider how the direction vector is chosen in the conjugate gradient method. Let $\mathbf{g}^{k+1} = \nabla f(\mathbf{x}^{k+1})$. Then the basic formula for the conjugate gradient direction is

$$\mathbf{d}^{k+1} = -\mathbf{g}^{k+1} + \beta \mathbf{d}^k \tag{9.8}$$

Thus, the current direction of search is a combination of the current negative gradient plus a scalar β times the previous direction of search. The crucial question is: how is the value of β to be determined? The criterion used is that successive directions of search should be conjugate. This means that $(\mathbf{d}^{k+1})^\mathsf{T}\mathbf{A}\mathbf{d}^k = 0$ for some specified matrix $\mathbf{A}$.

This apparently obscure choice of requirement can be shown to lead to desirable convergence properties for the conjugate gradient method. In particular, it has the property that the optimum of a positive definite quadratic function of n variables can be found in n or less steps. In the case of a quadratic, $\mathbf{A}$ is the matrix of coefficients of the squared and cross product terms. It can be shown that the requirement of conjugacy leads to a value for β given by

$$\beta = \frac{\left(\mathbf{g}^{k+1}\right)^\mathsf{T} \mathbf{g}^{k+1}}{\left(\mathbf{g}^{k}\right)^\mathsf{T} \mathbf{g}^{k}} \tag{9.9}$$

Now (9.4), (9.6), (9.8), and (9.9) lead to the conjugate gradient algorithm given by Fletcher and Reeves (1964) which has the form:

- Step 0: Input value for $\mathbf{x}^0$ and accuracy ε. Set $k = 0$ and compute $\mathbf{d}^k = -\nabla f(\mathbf{x}^k)$.
- Step 1: Determine s_k which is the value of s that minimizes $f(\mathbf{x}^k + s\mathbf{d}^k)$.
  Calculate $\mathbf{x}^{k+1}$ where $\mathbf{x}^{k+1} = \mathbf{x}^k + s_k\mathbf{d}^k$ and compute $\mathbf{g}^{k+1} = \nabla f(\mathbf{x}^{k+1})$.
  If norm$(\mathbf{g}^{k+1}) < \varepsilon$, then terminate with solution $\mathbf{x}^{k+1}$, else go to step 2.
- Step 2: Calculate new conjugate direction $\mathbf{d}^{k+1}$ where

$$\mathbf{d}^{k+1} = -\mathbf{g}^{k+1} + \beta\mathbf{d}^k \text{ and } \beta = (\mathbf{g}^{k+1})^\mathsf{T}\mathbf{g}^{k+1}/\{(\mathbf{g}^k)^\mathsf{T}\mathbf{g}^k\}$$

- Step 3: $k = k + 1$; go to step 1.

Note that in other forms of this algorithm the steps 1, 2, and 3 are repeated n times and then restarted with a steepest descent step from step 0. The following is a MATLAB function for this method.

```
function [x1,df,noiter] = mincg(f,derf,ftau,x,tol)
% Finds local min of a multivariable non-linear function in n variables
% using conjugate gradient method.
% Example call: res = mincg(f,derf,ftau,x,tol)
% f is a user defined multi-variable function,
% derf a user defined function of n first order partial derivatives.
% ftau is the line search function.
% x is a col vector of n starting values, tol gives required accuracy.
% x1 is solution, df is the gradient
```

```
% noiter is the number of iterations required.
% WARNING. Not guarenteed to work with all functions. For difficult
% problems the linear search accuracy may have to be adjusted.
global p1 d1
n = size(x);  noiter = 0;
% Calculate initial gradient
df = feval(derf,x);
% main loop
while norm(df)>tol
    noiter = noiter+1;
    df = feval(derf,x);
    d1 = -df;
    %Inner loop
    for inner = 1:n
        p1 = x;  tau = fminbnd(ftau,-10,10);
        % calculate new x
        x1 = x+tau*d1;
        % Save previous gradient
        dfp = df;
        % Calculate new gradient
        df = feval(derf,x1);
        % Update x and d
        d = d1; x = x1;
        % Conjugate gradient method
        beta = (df'*df)/(dfp'*dfp);
        d1 = -df+beta*d;
    end
end
```

Notice that the MATLAB function `fminbnd` is used in the function `mincg` to perform the single-variable minimization to find the best step value. It is important to note that the function `mincg` requires three input functions which must be supplied by the user. They are the function to be minimized, the partial derivatives of this function, and the line-search function. As implemented the function `mincg` requires the input functions to be user-defined functions, not anonymous functions. An example of the use of `mincg` follows.

The function to be minimized, which is taken from Styblinski and Tang (1990), is

$$f(x_1, x_2) = \left(x_1^4 - 16x_1^2 + 5x_1\right)/2 + \left(x_2^4 - 16x_2^2 + 5x_2\right)/2 \tag{9.10}$$

The function `f01` and the derivative of this function, `f01d` are defined as follows:

```
function f = f01(x)
f = 0.5*(x(1)^4-16*x(1)^2+5*x(1)) + 0.5*(x(2)^4-16*x(2)^2+5*x(2));

function f = f01d(x)
f = [0.5*(4*x(1)^3-32*x(1)+5); 0.5*(4*x(2)^3-32*x(2)+5)];
```

The MATLAB line-search function ftau2cg is defined as

```
function ftauv = ftau2cg(tau);
global p1 d1
q1 = p1+tau*d1;
ftauv = feval('f01',q1);
```

To test the mincg function we use the following simple MATLAB command:

```
>> [solution,gradient,iterations] = mincg('f01','f01d','ftau2cg',[1 -1]',.000005)
```

The results of executing these statements are

```
solution =
    -2.9035
    -2.9035

gradient =
  1.0e-006 *
     0.0156
    -0.2357

iterations =
      3
```

Note that

```
>> f = f01(solution)

f =
   -78.3323
```

This is also the minimum value of the function determined by using mincg. We provide both a three-dimensional and a contour plot of the function in Figs. 9.4 and 9.5. The latter includes a plot of the iterates and shows the path taken to reach the optimum solution from a particular starting point. The script used to obtain these graphs is

```
% e4s902.m
clf
[x,y] = meshgrid(-4.0:0.2:4.0,-4.0:0.2:4.0);
z = 0.5*(x.^4-16*x.^2+5*x)+0.5*(y.^4-16*y.^2+5*y);
figure(1)
surfl(x,y,z)
axis([-4 4 -4 4 -80 20])
xlabel('x1'), ylabel('x2'), zlabel('z')
x1=[1 2.8121 -2.8167 -2.9047 -2.9035];
y1=[0.5 -2.0304 -2.0295 -2.9080 -2.9035];
```

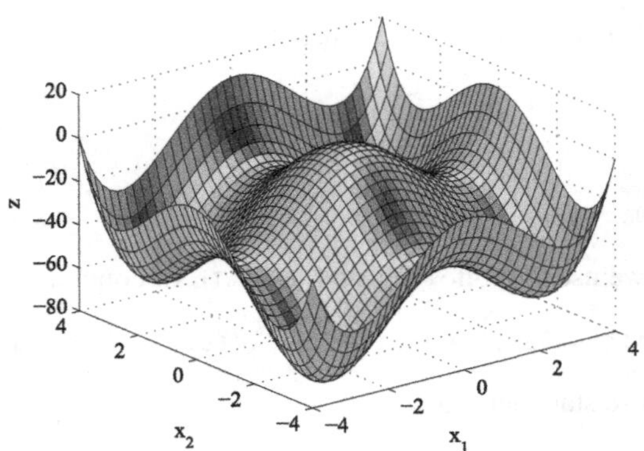

FIGURE 9.4

Three-dimensional plot of the Styblinski and Tang function.

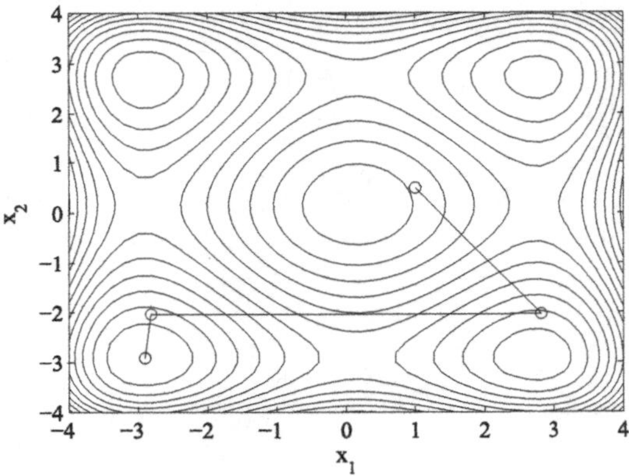

FIGURE 9.5

Contour plot of the Styblinski and Tang function, showing the location of four local minima. The conjugate gradient algorithm has found the one in the lower left corner. The search path taken by the algorithm is also shown.

```
figure(2)
contour(-4.0:0.2:4.0,-4.0:0.2:4.0,z,15);
xlabel('x1'), ylabel('x2')
hold on
plot(x1,y1,x1,y1,'o')
xlabel('x1'), ylabel('x2')
hold off
```

In script `e4s902.m` the vectors `x1` and `y1` contain the iterates for the conjugate gradient solution of the given function. These values were obtained by running a modified version of the `mincg` function separately. The minimum we have obtained is in fact the smallest of the four local minima that exist for this function, called the global minimum. However, this result was fortuitous; all that the conjugate gradient method is able to do is to find one of the four local minima and even this is not guaranteed for all problems. The conjugate gradient method, because of its small storage requirements, is one of the key algorithms used in neural network problems as part of the back propagation algorithm, but it has many other applications.

It should be noted that a MATLAB optimization toolbox is available and this provides a range of optimization procedures.

9.5 MOLLER'S SCALED CONJUGATE GRADIENT METHOD

In 1993 Moller, when working on optimization methods for neural networks, introduced a much improved version of the Fletcher's conjugate gradient method. Fletcher's conjugate gradient method uses a line search procedure to solve a single-variable minimization problem which is then used to find the optimum step to take in the chosen direction of search. The procedure used by Fletcher is a fragile, iterative, and computationally intensive process. In addition, the line search depends on a number of parameters which must be estimated by the user. Moller's paper (Moller, 1993) introduced a method which allowed the line search procedure to be replaced by a considerably simplified method for estimating an acceptable step size. However, using a simple estimation of the step size often fails and leads to non-stationary points. Moller noted that a simple approach to the problem fails because it only works for functions with positive definite matrices. Consequently, Moller suggested a method based on a combination of the Levenberg–Marquardt algorithm and the conjugate gradient algorithm. An outline of the algorithm is described in the following; for the details the reader is referred to Moller (1993).

Consider the n variable non-linear function $f(\mathbf{x})$. Moller introduces a scalar parameter λ_k which is adjusted at each iteration k after considering the sign of δ_k where

$$\delta_k = \mathbf{p}_k^\top \mathbf{H}_k \mathbf{p}_k$$

where $\mathbf{p}_k$ for $k = 1, 2, \ldots, n$ are a set of conjugate directions and $\mathbf{H}_k$ is the Hessian matrix of the function $f(\mathbf{x})$. If $\delta_k \geq 0$ then $\mathbf{H}_k$ is positive definite. However, since only first-order derivative information is known at each step of the conjugate gradient method, Moller suggests that the Hessian multiplied by $\mathbf{p}_k$ is approximated by:

$$\mathbf{s}_k = \frac{f'(\mathbf{x}_k + \sigma_k \mathbf{p}_k) - f'(\mathbf{x}_k)}{\sigma_k} \quad \text{for} \quad 0 < \sigma_k < 1$$

In practice the value of σ_k should be kept as small as possible for a good approximation. This expression in the limit tends to the true Hessian matrix multiplied by $\mathbf{p}_k$. The scalar λ_k is now introduced to regulate the approximation to the Hessian so ensuring it is positive definite, specifically by using the equation:

$$\mathbf{s}_k = \frac{f'(\mathbf{x}_k + \sigma_k \mathbf{p}_k) - f'(\mathbf{x}_k)}{\sigma_k} + \lambda_k \mathbf{p}_k \quad \text{for} \quad 0 < \sigma_k < 1$$

Thus, the value of λ_k is adjusted, and then we check the value of δ_k defined earlier using the approximation to the Hessian; if this is negative then the Hessian is no longer positive definite and the value of

λ_k is increased and s_k is checked again. This is repeated until the current estimate of the Hessian is positive definite. The key question is how should λ_k be adjusted to ensure the Hessian estimate becomes positive definite. Let the λ_k be increased to $\bar{\lambda}_k$; then

$$\bar{\mathbf{s}}_k = \mathbf{s}_k + (\bar{\lambda}_k - \lambda_k)\mathbf{p}_k$$

Now at any iteration k a new δ_k which we can denote as $\bar{\delta}_k$ can be computed from

$$\bar{\delta}_k = \mathbf{p}_k^{\mathsf{T}}\bar{\mathbf{s}}_k = \mathbf{p}_k^{\mathsf{T}}(\mathbf{s}_k + (\bar{\lambda}_k - \lambda_k)\mathbf{p}_k) = \mathbf{p}_k^{\mathsf{T}}\mathbf{s}_k + (\bar{\lambda}_k - \lambda_k)\mathbf{p}_k^{\mathsf{T}}\mathbf{p}_k$$

But $\mathbf{p}_k^{\mathsf{T}}\mathbf{s}_k$ is the original value of δ_k before λ_k was increased. So we have:

$$\bar{\delta}^k = \delta^k + (\bar{\lambda}_k - \lambda_k)\mathbf{p}_k^{\mathsf{T}}\mathbf{p}_k$$

Clearly we now require that new value of $\bar{\delta}^k$ be positive; hence, we require that

$$\delta_k + (\bar{\lambda}_k - \lambda_k)\mathbf{p}_k^{\mathsf{T}}\mathbf{p}_k > 0$$

This will be true if

$$\bar{\lambda}_k > \lambda_k - \frac{\delta_k}{\mathbf{p}_k^{\mathsf{T}}\mathbf{p}_k}$$

Moller suggests a reasonable choice of $\bar{\lambda}_k$ is:

$$\bar{\lambda}_k = 2\left(\lambda_k - \frac{\delta^k}{\mathbf{p}_k^{\mathsf{T}}\mathbf{p}_k}\right)$$

It is easily verified by back-substitution of this value for $\bar{\lambda}_k$ in our expression for $\bar{\delta}_k$ that

$$\bar{\delta}_k = -\delta_k + \lambda_k\mathbf{p}_k^{\mathsf{T}}\mathbf{p}_k$$

which, since δ_k was negative, λ_k is positive, and $\mathbf{p}_k^{\mathsf{T}}\mathbf{p}_k$ is a sum of squares, is clearly positive as required. The step size estimate is based on a quadratic approximation to the function being optimized at the current step and is calculated from:

$$\alpha_k = \frac{\mu_k}{\delta_k} = \frac{\mu_k}{\mathbf{p}_k^{\mathsf{T}}\mathbf{s}_k + \lambda_k\mathbf{p}_k^{\mathsf{T}}\mathbf{p}_k}$$

Here μ_k is the current negative gradient times the current direction of search $\mathbf{p}_k$. This gives the basis of the algorithm. However, an important issue is still to be decided, is how can the value of λ_k be safely and systematically varied. Moller provides a method based on a measure of how well the current quadratic approximation defined as f_q approximates the original function at the point considered. He does this by using the following definition:

$$\Delta_k = \frac{f(\mathbf{x}_k) - f(\mathbf{x}_k + \alpha_k\mathbf{p}_k)}{f(\mathbf{x}_k) - f_q(\alpha_k\mathbf{p}_k)}$$

By virtue of the fact that $f_q(\alpha_k, \mathbf{p}_k)$ is a quadratic approximation at the current iteration, this can be shown to be equivalent to:

$$\Delta_k = \frac{\delta_k^2(f(\mathbf{x}_k) - f(\mathbf{x}_k + \alpha_k \mathbf{p}_k))}{\mu_k^2}$$

Now if Δ_k is close to 1 then the quadratic approximation $f_q(\alpha_k, \mathbf{p}_k)$ must be close to $f(x_k + \alpha_k \mathbf{p}_k)$ and hence, a good local approximation to the function. This leads to the following steps for the adjustment of λ_k. Use the definition Δ_k described earlier as the quadratic approximation measure; more details can be found in Moller (1993). Then adjust λ_k as follows:

$$\text{If } \Delta_k > 0.75 \text{ then } \lambda_k = \lambda_k/4$$
$$\text{If } \Delta_k < 0.25 \text{ then } \lambda_k = \lambda_k + \frac{\delta_k(1 - \Delta_k)}{\mathbf{p}_k^\mathsf{T}\mathbf{p}_k}$$

These steps, together with any of the methods for generating conjugate gradient directions of search, provide an algorithm with a simple line search process. The outline of the major steps in Moller's algorithm are now given:

1. Choose initial approximation $\mathbf{x}_0$ and initial values for $\sigma_i < 10^{-4}$, $\lambda_i < 10^{-4}$, and $\bar{\lambda}_i = 0$. These values were suggested by Moller. Calculate initial negative gradient and assign it to $\mathbf{r}_1$, and assign $\mathbf{r}_1$ to the initial direction of search $\mathbf{p}_1$. Set $k = 1$.
2. Calculate second-order information. Specifically, values for: σ_k, $\bar{\mathbf{s}}_k$, and δ_k.
3. Scale δ_k using

$$\bar{\delta}_k = \delta_k + (\bar{\lambda}_k - \lambda_k)\mathbf{p}_k^\mathsf{T}\mathbf{p}_k$$

4. If $\delta_k < 0$ then make Hessian approximation positive definite using:

$$\bar{\delta}_k = -\delta_k + \lambda_k \mathbf{p}_k^\mathsf{T}\mathbf{p}_k$$

Set

$$\bar{\lambda}_k = 2\left(\lambda_k - \frac{\delta^k}{\mathbf{p}_k^\mathsf{T}\mathbf{p}_k}\right)$$

and

$$\bar{\lambda_k} = \lambda_k$$

5. Calculate step size from:

$$\alpha_k = \frac{\mu_k}{\delta_k}$$

6. Calculate factor to test goodness of quadratic fit Δ_k from

$$\Delta_k = \frac{\delta_k^2(f(\mathbf{x}_k) - f(\mathbf{x}_k + \alpha_k \mathbf{p}_k))}{\mu_k^2}$$

7. If $\Delta_k \geq 0$ then function can be reduced toward the minimum, so use

$$\mathbf{x}_{k+1} = \mathbf{x}_k + \alpha_k \mathbf{p}_k$$

Calculate new gradient

$$\mathbf{r}_{k+1} = -\nabla f(\mathbf{x}_{k+1})$$

Set $\bar{\lambda}_k = 0$. If $k \bmod N = 0$ then restart algorithm with

$$\mathbf{p}_{k+1} = \mathbf{r}_{k+1}$$

else calculate a new conjugate gradient direction.
Use some method for calculating the set of conjugate directions, see Fletcher and Reeves (1964) for example. A number of other methods are available.

$$\text{If } \Delta_k \geq 0.75 \text{ then } \lambda_k = 0.25\lambda_k$$

else

$$\bar{\lambda}_k = \lambda_k$$

8. If $\Delta_k < 0.25$ then increase scale parameter:

$$\lambda_k = \lambda_k + \delta_k(1 - \Delta_k)/\left(\mathbf{p}_k^\top \mathbf{p}_k\right)$$

9. If the gradient r_k is still not sufficiently close to zero then set $k = k + 1$ and go to Step 2 otherwise terminate and return optimum solution.

The following MATLAB function implements this method.

```
function [res, noiter] = minscg(f,derf,x,tol)
% Conjugate gradient optimization by Moller
% Finds local min of a multivariable non-linear function in n variables
% Example call: [res, noiter] = minscg(f,derf,x,tol)
% f is a user defined multi-variable function,
% derf a user defined function of n first order partial derivatives.
% x is a col vector of n starting values, tol gives required accuracy.
% res is solution, noiter is the number of iterations required.
lambda = 1e-8; lambdabar = 0; sigmac = 1e-5; sucess = 1;
deltastep = 0; [n m] = size(x);
% Calculate initial gradient
noiter = 0;
pv = -feval(derf,x); rv = pv;
while norm(rv)>tol
    noiter = noiter+1;
    if deltastep==0
        df = feval(derf,x);
```

```
    else
        df = -rv;
    end
    deltastep = 0;
    if sucess==1
        sigma = sigmac/norm(pv);
        dfplus = feval(derf,x+sigma*pv);
        stilda = (dfplus-df)/sigma;
        delta = pv'*stilda;
    end
    % Scale
    delta = delta+(lambda-lambdabar)*norm(pv)^2;
    if delta<=0
        lambdabar = 2*(lambda-delta/norm(pv)^2);
        delta = -delta+lambda*norm(pv)^2;
        lambda = lambdabar;
    end
    % Step size
    mu = pv'*rv; alpha = mu/delta;
    fv = feval(f,x);
    fvplus = feval(f,x+alpha*pv);
    delta1 = 2*delta*(fv-fvplus)/mu^2;
    rvold = rv; pvold = pv;
    if delta1>=0
        deltastep = 1;
        x1 = x+alpha*pv;
        rv = -feval(derf,x1);
        lambdabar = 0; sucess = 1;
        if rem(noiter,n) == 0
            pv = rv;
        else
            %Alternative conj grad direction generators may be used here
            % beta = (rv'*rv)/(rvold'*rvold);
            rdiff = rv-rvold;
            beta = (rdiff'*rv)/(rvold'*rvold);
            pv = rv+beta*pvold;
        end
        if delta1>=0.75
            lambda = 0.25*lambda;
        end
    else
        lambdabar = lambda;
        sucess = 0;
        x1 = x+alpha*pv;
```

```
    end
    if delta1<0.25
        lambda = lambda+delta*(1-delta1)/norm(pvold)^2;
    end
    x = x1;
end
res = x1;
```

We now apply the scaled conjugate gradient method to minimize two functions. The first function is the Styblinski–Tang function, see (9.10).

Using `minscg`, and the user-defined functions `f01` and `f01d` are given in Section 9.4, we have

```
>> [x, iterns] = minscg('f01','f01d',[1 -1]',.000005)

x =
     2.7468
    -2.9035

iterns =
      8
```

This is not the same solution as that determined by `mincg`. It is a local minimum value of the function but not the global minimum. Other initial values will lead to the global minimum.

The second problem is finding a minimum of Rosenbrock's function. This function is

$$f(x_1, x_2) = 100(x_2 - x_1^2)^2 + (1 - x_1)^2$$

The minimum of this function is at $x_1 = x_2 = 1$. We define the necessary anonymous functions, and solve the problem as follows:

```
>> fr = @(x) 100*(x(2)-x(1).^2).^2+(1-x(1)).^2;
>> frd = @(x) [-400*x(1).*(x(2)-x(1).^2)-2*(1-x(1)); 200*(x(2)-x(1).^2)];
>> [x, iterns] = minscg(fr,frd,[-1.2 1]',.0005)

x =
     1.0000
     1.0000

iterns =
    135
```

Note that a large number of iterations were required to solve this difficult problem.

9.6 CONJUGATE GRADIENT METHOD FOR SOLVING LINEAR SYSTEMS

We now apply the conjugate gradient algorithm to minimize a positive definite quadratic function which has the standard form

$$f(\mathbf{x}) = \left(\mathbf{x}^\mathsf{T}\mathbf{A}\mathbf{x}\right)/2 + \mathbf{p}^\mathsf{T}\mathbf{x} + q \tag{9.11}$$

Here $\mathbf{x}$ and $\mathbf{p}$ are n-component column vectors, $\mathbf{A}$ is an $n \times n$ positive definite symmetric matrix and q is a scalar. The minimum value of $f(\mathbf{x})$ is such that the gradient of $f(\mathbf{x})$ is zero. However, the gradient is easily found by direct differentiation as

$$\nabla f(\mathbf{x}) = \mathbf{A}\mathbf{x} + \mathbf{p} = \mathbf{0} \tag{9.12}$$

Thus, finding the minimum is equivalent to solving this system of linear equations which becomes, on letting $\mathbf{b} = -\mathbf{p}$,

$$\mathbf{A}\mathbf{x} = \mathbf{b} \tag{9.13}$$

Since we can use the conjugate gradient method to find the minimum of (9.11), then we can use it to solve the equivalent system of linear equations (9.13). The conjugate gradient method provides a powerful method for solving linear equation systems with positive definite symmetric matrices, and it follows quite closely the algorithm we have described for solving non-linear optimization problems. However, the line-search is greatly simplified and the value of gradient can be computed within the algorithm in this case. The algorithm takes the form:

1. Set $k = 0$: $\mathbf{x}^k = \mathbf{0}$, $\mathbf{g}^k = \mathbf{b}$, $\mu^k = \mathbf{b}^\mathsf{T}\mathbf{b}$, $\mathbf{d}^k = -\mathbf{g}^k$,
2. While system (9.13) is not satisfied
   $\mathbf{q}^k = \mathbf{A}\mathbf{d}^k$, $r^k = (\mathbf{d}^k)^\mathsf{T}\mathbf{q}^k$, $s^k = \mu^k/r^k$
   $\mathbf{x}^{k+1} = \mathbf{x}^k + s^k\mathbf{d}^k$, $\mathbf{g}^{k+1} = \mathbf{g}^k + s^k\mathbf{q}^k$
   $t^k = (\mathbf{g}^{k+1})^\mathsf{T}\mathbf{q}^k$, $\beta^k = t^k/r^k$
   $\mathbf{d}^{k+1} = -\mathbf{g}^{k+1} + \beta^k\mathbf{d}^k$, $\mu^{k+1} = \beta^k\mu^k$
   $k = k + 1$, end

Notice that the values of the gradient $\mathbf{g}$ and the step s are calculated directly and no MATLAB function or user-defined function is required.

The MATLAB function `solvercg` implements this algorithm and utilizes the stopping procedure suggested by Karmarkar and Ramakrishnan (1991). See this paper and also Golub and Van Loan (1989).

```
function xdash = solvercg(a,b,n,tol)
% Solves linear system ax = b using conjugate gradient method.
% Example call: xdash = solvercg(a,b,n,tol)
% a is an n x n positive definite matrix, b is a vector of n
% coefficients. tol is accuracy to which system is satisfied.
% WARNING Large, ill-cond. systems will lead to reduced accuracy.
xdash = [ ];  gdash = [ ];
ddash = [ ];  qdash = [ ];
```

```
q=[ ];
mxitr = n*n;
xdash = zeros(n,1);  gdash = -b;
ddash = -gdash; muinit = b'*b;
stop_criterion1 = 1;
k = 0;
mu = muinit;
% main stage
while stop_criterion1==1
    qdash = a*ddash;
    q = qdash; r = ddash'*q;
    if r==0
        error('r=0, divide by 0!!!')
    end
    s = mu/r;
    xdash = xdash+s*ddash;
    gdash = gdash+s*q;
    t = gdash'*qdash;  beta = t/r;
    ddash = -gdash+beta*ddash;
    mu = beta*mu; k = k+1;
    val = a*xdash;
    if ((1-val'*b/(norm(val)*norm(b)))<=tol) & (mu/muinit<=tol)
        stop_criterion1 = 0;
    end
    if k>mxitr
        stop_criterion1 = 0;
    end
end
```

The script e4s903.m generates a system of 10 equations with randomly selected elements on which this algorithm can be tested:

```
% e4s903.m
n = 10; tol = 1e-8;
A = 10*rand(n); b = 10*rand(n,1);
ada = A*A';
% To ensure a symmetric positive definite matrix.
sol = solvercg(ada,b,n,tol);
disp('Solution of system is:')
disp(sol)
accuracy = norm(ada*sol-b);
fprintf('Norm of residuals =%12.9f\n',accuracy)
```

Running script e4s903.m gives the following results:

```
Solution of system is:
    0.2527
   -0.2642
   -0.1706
    0.4284
    0.0017
   -0.1391
   -0.0231
   -0.0109
   -0.2310
    0.2928

Norm of residuals = 0.000000008
```

We note that the norm of the residuals is very small. For ill-conditioned matrices it is necessary to use some kind of pre-conditioner which reduces the condition number of the matrix; otherwise the method becomes too slow. Karmarkar and Ramakrishnan (1991) have used a pre-conditioned conjugate gradient method as part of an interior point algorithm to solve linear programming problems with 5000 rows and 333,000 columns.

MATLAB provides a range of iterative procedures based on conjugate gradient methods for solving $\mathbf{Ax} = \mathbf{b}$. These are the MATLAB functions `pcg`, `bicg`, and `cgs`.

9.7 METAHEURISTIC METHODS

We have seen how both calculus and gradient based methods often fail to find a global optimum. In this situation metaheuristic technique can be applied successfully. A heuristic technique is a problem specific approach that employs a practical method that often provides sufficient accuracy for the immediate goals. In the case of optimization, a heuristic method should find a good approximation to the global optimum solution, in contrast to a classical method that sometimes finds a local, rather than a global optimum solution. Metaheuristic methods are heuristic methods using a process that is problem-independent so that the methodology can be applied to a wide range of problems. Metaheuristic methods are usually stochastic in nature and are inspired by physical processes, such as gravity, or biological processes, such as bird flocking, or evolutionary processes, such as genetics.

9.8 SIMULATED ANNEALING

Simulated annealing is inspired by the physical process of annealing metal by cooling it slowly.

Here we provide a brief introduction to the ideas on which optimization using simulated annealing is based. The technique can be applied to large and difficult problems where we require the global optima and other techniques are inadequate. Even for relatively simple problems the technique can be slow.

If a metal is allowed to cool sufficiently slowly (metallurgically called post-annealing), its metallurgical structure is naturally able to find a minimum energy state for the system. If, however, the metal is cooled quickly, say by quenching in water, then this minimum energy state is not found. This concept of the natural process of finding a minimum energy state can be utilized to find the global optima of given non-linear functions. This optimization method is called simulated annealing.

The analogy is not perfect but the fast cooling process may be viewed as equivalent to finding a local minimum of a given non-linear function corresponding to the energy level, while the slow cooling corresponds to finding the ideal energy state or a global minimum of the function. This slow cooling process may be implemented using the Boltzmann probability distribution of energy states which plays a prominent part in thermodynamics and which has the form

$$P(E) = \exp(-E/kT)$$

where $P(E)$ is the probability of E, a particular energy state, k is Boltzmann's constant, and T is the temperature. This function is used to reflect the cooling process where a change in the energy level, which may be initially unfavorable, ultimately leads to a final minimum global energy state. This corresponds to the concept of moving out of the region of a local minimum of a non-linear function in the search for a global solution for the problem. This may require a temporary increase in the value of the objective function; that is, climbing out of the valley of a local minimum, although convergence to the global optimum may still occur if the adjustment to the temperature is slow enough. These ideas lead to an optimization algorithm used by Kirkpatrick et al. (1983) which has the following general structure.

Let $f(\mathbf{x})$ be the non-linear function to be minimized, where $\mathbf{x}$ is an n component vector. Then

Step 1. Set the iteration count k to zero, and set the temperature increment p count to zero. Chose an initial, trial solution $\mathbf{x}^{(k)}$ and an initial, arbitrary temperature T_p.
Step 2. Choose a second trial solution $\mathbf{x}^{(n)}$ close to $\mathbf{x}^{(k)}$.
Step 3. Compute $\Delta f = f(\mathbf{x}^{(n)}) - f(\mathbf{x}^{(k)})$ then:
If $\Delta f < 0$, then $\mathbf{x}^{(k+1)} = \mathbf{x}^{(n)}$.
If $\Delta f > 0$ and $\exp(-\Delta f/T_p) < r$, where r is a random number in the range 0 to 1, then $\mathbf{x}^{(k+1)} = \mathbf{x}^{(n)}$. Otherwise $\mathbf{x}^{(k+1)} = \mathbf{x}^{(k)}$; that is the initial trial value is unchanged.
Step 4. Set $k = k + 1$. Repeat from Step 2 until there is no significant change of function value.
Step 5. Lower the current temperature using an appropriate reduction process $T_{p+1} = g(T_p)$, set $p = p + 1$ and repeat from Step 2 until there is no further significant change in the function value from temperature reduction.

The key difficulties with this algorithm are choosing an initial temperature and a temperature reduction regime. This has generated many research papers but the details are not discussed here.

The MATLAB function `asaq` is an improved implementation of the preceding algorithm. It is based on a modified and simplified version of an algorithm described by Lester Ingber (1993). This uses an exponential cooling regime with some quenching to accelerate the convergence of the algorithm. The key parameters such as the values of `qf`, `tinit`, `maxstep` and the upper and lower bounds on the variables can be adjusted and may lead to some improvements in the convergence rate. A major change would be to use a different temperature adjustment regime and many alternatives have been suggested. The reader should view these parameter variations as an opportunity to experiment with simulated annealing.

```
function [fnew,xnew] = asaq(func,x,maxstep,qf,lb,ub,tinit)
% Determines optimum of a function using simulated annealing.
% Example call: [fnew,xnew]=asaq(func,x,maxstep,qf,lb,ub,tinit)
% func is the function to be minimized, x the initial approx.
% given as a column vector, maxstep the maximum number of main
% iterations, qf the quenching factor in range 0 to 1,
% Note: small value gives slow convergence, value close to 1 gives
% fast convergence, but may not supply global optimum,
% lb and ub are lower and upper bounds for the variables,
% tinit is the intial temperature value
% Suggested values for maxstep = 200, tinit = 100, qf = 0.9
% Initialisation
xold = x;
fold = feval(func,x);
n = length(x);
lk = n*10;
% Quenching factor q
q = qf*n;
% c values estimated
nv = log(maxstep*ones(n,1));
mv = 2*ones(n,1);
c = mv.*exp(-nv/n);
% Set values for tk
t0 = tinit*ones(n,1);
tk = t0;
% upper and lower bounds on x variables
% variables assumed to lie between -100 and 100
a = lb*ones(n,1);
b = ub*ones(n,1);
k = 1;
% Main loop
for mloop = 1:maxstep
    for tempkloop = 1:lk
        % Choose xnew as random neighbour
        fold = feval(func,xold);
        u = rand(n,1);
        y = sign(u-0.5).*tk.*((1+ones(n,1)./tk).^(abs((2*u-1))-1));
        xnew = xold+y.*(b-a);
        fnew = feval(func,xnew);
        % Test for improvement
        if fnew <= fold
            xold = xnew;
        elseif exp((fold-fnew)/norm(tk))>rand
            xold = xnew;
```

```
            end
        end
        % Update tk values
        tk = t0.*exp(-c.*k^(q/n));
        k = k+1;
end
tf = tk;
```

The results of using this function to optimize the Styblinski–Tang function, (9.10), follow. This function has been solved by the conjugate gradient method in Section 9.4.

```
>> fv = @(x) 0.5*(x(1)^4-16*x(1)^2+5*x(1)) +...
              0.5*(x(2)^4-16*x(2)^2+5*x(2));
>> for c = 1:20, [fnew(c),xnew(:,c)] = asaq(fv,[0 0].',200,0.9,-10,10,100); end
>> fmean = mean(fnew)

fmean =
  -71.9708

>> [fmin,idx] = min(fnew)

fmin =
  -78.3323

idx =
    19

>> xnew(:,idx)

ans =
   -2.9032
   -2.9031
```

Note that each run provides a different result and is not guaranteed to provide a global optimum unless the parameters are adjusted appropriately for the particular problem. The mean value of the 20 runs is −71.97, which is a relatively poor result, although the best (minimum) result is −78.3323, which compares well with the exact value of −78.3320. The corresponding values of **x** are −2.9031 and −2.9032 which again compare well with the exact values −2.9035, −2.9035.

Fig. 9.6 provides a plot of the variation in the function value for the final 40 iterations. It illustrates the behavior of the algorithm which allows both increases and decreases in the function value.

As a further illustration a contour plot showing only the final stages of the iteration, is given in Fig. 9.7.

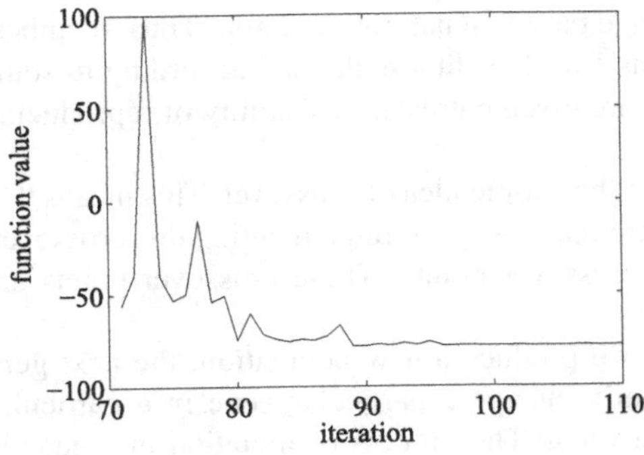

FIGURE 9.6

Graph showing the Styblinski–Tang function value for the final 40 iterations of the simulated annealing algorithm.

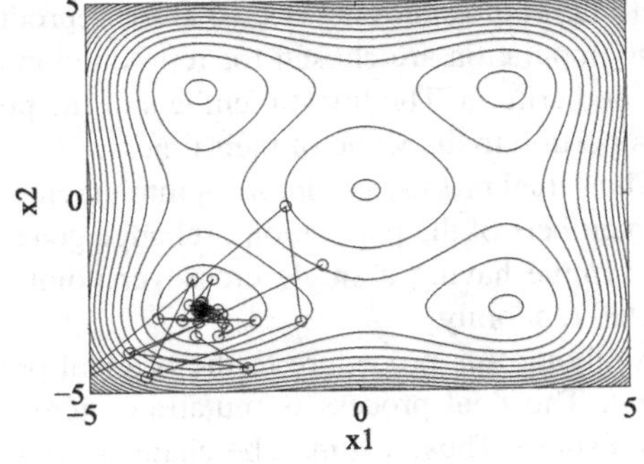

FIGURE 9.7

Contour plot of the Styblinski–Tang function. The final stages in the simulated annealing process are shown. Note how these values are concentrated in the lower left corner, close to the global minimum.

9.9 EVOLUTIONARY ALGORITHM

Most metaheuristic optimization methods use a population of *agents* to search for an optimum solution a predefined region – the search-space. These agents carry information concerning their position in the search space and, and in some cases, their velocities of exploration as well. They also interact with each other in various ways and comb the search-space looking for the global optimum solution. One class of metaheuristic algorithms are evolutionary algorithms. These algorithms provide a robust search procedure for solving difficult problems. The striking feature of these algorithms is that they are based on ideas derived from the science of genetics and the process of natural selection. This cross-fertilization from one field of science to another has led to stimulating and fruitful applications in many fields and particularly in computer science.

In this section we consider two evolutionary algorithms. The first is the genetic algorithm (GA) and we introduce the ideas on which it is based and explain how this is implemented for an optimization problem *in one variable*. The genetic algorithm comprises an initial population of agents, the size of which is generally related to the complexity of the problem under consideration. This population of agents is generated randomly and we can use the terminology of genetics to characterize it. Each member of the population corresponds to a *chromosome* and each element of the chromosome corresponds to a *gene*. Frequently the member of the population comprises a string of binary digits but it can be a decimal number or other coding that has an equivalent numeric value. This information provides the location of the member of the population in the search space and from that information the value of the function can be computed. The description of the GA that follows is a binary coded GA where trial.

From the initial population a new population must be developed and to do this we implement the analogue of specific genetic processes. These are:

1. selection based on fitness
2. recombination (implemented by crossover)
3. mutation

A set of chromosomes is selected at the reproduction stage based on natural selection. Thus, members of the population are chosen for reproduction on the basis of their fitness defined according to some specified criteria. The fittest members of the population are given a greater probability of reproducing in proportion to the value of their fitness.

The actual process of *mating* is implemented by using the simple idea of crossover. This means that two members of the population exchange genes. There are many ways of implementing this crossover, for example having a single crossover point or many crossover points. These crossover points are selected randomly.

Applying this procedure to the original population, we produce a new population; the next generation. The final process is mutation. Here we randomly change a particular gene in a particular chromosome. Thus, a 0 may be changed to a 1 or vice versa. The process of mutation in a genetic algorithm usually set to occur very rarely and hence, the probability of a change in a string is kept very low.

Having described the basic principles of a genetic algorithm, we now illustrate how it may be applied by considering a simple optimization problem and in so doing fill in some of the details to show how a genetic algorithm may be implemented. A manufacturer wishes to produce a container which consists of a hemisphere surmounted by a cylinder of fixed height. The height of the cylinder is fixed but the common radius of the cylinder and hemisphere may be varied between 2 and 4 units. The manufacturer wishes to find the radius value which *maximizes* the volume of the container. This is a simple problem and the optimum radius is obviously 4 units. However, it serves to illustrate how the genetic algorithm may be applied.

We can formulate this as a one-dimensional maximization problem by taking r as the common radius of the cylinder and hemisphere and h as the height of the cylinder. Taking $h = 2$ units leads to the formula

$$\text{Maximize } v = 2\pi r^3/3 + 2\pi r^2 \tag{9.14}$$

where $2 \le r \le 4$.

To solve this optimization problem we generate an initial set of strings of binary digits or bits to constitute the initial population of agents. The number of bits in each string – the string length – limits the accuracy with which we can find the solution to the problem so it must be chosen with care. In addition we must select the size of the initial population; again this must be chosen with care since a large initial population increases the time taken to implement the steps of the algorithm. A large population may be unnecessary since the algorithm automatically generates new members of the population in the process of searching the region.

Suppose an initial population of five chromosomes is generated, each with six genes, giving 0, 1, ..., $2^6 - 1$ possible values of the chromosome, as shown in Table 9.2. To aid the reader in the following discussion, we have labeled the five members of the population #1 to #5. Since we are interested in values of r in the range 2 to 4, we must transform the values of these binary strings to values in the range 2 to 4. In this example, the decimal equivalent, d, of the binary numbers, is in the range 0 to 63, and r is in the range 2 to 4, then the conversion from d to r is given by $r = 2d/63 + 2$. For example, when $d = 0$, $r = 2$ and when $d = 63$, $r = 4$. Carrying out this conversion, the decimal values, r, are shown in Table 9.2. As required, these values are in the range 2 to 4 and provide the initial population of values for r. However, these values tell us nothing about their fitness and to discover this we must judge them against some fitness criteria. In this case, the choice is easy since our objective is

Table 9.2 First generation of a GA

Population	binary	r	v	fitness %
Pop member #1	011100	2.8889	102.93	16.94
Pop member #2	111101	3.9365	225.12	37.04
Pop member #3	100100	3.1429	127.08	20.91
Pop member #4	000011	2.0952	46.85	7.71
Pop member #5	011101	2.9206	105.77	17.40
Total			607.75	100.00

Table 9.3 Generation of a new population

Population	binary	v	Population	binary	v
Pop member #2	111101	225.12	Pop member #2	111101	225.12
Pop member #2	111101	225.12	From #2 and #1	*11*1100	220.49
Pop member #1	011100	102.93	From #1 and #2	01*1101*	105.77
Pop member #5	011101	105.77	Pop member #5	011101	105.77
Pop member #1	011100	102.93	Pop member #1	011100	102.93
Total		761.89			760.10

to maximize the value of the function (9.14). We simply determine the values of the fitness function v for these values of r as shown in Table 9.2. Notice that at this stage that the total fitness is 607.75. The fittest member of the population string is member #2 which has a value $r = 3.9365$, with a fitness value $v = 225.1246$. Fortuitously, this is a very good result.

The next stage is reproduction, when the strings are selected according to their fitness so that there is a higher probability of more of the fittest chromosomes in the mating pool. However, this process of selection is more complex and is based on a process which simulates the use of a roulette wheel. The percentage of the roulette wheel that is allocated to a particular string is directly proportional to the fitness of the string. For each of the fitness values, the percentage fitness can be calculated from $100v_i/\sum v_i$, as shown in Table 9.2. Thus, conceptually, we spin a roulette wheel on which the population members #1 to #5 have the percentages of area 16.94%, 37.04%, 20.91%, 7.71%, and 17.40%, respectively, see Table 9.2. Thus, these chromosomes or strings have this biased chance of being selected. After a biased random choice that simulates the biased roulette wheel, the population members chosen for mating are shown in Table 9.3. Note that Population member #1 and Population member #2 have been favored by the selection process and duplicated. Note that this mating population, as we would expect, has a higher total fitness of 761.89, see Table 9.3.

We can now mate the members of this population but we mate only a proportion of them, in this case we have chosen 60% or 0.6. In this example the population size is 5 and $0.6 \times 5 = 3$. This number is rounded down to an even number, that is 2, since only an even number of members of the population can mate. Thus, two members of the population are randomly selected for mating. We see that two members of the original population, members #1 and #2, have mated by cross over to create two new members of the population. Cross over has been randomly chosen to take place after the second digit. In Table 9.3 the binary digits of #2 have been italicized to show the cross over more clearly. The new population fitness is shown in Table 9.3 and we see that the total fitness is 760.10. Notice that the total fitness has not improved and indeed, at this stage, we cannot expect improvements every time.

Table 9.4 Mutation in new population

Population	binary	v	binary	v
Pop member #2	111101	225.12	110101	189.80
From #2 and #1	111100	220.49	111100	220.49
From #1 and #2	011101	105.77	011101	105.77
Pop member #5	011101	105.77	011101	105.77
Pop member #1	011100	102.93	011110	108.66
Total		760.10		730.49

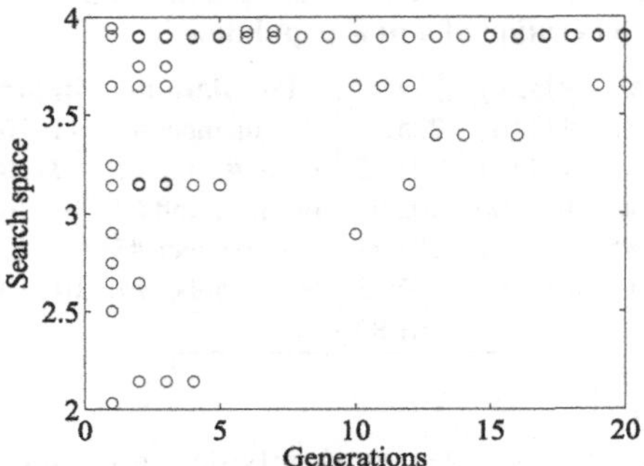

FIGURE 9.8

Genetic algorithm. Each member of the population is represented by o. Successive generations of the population concentrate toward the value 4 approximately.

Finally we perform mutation. The mutation rate is usually low and, in this case, we have chosen a mutation rate of 5%. This is again done randomly. Each bit is given a 5% chance of being changed. Thus we would expect, in 30 binary digits or bits, one or two bits to mutate. In this example two mutations have occurred; the third element of the first chromosome has changed from 1 to 0, and the fifth element of the last chromosome has changed from 0 to 1. In Table 9.4 the two mutated bits are underlined. Sometimes no mutation will occur. Note also the overall fitness is further reduced, but this is often the case. Part of the process is to encourage searching new areas. These areas may be areas of high fitness, but they may be areas of low fitness which will lower the overall fitness. Until they are searched this information is unknown. This completes the production of a new generation.

The process of selection based on fitness, reproduction, and mutation is now repeated using the new generation and subsequently repeated for many generations. If we do this for 20 generations we obtain $r = 3.8980$, and $v = 219.5219$. Since the exact solution is $r = 4$, this is a reasonable result. Fig. 9.8 gives a graphical representation of the progress of the genetic algorithm. It should be noted that each run of the genetic algorithm can produce a different result because of the random nature of the process. In addition the number of distinct values in the search space is limited by the chromosome length. In this example the chromosome length is 6 bits giving 2^6 or 64 divisions. Thus the range of r from 2 to 4 is divided into 64 divisions, each equal to 0.031746. If the chromosome length had been 8, then divisions in the range from 2 to 4 are equal to 0.007843.

There are many variants of the basic genetic algorithms. For example, Grey code rather than binary code can be used; selection can be implemented in many different ways; crossover can be changed to multi-point crossover or other alternatives. It is often reported that a genetic algorithm is slow in execution but it should be remembered that it is best applied to difficult problems where other methods fail, for example those which have multiple optima and where the global optimum is required. Since standard algorithms often fail in these cases, the extra time taken by the genetic algorithm is worthwhile.

In the description of the GA, we have applied the algorithm to a problem in one variable. Suppose we have a problem in three variables, and for simplicity we will assume that the search range in each variable is 0 to 10. We will also assume that each member of the population is represented by a string of 24 bits which is divided into three, 8 bits sub-strings. Thus, for example, the 24 bit string

11001010|01110001|00100111

is divided into the three sub-strings 11001010, 01110001, and 00100111. The sub-strings have decimal values of 202, 113, and 39. These convert to 7.9216, 4.4314, and 1.5294 and are the values of the three variables of the problem. Using these values, we can then determine the fitness of the member of the population and proceed as normal.

The genetic algorithm can be used to solve optimization problems for which the solutions are constrained to be members of a set of discrete values, rather than being continuous over a defined range. For example, suppose we wish to optimize the number of different classes of employees required for some purpose, and suppose the number of employees in a class could be between zero and 63. If the numbers of employees in a class is coded using 6 binary digits, the solution can only be an integer number in the range zero to 63. Non-integer solutions cannot be generated. A second example is more complicated. Suppose we wish to optimize the design of a steel framework that is required to carry a specific load. The members of the framework are subjected to internal loads and these loads will cause both internal stresses in the members and deflections of the framework, both of which must be limited for safe working. To simplify the following discussion we will assume that the way the beam reacts to loads can be consumed into a single measure of "performance". In practice the situation is more complicated and it is not possible to describe a beam's ability to carry load by a single figure. We optimize the framework by minimizing the weight of the steel required, thereby minimizing the cost. However, steel sections (or beams) are rolled in a limited range of sizes, and it would be uneconomical to have special sections rolled to suit the needs of an optimum design. Thus the framework must be constructed from standard sections. Suppose that eight sizes of steel section are available from which the framework will be designed. We will designate or reference the range of beam sizes 0 to 7 and in the genetic algorithm each beam in the framework will require 3 binary digits to specify it. Further suppose there are 10 beams in the framework so that 30 binary digits will be required for trial solution or design. For a particular beam, its reference number can be used to access, from a table, its weight and performance. Doing this for all the beams in the framework allows the framework weight and performance to be evaluated. Provided the performance is within safe limits, the GA optimization will proceed in the normal way seek the design with the lowest weight – the optimum design. This optimal design will, of course, only be composed of beams that are selected from the eight standard sizes.

It is beyond the scope of this book to give a detailed account of this field of study and the reader is referred to the excellent texts of Goldberg (1989) and Coley (1999). We now consider Differential Evolution, a more recent optimization method based on evolution.

9.10 DIFFERENTIAL EVOLUTION

The Differential Evolution (DE) algorithm is also an example of evolutionary programming. It was developed by Storn and Price (1997) to solve difficult global optimization problems, particularly when we seek the global optimum. The advantage of the DE algorithm is that it has a simple structure and is fast and robust. The DE algorithm uses mutation as the search mechanism and selection to direct the search towards a better solution. The principal difference between the DE algorithm and the GA is that the GA principally relies on crossover to obtain better solutions, whilst DE principally uses mutation as the primary search mechanism.

Consider a minimization problem in n_{var} dimensions. The search space is to be explored using n_{pop} trial solutions which are generated in the search space. The DE algorithm requires column vectors of n_{var} real elements, denoted by $\mathbf{x}_i$, where $i = 1, 2, ..., n_{pop}$. In the terminology of DE the vectors $\mathbf{x}_i$ are called target vectors. This initial population is uniformly randomly distributed within the lower and upper bounds specified for each dimension. We now describe the major steps in this process:

Mutation: To find a new approximation to $\mathbf{x}_i^{(k+1)}$ for the $(k+1)$th generation we select from a uniform random distribution three target vectors from the current, kth, generation $\mathbf{x}_{r1}^{(k)}$, $\mathbf{x}_{r2}^{(k)}$, and $\mathbf{x}_{r3}^{(k)}$ such that $r_1 \neq r_2 \neq r_3 \neq i$. The vector $\mathbf{v}_i^{(k+1)}$ is called the mutant vector and is defined as

$$\mathbf{v}_i^{(k+1)} = \mathbf{x}_{r1}^{(k)} + F(\mathbf{x}_{r2}^{(k)} - \mathbf{x}_{r3}^{(k)}), \quad i = 1, 2, ..., n_{pop} \tag{9.15}$$

where F is a user specified constant which controls the amplitude of the differential variation $(\mathbf{x}_{r2}^{(k)} - \mathbf{x}_{r3}^{(k)})$. The optimal range for this weighting or scaling factor F has been found to be in the range 0.5 to 1.0 although it can be in the range 0 to 2.

Crossover: The vector $\mathbf{u}_i^{(k+1)}$, called the candidate vector, is then obtained from

$$u_{ji}^{(k+1)} = \begin{cases} v_{ji}^{(k+1)} & \text{if } r_j \leq c_r \text{ or } j = s_i \\ x_{ji}^{(k)} & \text{if } r_j > c_r \text{ and } j \neq s_i \end{cases} \tag{9.16}$$

where $i = 1, 2, ..., n_{pop}$, $j = 1, 2, ..., n_{var}$, r_j is a random number chosen from a uniform distribution in the range 0 to 1, the crossover rate, c_r is a user supplied constant, also in the range 0 to 1, and s_i is a randomly chosen index in the range 1 to n_{var}. Thus (9.16) ensures that the candidate $\mathbf{u}_i^{(k+1)}$ will not be an exact copy of the target $\mathbf{x}_i^{(k)}$ and will get at least one parameter from the mutant vector $\mathbf{v}_i^{(k+1)}$.

Selection: The final step is to compute the next generation of trial values $\mathbf{x}_i^{(k+1)}$. This requires a selection process which, for the minimization of the objective function $f(\mathbf{x})$, is

$$\mathbf{x}_i^{(k+1)} = \begin{cases} \mathbf{u}_i^{(k+1)} & \text{if } f(\mathbf{u}_i^{(k+1)}) \leq f(\mathbf{x}_i^{(k)}) \\ \mathbf{x}_i^{(k)} & \text{if } f(\mathbf{u}_i^{(k+1)}) > f(\mathbf{x}_i^{(k)}) \end{cases} \tag{9.17}$$

where $i = 1, 2, ..., n_{pop}$. Using this new generation of trial solutions, the mutation, crossover, and selection process is repeated until convergence is achieved. Note that the user is only required to specify three parameters, the number of trial solutions, n_{pop}, together with F and c_r.

It is possible that one or more of the new trial values might lie outside the search space in one or more dimensions. If that is the case, then one option is to simply take the trial value back to the boundary of the search space. Alternatively, we might move the trial value as far inside the boundary as it is currently outside the boundary. Thus, for the jth variable in the ith trial solution, x_{ji}, we apply the following adjustment.

$$\text{if } x_{ji} > x_{j(max)}, \text{ then } x_{ji} = x_{j(max)}$$
$$\text{if } x_{ji} < x_{j(min)}, \text{ then } x_{ji} = x_{j(min)}$$

where $x_{j(max)}$ and $x_{j(min)}$ are the maximum and minimum values of the search space in the jth dimension. An alternative procedure to contain the trial values in the search space is

$$\text{if } x_{ji} > x_{j(max)}, \text{ then } x_{ji} = 2x_{j(max)} - x_{ji}$$
$$\text{if } x_{ji} < x_{j(min)}, \text{ then } x_{ji} = 2x_{j(min)} - x_{ji}$$

The differential evolution procedure is now illustrated by a simple example. One of many functions used to test optimization algorithms is the Alpine 2 function. This function is

$$f(\mathbf{x}) = \prod_{i=1}^{n_{var}} \sqrt{x_i}\sin(x_i)$$

where n_{var} is the number of variables or dimensions in the problem. The *maximum* of this function is $f(\mathbf{x}^*) = 2.808^{n_{var}}$ where $\mathbf{x}^* = [7.917, 7.917, ..., 7.917]$. The search space is usually limited for each variable in the range 0 to 10. Thus, if the problem has two variables, then the optimum of the function is given by $f(\mathbf{x}^*) = 2.808^2 = 7.8849$.

We now seek to maximize this function in three variables. Since DE seeks the minimum of a function, we seek to minimum of the negative of the Alpine 2 function; that is the test function we wish to minimize is $f(\mathbf{x}) = -\sqrt{x_1x_2x_3}\sin(x_1)\sin(x_2)\sin(x_3)$, We have chosen to make $c_r = 0.5$ and $F = 0.8$. We begin by randomly generating six trial solutions in the range 0 to 10, as shown in Table 9.5. The table also shows values of the test function for each trial solution. The process for generating the next set of trial vectors (for $\mathbf{x}_1$ and $\mathbf{x}_6$ only) is now described.

Initially, generate the parameters r_1, r_2, and r_3 from a uniform random distribution of integers in the range 1 to n_{pop}. In this case when $i = 1$, $r_1 = 3$, $r_2 = 2$, and $r_3 = 4$ were randomly chosen; when $i = 6$, $r_1 = 4$, $r_2 = 2$, and $r_3 = 5$ were randomly chosen. Thus, substituting in (9.15) gives

$$\mathbf{v}_1^{(1)} = \mathbf{x}_3^{(0)} + 0.8(\mathbf{x}_2^{(0)} - \mathbf{x}_4^{(0)}), \quad \mathbf{v}_6^{(1)} = \mathbf{x}_4^{(0)} + 0.8(\mathbf{x}_2^{(0)} - \mathbf{x}_5^{(0)})$$

The calculations are shown in Table 9.6. We now compare $\mathbf{v}$ and $\mathbf{x}$, as shown in (9.16). In Table 9.7 a zero in the $r > c_r$ column indicates that the random number is less than c_r, a one in the $r > c_r$ column indicates the random number was greater than c_r. It is seen that due to the numbers being chosen randomly, the second element $\mathbf{u}_1$ is derived from $\mathbf{x}_1$, similarly the second and third elements of $\mathbf{u}_6$ is taken from $\mathbf{x}_6$.

Having determined $\mathbf{u}_1$ and $\mathbf{u}_6$, we can compare $f(\mathbf{x}_1^{(0)})$ with $f(\mathbf{u}_1^{(0)})$, and $f(\mathbf{x}_6^{(0)})$ with $f(\mathbf{u}_6^{(0)})$. From Table 9.7 we see that $f(\mathbf{u}_1^{(0)}) < f(\mathbf{x}_1^{(0)})$ and so from (9.17) we have $\mathbf{x}_1^{(1)} = \mathbf{u}_1^{(0)}$. Also from Table 9.7 we see that $f(\mathbf{x}_6^{(0)}) < f(\mathbf{u}_6^{(0)})$ and hence, from (9.17), we have $\mathbf{x}_6^{(1)} = \mathbf{x}_6^{(0)}$.

Table 9.5 Differential Evolution: randomly generated trial solutions

	x_1	x_2	x_3	x_4	x_5	x_6
	8.1472	9.1338	2.7850	9.6489	9.5717	1.4189
	9.0579	6.3236	5.4688	1.5761	4.8538	4.2176
	1.2699	0.9754	9.5751	9.7059	8.0028	9.1574
$f(\mathbf{x})$	3.1749	0.0720	0.4592	0.7491	2.7637	−1.7015

Table 9.6 Showing $\mathbf{v}_1^{(1)} = \mathbf{x}_3^{(0)} + 0.8(\mathbf{x}_2^{(0)} - \mathbf{x}_4^{(0)})$, $\mathbf{v}_6^{(1)} = \mathbf{x}_4^{(0)} + 0.8(\mathbf{x}_2^{(0)} - \mathbf{x}_5^{(0)})$

x_3	x_2	x_4	v_1	x_4	x_2	x_5	v_6
2.7850	9.1338	9.6489	2.3729	9.6489	9.1338	9.5717	1.8077
5.4688	6.3236	1.5761	9.2668	1.5761	6.3236	4.8538	2.1309
9.5751	0.9754	9.7059	2.5907	9.7059	0.9754	8.0028	0.7385

Table 9.7 Crossover: generating u from x or v

	x_1	v_1	$r_1 > c_r$	u_1	x_6	v_6	$r_6 > c_r$	u_6
	8.1472	<u>2.3729</u>	0	2.3729	1.4189	<u>9.2986</u>	0	9.2986
	<u>9.0579</u>	9.2668	1	9.0579	<u>4.2176</u>	2.7519	1	4.2176
	1.2699	<u>2.5907</u>	0	2.5907	<u>9.1574</u>	4.0840	1	9.1574
$f(\mathbf{x})$	3.1749			0.9740	−1.7015			−0.5546

Having described the DE algorithm, we provide a MATLAB implementation of it as follows:

```
function [min_res,min_x] = diffevo(funk, nvar, x0_min,x0_max, F, C, npop, ngen)
% Differential Evolution
% x0_min and x0_max are column vectors specifying the minimum and maximum
% values of each variable in the search space
% F is user defined mutation parameter 0.85 suggested
% C is user defined crossover parameter 0.5 suggested
% nvar is the numer of variables or dimensions
% npop is the size of the population
% ngen is the maximum number of generations for the search
x_rng = x0_max-x0_min;
% Generate initial locations for population
x = zeros(nvar,npop);
x_max = zeros(nvar,npop); x_min = zeros(nvar,npop);
xold = zeros(nvar,npop); v = zeros(nvar,npop); u = zeros(nvar,npop);
r = zeros(nvar,npop); ri = zeros(nvar,npop);
for i = 1:npop
    x(:,i) = x0_min+x_rng.*rand(nvar,1);
    x_max(:,i) = x0_max;
    x_min(:,i) = x0_min;
end
```

```
xold = x;
for gen = 1:ngen
    % Mutation
    for i = 1:npop
        flg = 0;
        while flg == 0
            r = randi(npop);
            if r~=i
                r1 = r; flg = 1;
            end
        end
        flg = 0;
        while flg == 0
            r = randi(npop);
            if r~=i & r~=r1
                r2 = r;  flg = 1;
            end
        end
        flg = 0;
        while flg == 0
            r = randi(npop);
            if r~=i & r~=r1 & r~=r2
                r3 = r; flg = 1;
            end
        end
        v(:,i) = x(:,r1)+F*(x(:,r2)-x(:,r3));
    end  % End populations loop
    r = rand(nvar,npop);
    ri = randi(nvar,npop,1);
    % Crossover and selection
    for j = 1:npop
        % Mutation
        for i = 1:nvar
            u(i,j) = v(i,j)*(r(i,j)<=C | j==ri(i,1))+x(i,j)*(r(i,j)>C & j~=ri(i,1));
        end
        % Selection
        if feval(funk,u(:,j))<feval(funk,x(:,j))
            x(1:nvar,j) = u(1:nvar,j);
        end
    end % End populations loop
    % Limit x to search area.
    for i = 1:nvar
        for j = 1:npop
            if x(i,j) > x_max(i)
```

```
                x(i,j) = x_max(i)-0.5*(x_max(i)-xold(i,j));
            end
            if x(i,j) < x_min(i)
                x(i,j) = x_min(i)+0.5*(xold(i,j)-x_min(i));
            end
        end
    end
end  % End generations loop
for i = 1:npop
    res(i) = feval(funk,x(:,i));
end
[min_res, idx] = min(res);
min_x = x(:,idx);
```

The script e4s904.m determines the maximum value of the Alpine 2 function by determining the minimum of the negative of the function in 4 variables, we have:

```
%e4s904.m
clear all
format long
nvar = 4;
x0_min = zeros(nvar,1);
x0_max = 10*ones(nvar,1);
opt_f = -(2.808^nvar)
npop = 30;
ngen = 200;
min_x = zeros(nvar,1);
[min_res,min_x] = diffevo('alpine2', nvar, x0_min, x0_max, 0.8, 0.5, npop, ngen)
return
%-------------------------------------------------------
function f = alpine2(x)
nvar = length(x);
f = 1;
for i = 1:nvar
    f = f*sqrt(x(i))*sin(x(i));
end
f = -f;
end
```

Running script e4s904.m, we have

```
>> e4s904

opt_f =
  -62.171080298495987
```

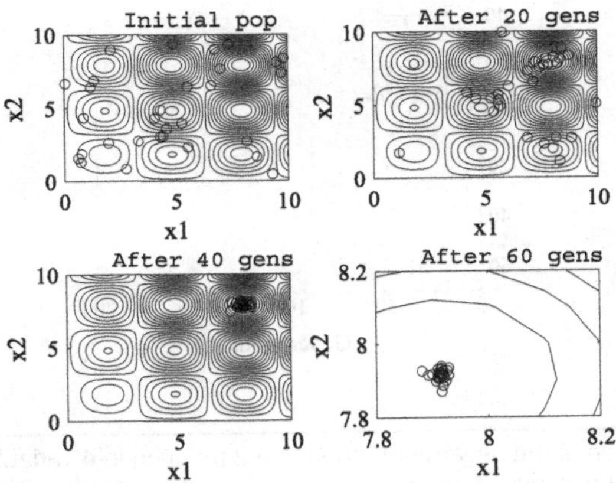

FIGURE 9.9

Contour plot of the Alpine 2 function showing the rapid convergence to the global maximum using Differential Evolution. The bottom right contour plot is greatly expanded.

```
min_res =
 -62.182698724243522

min_x =
    7.917085939985708
    7.917077617615944
    7.917060106742505
    7.917053640428087
```

Clearly, an accurate result is obtained after only 200 generations.

The ability of Differential Evolution to converge rapidly to a global minimum is illustrated in Fig. 9.9. This shows the search for the global minimum of the negative of the Alpine 2 function. The randomly generated initial population of 30 members rapidly converges to the global minimum. After 60 generations it is clear that the population has converged to the global minimum. More generations are required to obtain the value of the global minimum accurately.

Fig. 9.10 shows convergence to the minimum value of the negative of the Alpine 2 function with 4 variables, using the DE algorithm with a population of 30. The three plots show the maximum (worst), the mean, and the minimum (best) of the population. Convergence is to the exact value of the minimum, $-(2.808^{n_{var}}) = -62.1711$.

The power of the DE algorithm is illustrated in Table 9.8. The data is derived by using 20 runs of the MATLAB function `diffevo`. An optimization problem in 12 variables is demanding since there are approximately 4000 local minima. The DE algorithm provides a good result for this problem. With 2000 generations the mean value obtained for the optimum is of the right order. Increasing the number of generations to 4000 gives a much improved result.

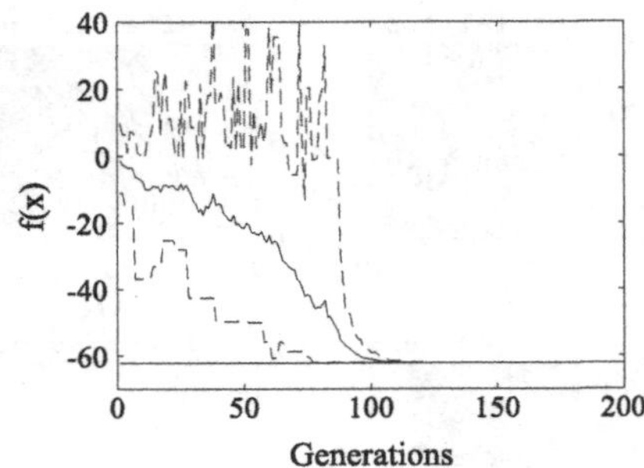

FIGURE 9.10

Graph showing the minimization of the negative of the Alpine 2 function in 4 variables. The plots show the maximum, mean, and minimum values of the population for 200 generations of the DE algorithm. The continuous line denotes the mean values and the dashed lines denote the maximum and minimum values. Convergence is to the exact solution, shown by the horizontal line.

Table 9.8 Optimization of Alpine 2 function

var	pop	gen	exact	mean	min
4	30	500	-6.217×10	-6.218×10	-6.218×10
8	30	1000	-3.865×10^3	-3.867×10^3	-3.867×10^3
12	30	2000	-2.403×10^5	-1.960×10^5	-2.404×10^5
12	30	4000	-2.403×10^5	-2.304×10^5	-2.404×10^5

The DE algorithm presented here is not the only form of differential evolution which has been shown to be of value. Storn and Price (1997) have described and classified some of the alternative forms. For example, one alternative is to replace (9.15) by

$$\mathbf{v}_i^{(k+1)} = \mathbf{x}_{r1}^{(k)} + F_1(\mathbf{x}_{r2}^{(k)} - \mathbf{x}_{r3}^{(k)}) + F_2(\mathbf{x}_{r4}^{(k)} - \mathbf{x}_{r5}^{(k)}) \tag{9.18}$$

Here $r_1 \neq r_2 \neq r_3 \neq r_4 \neq r_5 \neq i$, and, usually, $F_1 = F_2 = F$ and F is chosen to be in the range $0.5 \leq F \leq 1$.

9.11 CONSTRAINED NON-LINEAR OPTIMIZATION

In this section we consider the problem of optimizing a non-linear function, subject to one or more non-linear constraints. This problem can be expressed mathematically as follows:

$$\text{Minimize } f = f(\mathbf{x}) \quad \text{where } \mathbf{x}^\mathsf{T} = [x_1 \; x_2 \; \ldots \; x_n] \tag{9.19}$$

subject to the constraints

$$h_i(\mathbf{x}) = 0 \quad \text{where } i = 1, \ldots, p \tag{9.20}$$

Sometimes the minimization problem may have additional or alternative constraints that are of the form

$$g_j(\mathbf{x}) \geq b_j \quad \text{where} \quad j = 1, \ldots, q \tag{9.21}$$

To solve this problem we can use the Lagrange Multiplier Method. This method is not a purely numerical method, it requires the user to apply calculus, and the resulting equations are solved numerically. For large problems this is too onerous for the user, and for this reason it is not a practical method for solving this type of problem. However, it is theoretically important in the development of other, more practical, methods.

If constraints of the form of (9.21) are present, they must be converted to the form of (9.20) as follows. Let $\theta_j^2 = g_j(\mathbf{x}) - b_j$. If the constraint $g_j(\mathbf{x}) \geq b_j$ is violated, then θ_j^2 is negative and θ_j is imaginary. Thus, we have a requirement that for the constraints to be satisfied θ_j must be real. Thus the constraint equation (9.21) becomes

$$\theta_j^2 - g_j(\mathbf{x}) + b_j = 0 \quad \text{where} \quad j = 1, \ldots, q \tag{9.22}$$

This constraint equation is an equality constraint, as in (9.20).

To solve (9.19) we begin by forming the expression

$$L(\mathbf{x}, \boldsymbol{\theta}, \boldsymbol{\lambda}) = f(\mathbf{x}) + \sum_{i=1}^{p} \lambda_i h_i(\mathbf{x}) + \sum_{j=1}^{q} \lambda_{p+j}[\theta_j^2 - g_j(\mathbf{x}) + b_j] = 0 \tag{9.23}$$

The function L is called the Lagrange function and the scalar quantities λ_i are called the Lagrange multipliers. We now minimize this function using calculus: hence we take the following partial derivatives and set them to zero.

$$\begin{aligned}
\partial L/\partial x_k &= 0, \quad k = 1, \ldots, n \\
\partial L/\partial \lambda_r &= 0, \quad r = 1, \ldots, p+q \\
\partial L/\partial \theta_s &= 0, \quad s = 1, \ldots, q
\end{aligned}$$

We will find that when we set the differentials with respect to λ_r to zero, we force both $h_i(\mathbf{x})$ $(i = 1, 2, \ldots, n)$ and $\theta_j^2 - g_j(\mathbf{x}) + b_j$ $(j = 1, 2, \ldots, q)$ to be zero. Thus the constraints are satisfied. If these terms are zero then minimizing (9.23) is equivalent to minimizing (9.19) subject to (9.20) and (9.21). If we are dealing with a quadratic function with linear constraints then the resulting equations are all linear and relatively easy to solve.

Example 9.1. Consider the solution of a problem with a cubic function and quadratic constraints.

$$\text{Minimize } f = 2x + 3y - x^3 - 2y^2$$

subject to

$$\begin{aligned}
x + 3y - x^2/2 &\leq 5.5 \\
5x + 2y + x^2/10 &\leq 10 \\
x, y &\geq 0
\end{aligned}$$

To use the Lagrange method we change the form of the constraints to be equality constraints, thus:

$$\text{Minimize } f = 2x + 3y - x^3 - 2y^2$$
$$\text{subject to}$$
$$\theta_1^2 + x + 3y - x^2/2 - 5.5 = 0$$
$$\theta_2^2 + 5x + 2y + x^2/10 - 10 = 0$$
$$x, y \geq 0$$

Hence, forming L we have

$$L = 2x + 3y - x^3 - 2y^2 + \lambda_1(\theta_1^2 + x + 3y - x^2/2 - 5.5) + \lambda_2(\theta_2^2 + 5x + 2y + x^2/10 - 10)$$

Taking partial derivatives of L and setting them to zero gives

$$\partial L/\partial x = 2 - 3x^2 + \lambda_1(1 - x) + \lambda_2(5 + x/5) = 0 \tag{9.24}$$
$$\partial L/\partial y = 3 - 4y + 3\lambda_1 + 2\lambda_2 = 0 \tag{9.25}$$
$$\partial L/\partial \lambda_1 = \theta_1^2 + x + 3y - x^2/2 - 5.5 = 0 \tag{9.26}$$
$$\partial L/\partial \lambda_2 = \theta_2^2 + 5x + 2y + x^2/10 - 10 = 0 \tag{9.27}$$
$$\partial L/\partial \theta_1 = 2\lambda_1\theta_1 = 0 \tag{9.28}$$
$$\partial L/\partial \theta_2 = 2\lambda_2\theta_2 = 0 \tag{9.29}$$

If (9.28) and (9.29) are to be satisfied then there are four cases to consider as follows:

Case 1: $\theta_1^2 = \theta_2^2 = 0$. Then (9.12) through (9.27) become, with some rearrangement,

$$2 - 3x^2 + \lambda_1(1 - x) + \lambda_2(5 + x/5) = 0$$
$$3 - 4y + 3\lambda_1 + 2\lambda_2 = 0$$
$$x + 3y - x^2/2 - 5.5 = 0$$
$$5x + 2y + x^2/10 - 10 = 0$$

Case 2: $\lambda_1 = \theta_2^2 = 0$. Then (9.12) through (9.27) become, with some rearrangement,

$$2 - 3x^2 + \lambda_2(5 + x/5) = 0$$
$$3 - 4y + 2\lambda_2 = 0$$
$$\theta_1^2 + x + 3y - x^2/2 - 5.5 = 0$$
$$5x + 2y + x^2/10 - 10 = 0$$

Case 3: $\theta_1^2 = \lambda_2 = 0$. Then (9.12) through (9.27) become, with some rearrangement,

$$2 - 3x^2 + \lambda_1(1 - x) = 0$$
$$3 - 4y + 3\lambda_1 = 0$$
$$x + 3y - x^2/2 - 5.5 = 0$$
$$\theta_2^2 + 5x + 2y + x^2/10 - 10 = 0$$

Case 4: $\lambda_1 = \lambda_2 = 0$. Then (9.12) through (9.27) become, with some rearrangement,

$$2 - 3x^2 = 0$$
$$3 - 4y = 0$$
$$\theta_1^2 + x + 3y - x^2/2 - 5.5 = 0$$
$$\theta_2^2 + 5x + 2y + x^2/10 - 10 = 0$$

The solution of these sets of non-linear equations requires some iterative procedure. The MATLAB function `fminsearch` finds the minimum of a scalar function of several variables, given an initial estimate. The application of this function to this problem is illustrated for Case 1 in the MATLAB script `e4s905`. Since the right side of each equation is zero, when the solution is found, the function

$$[2 - 3x^2 + \lambda_1(1 - x) + \lambda_2(5 + x/5)]^2 + [3 - 4y + 3\lambda_1 + 2\lambda_2]^2 + \ldots$$
$$[x + 3y - x^2/2 - 5.5]^2 + [5x + 2y + x^2/10 - 10]^2$$

should equal zero. The MATLAB function `fminsearch` will choose values of x, y, λ_1, and λ_2 to minimize this expression and bring it very close to zero. This is generally referred to as unconstrained non-linear optimization. Thus we have converted a constrained optimization to an unconstrained one.

```
% e4s905.m
g = @(X) sqrt((2-3*X(1).^2+X(3).*(1-X(1))+X(4).*(5+X(1)/5)).^2 ...
    +(3-4*X(2)+3*X(3)+2*X(4)).^2+(X(1)+3*X(2)-X(1).^2/2-5.5).^2 ...
    +(5*X(1)+2*X(2)+X(1).^2/10-10).^2);
X = fminsearch(g, [1 1 1 1]);
x = X(1); y = X(2); f = 2*x+3*y-x^3-2*y^2;
lambda_1 = X(3); lambda_2 = X(4);
disp('Case 1')
disp(['x = ' num2str(x) ', y = ' num2str(y) ', f = ' num2str(f)])
disp(['lambda_1 = ' num2str(lambda_1) ...
    ', lambda_2 = ' num2str(lambda_2)])
```

Executing this script gives

```
Case 1
x = 1.2941, y = 1.6811, f = -0.18773
lambda_1 = 0.82718, lambda_2 = 0.62128
```

Cases 2, 3, and 4 can also be solved in a similar manner using the function `fminsearch`. However, the details of the MATLAB scripts are not shown; only the results are provided in Table 9.9.

Comparing the values of f computed for each case and shown in Table 9.9, it is clear that Case 1 gives the minimum solution. Fig. 9.11 shows the function, the constraints, and the four possible solutions. The optimal solution is not necessarily at the intersection of the constraint boundaries as it is in a linear system. All solutions are feasible but only Case 1 is the global minimum. Case 2 and Case 3 are local minima at the constraint boundary and Case 4 is a local maximum. It should be noted that often one or more solutions will not be feasible; that is they will not satisfy the constraints.

Table 9.9 Possible solutions of a minimization problem

Case	θ_1^2	θ_1^2	λ_1	λ_2	x	y	f
1	0	0	0.8272	0.6213	1.2941	1.6811	−0.1877
2	1.5654	0	0	0.8674	1.4826	1.1837	0.4552
3	0	2.3236	1.2270	0	0.8526	1.6703	0.5166
4	2.7669	4.3508	0	0	0.8165	0.7500	2.2137

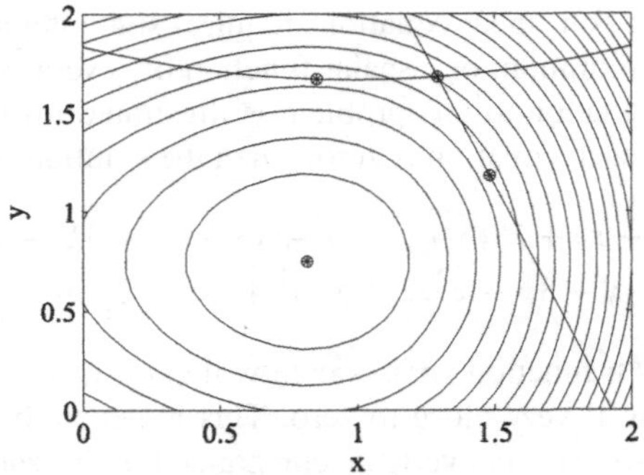

FIGURE 9.11

Graph showing the objective function and constraints for Example 9.1. The four solutions are also indicated.

9.12 THE SEQUENTIAL UNCONSTRAINED MINIMIZATION TECHNIQUE

We now give a brief introduction to a standard method for constrained optimization. The Sequential Unconstrained Minimization Technique (SUMT) method for constrained optimization converts the solution of a constrained optimization problem to the solution of a sequence of unconstrained problems. This method was developed by Fiacco and McCormick and others in the 1960s. See Fiacco and McCormick (1968, 1990).

Consider the optimization problem:

$$\text{Minimize } f(\mathbf{x}) \text{ subject to}$$
$$g_i(\mathbf{x}) \geq 0 \text{ for } i = 1, 2, \ldots, p$$
$$h_j(\mathbf{x}) = 0 \text{ for } j = 1, 2, \ldots, s$$

where $\mathbf{x}$ is an n component vector. By using barrier and penalty functions, the requirements of the constraints can be included with the function to be minimized so that the problem is converted to the unconstrained problem:

$$\text{Minimize} \quad f(\mathbf{x}) - r_k \sum_{i=1}^{p} \log_e(g_i(\mathbf{x})) + \frac{1}{r_k} \sum_{j=1}^{s} h_j(\mathbf{x})^2$$

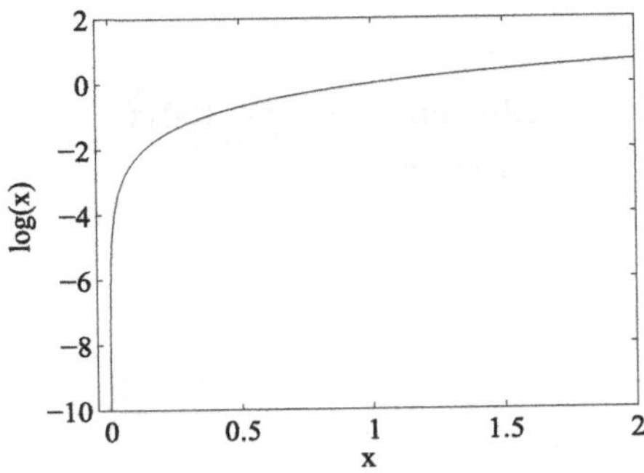

FIGURE 9.12

Graph of $\log_e(x)$.

Notice the effect of the added terms. The first term imposes a barrier at zero on the inequality constraints in that as the $g_i(\mathbf{x})$ approaches zero the function approaches minus infinity, thus imposing a substantial penalty. Fig. 9.12 illustrates this. The last term encourages the satisfaction of the equality constraints $h_j(\mathbf{x}) = 0$ since the smallest amount is added when all the constraints are zero; otherwise a substantial penalty is imposed. This means that this approach encourages the maintenance of the feasibility of the solution assuming we start with an initial solution which is within the feasible region of the inequality constraints. These methods are sometimes called interior point methods.

A sequence of problems are generated by starting with an arbitrarily large value for r_0 and then using $r_{k+1} = r_k/c$ where $c > 1$ and solving the resulting sequence of unconstrained optimization problems. The unconstrained minimization steps may of course present formidable difficulties for some problems. A simple stopping criterion is to examine the difference between the value of $f(\mathbf{x})$ between successive unconstrained optimizations. If the difference is less than a specified tolerance then stop the procedure.

There are various alternatives to this algorithm: for example, a reciprocal barrier function can be used instead of the preceding logarithmic function. The barrier term can be replaced by a penalty function term of the form

$$\sum_{i=1}^{p} \max(0, g_i(\mathbf{x}))^2$$

This term will add a substantial penalty if $g_i(\mathbf{x}) < 0$; otherwise, no penalty is applied. This method has the advantage that feasibility is not required and is called an exterior point method. However, the resulting unconstrained problems may present additional problems for the unconstrained minimization procedure. For more details on these methods see Lasdon et al. (1996).

Although purpose-built software is available for this method, a very simple illustration of its operation is given. The program shows some steps of the method for solving a constrained minimization problem notice that care must be taken in the choice of the initial r value and its reduction factor. To avoid the need for the derivative of the objective function, of the MATLAB function `fminsearch` is used in the solution of the following unconstrained problem. We first use the interior point method.

$$\text{Minimize } f = x_1^2 + 100x_2^2$$
$$\text{subject to}$$
$$4x_1 + x_2 \geq 6$$
$$x_1 + x_2 = 3$$
$$x_1, x_2 \geq 0$$

```
% e4s906.m
r = 10; x0 = [5 5]
while r>0.01
    fm = @(x) x(1).^2+100*x(2).^2-r*log(-6+4*x(1)+x(2)) ...
        +1/r*(x(1)+x(2)-3).^2-r*log(x(1))-r*log(x(2));
    x1 = fminsearch(fm,x0);
    r = r/5;
    x0 = x1
end
optval = x1(1).^2+100*x1(2).^2
```

Running script `e4s906.m` gives

```
x0 =
     5     5

x0 =
    3.6097    0.2261

x0 =
    2.1946    0.1035

x0 =
    2.2084    0.0553

x0 =
    2.7463    0.0377

x0 =
    2.9217    0.0317

optval =
    8.6366
```

This shows convergence to the optimum solution (3, 0) which satisfies the constraints. Using the exterior point method to solve the same problem we have:

```
% e4s907.m
r = 10; x0 = [5 5]
while r>0.01
    fm = @(x) x(1).^2+100*x(2).^2+1/r*min(0,(-6+4*x(1)+x(2))).^2 ...
     +1/r*(x(1)+x(2)-3).^2+1/r*min(0,x(1)).^2+1/r*min(0,x(2)).^2;
    x1 = fminsearch(fm,x0);
    r = r/5;
    x0 = x1
end
optval = x1(1).^2+100*x1(2).^2
```

Script `e4s907.m` gives:

```
x0 =
     5     5

x0 =
    0.9993    0.0040

x0 =
    1.4186    0.0094

x0 =
    2.1276    0.0213

x0 =
    2.7523    0.0275

x0 =
    2.9239    0.0293

optval =
    8.6350
```

Clearly the results are very similar.

9.13 SUMMARY

In this chapter we have introduced a number of more advanced areas of numerical analysis. The genetic algorithm, differential evolution, and simulated annealing are still topics for active research and development and are mainly used to tackle difficult optimization problems. The conjugate gradient method is a well established and widely used for problems which present a range of difficulties. In this chapter MATLAB functions are provided to allow the reader to experiment and explore the problems more deeply. However, it must be remembered that no optimization technique is guaranteed to solve all optimization problems. The structure of the algorithms is well reflected by the structure of the MATLAB functions.

The MathWorks Optimization toolbox provides a useful range of optimization functions which may be used in both education and research.

Note: For readers who a have particular interest in applying MATLAB to optimization problems, MATLAB provides an Optimization Toolbox.

9.14 PROBLEMS

9.1. Use the function `barnes` to Minimize $z = 5x_1 + 7x_2 + 10x_3$ subject to $x_1 + x_2 + x_3 \geq 4$, $x_1 + 2x_2 + 4x_3 \geq 5$, and $x_1, x_2, x_3 \geq 0$.

9.2. Maximize $p = 4y_1 + 5y_2$ subject to $y_1 + y_2 \leq 5$, $y_1 + 2y_2 \leq 7$, $y_1 + 4y_2 \leq 10$, and $y_1, y_2 \geq 0$. By introducing slack variables and subtracting one from each equality, write the constraints as equalities. Hence apply the function `barnes` to solve this problem. Notice that the optimum value of p for this problem is equal to the optimum value of z in Problem 9.1. Problem 9.2 is the dual of Problem 9.1. Thus, the optima of the objective function of Problem 9.1 and its dual, Problem 9.2, should be equal.

9.3. Maximize $z = 2u_1 - 4u_2 + 4u_3$ subject to $u_1 + 2u_2 + u_3 \leq 30$, $u_1 + u_2 = 10$, $u_1 + u_2 + u_3 \geq 8$, and $u_1, u_2, u_3 \geq 0$ using the function `barnes`.

Hint: Remember to use slack variables to ensure that the main constraints are equalities.

9.4. Use the function `mincg`, with tolerance 0.005, to minimize Rosenbrock's function

$$f(x, y) = 100\left(x^2 - y\right)^2 + (1 - x)^2$$

starting with the initial approximation $x = 0.5$, $y = 0.5$ and using a line-search accuracy 10 times the machine precision in the MATLAB function `fminsearch`. To obtain an impression of how this function varies, plot it, using MATLAB, in the range $0 \leq x \leq 2, 0 \leq y \leq 2$.

9.5. Use the function `mincg`, with tolerance 0.00005, to minimize the five-variable function

$$\begin{aligned} z = {} & 0.5(x_1^4 - 16x_1^2 + 5x_1) + 0.5(x_2^4 - 16x_2^2 + 5x_2) + \dots \\ & + (x_3 - 1)^2 + (x_4 - 1)^2 + (x_5 - 1)^2 \end{aligned}$$

Use $x_1 = 1$, $x_2 = 2$, $x_3 = 0$, $x_4 = 2$, and $x_5 = 3$ for the starting values in `mincg`. Experiment further with other starting values.

9.6. Use the function `solvercg` to solve the matrix equation $\mathbf{Ax} = \mathbf{b}$ where

$$\mathbf{A} = \begin{bmatrix} 5 & 4 & 1 & 1 \\ 4 & 5 & 1 & 1 \\ 1 & 1 & 4 & 2 \\ 1 & 1 & 2 & 4 \end{bmatrix} \quad \mathbf{b} = \begin{bmatrix} 1 \\ 2 \\ 3 \\ 4 \end{bmatrix}$$

Check the accuracy of the solution by finding the value of `norm(b-Ax)`.

9.7. Maximize the function $z(x, y) = 1/\{(x-1)^2 + 2\} + 1/\{(y-1)^2 + 2\}$ in the range $x, y = 0$ to 2 using the function `diffevo`. Use different initial population sizes, coefficients C and F, and numbers of generations. Notice that this is not a simple exercise since for each set of conditions it is necessary to solve the problem several times to take account of the random nature of the process. Given that the optimum value of the function is 0.5, plot the error in the optimum value of the function for each run under a particular set of parameters. Then change one of the parameters and repeat the process. Differences in the error plots may or may not be discernible.

9.8. Plot the function $z = x^2 + y^2$ in the range $-2 \le x \le 2$ and $-2 \le y \le 2$. The differential evolution function with appropriate parameters

9.9. Use the function `golden`, given in Section 9.3 to minimize the single-variable function: $y = e^{-x}\cos(3x)$ for x in the range 0 to 2. Use a tolerance of 0.00001. You should check your result by using the MATLAB function `fminsearch` to minimize the same function in the same range. As a further confirmation you might plot the function in the range 0 to 4 using the MATLAB function `fplot`.

9.10. Use the simulated annealing function `asaq` (with the same values of parameters used in the call of the function in Section 9.8) to minimize the two-variable function f, where

$$f = (x_1 - 1)^2 + 4(x_2 + 3)^2$$

Compare your result with the exact answer which is clearly, $x_1 = 1$ and $x_2 = -3$. This function has only one minimum in the region considered. As a more demanding test, use the same call as used for the preceding function to minimize the function f where

$$f = 0.5\left(x_1^4 - 16x_1^2 + 5x_1\right) + 0.5\left(x_2^4 - 16x_2^2 + 5x_2\right) - 10\cos\{4(x_1 + 2.9035)\}\cos\{4(x_2 + 2.9035)\}$$

The global optimum for this problem is $x_1 = -2.9035$ and $x_2 = -2.9035$. Try several runs of function `asaq` for this problem. All runs may not provide the global optimum since this problem has many local optima.

9.11. A method for solving a system of non-linear equations is to re-express them as an optimization problem. Consider the system of equations:

$$2x - \sin((x+y)/2) = 0$$
$$2y - \cos((x-y)/2) = 0$$

These can be rewritten as

$$\text{Minimize } z = (2x - \sin((x+y)/2))^2 + (2y - \cos((x-y)/2))^2$$

Use the MATLAB function `minscg` to minimize this function use the starting point $x = 10$ and $y = -10$.

9.12. Write a MATLAB script that provides three-dimensional plots for the function $z = f(x, y)$ defined as follows:

$$z = f(x, y) = (1-x)^2 e^{-p} - pe^{-p} - e^{(-(x+1)^2 - y^2)}$$

where p is defined by

$$p = x^2 + y^2$$

for x and y in the range $x = -4 : 0.1 : 4$ and $y = -4 : 0.1 : 4$. The script should use the MATLAB `surf` and `contour` functions to provide separate three-dimensional and contour plots. Use the MATLAB function `ginput` to select and assign to an appropriate matrix three points which appear to be optimal on the contour plot. Use the function $z = f(x, y)$ defined earlier to find the values of z for these points. Then find the maximum and minimum of these z values using the MATLAB functions `max` and `min`. Finally, approximate the global minimum and maximum of the function.

An alternative method of finding the minimum is to use the MATLAB function `fminsearch` in the form `x = fminsearch(funxy,xv)`, where `funxy` is an anonymous function or user-defined function given by the user and `xv = [-4 4]` is a vector of initial approximations to the location of the minimum. Experiment with different initial approximations to see if your results vary.

9.13. Solve the minimization problem described in Problem 9.12 using differential evolution.

9.14. Write a MATLAB script to minimize $f(x) = x_1^4 + x_2^2 + x_1$ subject to $4x_1^3 + x_2 \geq 6$, $x_1 + x_2 = 3$, and $x_1, x_2 \geq 0$. Use the sequential unconstrained minimization technique with a logarithmic barrier function and an initial approximation vector of $x = [5,\ 5]$. Use an initial value for the parameter r_0 of 10, a reduction parameter $c = 5$ and $r_{k+1} = r_k/c$. Continue iterations while r_k is greater than 0.0001.

9.15. Write a MATLAB script to solve the constrained optimization problem described in Problem 9.14. However, use the penalty function of the form $[\min(0,\ g_i(\mathbf{x}))]^2$ instead of the logarithmic barrier function where the $g_i(\mathbf{x})$ are the greater than or equal to the constraints. Use the same initial starting point and values for r_0 and c. Compare the solution you find with this method to that which was achieved with Problem 9.14.

9.16. Solve the Rosenbrock's two variable optimization problem:

$$\text{Minimize} \quad f(\mathbf{x}) = 100(x_2 - x_1^2)^2 + (1 - x_1)^2$$

Using the MATLAB function `asaq` with initial approximation [−1.2 1]. With quenching factor 0.9, the upper and lower bounds for the variables given by −10 and 10, respectively, initial temperature value of 100 and the maximum number of main iterations equal to 800. The solution to this problem is [1 1].

9.17. Using the MATLAB function given for the method of differential evolution, optimize the following functions:

(a) Rastrigin's function:

$$f(\mathbf{x}) = An + \sum_{i=1}^{n} [x_i^2 - A\cos(2\pi x_i)]$$

where $A = 10$. The region in which the function is defined is for x_i values in the range: $-5.12 \leq x_i \leq 5.12$ for all i. The global optimum is given by $f((x)) = 0$, where $\mathbf{x} = [0\,0\,0\ \ldots\ 0]$. Minimize this function for $n = 4$, 6, and 8.

(b) Easom's function is the two variable problem:

$$f(x, y) = -\cos(x)\cos(y)\exp\left[-\left((x-\pi)^2 + (y-\pi)^2\right)\right]$$

Determine the minimum of this function, searching in the range $-50 \leq x,\ y \leq 50$. The global solution of this problem is given by $f(\pi, \pi) = 1$. Increase the search range to $-60 \leq x,\ y \leq 60$ and then $-80 \leq x,\ y \leq 80$ and finally $-100 \leq x,\ y \leq 100$. Does this increase the difficulty of minimizing the function?

CHAPTER

10

APPLICATIONS OF THE SYMBOLIC TOOLBOX

Abstract

The Symbolic Toolbox provides an extensive list of functions for the symbolic manipulation of symbolic expressions and equations. In this chapter, we illustrate by examples the use of a wide range functions used for symbolic manipulation.

10.1 INTRODUCTION TO THE SYMBOLIC TOOLBOX

Originally, the MATLAB Symbolic Tool box used the Maplesoft symbolic software to carry out the symbolic operations and pass the results to MATLAB. However, since late 2008 MATLAB has used MuPAD for its symbolic engine. This change has caused minor changes in the way results are presented, but the results are generally equivalent. Also, from 2016, the MATLAB live editor can be used with the symbolic toolbox.

Since we are also using the Symbolic Toolbox in the field of numerical analysis we begin by giving some examples of the beneficial application of this toolbox. It provides:

1. the symbolic first derivative of a given single-variable non-linear function which, for example, is required by Newton's method for the solution of single-variable non-linear equation, see Chapter 3,
2. the Jacobian for a system of non-linear simultaneous equations, see Chapter 3,
3. the symbolic gradient vector of a given non-linear function which is required, or example, for the conjugate gradient method for minimizing a non-linear function, see Chapter 9.

An important feature of the Symbolic Toolbox is that it allows an extra dimension of experimentation. For example, a user can test a given numerical algorithm by solving a test problem symbolically, providing it has a solution in the closed form, and compare this exact solution with a numerical solution. In addition, the study of the accuracy of computations can be enhanced using the Symbolic Toolbox variable-precision arithmetic feature. This feature allows the user to perform certain computations to an unlimited precision.

In Sections 10.2 through 10.14 we provide an introduction to some of the Symbolic Toolbox features but it is not our intention to provide details of all the features. In Section 10.15 we describe applications of the Symbolic Toolbox to specific numerical algorithms.

Numerical Methods. https://doi.org/10.1016/B978-0-12-812256-3.00019-1

10.2 SYMBOLIC VARIABLES AND EXPRESSIONS

The first key point to note is that symbolic variables and expressions are different from the standard variables and expressions of MATLAB, and we must distinguish clearly between them. Symbolic variables and symbolic expressions are not required to have numerical values but rather define a structural relationship between symbolic variables, i.e. an algebraic expression.

To define any variable as a symbolic variable the `sym` function must be used thus:

```
>> x = sym('x')

x =
x

>> d1 = sym('d1')

d1 =
d1
```

Alternatively, we can use the statement `syms` to define any number of symbolic variables. Thus, for example

```
>> syms a b c d3
```

provides four symbolic variables `a`, `b`, `c`, and `d3`. Note that there is no output to the screen. The key word `syms` is a useful short-cut for the definition of variables and we have used this approach in this text. To check which variables have been declared as symbolic we can use the standard `whos` command. Thus if we use this command after the `syms` declaration, above we obtain

```
>> whos
  Name      Size            Bytes  Class    Attributes

  a         1x1                 8  sym
  b         1x1                 8  sym
  c         1x1                 8  sym
  d1        1x1                 8  sym
  d3        1x1                 8  sym
  x         1x1                 8  sym
```

Once variables have been defined as symbolic then expressions can be written using them directly in MATLAB and they will be treated as symbolic expressions. For example, once `x` has been defined as a symbolic variable, the statement

```
>> syms x
>> 1/(1+x)
```

produces the symbolic expression

```
ans =
1/(x + 1)
```

To set up a symbolic matrix we first define any symbolic variables involved in the matrix. Then we enter the statement which defines the matrix in terms of these symbolic variables in the usual manner. On execution the matrix will be displayed as shown below.

```
>> syms x y
>> d = [x+1 x^2 x-y;1/x 3*y/x 1/(1+x);2-x x/4 3/2]

d =
[ x + 1,      x^2,       x - y]
[   1/x, (3*y)/x, 1/(x + 1)]
[ 2 - x,      x/4,         3/2]
```

Note that d is automatically made symbolic by the assignment of a symbolic expression. We can address individual elements or specific rows and columns in the usual way, thus:

```
>> d(2,2)

ans =
(3*y)/x

>> c = d(2,:)

c =
[ 1/x, (3*y)/x, 1/(x + 1)]
```

We now consider the manipulation of a symbolic expression. First we set up a symbolic expression thus:

```
>> e = (1+x)^4/(1+x^2)+4/(1+x^2)

e =
(x + 1)^4/(x^2 + 1) + 4/(x^2 + 1)
```

To see more clearly what this expression represents we can use the function pretty to get a more conventional layout of the function thus:

```
>> pretty(e)

          4
   (x + 1)      4
   -------- + ------
     2          2
    x  + 1     x  + 1
```

Not very pretty! We can simplify the symbolic expression e using the function simplify:

```
>> simplify(e)

ans =
x^2 + 4*x + 5
```

We can expand expressions using `expand` thus:

```
>> syms x p; p = expand((1+x)^4)

p =
x^4 + 4*x^3 + 6*x^2 + 4*x + 1
```

Note the layout of this and other expressions may vary slightly from one computer platform to another. The expression for `p`, in turn, can be rearranged into a nested form using the function `horner` thus:

```
>> horner(p)

ans =
x*(x*(x*(x + 4) + 6) + 4) + 1
```

We may factorize expressions using the function `factor`. Declaring `a`, `b`, and `c` as symbolic variables, then

```
>> syms a b c
>> factor(a^3+b^3+c^3-3*a*b*c)

ans =
[ a + b + c, a^2 - a*b - a*c + b^2 - b*c + c^2]
```

This has produced 2 factors.

The function `simplify`, as its name suggests, attempts to simplify expressions. Thus

```
>> syms a b
>> simplify(a^3+3*a^2*b+3*b^2*a+b^3)

ans =
(a + b)^3

>> simplify((1+a)/(1+a)^2)

ans =
1/(a + 1)
```

When manipulating algebraic, trigonometric, and other expressions, it is important to be able to substitute an expression or constant for any given variable. For example,

```
>> syms u v w
>> fmv = pi*v*w/(u+v+w)

fmv =
(pi*v*w)/(u + v + w)
```

Now we substitute for various variables in this expression. The following statement substitutes the symbolic expression `2*v` for the variable `u`:

```
>> subs(fmv,u,2*v)

ans =
(v*w*pi)/(3*v + w)
```

The following statement substitutes the symbolic constant `1` for the variable `v` in the previous result held in `ans`:

```
>> subs(ans,v,1)

ans =
(pi*w)/(w + 3)
```

Finally, we substitute the symbolic constant `1` for `w` to give

```
>> subs(ans,w,1)

ans =
    pi/4
```

As a further example of the use of the `subs` function, consider the statements:

```
>> syms y
>> f = 8019+20412*y+22842*y^2+14688*y^3+5940*y^4 ...
                                 +1548*y^5+254*y^6+24*y^7+y^8;
```

Now to substitute `x-3` for `y` we use:

```
>> subs(f,y,x-3)

ans =
20412*x + 22842*(x - 3)^2 + 14688*(x - 3)^3 + 5940*(x - 3)^4
   + 1548*(x - 3)^5 + 254*(x - 3)^6 + 24*(x - 3)^7 + (x - 3)^8 - 53217
```

Using the `collect` function to collect terms having the same power of `x`, we obtain a major simplification as follows:

```
>> collect(ans)

ans =
x^8 + 2*x^6
```

The process of rearranging and simplifying algebraic and transcendental expressions is difficult and the Symbolic Toolbox can be both powerful and frustrating. Powerful because, as we have shown, it is capable of simplifying complex expressions; frustrating because it sometimes fails on relatively simple problems.

10.3 VARIABLE PRECISION ARITHMETIC IN SYMBOLIC CALCULATIONS

In symbolic calculations involving numerical values, the function `vpa`, standing for variable precision arithmetic, may be used to obtain any number of decimal places. It should be noted that the result of using this function is a symbolic constant, not a numerical value. Thus to provide $\sqrt{6}$ to 100 places we write `vpa(sqrt(6),100)`. Clearly the accuracy here is not restricted to 16 decimal places as in ordinary arithmetic calculations. A nice illustration of this feature is given below where we provide an implementation of the famous algorithm of the Borwein's which amazingly quadruples the number of accurate decimal places of π at each iteration!We include script `e4sX01.m` for illustration only; MATLAB can give π to as many digits as required by writing, for example, `vpa(pi,100)`.

```
% Script e4sX01.m  Borwein iteration for pi
n = input('Enter number of iterations');
n = n-1;
y0 = sqrt(2)-1; a0 = 6-4*sqrt(2);
np = 4;
for k = 0:n
    yv = (1-y0^4)^0.25; y1 = (1-yv)/(1+yv);
    a1 = a0*(1+y1)^4-2.0^(2*k+3)*y1*(1+y1+y1^2);
    rpval = a1;  pval = vpa(1/rpval,np)
    a0 = a1; y0 = y1; np = 4*np;
end
```

The results of four iterations are given below

```
enter n 3

n =
     3

pval =
3.142

pval =
3.141592653589793

pval =
3.1415926535897932384626433832795028841971693993751058209749445923

pval =
3.1415926535897932384626433832795028841971693993751058209749445923
0781640628620899862803482534211706798214808651328230664709384460
9550582231725359408128481117450284102701938521105559644622948954
9303819644288109756659334461284756482337867831652712019091456491

pval-vpa(pi,100)
```

```
ans =
2.0908055377181642372760353024684e-108
```

A very small difference!

Theoretically, the function vpa can be utilized to give results to any number of decimal places. An excellent introduction to the computation of π is given by Bailey (1988).

10.4 SERIES EXPANSION AND SUMMATION

In this section we consider both the development of series approximations for functions and the summation of series.

We begin by showing how a symbolic function can be expanded in the form of a Taylor series using the MATLAB function taylor(f,n). This provides the $(n-1)$th-degree polynomial approximation for the symbolically defined function f. If the taylor function has only one parameter, then it provides the fifth-degree polynomial approximation for that function. Consider the following examples:

```
>> syms x
>> taylor(cos(exp(x)),'Order',4)

ans =
- (cos(1)*x^3)/2 + (- cos(1)/2 - sin(1)/2)*x^2 - sin(1)*x + cos(1)
```

This provides an expansion to x^3.

```
>> s = taylor(exp(x),'Order',8)

s =
x^7/5040 + x^6/720 + x^5/120 + x^4/24 + x^3/6 + x^2/2 + x + 1
```

The series expansion for the exponential function can be summed using the function symsum with $x = 0.1$. To use this function we must know the form of the general term. In this case the series definition is

$$e^{0.1} = \sum_{r=1}^{\infty} \frac{0.1^{r-1}}{(r-1)!} \text{ or } \sum_{r=1}^{\infty} \frac{0.1^{r-1}}{\Gamma(r)}$$

Thus, to sum the first eight terms, that is 1 to 8, we have

```
>> syms r
>> symsum((0.1)^(r-1)/gamma(r),1,8)

ans =
55700614271/50400000000
```

This can then be evaluated using the function double thus:

```
>> double(ans)

ans =
     1.1052
```

In this example, a simple alternative is to use the function subs to replace x by 0.1 in the symbolic function represented by s, and then use double to evaluate it, as follows:

```
>> double(subs(s,x,0.1))

ans =
     1.1052
```

This gives a good approximation since we can see that, as a non-symbolic process,

```
>> exp(0.1)

ans =
     1.1052
```

The function symsum can be used to perform many different summations using different combinations of parameters. The following examples illustrate the different cases. To sum the series

$$S = 1 + 2^2 + 3^2 + 4^2 + \ldots + n^2 \tag{10.1}$$

for 1 to n we proceed as follows:

```
>> syms r n
>> symsum(r*r,1,n)

ans =
(n*(2*n + 1)*(n + 1))/6
```

Another example is to sum the series

$$S = 1 + 2^3 + 3^3 + 4^3 + \ldots + n^3 \tag{10.2}$$

Here we use

```
>> symsum(r^3,1,n)

ans =
(n^2*(n + 1)^2)/4
```

Infinity can be used as an upper limit. As an example of this, consider the following infinite sum:

$$S = 1 + \frac{1}{2^2} + \frac{1}{3^2} + \frac{1}{4^2} + \ldots \frac{1}{r^2} + \ldots$$

Summing an infinity of terms, we have

```
>> symsum(1/r^2,1,inf)

ans =
pi^2/6
```

This is an interesting series and is a particular case of the Riemann Zeta function (implemented in MATLAB by `zeta(k)`) which gives the sum of the following series

$$\zeta(k) = 1 + \frac{1}{2^k} + \frac{1}{3^k} + \frac{1}{4^k} + \ldots + \frac{1}{r^k} + \ldots \tag{10.3}$$

For example:

```
>> zeta(2)

ans =
    1.6449

>> zeta(3)

ans =
    1.2021
```

A further interesting example is a summation involving the gamma function written $\Gamma(r)$, where $\Gamma(r) = 1 \times 2 \times 3 \times \ldots \times (r-2)(r-1) = (r-1)!$ for integer values of r. This function is implemented in MATLAB by `gamma(r)`. For example, to sum the following series to infinity

$$S = 1 + \frac{1}{1} + \frac{1}{2!} + \frac{1}{3!} + \ldots + \frac{1}{r!} + \ldots$$

we use the MATLAB statement

```
>> symsum(1/gamma(r),1,inf)

ans =
exp(1)

>> vpa(ans,100)

ans =
2.718281828459045235360287471352662497757247093699959574966967627
724076630353547594571382178525166427
```

Notice that the use of the `vpa` function leads to an interesting evaluation of e to a large number of decimal places.

A further example of summation is given by

$$S = 1 + \frac{1}{1!} + \frac{1}{(2!)^2} + \frac{1}{(3!)^2} + \frac{1}{(4!)^2} + \ldots$$

In MATLAB, this becomes

```
>> symsum(1/gamma(r)^2,1,inf)

ans =
hypergeom([], 1, 1)

>> vpa(hypergeom([], 1, 1),20)

ans =
2.2795853023360668 21
```

Gauss's hypergeometric function is described by Abramowitz and Stegun (1965) or Olver et al. (2010). It can be evaluated under certain conditions; use MATLAB `help hypergeom` for further information.

10.5 MANIPULATION OF SYMBOLIC MATRICES

Some of the MATLAB functions, such as `eig`, that can be applied to numerical matrices can also be applied directly to symbolic matrices. However, these features must be used with care for two reasons. Firstly, manipulating large symbolic matrices can be a very slow process. Secondly, the symbolic results derived from such operations can be of such algebraic complexity that it is difficult or almost impossible to obtain insight into the meaning or structure of the matrix.

We will begin by finding the eigenvalues of a simple 4×4 matrix expressed in terms of two symbolic variables.

```
>> syms a b
>> Sm = [a b 0 0;b a b 0;0 b a b;0 0 b a]

Sm =
[ a, b, 0, 0]
[ b, a, b, 0]
[ 0, b, a, b]
[ 0, 0, b, a]

>> eig(Sm)

ans =
 a - b/2 - (5^(1/2)*b)/2
 a - b/2 + (5^(1/2)*b)/2
 a + b/2 - (5^(1/2)*b)/2
 a + b/2 + (5^(1/2)*b)/2
```

In this problem the expressions for the eigenvalues are quite simple. In contrast, we will now consider an example which appears to be equally simple in its original structure but develops into a problem with eigenvalues that are much more complicated.

```
>> syms A p
>> A = [1 2 3;4 5 6;5 7 9+p]

A =
[ 1, 2,     3]
[ 4, 5,     6]
[ 5, 7, p + 9]
```

Inspection of this matrix reveals that if $p = 0$, the matrix is singular. Evaluating the determinant of the matrix, we have

```
>> det(A)

ans =
-3*p
```

From this simple result it can immediately be seen that as p tends to zero, the determinant of the matrix tends to zero, indicating that the matrix is singular. Inverting the matrix `A` gives

```
>> B = inv(A)

B =
[    -(5*p + 3)/(3*p), (2*p - 3)/(3*p),  1/p]
[ (2*(2*p + 3))/(3*p),  -(p - 6)/(3*p), -2/p]
[                -1/p,            -1/p,  1/p]
```

This is much more difficult to interpret, although it can be seen that as p tend to zero, each element of the inverse matrix tends to infinity; i.e., the inverse does not exist. Note that every element of this inverse matrix is a function of p whereas only one element of the original matrix is a function of p. Finally, we can compute the eigenvalues of the original matrix using the statement `v = eig(A)`. The value of the symbolic object `v` is not shown here because it is so long and complicated. We can find out how many characters (including spaces) that are required to express the three eigenvalues symbolically as follows:

```
>>n = length(char(v))

n =
1722
```

This output is very difficult to read, let alone understand. Using the pretty print facility (`pretty`) improves the situation, but the output is still requires 106 character spaces per line – far more than can be displayed on this page.

The following scripts compute the eigenvalues both symbolically and numerically. In each case the parameter p is varied from 0 to 2 in steps of 0.1. However, only the eigenvalues which correspond to $p = 0.9$ and 1.9 are displayed. The script `e4sX02.m` determines the eigenvalues symbolically, and then the values of p are substituted into the symbolic eigenvalue expressions using the function subs; then the function double is used to provide the numerical results.

```
% e4sX02.m
disp('Script 1; Symbolic - numerical solution')
c = 1; v = zeros(3,21);
tic
syms a p u w
a = [1 2 3;4 5 6;5 7 9+p];
w = eig(a);  u = [ ];
for s = 0:0.1:2
    u = [u,subs(w,p,s)];
end
v = sort(real(double(u)));
toc
v(:,[10 20])
```

Running script e4sX02.m gives

```
Script 1; Symbolic - numerical solution
Elapsed time is 0.423691 seconds.

ans =
   -0.4255   -0.4384
    0.3984    0.7854
   15.9270   16.5530
```

An alternative approach is to find the eigenvalues of the same matrix by substituting numerical values of p into the numerical matrix and then computing the eigenvalues. The script e4sX03.m carries out this process. Again, only the eigenvalues for $p = 0.9$ and 1.9 are displayed.

```
% e4sX03.m
disp('Script 2: Numerical solution')
c = 1; v = zeros(3,21);
tic
for p = 0:.1:2
    a = [1 2 3;4 5 6;5 7 9+p];
    v(:,c) = sort(eig(a));
    c = c+1;
end
toc
v(:,[10 20])
```

Running script e4sX03.m gives

```
Script 2: Numerical solution
Elapsed time is 0.000415 seconds.
```

```
ans =
   -0.4255   -0.4384
    0.3984    0.7854
   15.9270   16.5530
```

As expected, the methods give identical results and show that the eigenvalues are real. Note that the symbolic approach is much slower.

We conclude this section with an example which illustrates the advantage of the symbolic approach for certain problems. We wish to find the eigenvalues of a matrix which can be generated using the MATLAB statement `gallery(5)`. We will begin by finding the eigenvalues of this matrix in a non-symbolic way.

```
>> B = gallery(5)

B =
          -9          11         -21          63        -252
          70         -69         141        -421        1684
        -575         575       -1149        3451      -13801
        3891       -3891        7782      -23345       93365
        1024       -1024        2048       -6144       24572

>> format long e
>> eig(B)

ans =
     -3.472940132398842e-02 + 2.579009841174434e-02i
     -3.472940132398842e-02 - 2.579009841174434e-02i
      1.377760760018629e-02 + 4.011025813393478e-02i
      1.377760760018629e-02 - 4.011025813393478e-02i
      4.190358744843689e-02 + 0.000000000000000e+00i
```

The eigenvalues appear to be small. One eigenvalue is real and the remainder are complex conjugate pairs. However, using the symbolic approach we have

```
>> A = sym(gallery(5))

A =
[   -9,    11,   -21,     63,   -252]
[   70,   -69,   141,   -421,   1684]
[ -575,   575, -1149,   3451, -13801]
[ 3891, -3891,  7782, -23345,  93365]
[ 1024, -1024,  2048,  -6144,  24572]

>> eig(A)
```

```
ans =
 0
 0
 0
 0
 0
```

How do we verify which of these two solutions is correct? If we rearrange the eigenvalue problem into the form given by (2.41), that is,

$$(\mathbf{B} - \lambda\mathbf{I})\,\mathbf{x} = 0$$

then the eigenvalues are the roots of $|\mathbf{A} - \lambda\mathbf{I}| = 0$. We can find these roots symbolically in MATLAB thus:

```
>> syms lambda
>> D = A-lambda*sym(eye(5));
>> det(D)

ans =
-lambda^5
```

We have shown that $|\mathbf{A} - \lambda\mathbf{I}| = -\lambda^5$ and hence the eigenvalues are the roots of $-\lambda^5 = 0$, i.e., zero. Here the Symbolic Toolbox has revealed the true solution.

10.6 SYMBOLIC METHODS FOR THE SOLUTION OF EQUATIONS

The function available in MATLAB to solve symbolic equations is `solve`. This function is most useful for solving polynomials since it provides expressions for all roots. To use `solve` we must set up the expression for the equation we wish to solve in terms of a symbolic variable. For example,

```
>> syms x
>> f = x^3-7/2*x^2-17/2*x+5

f =
x^3 - (7*x^2)/2 - (17*x)/2 + 5

>> solve(f)

ans =
  -2
 1/2
   5
```

The following example illustrates the solution of a system of two simultaneous equations in two variables. In this example the two equations must be converted from string to symbolic form using the function `str2sym`.

```
>> syms x y
>> [x y] = solve(str2sym('x^2+y^2=a'),str2sym('x^2-y^2=b'))
```

This gives the four solutions:

```
x =
 -(2^(1/2)*(a + b)^(1/2))/2
  (2^(1/2)*(a + b)^(1/2))/2
 -(2^(1/2)*(a + b)^(1/2))/2
  (2^(1/2)*(a + b)^(1/2))/2

y =
 -(2^(1/2)*(a - b)^(1/2))/2
 -(2^(1/2)*(a - b)^(1/2))/2
  (2^(1/2)*(a - b)^(1/2))/2
  (2^(1/2)*(a - b)^(1/2))/2
```

A simple check of this solution may be obtained by inserting these solutions back in the original equations thus:

```
>> x.^2+y.^2, x.^2-y.^2

ans =
 a
 a
 a
 a

ans =
 b
 b
 b
 b
```

Hence all four solutions satisfy the equations simultaneously.

In practice it is rarely possible to determine the symbolic solutions of general single-variable or multi-variable non-linear equations and, where possible, a numeric approximation is provided. For example:

```
>> f = sin(x^3+log(x));
>> solve(f)
Warning: Unable to solve symbolically. Returning a numeric approximation instead.
> In solve (line 304)

ans =
98.395389162103351442766673981852
```

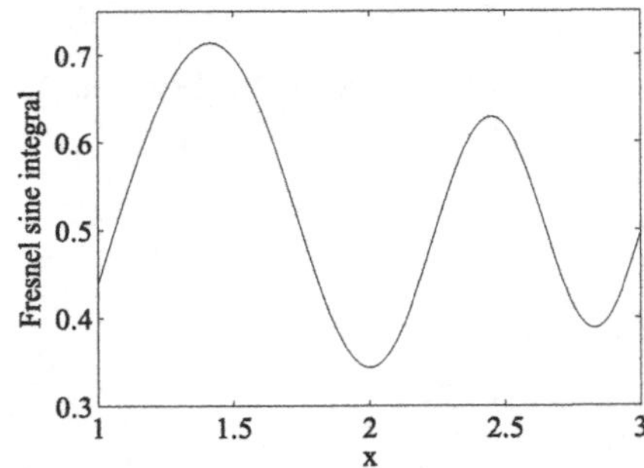

FIGURE 10.1

Plot of the Fresnel sine integral in the range $x = 1$ to $x = 3$.

10.7 SPECIAL FUNCTIONS

The MATLAB Symbolic Toolbox provides the user with access to a large number special functions and polynomials. One of these functions is the Fresnel sine integral. This is defined by

$$S(x) = \int_0^x \sin(t^2)\,dt$$

To evaluate this integral for $x = 4.2$ we use the function `fresnels` as follows:

```
>> x = 4.2; y = fresnels(x)

y =
    0.5632
```

Fig. 10.1 provides a plot of this function using the following commands:

```
>> x=1:.01:3; y = fresnels(x);
>> plot(x,y)
>> xlabel('x'), ylabel('Fresnel sine integral')
```

Another interesting function is the logarithmic integral defined by

$$\mathrm{li}(x) = \int_0^x \frac{dt}{\log_e(t)}$$

To evaluate the rounded value for this integral for $x = 1000$, 10,000 and 100,000 we use the `logint` as follows:

```
>> y = round(logint([1000 10000 100000]))

y =
         178          1246          9630
```

The logarithmic integral can be used to estimate the number of primes below a particular value, and the estimation becomes more accurate as the number of primes increases. To find the number of primes below 1000, 10,000 and 100,000 we use the `primes` function to list the primes below 1000, 10,000 and 100,000 as follows:

```
>> p1 = length(primes(1000)); p2 = length(primes(10000));
>> p3 = length(primes(100000)); p = [p1 p2 p3]

p =
         168          1229          9592
```

Taking the ratio between the values of the logarithmic integral `y` and the number of primes `p`, we have

```
>> p./y

ans =
    0.9438    0.9864    0.9961
```

Notice that the ratio of the value of the logarithmic integral and the number of primes less than N tends to 1 as N tends to infinity.

Two other important functions are the Dirac and Heaviside functions. Information about them can be obtained using the command `help` in the usual way.

The Dirac delta, or impulse, function, $\delta(x)$, is defined as follows:

$$\delta(x - x_0) = 0 \text{ where } x \neq x_0 \tag{10.4}$$

$$\int_{-\infty}^{\infty} f(x)\,\delta(x - x_0)\,dx = f(x_0) \tag{10.5}$$

If $f(x) = 1$, then

$$\int_{-\infty}^{\infty} \delta(x - x_0)\,dx = 1 \tag{10.6}$$

By (10.4), the delta function only exists at $x = x_0$. Furthermore, from (10.6), the area under the delta function is unity. This function is implemented in MATLAB by `dirac(x)`.

The Heaviside, or unit step, function is defined as follows:

$$u(x - x_0) = \begin{cases} 1, & \text{for } x - x_0 > 0 \\ 0.5, & \text{for } x - x_0 = 0 \\ 0, & \text{for } x - x_0 < 0 \end{cases} \tag{10.7}$$

This function is implemented in MATLAB by `heaviside(x)`. As an example, consider

```
>> heaviside(2)

ans =
     1
```

In Sections 10.8, 10.10, and 10.14, we consider various aspects of symbolic manipulation and we illustrate some of the properties of the Dirac and Heaviside functions.

10.8 SYMBOLIC DIFFERENTIATION

Essentially any function can be differentiated symbolically but the process is not always easy. For example, the following function given by Swift (1977) is tedious to differentiate by hand:

$$f(x) = \sin^{-1}\left(\frac{e^x \tan x}{\sqrt{x^2+4}}\right)$$

Differentiating this function symbolically using the MATLAB function `diff` gives

```
>> syms x
>> diff(asin(exp(x)*tan(x)/sqrt(x^2+4)))

ans =
((exp(x)*(tan(x)^2 + 1))/(x^2 + 4)^(1/2) +
(exp(x)*tan(x))/(x^2 + 4)^(1/2) -
(x*exp(x)*tan(x))/(x^2 + 4)^(3/2))/
(1 - (exp(2*x)*tan(x)^2)/(x^2 + 4))^(1/2)

>> pretty(ans)
 /                  2                                               \
 | exp(x) (tan(x)  + 1)    exp(x) tan(x)    x exp(x) tan(x) |
 | -------------------- + ------------- - --------------- |
 |               2                     2               2      3/2   |
 \     sqrt(x  + 4)           sqrt(x  + 4)       (x  + 4)         /

                  /                        2 \
                  |      exp(2 x) tan(x)     |
           /sqrt| 1 - ----------------     |
                  |              2             |
                  \             x  + 4        /
```

We now explain specifically how this is achieved. Differentiation is performed on a specified function with respect to a specific variable. When using the Symbolic Toolbox, the variables or variable of the expression must be defined as symbolic. Once this is done, the expression will be treated as symbolic and differentiation may be performed. Consider the following example:

```
>> syms z k
>> f = k*cos(z^4);
>> diff(f,z)

ans =
-4*k*z^3*sin(z^4)
```

Note that the differentiation is performed with respect to `z`, the variable indicated in the second parameter. Differentiation may be performed with respect to `k` as follows:

```
>> diff(f,k)

ans =
cos(z^4)
```

If the variable with respect to which the differentiation is being performed is not indicated, MATLAB chooses the variable name alphabetically closest to x.

Higher-order differentiation can also be performed by including an additional integer parameter which indicates the order of the differentiation (4 in this case) thus:

```
>> syms n
>> diff(k*z^n,4)

ans =
k*n*z^(n - 4)*(n - 1)*(n - 2)*(n - 3)
```

The following example illustrates how we can obtain a standard numerical value from our symbolic differentiation:

```
>> syms x
>> f = x^2*cos(x);
>> df = diff(f)

df =
2*x*cos(x) - x^2*sin(x)
```

We can now substitute a numerical value for x thus:

```
>> subs(df,x,0.5)

ans =
cos(1/2) - sin(1/2)/4
```

Finally, we consider the symbolic differentiation of the Heaviside, or unit step, function, thus

```
diff(heaviside(x))

ans =
dirac(x)
```

This is as expected since the differentiation of the Heaviside function is zero for all x except when $x = 0$. This is the definition of the Dirac function.

10.9 SYMBOLIC PARTIAL DIFFERENTIATION

Partial derivatives of any multi-variable function can be found by differentiating with respect to each variable in turn. As an example, we set up a symbolic function of three variables, assigned to fmv, as follows:

```
>> syms u v w
>> fmv = u*v*w/(u+v+w)

fmv =
(u*v*w)/(u + v + w)

>> pretty(fmv)
    u v w
  ---------
  u + v + w
```

Now we differentiate with respect to u, v, and w in turn:

```
>> d = [diff(fmv,u) diff(fmv,v) diff(fmv,w)]

d =
[ (v*w)/(u + v + w) - (u*v*w)/(u + v + w)^2,
          (u*w)/(u + v + w) - (u*v*w)/(u + v + w)^2,
                 (u*v)/(u + v + w) - (u*v*w)/(u + v + w)^2]
```

To obtain mixed partial derivatives, we simply differentiate with respect to the second variable the answer obtained after differentiation with respect to the first. For example,

```
>> diff(d(3),u)

ans =
v/(u + v + w) - (u*v)/(u + v + w)^2
                    - (v*w)/(u + v + w)^2 + (2*u*v*w)/(u + v + w)^3
```

To clarify the structure of the above expression we use the function `pretty` thus:

```
>> pretty(ans)
      v             u v               v w             2 u v w
  --------- - ------------- - ------------- + -------------
  u + v + w               2               2               3
              (u + v + w)     (u + v + w)     (u + v + w)
```

This provides the mixed second-order partial derivative with respect to w and then u, since d3 is the derivative of the function fmv with respect to w.

10.10 SYMBOLIC INTEGRATION

The integration process presents more difficulties than differentiation because not all functions can be integrated to produce an algebraic expression. Even functions that can be integrated to produce an algebraic expression often require considerable skill and experience such an expression.

To use the Symbolic Toolbox, we begin by defining the symbolic expression `f`. Then the function `int(f,a,b)` performs symbolic integration where `a` and `b` are the lower and upper limits, respectively. This results in a symbolic constant. If the upper and lower limits are omitted, the result is an expression which is the formula for the indefinite integral. In either case, if the integral cannot be evaluated, which frequently happens, the original function is returned. It must be stressed that many integrals can only be evaluated numerically.

Consider the following indefinite integral:

$$I = \int u^2 \cos u \, du$$

In MATLAB this becomes

```
>> syms u
>> f = u^2*cos(u); int(f)

ans =
sin(u)*(u^2 - 2) + 2*u*cos(u)
```

We note that the result is a formula as expected and not a numerical value. However, if upper and lower limits are specified, we obtain a symbolic constant. For example,

$$y = \int_0^{2\pi} e^{-x/2} \cos(100x) \, dx$$

Thus we have

```
>> syms x, res = int(exp(-x/2)*cos(100*x),0,2*pi)

res =
2/40001 - (2*exp(-pi))/40001
```

We can obtain a numerical value for this using the `vpa` function as follows:

```
>> vpa(res)

ans =
0.000047838108134108034810408852920091
```

This result confirms the numerical solution given in Section 4.11.

The following examples require infinite limits. Limits of this type are easily accommodated using the symbol `inf` and `-inf`. Consider the following integrals:

$$y = \int_0^{\infty} \log_e \left(1 + e^{-x}\right) dx \text{ and } y = \int_{-\infty}^{\infty} \frac{dx}{(1 + x^2)^2}$$

These can be evaluated as follows:

```
>> syms x, int(log(1+exp(-x)),0,inf)

ans =
pi^2/12

>> syms x, f = 1/(1+x^2)^2;
>> int(f,-inf,inf)

ans =
pi/2
```

These results confirm those given by the numerical integration in Section 4.8.

We now consider what happens when an integral cannot be evaluated by symbolic means. In this case we must resort to numerical procedures to find an approximate numerical value for our integral.

```
>> p = sin(x^3);
>> int(p)

ans =
int(sin(x^3), x)
```

Note that the integral cannot be evaluated symbolically. If we now insert upper and lower limits for this integral, we have

```
>> res = int(p,0,1)

res =
hypergeom([2/3], [3/2, 5/3], -1/4)/4

>> vpa(res)

ans =
0.23384524559381660956535064328083
```

Note that the solution is returned in terms of Gauss's hypergeometric function; see Abramowitz and Stegun (1965) or Olver et al. (2010). It can be evaluated under certain conditions; see MATLAB `help hypergeom`.

We now consider two interesting examples which raise new issues in symbolic processing. Consider the two integrals

$$\int_0^\infty e^{-x}\log_e(x)\,dx \text{ and } \int_0^\infty \frac{\sin^4(mx)}{x^2}\,dx$$

We may evaluate the first integral as follows:

```
>> syms x; int(exp(-x)*log(x),0,inf)

ans =
-eulergamma

>> y = vpa(-eulergamma,20)

y =
-0.57721566490153286061
```

In MATLAB, `eulergamma` is Euler's constant and is defined by

$$C = \lim_{p\to\infty}\left[-\log_e p + \frac{1}{2} + \frac{1}{3} + \ldots + \frac{1}{p}\right] = 0.577215\ldots$$

Thus, MATLAB shows how Euler's constant arises and allows us to evaluate it to any specified number of significant figures.

Now, considering the second integral,

```
>> syms m, int(sin(m*x)^4/x^2,0,inf)

ans =
piecewise(0 < m, (pi*m)/4, in(m, 'real'), (pi*abs(m))/4, ~in(m, 'real'),
                                                int(sin(x*m)^4/x^2, x, 0, Inf))
```

This complicated MATLAB result attempts to provide, where possible, an evaluation of the integral for various ranges m. Finally, we integrate the Dirac function from $-\infty$ to ∞

```
>> int(dirac(x),-inf, inf)

ans =
1
```

This result accords with the definition of the Dirac or delta function, see (10.6).

We now consider the integral of the Heaviside function from −5 to 3.

```
>> int(heaviside(x),-5,3)

ans =
3
```

From the definition of the Heaviside function, (10.7), we see that this result is as expected.

Symbolic integration can be performed for a function of two variables by repeated application of the `int` function. Consider the double integral defined by (4.49) and repeated here for convenience:

$$\int_{x^2}^{x^4} dy \int_1^2 x^2 y\, dx$$

This can be evaluated symbolically thus:

```
>> syms x y; f = x^2*y;
>> int(int(f,y,x^2,x^4),x,1,2)

ans =
6466/77
```

which confirms the numerical result given on Section 4.15.2.

10.11 SYMBOLIC SOLUTION OF ORDINARY DIFFERENTIAL EQUATIONS

The Symbolic Toolbox can be used to solve, symbolically, first-order differential equations, systems of first-order differential equations or higher-order differential equations, together with any initial conditions provided. This symbolic solution of differential equations is implemented in MATLAB using the function `dsolve`, and its use is illustrated by a range of examples.

It is important to note that this approach only provides a symbolic solution if one exists. If no solution exists, then the user should apply one of the numerical techniques provided in MATLAB, such as `ode45`.

The general form of a call of the function `dsolve` to solve a differential equation system is

```
sol = dsolve('de1, de2, de3, ... , den, in1, in2, in3, ... , inn');
```

The independent variable is assumed to be `t` unless given by an optional final parameter of `dsolve`. The parameters `de1`, `de2`, `de3` up to `den` stand for the individual differential equations. These must be written in symbolic form using symbolic variables, standard MATLAB operators and the symbols `D`, `D2`, `D3`, `D4`, ..., etc., which represent the first-, second-, third-, and higher-order differential operators respectively. The parameters `in1`, `in2`, `in3`, `in4`, and so on, represent the initial conditions for the differential equations, if these conditions are required. An example of how these initial conditions should be written, assuming an dependent variable `y`, is

```
y(0) = 1,  Dy(0) = 0,  D2y(0) = 9.1
```

which means that the value of y is 1, $dy/dt = 0$ and $d^2y/dt^2 = 9.1$ when $t = 0$.

The solution is returned to `sol` as a MATLAB structure, and consequently the names of the dependent variables must be used to indicate the individual components. For example, if `g` and `y` are two dependent variables for the differential equation, `sol.y` gives the solution for the dependent variable `y` and `sol.g` gives the solution for the dependent variable `g`.

To illustrate these points we consider some examples. Consider a first-order differential equation:

$$\left(1+t^2\right)\frac{dy}{dt} + 2ty = \cos t$$

We may solve this without initial conditions using `dsolve` as follows:

```
>> s = dsolve('(1+t^2)*Dy+2*t*y=cos(t)')

s =
-(C4 - sin(t))/(t^2 + 1)
```

Notice that the solution contains the arbitrary constant C4. If we now solve the same equation using an initial condition, we proceed as follows:

```
>> s = dsolve('(1+t^2)*Dy+2*t*y=cos(t),y(0)=0')

s =
sin(t)/(t^2 + 1)
```

Note that in this case there is no arbitrary constant.

We now solve a second-order system

$$\frac{d^2y}{dx^2} + y = \cos 2x$$

with the initial conditions $y = 0$ and $dy/dx = 1$ at $x = 0$. To solve this differential equation dsolve has the form below, where x is the independent variable.

```
>> dsolve('D2y+y=cos(2*x), Dy(0)=1, y(0)=0','x')

ans =
(2*cos(x))/3 + sin(x) + sin(x)*(sin(3*x)/6 + sin(x)/2) -
  (2*cos(x)*(6*tan(x/2)^2 - 3*tan(x/2)^4 + 1))/(3*(tan(x/2)^2 + 1)^3)

>> simplify(ans)

ans =
(5*cos(x))/3 + sin(x) - (8*cos(x/2)^4)/3 + 1
```

We now solve a fourth-order differential equation:

$$\frac{d^4y}{dt^4} = y$$

with initial conditions

$$y = 1,\ dy/dt = 0,\ d^2y/dt^2 = -1,\ d^3y/dt^3 = 0 \text{ when } t = \pi/2.$$

We again use dsolve. In this example D4 stands for the fourth derivative operator, with respect to t, and so on.

```
>> dsolve('D4y=y, y(pi/2)=1, Dy(pi/2)=0, D2y(pi/2)=-1, D3y(pi/2)=0')

ans =
sin(t)
```

However, we note that if we try to solve the apparently simple problem

$$\frac{dy}{dx} = \frac{e^{-x}}{x}$$

with the initial condition $y = 1$ when $x = 1$, difficulties arise. Applying `dsolve`, we have

```
>> dsolve('Dy=exp(-x)/x, y(1)=1', 'x')

ans =
1 - expint(x) - ei(-1)
```

The function `expint(z)`, where z is real or complex, is the exponential integral function defined by

$$E_1(z) = -\int_z^{\infty} \frac{\exp(-t)}{t} dt \quad z \neq 0$$

$E_1(z)$ assumes its principal value. In fact $E_1(z)$ is a particular case of the more general $E_n(z)$ function.

Details of this mathematical function can be found in Abramowitz and Stegun (1965) and Olver et al. (2010). The function `ei(z)` is the one-argument exponential integral function and is defined, for $x > 0$ by

$$\mathrm{Ei}(x) = -\int_{-x}^{\infty} \frac{\exp(-t)}{t} dt = \int_{-\infty}^{x} \frac{\exp(t)}{t} dt \quad x > 0$$

The functions E_1 and Ei are related. Thus

$$E_1(x) = \int_x^{\infty} \frac{\exp(-t)}{t} dt = -\mathrm{Ei}(-x) \quad x > 0$$

and

$$E_1(-x) = -\mathrm{Ei}(x) - \jmath\pi$$

We can illustrate these relationships. For example, if $x = 3$,

```
>> x = 3;
>> [expint(x) -ei(-x)]

ans =
    0.0130    0.0130

>> [expint(-x) -ei(x)-j*pi]

ans =
  -9.9338 - 3.1416i  -9.9338 - 3.1416i
```

Returning to the solution of the differential equation, we can evaluate the `ei(-1)` as follows:

```
>> ei(-1)

ans =
-0.2194
```

Thus, for example, when $x = 2$

```
>> x = 2;
>> 1 - expint(x) - ei(-1)

ans =
     1.1705
```

We will now attempt to solve the following apparently simple differential equation:

$$\frac{dy}{dx} = \cos(\sin x)$$

Applying `dsolve` we have

```
>> dsolve('Dy=cos(sin(x))','x')

ans =
C14 + int(cos(sin(x)), x, 'IgnoreSpecialCases', true,
'IgnoreAnalyticConstraints', true)
```

In this case `dsolve` fails to solve the equation.

The differential equation may also contain constants represented as symbols. For example, if we wish to solve the equation

$$\frac{d^2x}{dt^2} + \frac{a}{b}\sin t = 0$$

with initial conditions $x = 1$ and $dx/dt = 0$ when $t = 0$. Thus

```
>> syms x t a b
>> x = dsolve('D2x+(a/b)*sin(t)=0,x(0)=1,Dx(0)=0')

x =
(a*sin(t))/b - (a*t)/b + 1
```

Note how the variables a and b appear in the solution as expected.

As an example of solving two simultaneous differential equations, we note that this differential equation may be rewritten as

$$\frac{du}{dt} = -\frac{a}{b}\sin t$$

$$\frac{dx}{dt} = u$$

Using the same initial conditions, `dsolve` may be applied to solve these equations by writing

```
>> syms u
>> [u x] = dsolve('Du+(a/b)*sin(t)=0,Dx=u,x(0)=1,u(0)=0')
```

```
u =
(a*cos(t))/b - a/b

x =
(a*sin(t))/b - (a*t)/b + 1
```

The same solution as that obtained from `dsolve` applied directly to solve the second-order differential equation.

The script `e4sX04.m` provides an interesting comparison between the symbolic and numerical approach. It consists of a script and the output from the script. The script compares the use of `dsolve` for the symbolic solution of a differential equation, and the use of `ode45` for the numerical solution of the same differential equation. Note that the symbolic solution is obtained in two ways: by solving the second-order equation directly and by separating it into two first-order simultaneous equations. Both approaches provide the same solution.

```
% e4sX04.m  Simultaneous first order differential equations
% dx/dt = y, Dy = 3*t-4*x.
% Using dsolve this becomes
syms y t x
x = dsolve('D2x+4*x=3*t','x(0)=0', 'Dx(0)=1')
tt = 0:0.1:5; p = subs(x,t,tt); pp = double(p);
% Plot the symbolic solution to the differential equ'n
plot(tt,pp,'r')
hold on
xlabel('t'), ylabel('x')
sol = dsolve('Dx=y','Dy=3*t-4*x', 'x(0)=0', 'y(0)=1');
sol_x = sol.x, sol_y = sol.y
fv = @(t,x) [x(2); 3*t-4*x(1)];
options = odeset('reltol', 1e-5,'abstol',1e-5);
tspan = [0 5]; initx = [0 1];
[t,x] = ode45(fv,tspan,initx,options);
plot(t,x(:,1),'k+');
axis([0 5 0 4])
```

Executing script `e4sX04.m` gives the symbolic solution

```
x =
 (3*t)/4 + sin(2*t)/8

sol_x =
(3*t)/4 + sin(2*t)/8

sol_y =
cos(2*t)/4 + 3/4
```

This script also provides the graph of the symbolic solution with alternate numerical solution values also plotted, Fig. 10.2. Note how the numerical solution is consistent with the symbolic one.

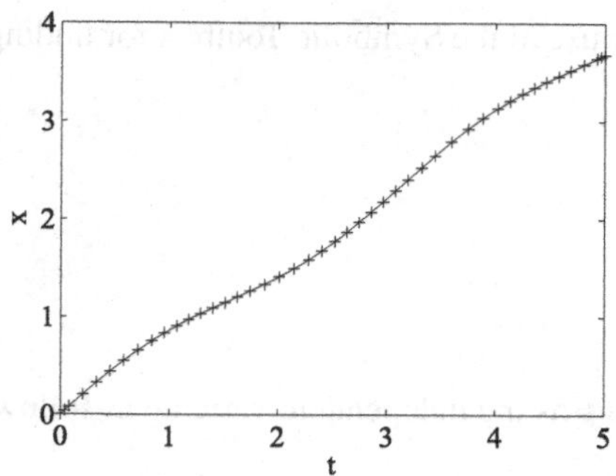

FIGURE 10.2

Symbolic solution and numerical solution indicated by +.

10.12 THE LAPLACE TRANSFORM

The Symbolic Toolbox allows the symbolic determination of the Laplace transforms of many functions. The Laplace transform is used to transform a linear differential equation into an algebraic equation in order to simplify the process of obtaining the solution. It can also be used to transform a system of *differential* equations into a system of *algebraic* equations. The Laplace transform maps a continuous function $f(t)$ in the t domain, where $f(t) = 0$ if $t < 0$, into the function $F(s)$ in the s domain. The domain s is complex and s can be expresses as $s = \sigma + \jmath\omega$ where σ and ω are real numbers. The Laplace transform is related to the Fourier transform, but whereas the Fourier transform of a function is a complex function of a real variable (frequency), the Laplace transform of a function is a complex function of a complex variable.

Let $f(t)$ have a finite origin which can be assumed to be at $t = 0$. In this case we can write

$$F(s) = \int_0^\infty f(t)e^{-st}\, dt \tag{10.8}$$

where $f(t)$ is a given function defined for all positive values of t and $F(s)$ is the Laplace transform of $f(t)$. This transform is called the one-sided Laplace transform. The parameter s is restricted so that the integral converges, and it should be noted that for many functions the Laplace transform does not exist. In general, computation of inverse Laplace transforms is difficult and requires techniques from complex analysis. The simplest inversion formula is given by the Bromwich integral as follows:

$$f(t) = \frac{1}{\jmath 2\pi} \int_{\sigma_0 - \jmath\infty}^{\sigma_0 + \jmath\infty} F(s)e^{st}\, ds \tag{10.9}$$

where $\jmath = \sqrt{-1}$, σ_0 is any real value, such that the contour $\sigma_0 - \jmath\omega$, for $-\infty < \omega < \infty$ is in the region of convergence of $F(s)$. In practice (10.9) is not used to compute the inverse transform. The Laplace transform of $f(t)$ may be denoted by the operator $\mathcal{L}$. Thus

$$F(s) = \mathcal{L}[f(t)] \text{ and } f(t) = \mathcal{L}^{-1}[F(s)]$$

We now give examples of the use of the Symbolic Toolbox for finding Laplace transforms of certain functions.

```
>> syms t
>> laplace(t^4)

ans =
24/s^5
```

We can use a variable other than `t` as the independent variable as follows:

```
>> syms x; laplace(heaviside(x))

ans =
1/s
```

In this brief introduction to the Laplace transform we will not discuss its properties but merely state the following results:

$$\mathcal{L}\left[\frac{df}{dt}\right] = sF(s) - f(0)$$

$$\mathcal{L}\left[\frac{d^2 f}{dt^2}\right] = s^2 F(s) - sf(0) - f^{(1)}(0)$$

where $f(0)$ and $f^{(1)}(0)$ are the values of $f(t)$ and its first derivative when $t = 0$. This pattern is continued for higher derivatives.

Suppose we wish to solve the following differential equation:

$$\ddot{y} - 3\dot{y} + 2y = 4t + e^{3t}, \ \ y(0) = 1, \ \ \dot{y}(0) = -1 \tag{10.10}$$

where the dot notation denotes differentiation with respect to time. Taking the Laplace transform of (10.10), we have

$$s^2 Y(s) - sy(0) - y^{(1)}(0) - 3\{sY(s) - y(0)\} + 2Y(s) = \mathcal{L}[4t + e^{3t}] \tag{10.11}$$

Now, from the definition of the Laplace transform or from tables, we can determine $\mathcal{L}[4t + e^{3t}]$. Here we use the Symbolic Toolbox to determine the required transform thus:

```
>> syms s t
>> laplace(4*t+exp(3*t))

ans =
1/(s - 3) + 4/s^2
```

Substituting this result in (10.11) and rearranging gives

$$(s^2 - 3s + 2)Y(s) = \frac{4}{s^2} + \frac{1}{s-3} - 3y(0) + sy(0) + y^{(1)}(0)$$

Applying the initial conditions and further rearrangement, we have

$$Y(s) = \left(\frac{1}{s^2 - 3s + 2}\right)\left(\frac{4}{s^2} + \frac{1}{s-3} - 4 + s\right)$$

To obtain the solution $y(t)$, we must determine the inverse transform of this equation. It has already been stated that in practice (10.9) is not used to compute the inverse transform. The usual procedure is to rearrange the transform into one that can be recognized in tables of Laplace transforms, and typically the method of partial fractions is used for this purpose. However, the MATLAB Symbolic Toolbox allows us to avoid this task and determine inverse transforms using the `ilaplace` statement thus:

```
>> ilaplace((4/s^2+1/(s-3)-4+s)/(s^2-3*s+2))

ans =
2*t - 2*exp(2*t) + exp(3*t)/2 - exp(t)/2 + 3
```

Thus $y(t) = 2t + 3 + 0.5(e^{3t} - e^t) - 2e^{2t}$.

10.13 THE Z-TRANSFORM

In the solution of difference equations representing discrete systems, the Z-transform plays a similar role to that of the Laplace transform in the solution of differential equations. The Z-transform is defined by

$$F(z) = \sum_{n=0}^{\infty} f_n z^{-n} \tag{10.12}$$

where f_n is a sequence of data beginning at f_0. The function $F(z)$ is called the unilateral or single sided Z-transform of f_n and is denoted $\mathcal{Z}[f_n]$. Thus

$$F(z) = \mathcal{Z}[f_n]$$

The inverse transform is denoted by $\mathcal{Z}^{-1}[F(z)]$. Thus

$$f_n = \mathcal{Z}^{-1}[F(z)]$$

Like the Laplace transform, the Z-transform has many important properties. These properties will not be discussed here, but we do provide the following important results:

$$\mathcal{Z}\left[f_{n+k}\right] = z^k F(z) - \sum_{m=0}^{k-1} z^{k-m} f_m \tag{10.13}$$

$$\mathcal{Z}\left[f_{n-k}\right] = z^{-k} F(z) + \sum_{m=1}^{k} z^{-(k-m)} f_{(-m)} \tag{10.14}$$

These are the left, and right, shifting properties respectively.

We can use the Z-transform to solve *difference* equations in much the same way as we use the Laplace transform to solve *differential* equations. For example, consider the following difference equation:

$$6y_n - 5y_{n-1} + y_{n-2} = \frac{1}{4^n}, \quad n \geq 0 \tag{10.15}$$

Here n is an integer and y_n is a sequence of data values beginning at y_0. However, when $n = 0$ in (10.15), we require the values of y_{-1} and y_{-2} to be specified. These are *initial conditions* and play a similar role to the initial conditions in a differential equation. Let the initial conditions be $y_{-1} = 1$ and $y_{-2} = 0$. Taking $k = 1$ in (10.14), we have

$$\mathcal{Z}\left[y_{n-1}\right] = z^{-1}Y(z) + y_{-1}$$

and taking $k = 2$ in (10.14), we have

$$\mathcal{Z}\left[y_{n-2}\right] = z^{-2}Y(z) + z^{-1}y_{-1} + y_{-2}$$

Taking the Z-transform of (10.15), and substituting for the transformed values of y_{-1} and y_{-2}, we have

$$6Y(z) - 5\{z^{-1}Y(z) + y_{-1}\} + \{z^{-2}Y(z) + z^{-1}y_{-1} + y_{-2}\} = \mathcal{Z}\left[\frac{1}{4^n}\right] \tag{10.16}$$

We could use the basic definition of the Z-transform, or tabulated relationships to determine the Z-transform of the right side of this equation. However, the MATLAB Symbolic Toolbox gives the Z-transform of a function thus:

```
>> syms z n
>> ztrans(1/4^n)

ans =
z/(z - 1/4)
```

Substituting this result in (10.16), we have

$$(6 - 5z^{-1} + z^{-2})Y(z) = \frac{4z}{4z - 1} - z^{-1}y_{-1} - y_{-2} + 5y_{-1}$$

Substituting for y_{-1} and y_{-2} gives

$$Y(z) = \left(\frac{1}{6 - 5z^{-1} + z^{-2}}\right)\left(\frac{4z}{z - 1} - z^{-1} + 5\right)$$

We can determine y_n by taking the inverse of the Z-transform of $Y(z)$. Using the MATLAB function `iztrans` we have

```
>> iztrans((4*z/(4*z-1)-z^(-1)+5)/(6-5*z^(-1)+z^(-2)))

ans =
(5*(1/2)^n)/2 - 2*(1/3)^n + (1/4)^n/2
```

Thus

$$y_n = \frac{5}{2}\left(\frac{1}{2}\right)^n - 2\left(\frac{1}{3}\right)^n + \frac{1}{2}\left(\frac{1}{4}\right)^n$$

Evaluating this solution for $n = -2$ and -1 shows that it satisfies the initial conditions.

10.14 FOURIER TRANSFORM METHODS

Fourier analysis transforms data or functions from the time or spacial domain into the frequency domain. In this description, we will transform from the x domain into the ω domain.

The Fourier series transforms a periodic function in the x domain into corresponding discrete values in the frequency domain; these discrete values are the Fourier coefficients. In contrast, the Fourier transform takes a non-periodic and continuous function in the x domain and transforms it into an infinite and continuous function in the frequency domain.

The Fourier transform of a function $f(x)$ is given by

$$F(s) = \mathcal{F}[f(x)] = \int_{-\infty}^{\infty} f(x)e^{-sx}\,dx \tag{10.17}$$

where $s = \jmath\omega$, i.e., s is imaginary. Thus we may write

$$F(\omega) = \mathcal{F}[f(x)] = \int_{-\infty}^{\infty} f(x)e^{-\jmath\omega x}\,dx \tag{10.18}$$

$F(\omega)$ is complex and is called the frequency spectrum of $f(x)$. The inverse Fourier transform is given by

$$f(x) = \mathcal{F}^{-1}[F(\omega)] = \frac{1}{2\pi}\int_{-\infty}^{\infty} F(w)e^{\jmath\omega x}\,d\omega \tag{10.19}$$

Here we use the operators $\mathcal{F}$ and $\mathcal{F}^{-1}$ to indicate the Fourier transform and the inverse Fourier transform, respectively. Not all functions have a Fourier transform, and certain conditions must be met if the Fourier transform is to exist (see Bracewell, 1978). The Fourier transform has important properties that we introduce as appropriate.

We will begin by using the MATLAB symbolic function `fourier` to determine the Fourier transform of $\cos(3x)$.

```
>> syms x, y = fourier(cos(3*x))

y =
pi*(dirac(w - 3) + dirac(w + 3))
```

This Fourier transform pair is shown diagrammatically in Fig. 10.3. The Fourier transform tells us that the frequency spectrum of this cosine function consists of two infinitely narrow components, one at $\omega = 3$ and one at $\omega = -3$. (For a description of the Dirac function, see Section 10.7.) The MATLAB function `ifourier` implements the symbolic inverse Fourier transform. Thus

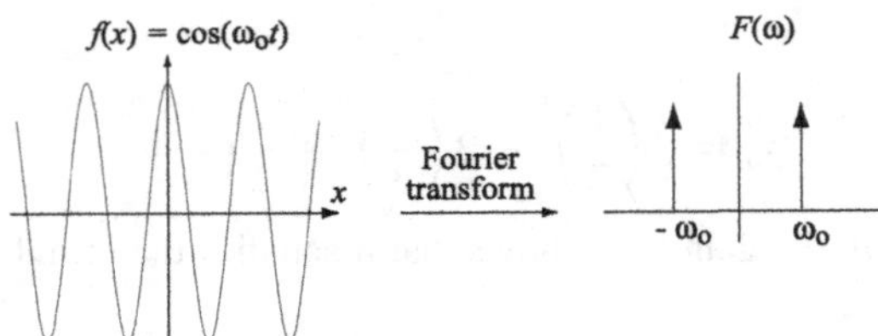

FIGURE 10.3

The Fourier transform of a cosine function.

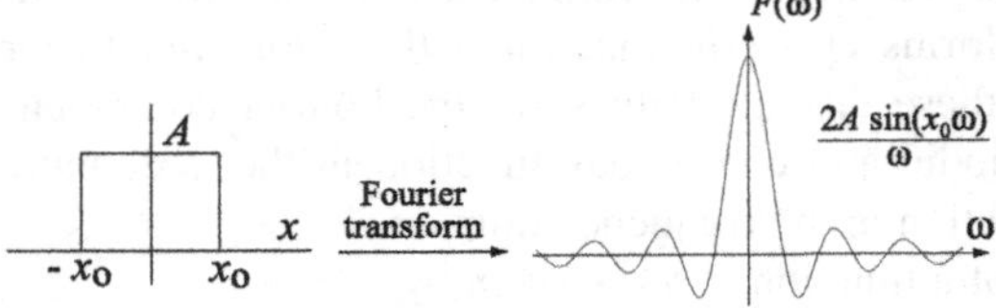

FIGURE 10.4

The Fourier transforms of a "top-hat" function.

```
>> z = ifourier(y)

z =
1/(2*exp(x*3i)) + exp(x*3i)/2

>> simplify(z)

ans =
cos(3*x)
```

As a second example of the use of the Fourier transform, consider the transform of the function shown in Fig. 10.4 which has a unit value in the range $-2 < x < 2$ and zero elsewhere. We show below how this function has been constructed from two Heaviside functions. (The Heaviside function is described in Section 10.7.)

```
>> syms x
>> fourier(heaviside(x+2) - heaviside(x-2))

ans =
(cos(2*w)*1i + sin(2*w))/w - (cos(2*w)*1i - sin(2*w))/w
```

This expression can be simplified thus

```
>> simplify(ans)

ans =
(2*sin(2*w))/w
```

Note that the original function, which in the x domain is limited to the range $-2 < x < 2$, has a frequency spectrum that is continuous between $-\infty < \omega < \infty$. This is shown in Fig. 10.4.

We now illustrate the use of the Fourier transform in the solution of a partial differential equation. Consider the equation

$$\frac{\partial u}{\partial t} = \frac{\partial^2 u}{\partial x^2} \quad (-\infty < x < \infty,\ t > 0) \tag{10.20}$$

subject to the initial condition

$$u(x, 0) = \exp(-a^2 x^2) \text{ where } a = 0.1$$

It can be proved that

$$\mathcal{F}\left[\frac{\partial^2 u}{\partial x^2}\right] = -\omega^2 \mathcal{F}[u] \text{ and } \mathcal{F}\left[\frac{\partial u}{\partial t}\right] = \frac{\partial}{\partial t}\{\mathcal{F}[u]\}$$

Thus, taking the Fourier transform of (10.20), we have

$$\frac{\partial}{\partial t}\{\mathcal{F}[u]\} + \omega^2 \mathcal{F}[u] = 0$$

Solving this first-order differential equation for $\mathcal{F}[u]$ gives

$$\mathcal{F}[u] = A \exp(-\omega^2 t) \tag{10.21}$$

To determine the constant A we must apply the initial conditions. We begin by using MATLAB to find the Fourier transform of the initial conditions thus:

```
>> syms x y z w
>> z = fourier(exp(-x^2/100))

z =
(10*pi^(1/2))*exp(-25*w^2)
```

Hence

$$\mathcal{F}[u(x, 0)] = \sqrt{100\pi} \exp(-25\omega^2)$$

Comparing this equation with (10.21) when $t = 0$, we see that

$$A = \sqrt{100\pi} \exp(-25\omega^2)$$

Substituting this result in (10.21), we have

$$\mathcal{F}[u] = \sqrt{100\pi} \exp(-25\omega^2) \exp(-\omega^2 t)$$

Taking the inverse transform of this equation gives

$$u(x,t) = \mathcal{F}^{-1}\left[\sqrt{100\pi}\exp(-25\omega^2)\exp(-\omega^2 t)\right]$$

Suppose we require a solution when $t = 4$. Using MATLAB to compute the inverse Fourier transform, we have

```
>> y = z*exp(-4*w^2)

y =
(10*pi^(1/2))*exp(-29*w^2)

>> ifourier(y)

ans =
(5*29^(1/2)*exp(-x^2/116))/29
```

Thus, when $t = 4$, the solution of (10.20) is

$$u(x,4) = \frac{5\sqrt{29}}{29}\exp(-x^2/116)$$

Consider now the Fourier transform of the Heaviside or step function.

```
>> syms x
>> fourier(heaviside(x))

ans =
pi*dirac(w) - 1i/w
```

Thus, the real part of the Fourier transform of the Heaviside function is π times the Dirac function at $\omega = 0$ and the imaginary part is $-1/\omega$, which tends to plus and minus infinity when $\omega = 0$. The step function has many applications. For example, if we require the Fourier transform of the function

$$f(x) = \begin{cases} e^{-2x} & x \geq 0 \\ 0 & x < 0 \end{cases}$$

using the Heaviside or unit step function, we can rewrite this as

$$f(x) = u(x)e^{-2x} \quad \text{for all } x$$

where $u(x)$ is the Heaviside function. The MATLAB implementation is

```
>> syms x
>> fourier(heaviside(x)*exp(-2*x))

ans =
1/(2 + w*1i)
```

10.15 LINKING SYMBOLIC AND NUMERICAL PROCESSES

Symbolic algebra can be used to ease the burden for the user in the numerical solution process. To illustrate this we show how the Symbolic Toolbox can be used in a version of Newton's method for solving a non-linear equation which only requires the user to supply the function itself. The usual implementation of this algorithm, (see Section 3.7), requires the user to supply the first-order derivative of the function as well as the function itself. The modified algorithm takes the following form of the MATLAB function fnewtsym:

```
function [res, it] = fnewtsym(func,x0,tol)
% Finds a root of f(x) = 0 using Newton's method
% using the symbolic toolbox
% Example call: [res, it] = fnewtsym(func,x,tol)
% The user defined function func is the function f(x) which must
% be defined as a symbolic function.
% x is an initial starting value, tol is required accuracy.
it = 1; syms dfunc x
% Now perform the symbolic differentiation:
dfunc = diff(sym(func));
d = double(subs(func,x,x0)/subs(dfunc,x,x0));
while abs(d)>tol
    x1 = x0-d; x0 = x1;
    d = double(subs(func,x,x0)/subs(dfunc,x,x0));
    it = it+1;
end
res = x0;
```

Notice the use of the subs and double functions so that a numerical value is returned. To illustrate the use of fnewtsym we will solve $\cos x - x^3 = 0$ to find the root closest to 1 with an accuracy to four decimal places.

```
>> [r,iter] = fnewtsym('cos(x)-x^3',1,0.00005)

r =

    0.8655


iter =
     4
```

These results are identical to those obtained using the function fnewton, (see Section 3.7), but here the user is not required to provide the derivative of the function.

The following examples provide further illustrations of how the Symbolic Toolbox can help users by performing routine, tedious, and sometimes difficult tasks required by some numerical methods. We have seen how the single-variable Newton's method can be modified by using symbolic differentiation and we now extend this to give the symbolic version of Newton's multi-variable method to solve a system of equations where the use of symbolic functions provides an even greater saving for the user. The equations solved in this example are

$$x_1 x_2 = 2$$
$$x_1^2 + x_2^2 = 4$$

The MATLAB function `newtmvsym` takes the form

```
function [x1,fr,it] = newtmvsym(x,f,n,tol)
% Newton's method for solving a system of n non-linear equations
% in n variables. This version is restricted to two variables
% Example call: [xv,it] = newtmvsym(x,f,n,tol)
% Requires an initial approximation column vector x. tol is
% required accuracy.
% User must define functions f, the system equations
% xv is the solution vector, parameter it is number of iterations.
syms a b
xv = sym([a b]); it = 0;
fr = double(subs(f,xv,x));
while norm(fr)>tol
    Jr = double(subs(jacobian(f,xv),xv,x));
    x1 = x-(Jr\fr')'; x = x1;
    fr = double(subs(f,xv,x1));
    it = it+1;
end
```

Notice how this function uses the symbolic Jacobian function in the line:

```
Jr = double(subs(jacobian(f,xv),xv,x));
```

This ensures that the user is not required to find the four partial derivatives required in this case. This function is embedded in script `e4sX05.m`:

```
% e4sX05.m.  Script for running newtonmvsym.m
syms a b
x = sym([a b]);
format long
f = [x(1)*x(2)-2,x(1)^2+x(2)^2-4];
[x1,fr,it] = newtmvsym([1 0],f,2,.000000005)
```

Running script `e4sX05.m` gives

```
x1 =
   1.414244079951220   1.414183044794970
```

```
fr =
   1.0e-08 *
  -0.093132261519485   0.186264506804511

it =
    14
```

We note that this result provides an accurate solution for the given problem in two variables.

It is interesting to note that in a similar way we can write a script for the conjugate gradient method that enables the user to avoid having to provide the first-order partial derivatives of the function to be minimized. This script uses the following fragment of MATLAB code

```
n = 4;
for i = 1:n, dfsymb(i) = diff(sym(f),xv(i)); end
df = double(subs(dfsymb,xv(1:n),x(1:n)'));
```

to obtain the gradient of the function where required. To run this modified function a script similar to that given above for the multi-variable Newton method must be written. This script defines the non-linear function to be optimized and defines a symbolic vector `x`.

10.16 SUMMARY

We have introduced a wide range of symbolic functions and shown how they can be applied to such standard mathematical processes as integration, differentiation, expansion, and simplification. We have also shown how symbolic methods can sometimes be directly linked to numerical procedures with good effect. For those with access to the Symbolic Toolbox, care must be taken to choose the appropriate and efficient method, symbolic or numerical, for the problem.

10.17 PROBLEMS

10.1. Using the appropriate MATLAB symbolic function, rearrange the following expression as a polynomial in x:

$$\left(x-\frac{1}{a}-\frac{1}{b}\right)\left(x-\frac{1}{b}-\frac{1}{c}\right)\left(x-\frac{1}{c}-\frac{1}{a}\right)$$

10.2. Using the appropriate MATLAB symbolic function, multiply the polynomials

$$f(x)=x^4+4x^3-17x^2+27x-19 \text{ and } g(x)=x^2+12x-13$$

and simplify the resulting expression. Then arrange the solution in nested form.

10.3. Using the appropriate MATLAB symbolic function, expand the following functions:
(i) $\tan(4x)$ in terms of powers of $\tan(x)$
(ii) $\cos(x + y)$ in terms of $\cos(x)$, $\cos(y)$, $\sin(x)$, $\sin(y)$
(iii) $\cos(3x)$ in terms of powers of $\cos(x)$
(iv) $\cos(6x)$ in terms of powers of $\cos(x)$

10.4. Using the appropriate MATLAB symbolic function, expand $\cos(x + y + z)$ in terms of $\cos(x)$, $\cos(y)$, $\cos(z)$, $\sin(x)$, $\sin(y)$, $\sin(z)$.

10.5. Using the appropriate MATLAB symbolic function, expand the following functions in ascending powers of x up to x^7:
(i) $\sin^{-1}(x)$
(ii) $\cos^{-1}(x)$
(iii) $\tan^{-1}(x)$

10.6. Using the appropriate MATLAB symbolic function, expand $y = \log_e(\cos(x))$ in ascending powers of x up to x^{12}.

10.7. The first three terms of a series are

$$\frac{4}{1\cdot 2\cdot 3}\left(\frac{1}{3}\right)+\frac{5}{2\cdot 3\cdot 4}\left(\frac{1}{9}\right)+\frac{6}{3\cdot 4\cdot 5}\left(\frac{1}{27}\right)+\ldots$$

Sum this series to n terms using the MATLAB function `symsum`.

10.8. Using the appropriate MATLAB symbolic function, sum the first hundred terms of the series whose kth term is k^{10}.

10.9. Using the appropriate MATLAB symbolic function, verify that

$$\sum_{k=1}^{\infty} k^{-4} = \frac{\pi^4}{90}$$

10.10. For the matrix

$$\mathbf{A} = \begin{bmatrix} 1 & a & a^2 \\ 1 & b & b^2 \\ 1 & c & c^2 \end{bmatrix}$$

determine $\mathbf{A}^{-1}$ symbolically using the MATLAB function `inv` and, using the function `factor`, express it in the following form

$$\begin{bmatrix} \dfrac{cb}{(a-c)(a-b)} & \dfrac{-ac}{(b-c)(a-b)} & \dfrac{ab}{(b-c)(a-c)} \\ \dfrac{-(b+c)}{(a-c)(a-b)} & \dfrac{(a+c)}{(b-c)(a-b)} & \dfrac{-(a+b)}{(b-c)(a-c)} \\ \dfrac{1}{(a-c)(a-b)} & \dfrac{-1}{(b-c)(a-b)} & \dfrac{1}{(b-c)(a-c)} \end{bmatrix}$$

10.11. Using the appropriate MATLAB symbolic function, show that characteristic equation of the matrix

$$\mathbf{A} = \begin{bmatrix} a_1 & a_2 & a_3 & a_4 \\ 1 & 0 & 0 & 0 \\ 0 & 1 & 0 & 0 \\ 0 & 0 & 1 & 0 \end{bmatrix}$$

is

$$\lambda^4 - a_1\lambda^3 - a_2\lambda^2 - a_3\lambda - a_4 = 0$$

10.12. The rotation matrix $\mathbf{R}$ is given below. Define it symbolically and find its second and fourth powers using MATLAB.

$$\mathbf{R} = \begin{bmatrix} \cos\theta & \sin\theta \\ -\sin\theta & \cos\theta \end{bmatrix}$$

10.13. Solve, symbolically, the general cubic equation having the form $x^3 + 3hx + g = 0$. *Hint*: Use the MATLAB function `subexpr` to simplify your result.

10.14. Using the appropriate MATLAB symbolic function, solve the cubic equation $x^3 - 9x + 28 = 0$.

10.15. Using the appropriate MATLAB symbolic functions, find the roots of $z^6 = 4\sqrt{2}(1 + \jmath)$ and plot these roots in the complex plane. *Hint*: Convert the answer using `double`.

10.16. Using the appropriate MATLAB symbolic function, differentiate the following function with respect to x:

$$y = \log_e \left\{ \frac{(1-x)(1+x^3)}{1+x^2} \right\}$$

10.17. Given that Laplace's equation is

$$\frac{\partial^2 z}{\partial x^2} + \frac{\partial^2 z}{\partial y^2} = 0$$

use the appropriate MATLAB symbolic function to verify that this equation is satisfied by the following functions:
(i) $z = \log_e(x^2 + y^2)$
(ii) $z = e^{-2y}\cos(2x)$

10.18. Using the appropriate MATLAB symbolic function, verify that $z = x^3 \sin y$ satisfies the following conditions:

$$\frac{\partial^2 z}{\partial x \partial y} = \frac{\partial^2 z}{\partial y \partial x} \quad \text{and} \quad \frac{\partial^{10} z}{\partial x^4 \partial y^6} = \frac{\partial^{10} z}{\partial y^6 \partial x^4} = 0$$

10.19. Using the appropriate MATLAB symbolic functions, determine the integrals of the following functions and then differentiate the result to recover the original function:

$$\text{(i)}\ \frac{1}{(a + fx)(c + gx)} \qquad \text{(ii)}\ \frac{1 - x^2}{1 + x^2}$$

10.20. Using the appropriate MATLAB symbolic function, determine the following indefinite integrals:

$$\text{(i)} \int \frac{1}{1+\cos x+\sin x}\,dx \quad \text{(ii)} \int \frac{1}{a^4+x^4}\,dx$$

10.21. Using the appropriate MATLAB symbolic function, verify the following result:

$$\int_0^\infty \frac{x^3}{e^x-1}\,dx = \frac{\pi^4}{15}$$

10.22. Using the appropriate MATLAB symbolic function, evaluate the following integrals:

$$\text{(i)} \int_0^\infty \frac{1}{1+x^6}\,dx \quad \text{(ii)} \int_0^\infty \frac{1}{1+x^{10}}\,dx$$

10.23. Using the appropriate MATLAB symbolic function, evaluate the integral

$$\int_0^1 \exp(-x^2)\,dx$$

by expanding $\exp(-x^2)$ in an ascending series of powers of x and then integrating term by term to obtain a series approximation. Expand the series to both x^6 and x^{14} and hence find two approximations to the integral. Compare the accuracy of your results. The evaluation of the integral to 10 decimal places is 0.7468241328.

10.24. Using the appropriate MATLAB symbolic function, verify the following result:

$$\int_0^\infty \frac{\sin(x^2)}{x}\,dx = \frac{\pi}{4}$$

10.25. Using the appropriate MATLAB symbolic function, evaluate the integral

$$\int_0^1 \log_e(1+\cos x)\,dx$$

by expanding $\log_e(1+\cos x)$ in an ascending series of powers of x and then integrating term by term to obtain a series approximation. Expand the series up to the term in x^4. Compare the accuracy of your result. The solution to 10 decimal places is 0.6076250333.

10.26. Using the appropriate MATLAB symbolic function, evaluate the following repeated integral:

$$\int_0^1 dy \int_0^1 \frac{1}{1-xy}\,dx$$

10.27. Using the appropriate MATLAB symbolic function, solve the following differential equation which arises in the study of consumer behavior:

$$\frac{d^2y}{dt^2} + (bp+aq)\frac{dy}{dt} + ab(pq-1)y = cA$$

Also find the solution in the case where $p = 1$, $q = 2$, $a = 2$, $b = 1$, $c = 1$, and $A = 20$. *Hint*: Use the MATLAB function `subs`.

10.28. Using the appropriate MATLAB symbolic function, solve the following pair of simultaneous differential equations

$$2\frac{dx}{dt} + 4\frac{dy}{dt} = \cos t$$
$$4\frac{dx}{dt} - 3\frac{dy}{dt} = \sin t$$

10.29. Using the appropriate MATLAB symbolic function, solve the following differential equation:

$$(1 - x^2)\frac{d^2y}{dx^2} - 2x\frac{dy}{dx} + 2y = 0$$

10.30. Using the appropriate MATLAB symbolic function, solve the following differential equations using the Laplace transform:

(i) $\dfrac{d^2y}{dt^2} + 2y = \cos(2t)$, $y = -2$ and $\dfrac{dy}{dt} = 0$ when $t = 0$

(ii) $\dfrac{dy}{dt} - 2y = t$, $y = 0$ when $t = 0$

(iii) $\dfrac{d^2y}{dt^2} - 3\dfrac{dy}{dt} + y = \exp(-2t)$, $y = -3$ and $\dfrac{dy}{dt} = 0$ when $t = 0$

(iv) $\dfrac{dq}{dt} + \dfrac{q}{c} = 0$, $q = V$ when $t = 0$

10.31. Solve the following difference equations symbolically using the Z-transform:

(i) $y_n + 2y_{n-1} = 0$, $y_{-1} = 4$

(ii) $y_n + y_{n-1} = n$, $y_{-1} = 10$

(iii) $y_n - 2y_{n-1} = 3$, $y_{-1} = 1$

(iv) $y_n - 3y_{n-1} + 2y_{n-2} = 3(4^n)$, $y_{-1} = -3$, $y_{-2} = 5$

10.32. A formula for the computation of π is given by the infinite series developed by the Chudnovsky brothers in 1987 (see Chudnovsky and Chudnovsky, 1989).

$$\frac{1}{\pi} = 12\sum_{k=0}^{\infty}(-1)^k\frac{(6k)!(13591409 + 545140134k)}{(3k)!(k!)^3(640320)^{3k+3/2}}$$

Use the symbolic toolbox and in particular the functions `vpa` and `sym` to compute various values of π to using 2, 3, 4, 5, 6, ..., 10 terms in the series. For each of these estimations, plot the number of decimal places achieved.

APPENDIX

MATRIX ALGEBRA

The aim of this appendix is to give a brief overview of matrix algebra, which covers a number of issues referred to in the main text of this book. This appendix includes an introduction to matrix properties, operators, and classes.

A.1 INTRODUCTION

Since many MATLAB functions and operators act on matrices and arrays, it is important that the user of MATLAB should feel at ease with matrix notation and matrix algebra. MATLAB is an ideal environment in which to experiment and learn matrix algebra. While MATLAB cannot provide a formal proof of any relationship, it does allow the user to verify results and rapidly gain experience in matrix manipulation. In this appendix only definitions and results are provided. For proofs and further explanation, it is recommended that the reader consult Golub and Van Loan (1989).

A.2 MATRICES AND VECTORS

A matrix is a rectangular array of elements which in itself cannot be evaluated. An element of a matrix can be a real or complex number, an algebraic expression, or another matrix. Normally matrices are enclosed in square brackets, round brackets, or braces. In this text square brackets are used. A complete matrix is denoted by an emboldened character. For example,

$$\mathbf{A} = \begin{bmatrix} 3 & -2 \\ -2 & 4 \end{bmatrix}, \quad \mathbf{B} = \begin{bmatrix} \mathbf{A} & \mathbf{A} & 2\mathbf{A} \\ \mathbf{A} & -\mathbf{A} & \mathbf{A} \end{bmatrix}$$

$$\mathbf{x} = \begin{bmatrix} 11 \\ -3 \\ 7 \end{bmatrix}, \quad \mathbf{e} = \begin{bmatrix} (2+3\imath) & \left(p^2+q\right) & (-4+7\imath) & (3-4\imath) \end{bmatrix}$$

and hence

$$\mathbf{B} = \begin{bmatrix} 3 & -2 & 3 & -2 & 6 & -4 \\ -2 & 4 & -2 & 4 & -4 & 8 \\ 3 & -2 & -3 & 2 & 3 & -2 \\ -2 & 4 & 2 & -4 & -2 & 4 \end{bmatrix}$$

where $\imath = \sqrt{(-1)}$. In the above examples $\mathbf{A}$ is a 2×2 square matrix with two rows and two columns of real coefficients. It also has the property of being a symmetric matrix, see Section A.7. The matrix

B is built up from the matrix **A** and so **B** is a 4×6 real matrix. The matrix **x** is a 3×1 matrix and is usually called a column vector and **e** is a 1×4 complex matrix, usually called a row vector. Note that **e** has the algebraic expression $p^2 + q$ for its second element. In this vector each element is enclosed in round brackets to clarify its structure. Enclosing an element in round brackets is not a requirement.

If we wish to refer to a particular element in a matrix, we use subscript notation: the first subscript denotes the row, the second the column. In the case of the row and column vectors it is conventional to use a single subscript. Thus, in the examples above,

$$a_{21} = -2, \quad b_{25} = -4, \quad x_2 = -3, \quad e_4 = 3 - 4\imath$$

Note also that although **A** and **B** are upper case letters it is conventional to refer to their elements by lower case letters. In general the element in the ith row and jth column of **A** is denoted by a_{ij}.

A.3 SOME SPECIAL MATRICES

The identity matrix. The identity matrix, denoted by **I**, has unit values along the leading diagonal and zeros elsewhere. The leading diagonal is the diagonal of elements from the top left to the bottom right of the matrix. For example,

$$\mathbf{I}_2 = \begin{bmatrix} 1 & 0 \\ 0 & 1 \end{bmatrix}, \quad \mathbf{I}_3 = \begin{bmatrix} 1 & 0 & 0 \\ 0 & 1 & 0 \\ 0 & 0 & 1 \end{bmatrix}$$

The subscript indicating the size of the matrix is usually omitted. The identity matrix behaves rather like the scalar quantity 1. In particular, pre- or post-multiplying a matrix by **I** does not change it.

The diagonal matrix. This matrix is square and has non-zero elements *only* along the leading diagonal. Thus

$$\mathbf{A} = \begin{bmatrix} 4 & 0 & 0 & 0 \\ 0 & -2 & 0 & 0 \\ 0 & 0 & 0 & 0 \\ 0 & 0 & 0 & 9 \end{bmatrix}, \quad \mathbf{B} = \begin{bmatrix} 12 & 0 & 0 \\ 0 & -2 & 0 \\ 0 & 0 & -6 \end{bmatrix}$$

The tridiagonal matrix. This matrix is square and has non-zero elements along the leading diagonal and the diagonals immediately above and below it. Thus, using "x" to denote a non-zero element,

$$\mathbf{A} = \begin{bmatrix} x & x & 0 & 0 & 0 \\ x & x & x & 0 & 0 \\ 0 & x & x & x & 0 \\ 0 & 0 & x & x & x \\ 0 & 0 & 0 & x & x \end{bmatrix}$$

Triangular and Hessenberg matrices. A lower triangular matrix has non-zero elements only on and below the leading diagonal. An upper triangular matrix has non-zero elements on and above the leading diagonal. The Hessenberg matrix is similar to the triangular matrix except that in addition it has non-zero elements on the diagonals adjacent to the leading diagonal.

$$\begin{bmatrix} x & x & x & x & x \\ 0 & x & x & x & x \\ 0 & 0 & x & x & x \\ 0 & 0 & 0 & x & x \\ 0 & 0 & 0 & 0 & x \end{bmatrix}, \begin{bmatrix} x & 0 & 0 & 0 & 0 \\ x & x & 0 & 0 & 0 \\ x & x & x & 0 & 0 \\ x & x & x & x & 0 \\ x & x & x & x & x \end{bmatrix}, \begin{bmatrix} x & x & x & x & x \\ x & x & x & x & x \\ 0 & x & x & x & x \\ 0 & 0 & x & x & x \\ 0 & 0 & 0 & x & x \end{bmatrix}$$

The first matrix is upper triangular, the second is lower triangular and the last is an upper Hessenberg matrix.

A.4 DETERMINANTS

The determinant of **A** is written |**A**| or det(**A**). For a 2 × 2 array we define its determinant as follows:

$$\text{If } \mathbf{A} = \begin{bmatrix} a_{11} & a_{12} \\ a_{21} & a_{22} \end{bmatrix} \text{ then } \det(\mathbf{A}) = \begin{vmatrix} a_{11} & a_{12} \\ a_{21} & a_{22} \end{vmatrix} = a_{11}a_{22} - a_{21}a_{12} \tag{A.1}$$

In general for an $n \times n$ array, **A**, cofactors $C_{ij} = (-1)^{i+j}\,\Delta_{ij}$ can be defined. In this definition Δ_{ij} is the determinant formed from **A** when the elements of the ith row and jth column are deleted. Δ_{ij} is called the minor of **A**. Then

$$\det(\mathbf{A}) = \sum_{k=1}^{n} a_{ik}C_{ik} \text{ for any } i = 1, 2, ..., n \tag{A.2}$$

This is known as an expansion along the ith row. Frequently the first row is used. This equation replaces the problem of evaluating one $n \times n$ determinant **A** by the evaluation of n, $(n-1) \times (n-1)$ determinants. The process can be continued until the cofactors are reduced to 2 × 2 determinants. Then the formula (A.1) is used. This is the formal definition for the determinant of **A** but it is not a computationally efficient procedure.

A.5 MATRIX OPERATIONS

Matrix transposition. In this operation the rows and columns of a matrix are interchanged or transposed. The transposition of a real matrix **A** is denoted by $\mathbf{A}^\top$. For example,

$$\mathbf{A} = \begin{bmatrix} 1 & -2 & 4 \\ 2 & 1 & 7 \end{bmatrix}, \mathbf{A}^\top = \begin{bmatrix} 1 & 2 \\ -1 & 1 \\ 4 & 7 \end{bmatrix}, \mathbf{x} = \begin{bmatrix} 1 \\ 2 \\ 3 \end{bmatrix}, \mathbf{x}^\top = \begin{bmatrix} 1 & 2 & 3 \end{bmatrix}$$

Note that a square matrix remains square when it is transposed and a column vector transposes into a row vector and vice versa.

Matrix addition and subtraction. This is done by adding or subtracting corresponding elements in the matrices. Thus

$$\begin{bmatrix} 1 & 3 \\ -4 & 5 \end{bmatrix} + \begin{bmatrix} 5 & -4 \\ 6 & 6 \end{bmatrix} = \begin{bmatrix} 6 & -1 \\ 2 & 11 \end{bmatrix}, \begin{bmatrix} -4 \\ 6 \\ 11 \end{bmatrix} - \begin{bmatrix} 3 \\ -3 \\ 2 \end{bmatrix} = \begin{bmatrix} -7 \\ 9 \\ 9 \end{bmatrix}$$

It is apparent that only matrices with the same number of rows and the same number of columns can be added and subtracted. In general, if $\mathbf{A} = \mathbf{B} + \mathbf{C}$ then $a_{ij} = b_{ij} + c_{ij}$.

Scalar multiplication. Every element of a matrix is multiplied by a scalar quantity. Thus if $\mathbf{A} = s\mathbf{B}$, where s is a scalar, then $a_{ij} = sb_{ij}$.

Matrix multiplication. We can only multiply two matrices **B** and **C** together if the number of columns in **B** is equal to the number of rows in **C**. Such matrices are said to be conformable. If **B** is a $p \times q$ matrix and **C** is a $q \times r$ matrix, then we can determine the product **BC** and the result, **A**, will be a $p \times r$ matrix. Because the order of matrix multiplication is important, we say that **B** pre-multiplies **C** or **C** post-multiplies **B**. The elements of **A** are determined from the following relationship:

$$a_{ij} = \sum_{k=1}^{n} b_{ik}c_{kj} \text{ for } i = 1, 2, \ldots, n; \; j = 1, 2, \; \ldots, n$$

For example,

$$\begin{bmatrix} 2 & -3 & 1 \\ -5 & 4 & 3 \end{bmatrix} \begin{bmatrix} -6 & 4 & 1 \\ -4 & 2 & 3 \\ 3 & -7 & -1 \end{bmatrix}$$

$$= \begin{bmatrix} 2(-6) + (-3)(-4) + 1(3) & 2(4) + (-3)2 + 1(-7) & 2(1) + (-3)3 + 1(-1) \\ (-5)(-6) + 4(-4) + 3(3) & (-5)4 + 4(2) + 3(-7) & (-5)1 + 4(3) + 3(-1) \end{bmatrix}$$

$$= \begin{bmatrix} 3 & -5 & -8 \\ 23 & -33 & 4 \end{bmatrix}$$

Note that the product of a 2×3 and 3×3 matrix is a 2×3 matrix. Consider four further examples of matrix multiplication:

$$\begin{bmatrix} 1 & 2 \\ 3 & 4 \end{bmatrix} \begin{bmatrix} 5 & 6 \\ 3 & 2 \end{bmatrix} = \begin{bmatrix} 11 & 10 \\ 27 & 26 \end{bmatrix}, \quad \begin{bmatrix} 5 & 6 \\ 3 & 2 \end{bmatrix} \begin{bmatrix} 1 & 2 \\ 3 & 4 \end{bmatrix} = \begin{bmatrix} 23 & 34 \\ 9 & 14 \end{bmatrix}$$

$$\begin{bmatrix} 1 & 2 & 3 \end{bmatrix} \begin{bmatrix} -4 \\ 3 \\ 3 \end{bmatrix} = 11, \quad \begin{bmatrix} -4 \\ 3 \\ 3 \end{bmatrix} \begin{bmatrix} 1 & 2 & 3 \end{bmatrix} = \begin{bmatrix} -4 & -8 & -12 \\ 3 & 6 & 9 \\ 3 & 6 & 9 \end{bmatrix}$$

In the above examples it is seen that while the 2×2 matrices can be multiplied in either order, the product is different. This is an important observation and in general $\mathbf{BC} \neq \mathbf{CB}$. Note also that multiplying a row by a column vector gives a scalar whereas multiplying a column by a row results in a matrix.

Matrix inversion. The inverse of a square matrix $\mathbf{A}$ is written $\mathbf{A}^{-1}$ and is defined by

$$\mathbf{A}\mathbf{A}^{-1} = \mathbf{A}^{-1}\mathbf{A} = \mathbf{I}$$

The formal definition of $\mathbf{A}^{-1}$ is

$$\mathbf{A}^{-1} = \text{adj}\,(\mathbf{A})\,/\det(\mathbf{A}) \tag{A.3}$$

where adj($\mathbf{A}$) is the adjoint of $\mathbf{A}$. The adjoint of $\mathbf{A}$ is given by

$$\text{adj}\,(\mathbf{A}) = \Delta^{\mathsf{T}}$$

where $\mathbf{D}$ is a matrix composed of the cofactors Δ_{ij} of $\mathbf{A}$. Using (A.3) is not an efficient way to compute an inverse.

A.6 COMPLEX MATRICES

A matrix can have elements that are complex and such a matrix can be expressed in terms of two real matrices. Thus

$$\mathbf{A} = \mathbf{B} + \imath\mathbf{C} \quad \text{where } \imath = \surd(-1)$$

where $\mathbf{A}$ is complex and $\mathbf{B}$ and $\mathbf{C}$ are real matrices. Conjugation of $\mathbf{A}$ is normally denoted by $\mathbf{A}^*$ and is equal to

$$\mathbf{A}^* = \mathbf{B} - \imath\mathbf{C}$$

Matrix $\mathbf{A}$ can be transposed so that

$$\mathbf{A}^{\mathsf{T}} = \mathbf{B}^{\mathsf{T}} + \imath\mathbf{C}^{\mathsf{T}}$$

Matrix $\mathbf{A}$ can be transposed *and* conjugated at the same time and this is denoted by $\mathbf{A}^{\mathsf{H}}$ and called the Hermitian transpose. Thus

$$\mathbf{A}^{\mathsf{H}} = \mathbf{B}^{\mathsf{T}} - \imath\mathbf{C}^{\mathsf{T}}$$

For example,

$$\mathbf{A} = \begin{bmatrix} 1-\imath & -2-3\imath & 4\imath \\ 2 & 1+2\imath & 7+5\imath \end{bmatrix}, \quad \mathbf{A}^* = \begin{bmatrix} 1+\imath & -2+3\imath & -4\imath \\ 2 & 1-2\imath & 7-5\imath \end{bmatrix}$$

$$\mathbf{A}^{\mathsf{T}} = \begin{bmatrix} 1-\imath & 2 \\ -2-3\imath & 1+2\imath \\ 4\imath & 7+5\imath \end{bmatrix}, \quad \mathbf{A}^{\mathsf{H}} = \begin{bmatrix} 1+\imath & 2 \\ -2+3\imath & 1-2\imath \\ -4\imath & 7-5\imath \end{bmatrix}$$

It is important to note that the MATLAB expression `A'` gives the conjugation and transposition of `A` when applied to a complex matrix, i.e., it is equivalent to $\mathbf{A}^{\mathsf{H}}$. However, `A.'` gives ordinary transposition which corresponds to $\mathbf{A}^{\mathsf{T}}$.

A.7 MATRIX PROPERTIES

The real square matrix **A** is

$$\text{symmetric if: } \mathbf{A}^{\mathsf{T}} = \mathbf{A}$$

$$\text{skew-symmetric if: } \mathbf{A}^{\mathsf{T}} = -\mathbf{A}$$

$$\text{orthogonal if: } \mathbf{A}^{\mathsf{T}} = \mathbf{A}^{-1}$$

$$\text{nilpotent if: } \mathbf{A}^{p} = \mathbf{0}, \text{ where } p \text{ is a positive integer}$$

$$\text{idempotent if: } \mathbf{A}^{2} = \mathbf{A}$$

The complex square matrix $\mathbf{A} = \mathbf{B} + \imath\mathbf{C}$ is

$$\text{Hermitian if: } \mathbf{A}^{\mathsf{H}} = \mathbf{A}$$

$$\text{unitary if: } \mathbf{A}^{\mathsf{H}} = \mathbf{A}^{-1}$$

A.8 SOME MATRIX RELATIONSHIPS

If **P**, **Q**, and **R** are matrices such that

$$\mathbf{W} = \mathbf{P}\mathbf{Q}\mathbf{R}$$

then

$$\mathbf{W}^{\mathsf{T}} = \mathbf{R}^{\mathsf{T}}\mathbf{Q}^{\mathsf{T}}\mathbf{P}^{\mathsf{T}} \tag{A.4}$$

and

$$\mathbf{W}^{-1} = \mathbf{R}^{-1}\mathbf{Q}^{-1}\mathbf{P}^{-1} \tag{A.5}$$

If **P**, **Q**, and **R** are complex, then (A.5) is still valid and (A.4) becomes

$$\mathbf{W}^{\mathsf{H}} = \mathbf{R}^{\mathsf{H}}\mathbf{Q}^{\mathsf{H}}\mathbf{P}^{\mathsf{H}} \tag{A.6}$$

A.9 EIGENVALUES

Consider the eigenvalue problem

$$\mathbf{A}\mathbf{x} = \lambda\mathbf{x}$$

If **A** is an $n \times n$ symmetric matrix, then there are n real eigenvalues, λ_i, and n real eigenvectors, $\mathbf{x}_i$, that satisfy this equation. If **A** is an $n \times n$ Hermitian matrix, then there are n real eigenvalues, λ_i, and n complex eigenvectors, $\mathbf{x}_i$, that satisfy the eigenvalue problem. The polynomial in λ given by $\det(\mathbf{A} - \lambda\mathbf{I}) = 0$ is called the characteristic equation. The roots of this polynomial are the eigenvalues of **A**. The sum of the eigenvalues of **A** equals trace(**A**) where trace(**A**) is defined as the sum of the elements on the leading diagonal of **A**. The product of the eigenvalues of **A** equals det(**A**).

It is interesting to note that if we define **C** as

$$\mathbf{C} = \begin{bmatrix} -p_1/p_0 & -p_2/p_0 & \dots & -p_{n-1}/p_0 & -p_n/p_0 \\ 1 & 0 & \dots & 0 & 0 \\ 0 & 1 & \dots & 0 & 0 \\ \vdots & \vdots & & \vdots & \vdots \\ 0 & 0 & \dots & 1 & 0 \end{bmatrix}$$

then the eigenvalues of **C** are the roots of the polynomial

$$p_0x^n + p_1x^{n-1} + \dots + p_{n-1}x + p_n = 0$$

The matrix **C** is called the companion matrix.

A.10 DEFINITION OF NORMS

The p-norm for the vector **v** is defined thus:

$$||\mathbf{v}||_p = \left(|v_1|^p + |v_2|^p + \dots + |v_n|^p\right)^{1/p} \tag{A.7}$$

The parameter p can take any value but only three values are commonly used. If $p = 1$ in (A.7), we have the 1-norm, $||\mathbf{v}||_1$, thus

$$||\mathbf{v}||_1 = |v_1| + |v_2| + \dots + |v_n| \tag{A.8}$$

If $p = 2$ in (A.7), we have the 2-norm or Euclidean norm of the vector **v** which is written $||\mathbf{v}||$ or $||\mathbf{v}||_2$ and is defined thus:

$$||\mathbf{v}||_2 = \sqrt{v_1^2 + v_2^2 + \dots + v_n^2} \tag{A.9}$$

Note that it is not necessary to take the modulus of the elements because in this case each element value is squared. The Euclidean norm is also called the length of the vector. These names arise from the fact that in two- or three-dimensional Euclidean space a vector of two or three elements is used to specify a position in space. The distance from the origin to the specified position is identical to the Euclidean norm of the vector.

If p tends to infinity in (A.7), we have $||\mathbf{v}||_\infty = \max(|v_1|, |v_2|, \dots, |v_n|)$, the infinity norm. At first sight this might appear inconsistent with (A.7). However, when p tends to infinity, the modulus of each element is raised to a very large power and the largest element will dominate the summation.

These functions are implemented in MATLAB; `norm(v,1)`, `norm(v,2)` (or `norm(v)`), and `norm(v,inf)` return the 1, 2, and infinity norms of the vector `v`, respectively.

A.11 REDUCED ROW ECHELON FORM

The RREF of a matrix also has an important role to play in the theoretical understanding of linear algebra. A matrix is transformed into its RREF when the following conditions have been met:

1. all zero rows, if they exist, are at the bottom of the matrix,
2. the first non-zero element in every non-zero row is unity,
3. for each non-zero row, the first non-zero element appears to the right of the first non-zero element of the preceding row,
4. for any column in which the first non-zero element of a row appears, all other elements are zero.

The RREF is determined by using a finite sequence of elementary row operations. The RREF is a standard form and it is the most fundamental form of a matrix that can be achieved using elementary row operations alone.

For a system of equations $\mathbf{Ax} = \mathbf{b}$ we can define the augmented matrix $[\mathbf{A}\ \mathbf{b}]$. If this matrix is transformed to its RREF, the following may be deduced:

1. If $[\mathbf{A}\ \mathbf{b}]$ is derived from an inconsistent system (i.e., no solution exists) the RREF has a row of the form $[0 \ldots\ 0\ 1]$.
2. If $[\mathbf{A}\ \mathbf{b}]$ is derived from a consistent system with an infinity of solutions, then the number of columns of the coefficient matrix is greater than the number of non-zero rows in the RREF; otherwise there is a unique solution and it appears in the last (augmented) column of the RREF.
3. A zero row in the RREF indicates that the original set of equations contained equations with redundant information, i.e., information contained in other equations of the system.

In computing the RREF, numerical problems can arise that are common to other procedures that use elementary row operations; see Section 2.6.

A.12 DIFFERENTIATING MATRICES

The rules for matrix differentiation are essentially the same as those for scalars, but care must be taken to ensure that the order of the matrix operations is maintained. The process is illustrated by the following example: differentiate $f(\mathbf{x}) = \mathbf{x}^{\mathsf{T}}\mathbf{Ax}$ where $\mathbf{x}$ is a column vector with n elements, $(x_1, x_2, x_3, \ldots, x_n)^{\mathsf{T}}$, and $\mathbf{A}$ has elements a_{ij} for $i, j = 1, 2, \ldots, n$. We note first that any matrix associated with a quadratic form must be symmetric. Hence the matrix $\mathbf{A}$ is symmetric. We require the gradient of $f(\mathbf{x})$, i.e. $\nabla f(\mathbf{x})$. The gradient consists of all the first-order partial derivatives of $f(\mathbf{x})$ with respect to each component of the vector $\mathbf{x}$. Now multiplying out the terms of $f(\mathbf{x})$ we have $f(\mathbf{x})$ expressed in component terms as

$$f(\mathbf{x}) = \sum_{i=1}^{n} a_{ij}x_i^2 + \sum_{i=1}^{n} \sum_{j=1, j\neq i}^{n} a_{ij}x_i x_j$$

However we note that since **A** is symmetric $a_{ij} = a_{ji}$ and consequently the terms $a_{ij}x_ix_j + a_{ji}x_ix_j$ can be written as $2a_{ij}x_ix_j$. Hence

$$\frac{\partial f(\mathbf{x})}{\partial x_i} = 2a_{ii}x_i + 2\sum_{j=1,\ j\neq i}^{n} a_{ij}x_j$$

This is of course equivalent to the matrix form

$$\nabla f(\mathbf{x}) = 2\mathbf{A}\mathbf{x}$$

and this provides the standard matrix result where **x** is a column vector.

A.13 SQUARE ROOT OF A MATRIX

In order to have a square root, a matrix must be square. If **A** is a square matrix and **BB** = **A**, then **B** is the square root of **A**. If **A** is singular it may not have a square root.

The square matrix **A** can be factorized to give $\mathbf{A} = \mathbf{X}\mathbf{D}\mathbf{X}^{-1}$ where **D** is a diagonal matrix comprising the n eigenvalues of **A** and **X** is an $n \times n$ array of the eigenvectors of **A**. We can expand this expression for **A** to give

$$\mathbf{A} = (\mathbf{X}\mathbf{D}^{1/2}\mathbf{X}^{-1})(\mathbf{X}\mathbf{D}^{1/2}\mathbf{X}^{-1})$$

Since

$$\mathbf{A} = \mathbf{B}\mathbf{B}$$

then

$$\mathbf{B} = \mathbf{X}\mathbf{D}^{1/2}\mathbf{X}^{-1}$$

The square root of the diagonal matrix of eigenvalues, **D**, is determined by taking the square root of each diagonal element, i.e. each eigenvalue. Any number, real or complex, will have one positive and one negative square root. Thus to determine the square root of **D** (and hence **A**) we must consider every combination of the positive and negative square roots of the eigenvalues. This gives 2^n possible combinations and hence there are 2^n expressions for $\mathbf{D}^{1/2}$. This will lead to 2^n different square root matrices, **B**. If $\mathbf{D}^{1/2}$ comprises all the positive roots then the resulting square root matrix is called the principal square root. This matrix is unique. Consider the following example. If

$$\mathbf{A} = \begin{bmatrix} 31 & 37 & 34 \\ 55 & 67 & 64 \\ 91 & 115 & 118 \end{bmatrix}$$

then, taking the $2^3 = 8$ combinations of square roots, we obtain the following square roots of **A**. Note that $\mathbf{B}_0$ is the principal square root.

$$\mathbf{B}_{0/1} = \pm \begin{bmatrix} 2.9798 & 2.9296 & 1.8721 \\ 4.3357 & 5.0865 & 3.9804 \\ 5.0313 & 7.1413 & 8.9530 \end{bmatrix} \qquad \mathbf{B}_{2/3} = \pm \begin{bmatrix} 1.0000 & 2.0000 & 3.0000 \\ 3.0000 & 4.0000 & 5.0000 \\ 8.0000 & 9.0000 & 7.0000 \end{bmatrix}$$

$$\mathbf{B}_{4/5} = \pm \begin{bmatrix} 2.8115 & 3.0713 & 1.8437 \\ 4.5426 & 4.9123 & 4.0153 \\ 4.9594 & 7.2019 & 8.9408 \end{bmatrix} \qquad \mathbf{B}_{6/7} = \pm \begin{bmatrix} 1.1683 & 1.8583 & 3.0284 \\ 2.7931 & 4.1742 & 4.9651 \\ 8.0719 & 8.9395 & 7.0121 \end{bmatrix}$$

Multiplying any one of these matrices by itself will result in the original matrix **A**.

APPENDIX

ERROR ANALYSIS

B

All numerical processes are subject to error. Errors may be of the following types:

1. truncation errors that are inherent in the numerical algorithm,
2. rounding errors due to the necessity to work to a finite number of significant figures,
3. errors due to inaccurate input data,
4. simple human errors in coding, which should not happen but does!

Examples of the errors described in 1 can be found throughout this text, see, for example, Chapters 3, 4, and 5. Here we consider the implications of the type of errors described in 2 and 3 above. Errors of the type described in 4 are outside the scope of this text.

B.1 INTRODUCTION

Error analysis estimates the error in some computation caused by errors in some previous process. The previous process may be some experimentation, observation, or rounding in a calculation. Generally, we require an upper estimate of the error that can arise when circumstances conspire to be at their worst! We now illustrate this by a specific example. Suppose that $a = 4 \pm 0.02$ (which implies an error of $\pm 0.5\%$) and $b = 2 \pm 0.03$ (which implies an error of $\pm 1.5\%$); then the highest value of a/b results when we divide 4.02 by 1.97 (to give 2.041) and the lowest value of a/b results when we divide 3.98 by 2.03 (to give 1.960). Thus, compared with the nominal value of a/b $(= 2)$ we see that the extremes are 2.05% above and 2.0% below the nominal value.

A particular aspect of error analysis is to determine how sensitive a particular calculation is to an error in a specific parameter. Thus, we deliberately modify the value of a parameter to determine how sensitive the final answer is to changes in that parameter. For example, consider the following equation:

$$a = 100\frac{\sin\theta}{x^3}$$

If a is evaluated for $\theta = 70$ and $x = 3$, then $a = 3.4803$. If θ is increased by 10%, then $a = 3.6088$, an increase of 3.69%. If θ is decreased by 10%, then $a = 3.3$, a decrease of 5.18%. Similarly, if we independently increase x by 10%, then $a = 2.6148$ which is a decrease of 24.8%. If we decrease x by 10%, then $a = 4.7741$, an increase of 37.17%. Clearly, the value of a is much more sensitive to small changes in x than in θ.

B.2 ERRORS IN ARITHMETIC OPERATIONS

More usually, each of the independent variables has a specified error and we wish to find the overall error in a calculation. We now consider how we can estimate the errors that arise from the standard arithmetic operations. Let x_a, y_a, and z_a be approximations to the exact values x, y, and z, respectively. Let the errors in x, y, and z be x_ε, y_ε, and z_ε, respectively. Then these are given by

$$x_\varepsilon = x - x_a, \quad y_\varepsilon = y - y_a, \quad z_\varepsilon = z_a$$

Hence

$$x = x_\varepsilon + x_a, \quad y = y_\varepsilon + y_a, \quad z = z_\varepsilon + z_a$$

If $z = x \pm y$, then

$$z = (x_a + x_\varepsilon) \pm (y_a + y_\varepsilon) = (x_a \pm y_a) + (x_\varepsilon \pm y_\varepsilon)$$

Now $z_a = x_a \pm y_a$ and hence from the above definitions, $z_\varepsilon = x_\varepsilon \pm y_\varepsilon$. Normally we are concerned with the maximum possible error and since x_ε and y_ε may be positive or negative quantities, then

$$\max(|z_\varepsilon|) = |x_\varepsilon| + |y_\varepsilon|$$

Consider now the process of multiplication. If $z = xy$, then

$$z = (x_a + x_\varepsilon)(y_a + y_\varepsilon) = x_a y_a + x_\varepsilon y_a + y_\varepsilon x_a + x_\varepsilon y_\varepsilon \tag{B.1}$$

Assuming the errors are small, we can neglect the product of errors in the above equation. It is convenient to work in terms of relative error, where the relative error in x, x_ε^R is given by

$$x_\varepsilon^R = x_\varepsilon / x \approx x_\varepsilon / x_a$$

Thus, dividing (B.1) by $z_a = x_a y_a$ we have

$$\frac{(z_a + z_\varepsilon)}{z_a} = 1 + \frac{x_\varepsilon}{x_a} + \frac{y_\varepsilon}{y_a}$$

or

$$\frac{z_\varepsilon}{z_a} = \frac{x_\varepsilon}{x_a} + \frac{y_\varepsilon}{y_a} \tag{B.2}$$

(B.2) can be written

$$z_\varepsilon^R = x_\varepsilon^R + y_\varepsilon^R$$

Again, we want to estimate the worst case error in z and since the error in x and y may be positive or negative, we have

$$\max\left(\left|z_\varepsilon^R\right|\right) = \left|x_\varepsilon^R\right| + \left|y_\varepsilon^R\right| \tag{B.3}$$

It can easily be shown that if $z = x/y$, the maximum relative error in z is also given by (B.3). This proof is left as an exercise for the reader.

A more general approach to error analysis is to use a Taylor series. Thus if $y = f(x)$ and $y_a = f(x_a)$, then we can write

$$y = f(x) = f(x_a + x_\varepsilon) = f(x_a) + x_\varepsilon f'(x_a) + \dots$$

Now

$$y_\varepsilon = y - y_a = f(x) - f(x_a)$$

Hence

$$y_\varepsilon = x_\varepsilon f'(x_a)$$

For example, consider $y = \sin\theta$ where $\theta = \pi/3 \pm 0.08$. Thus $\theta_\varepsilon = \pm 0.08$. Hence

$$y_\varepsilon = \theta_\varepsilon \frac{d}{d\theta}\{\sin(\theta)\} = \theta_\varepsilon \cos(\pi/3) = 0.08 \times 0.5 = 0.04.$$

B.3 ERRORS IN THE SOLUTION OF LINEAR EQUATION SYSTEMS

We now consider the problem of estimating the error in the solution of a set of linear equations, $\mathbf{Ax} = \mathbf{b}$. For this analysis we must introduce the concept of a matrix norm.

The formal definition of a matrix p-norm is

$$\|\mathbf{A}\|_p = \max \frac{\|\mathbf{Ax}\|_p}{\|\mathbf{x}\|_p} \quad \text{if } x \neq 0$$

where $\|\mathbf{x}\|_p$ is the vector norm defined in Section A.10. In practice, matrix norms are not computed using this definition directly. For example the 1-norm, 2-norm, and the infinity-norm are computed as follows:

$$||\mathbf{A}||_1 = \text{maximum absolute column sum of } \mathbf{A}$$
$$||\mathbf{A}||_2 = \text{maximum singular value of } \mathbf{A}$$
$$||\mathbf{A}||_\infty = \text{maximum absolute row sum of } \mathbf{A}$$

Having defined the matrix norm we now consider the solution of the equation system

$$\mathbf{Ax} = \mathbf{b}$$

Let the exact solution of this system be $\mathbf{x}$ and the computed solution be $\mathbf{x}_c$. Then we may define the error as

$$\mathbf{x}_e = \mathbf{x} - \mathbf{x}_c$$

We can also define the residual $\mathbf{r}$ as

$$\mathbf{r} = \mathbf{b} - \mathbf{A}\mathbf{x}_c$$

We note that large residuals are indicative of inaccuracies but small residuals do not guarantee accuracy. For example, consider the case where

$$\mathbf{A} = \begin{bmatrix} 2 & 1 \\ 2+\varepsilon & 1 \end{bmatrix}, \quad \mathbf{b} = \begin{bmatrix} 3 \\ 3+\varepsilon \end{bmatrix}$$

The exact solution of $\mathbf{Ax} = \mathbf{b}$ (with a residual $\mathbf{r} = \mathbf{0}$) is

$$\mathbf{x} = \begin{bmatrix} 1 \\ 1 \end{bmatrix}$$

However, if we consider the very poor approximation

$$\mathbf{x}_c = \begin{bmatrix} 1.5 \\ 0 \end{bmatrix}$$

then the residual is

$$\mathbf{b} - \mathbf{A}\mathbf{x}_c = \begin{bmatrix} 0 \\ -0.5\varepsilon \end{bmatrix}$$

If $\varepsilon = 0.00001$, then the residual is very small, even though the solution is very inaccurate.

To obtain a formula which provides bounds on the relative error of the computed value, $\mathbf{x}_c$ we proceed as follows:

$$\mathbf{r} = \mathbf{b} - \mathbf{A}\mathbf{x}_c = \mathbf{Ax} - \mathbf{A}\mathbf{x}_c = \mathbf{A}\mathbf{x}_\varepsilon \tag{B.4}$$

From (B.4) we have

$$\mathbf{x}_\varepsilon = \mathbf{A}^{-1}\mathbf{r}$$

Taking the norms of this equation we have

$$\|\mathbf{x}_\varepsilon\| = \left\|\mathbf{A}^{-1}\mathbf{r}\right\| \tag{B.5}$$

We can choose to use any p-norm and in the analysis that follows the subscript p is omitted. A property of norms is that $||\mathbf{AB}|| \leq ||\mathbf{A}||\,||\mathbf{B}||$. Thus we have, from (B.5)

$$\|\mathbf{x}_\varepsilon\| \leq \left\|\mathbf{A}^{-1}\right\| \|\mathbf{r}\| \tag{B.6}$$

But $\mathbf{r} = \mathbf{A}\mathbf{x}_\varepsilon$ and so

$$\|\mathbf{r}\| \leq \|\mathbf{A}\| \;\; \|\mathbf{x}_\varepsilon\|$$

Therefore

$$\frac{\|\mathbf{r}\|}{\|\mathbf{A}\|} \leq \|\mathbf{x}_\varepsilon\|$$

Combining this equation with (B.6) we have

$$\frac{\|\mathbf{r}\|}{\|\mathbf{A}\|} \leq \|\mathbf{x}_\varepsilon\| \leq \left\|\mathbf{A}^{-1}\right\| \ \|\mathbf{r}\| \tag{B.7}$$

Now since $\mathbf{x} = \mathbf{A}^{-1}\mathbf{b}$ we have similarly

$$\frac{\|\mathbf{b}\|}{\|\mathbf{A}\|} \leq \|\mathbf{x}\| \leq \left\|\mathbf{A}^{-1}\right\| \|\mathbf{b}\| \tag{B.8}$$

If none of the terms in the above equation are zero, we can take reciprocals to give

$$\frac{1}{\left\|\mathbf{A}^{-1}\right\| \|\mathbf{b}\|} \leq \frac{1}{\|\mathbf{x}\|} \leq \frac{\|\mathbf{A}\|}{\|\mathbf{b}\|} \tag{B.9}$$

Multiplying the corresponding terms in (B.7) and (B.9) gives

$$\frac{1}{\|\mathbf{A}\| \left\|\mathbf{A}^{-1}\right\|} \frac{\|\mathbf{r}\|}{\|\mathbf{b}\|} \leq \frac{\|\mathbf{x}_\varepsilon\|}{\|\mathbf{x}\|} \leq \|\mathbf{A}\| \left\|\mathbf{A}^{-1}\right\| \frac{\|\mathbf{r}\|}{\|\mathbf{b}\|} \tag{B.10}$$

This equation gives error bounds for the relative error in the computation which are directly computable. The condition number of $\mathbf{A}$is given by cond($\mathbf{A}$, p) $= ||\mathbf{A}||_p||\mathbf{A}^{-1}||_p$. Hence (B.10) can be re-written in terms of cond($\mathbf{A}$, p). When $p = 2$, cond($\mathbf{A}$) is the ratio of the largest singular value of $\mathbf{A}$ to the smallest.

We now show how (B.10) can be used to estimate the relative error in the solution of $\mathbf{Ax} = \mathbf{b}$ when $\mathbf{A}$ is the Hilbert matrix. We have chosen the Hilbert matrix because its condition number is large and its inverse is known and so we can compute the actual error in the computation of $\mathbf{x}$. The following MATLAB script evaluates (B.10) for a specific Hilbert matrix using the 2-norm.

```
n = 6, format long
a = hilb(n); b = ones(n,1);
xc = a\b;
x = invhilb(n)*b;
exact_x = x';
err = abs((xc-x)./x);
nrm_err = norm(xc-x)/norm(x)
r = b-a*xc;
L_Lim = (1/cond(a))*norm(r)/norm(b)
U_Lim = cond(a)*norm(r)/norm(b)
```

Running this script gives

```
n =
   6

nrm_err =
3.316798106133016e-11

L_Lim =
3.351828310510846e-21

U_Lim =
7.492481073232495e-07
```

We see that the norm of the actual relative error, 3.316×10^{-11}, lies between the bounds 3.35×10^{-21} and 7.49×10^{-7}.

Solutions to Selected Problems

CHAPTER 1

1.1. (a) Since some x are negative the corresponding square roots are imaginary and $\imath = \sqrt{-1}$ is used. (b) In executing `x./y` the divide by zero produces the symbol ∞ and a warning.

1.2. (b) Note that `t2` is identical to `c` but `t1` is not since the `sqrt` function gives the square root of the individual elements of `c`.

1.4. $x = 2.4545$, $y = 1.4545$, $z = -0.2727$. Note when using the `/` operator the solution is given by `x=b'/a'`.

1.8. The plot does not truly represent the function $\cos(x^3)$ because there are insufficient plotting points.

1.9. Function `fplot` automatically adjusts to provide a smoother plot. However, changing `x` to `-2:0.01:2` gives a similar quality graph using the function `plot`.

1.12. $x = 1.6180$.

1.14. Using $x_1 = 1$, $x_2 = 2$, ..., $x_6 = 6$, a suitable script is

```
n = 6; x = 1:n;
for j = 1:n,
    p(j) = 1;
    for i = 1:n
        if i~=j
            p(j) = p(j)*x(i);
        end
    end
end
p
```

1.15. A suitable script is

```
x = 0.82; tol = 0.005; s = x; i = 2; term = x;
while abs(term)>tol
    term = -term*x; s = s+term/i; i = i+1;
end
s, log(1+x)
```

1.17. The form of the function is

```
function [x1,x2] = funct1(a,b,c)
d = b*b-4*a*c;
if d==0
    x1 = -b/(2*a); x2 = x1;
else
```

```
    x1 = (-b+sqrt(d))/(2*a); x2 = (-b-sqrt(d))/(2*a);
end
```

1.18. A possible script is

```
function [x1,x2] = funct2(a,b,c)
if a~= 0
    %as in problem 1.17
else
    disp('warning only one root'); x1 = -c/b; x2 = x1;
end
```

1.19. The graph provides an initial approximation of 1.5. Use the function call `fzero('funct3',1.5)` to obtain the root as 1.2512.

1.20. A possible script is

```
x=[ ]; x(1) = 1873;
c = 1; xc = x(1);
while xc>1
    if (x(c)/2)==floor(x(c)/2)
        x(c+1) = (x(c))/2;
    else
        x(c+1) = 3*x(c)+1;
    end
    xc = x(c+1); c = c+1;
    if c>1000
        break
    end
end
plot(x)
```

Try different values for `x(1)`. For example 1173, 1409, etc.

1.21. A possible script is

```
x = -4:0.1:4; y = -4:0.1:4;
[x,y] = meshgrid(-4:0.1:4,-4:0.1:4);
p = x.^2+y.^2;
z = (1-x.^2).*exp(-p)-p.*exp(-p)-exp(-(x+1).^2-y.^2);
subplot(3,1,1)
mesh(x,y,z)
xlabel('x'), ylabel('y'), zlabel('z')
title('mesh')
subplot(3,1,2)
surf(x,y,z)
xlabel('x'), ylabel('y'), zlabel('z')
title('surf')
```

```
subplot(3,1,3)
mesh(x,y,z)
xlabel('x'), ylabel('y'), zlabel('z')
title('contour')
```

1.22. A possible script is

```
clf
a = 11; b= 6;
t = -20:0.1:20;
% Cycloid
x = a*(t-sin(t));y=a*(1-cos(t));
subplot(3,1,1), plot(x,y)
xlabel('x-xis'), ylabel('y-xis'), title('Cycloid')
% witch of agnesi
x1 = 2*a*t;y1=2*a./(1+t.^2);
subplot(3,1,2), plot(x1,y1)
xlabel('x-xis'), ylabel('y-xis')
title('witch of agnesi')
% Complex structure
x2 = a*cos(t)-b*cos(a/b*t);
y2 = a*sin(t)-b*sin(a/b*t);
subplot(3,1,3), plot(x2,y2)
xlabel('x-xis'), ylabel('y-xis')
title('Complex structure')
```

1.23. A possible function is

```
function r = zetainf(s,acc)
sum = 0; n = 1; term = 1+acc;
while abs(term)>acc
    term = 1/n.^s;
    sum = sum +term;
    n = n+1;
end
r = sum;
```

1.24. A possible function is

```
function res = sumfac(n)
sum = 0;
for i = 1:n
    sum = sum+i^2/factorial(i);
end
res = sum;
```

1.26. A possible script is

```
rho1 = [zeros(2), eye(2); eye(2), zeros(2)]
rho2 = [zeros(2), i*eye(2); -i*eye(2), zeros(2)]
rho3 = [eye(2), zeros(2); zeros(2), -eye(2)]
q1 = [zeros(4)  rho1;-rho1  zeros(4)]
q1 = [zeros(4)  rho2;-rho2  zeros(4)]
q1 = [zeros(4)  rho3;-rho3  zeros(4)]
```

1.27. A possible script is

```
x = -4:0.001:4;
y = 1./(((x+2.5).^2).*((x-3.5).^2));
plot(x,y)
ylim([0,20])
xlim([-3,-2])
```

1.28. A possible script is

```
y = @(x)x.^2.*cos(1+x.^2);
y1 = @(x) (1+exp(x))./(cos(x)+sin(x));
x = 0:0.1:2;
subplot(1,2,1), plot(x,y(x))
xlabel('x'), ylabel('y')
subplot(1,2,2), plot(x,y1(x))
xlabel('x'), ylabel('y')
```

1.39. The value of the summation to a high degree of accuracy is $I = 0.538079506912768$. Summing the first 10 terms gives 0.53807950. The eleventh term is equal to 1.3125×10^{-08}.

CHAPTER 2

2.1. Note the large error in the inverse of the square of the Hilbert matrix when $n = 6$.

2.2. For $n = 3$, 4, 5, and 6, the answers are 2.7464×10^5, 2.4068×10^8, 2.2715×10^{11}, and 2.2341×10^{14}, respectively. The large errors in Problem 2.1 arise from the fact that the Hilbert matrix is very ill-conditioned, as shown by these results.

2.3. For example, with $n = 5$, $a = 0.2$, $b = 0.1$; thus $a + 2b < 1$ and maximum error in the matrix coefficients is 1.0412×10^{-5}. Taking $n = 5$, $a = 0.3$, $b = 0.5$; thus $a + 2b > 1$ and after 10 terms maximum error in matrix coefficients is 10.8770. After 20 terms maximum error is 50.5327, clearly diverging.

2.4. Eigenvalues are 5, $2 + 2\imath$, and $2 - 2\imath$. Thus taking $\lambda = 5$ in the matrix $(\mathbf{A} - \lambda\mathbf{I})$ and finding the RREF gives

$$\mathbf{p} = \begin{bmatrix} 1 & 0 & -1.3529 \\ 0 & 1 & 0.6471 \\ 0 & 0 & 0 \end{bmatrix}$$

Hence $\mathbf{px} = \mathbf{0}$. Solving this gives $x_1 = 1.3529x_3$, $x_2 = -0.6471x_3$, and x_3 is arbitrary.

2.6. $\mathbf{x}^\top = [0.9500\ 0.9811\ 0.9727]$. All methods give identical solution. Note if `[q,r] = qr(a)` and `y = q'*b;`, then `x = r(1:3,1:3)\y(1:3)`.

2.7. The solution is $[0\ 0\ 0\ 0 \ldots n+1]$.

2.10. For $n = 20$ condition number is 178.0643; theoretical condition number is 162.1139. For $n = 50$ condition number is 1053.5; theoretical condition number is 1013.2.

2.11. Right-hand vectors are

$$\begin{bmatrix} 0.0484 + 0.4447\imath \\ -0.3962 + 0.4930\imath \\ 0.4930 + 0.3962\imath \end{bmatrix} \begin{bmatrix} 0.0484 - 0.4447\imath \\ -0.3962 - 0.4930\imath \\ 0.4930 - 0.3962\imath \end{bmatrix} \begin{bmatrix} 0.4082 \\ 0.8165 \\ 0.4082 \end{bmatrix}$$

The corresponding eigenvalues are $2 + 4\imath$, $2 - 4\imath$, and 1. The left-hand vectors are obtained by using the function `eig` on the transposed matrix.

2.12. (i) Largest eigenvalue is 242.9773; (ii) Eigenvalue nearest 100 is 112.1542; (iii) Smallest eigenvalue is 77.6972.

2.14. For $n = 5$, largest eigenvalue is 12.3435, smallest eigenvalue is 0.2716. For $n = 50$, largest eigenvalue is 1.0337×10^3, smallest eigenvalue is 0.2502.

2.15. Using the function `roots` we compute eigenvalues 22.9714, -11.9714, $1.0206 \pm 0.0086\imath$, $1.0083 \pm 0.0206\imath$, $0.9914 \pm 0.0202\imath$, $0.9798 \pm 0.0083\imath$. Using the function `eig` we have 22.9714, -11.9714, 1, 1, 1, 1, 1, 1, 1, 1. This is a more accurate solution.

2.16. Both `eig` and `roots` give results that only differ by less than 1×10^{-10}. Eigenvalues are 242.9773, 77.6972, 112.1542, 167.4849, 134.6865.

2.17. Sum of eigenvalues is 55, product of eigenvalues is 1.

2.18. $c = 0.641n^{1.8863}$.

2.19. A suitable function is

```
function appinv = invapprox(A,k)
ev = eig(A);
evm = max(ev);
if abs(evm)>1
    disp('Method fails')
    appinv = eye(size(A));
else
    appinv = eye(size(A));
    for i = 1:k
        appinv = appinv+A^i;
    end
end
```

2.20. The MATLAB operator gives a much better result. A suitable function is

```
function [res1,res2, nv1,nv2] = udsys(A,b)
newA = A'*A; newb = A'*b;
x1 = inv(newA)*newb;
nv1 = norm(A*x1-b);
x2 = A\b;
```

```
nv2 = norm(A*x2-b);
res1 = x1; res2 = x2;
```

2.22. Exact solution is

$$\mathbf{x} = [-12.5\ \ -24\ \ -34\ \ -42\ \ -47.5\ \ -50\ \ -49\ \ -44\ \ -34.5\ \ -20]^{\mathrm{T}}.$$

The Gauss–Seidel method requires 149 iterations and the Jacobi method requires 283 iterations to give the result to the required accuracy.

CHAPTER 3

3.2. The solution is 27.8235.

3.3. The solutions are -2 and 1.6344.

3.4. For $c = 5$, with the initial approximation 1.3 or 1.4, the root 1.3735 is obtained after two or three iterations. When $c = 10$, with the initial approximation 1.4, the root 1.4711 is obtained after five iterations. With initial approximation 1.3, convergence is to 130.2764 after 25 iterations. This is a root but the discontinuity in the function has degraded the performance of the Newton algorithm.

3.5. Schroder's method provides the solution $x = 1.0$ in only one iteration but Newton's method gives $x = 0.9991$ and requires 36 iterations. The solution obtained by Schroder's method is more accurate.

3.6. The equation can be rearranged into the form $x = \exp(x/10)$. Iteration gives $x = 1.1183$. There may be other successful rearrangements.

3.7. The solution is $E = 0.1280$.

3.8. The answers are 2.1602×10^{-16} and 7.1861×10^{-17} for initial values 1 and -1.5, respectively. The exact solution is clearly 0 but this is a difficult problem.

3.9. The three answers are 1.4299, 1.4468, and 1.4458 which are obtained for four, five, and six terms, respectively. These answers are converging to the correct answer.

3.10. Both approaches give identical results, $x = 8.2183$, $y = 2.2747$. The single variable function is $x/5 - \cos x = 2$. Alternatively the following call can be used

```
newtonmv([1 1]','p310','p310d',2,1e-4)
```

It requires the functions and derivatives to be defined thus:

```
function v = p310(x)
v = zeros(2,1);
v(1) = exp(x(1)/10)-x(2);
v(2) = 2*log(x(2))-cos(x(1))-2;

function vd = p310d(x)
vd = zeros(2,2);
vd(1,:) = [exp(x(1)/10)/10 -1];
vd(2,:) = [sin(x(1)) 2/x(2)];
```

3.11. The solution given by `broyden` is $x = 0.1605$, $y = 0.4931$.

3.12. A solution is $x = 0.9397$, $y = 0.3420$. The MATLAB function `newtonmv` requires 7 iterations and `broyden` requires 33.

3.14. The five roots are 1, $-\imath, \imath, -\sqrt{2}, \sqrt{2}$.

3.15. Solution is $x = -0.1737 - 0.9848\imath, 0.9397 + 0.3420\imath$, and $-0.7660 + 0.6428\imath$. This is identical to the exact answer.

3.16. The MATLAB function required is

```
function v = jarrett(f,x1,x2,tol)
gamma = 0.5; d = 1;
while abs(d)>to
    f2 = feval(f,x2);f1=feval(f,x1);
    df = (f2-f1)/(x2-x1); x3 = x2-f2/df; d = x2-x3;
    if f1*f2>0
        x2 = x1; f2 = gamma*f1;
    end
    x2 = x3
end
```

3.17. The third-order method provides the required accuracy after seven iterations. The second-order method requires ten iterations.

3.18. The graphs show that for $c = 2.8$ there is convergence to a single solution, for $c = 3.25$ the iteration oscillates between two values, for $c = 3.5$ the iteration oscillates between four values and for $c = 3.8$ there is chaotic oscillation between many values.

3.20. Here is an example with p and q chosen to give real roots:

```
>> p=2.5; q = -1; if p^3/q^2>27/4, r = roots([1 0 -p -q]), end

r =

   -1.7523
    1.3200
    0.4323
```

3.21. Commands to solve this problem are

```
>> y1 = roots([1 0 6 -60 36]);
>> y = y1(3:4)';
>> x = 6./y;
>> z = 10-x-y

z =
    4.9646    0.0446
```

3.22. A script to solve this problem is

```
c1 = (sinh(x)+sin(x))./(2*x);
c3 = (sinh(x)-sin(x))./(2*x.^3);
```

```
fzero(@ (x) c1^2-x.^4*c3.^2,5)
fzero(@ (x) c1^2-x.^4*c3.^2,30)
```

3.23. A script to solve this problem is

```
p = 3; q = 4;
pc = [1 0 3*p -2*q];
xv = roots(pc)
%check using Cardano formula
c1 = q+sqrt(q^2+p^3)
c2 = q-sqrt(q^2+p^3)
r1 = nthroot(c1,3)
r2 = nthroot(c2,3)
x1 = r1+r2
%Complex roots use cubes of unity
w1 = (-1+i*sqrt(3))/2;w2=(-1-i*sqrt(3))/2;
x2 = w1*r1+w2*r2
x3 = w2*r1+w1*r2
```

3.24. A script to solve this problem is

```
a = 1; b = 2; c = 3;
%polynomial coefficients are
pc(1) = (a+b+c); pc(2) = 3-(a*(b+c)+b*(a+c)+c*(a+b));
pc(3) = 3*a*b*c-2*(a+b+c); pc(4) = b*c+a*c+a*b;
sol = roots(pc)
polyval(pc,sol)
```

3.25. Since $\sin(3\theta) = -4\sin^3\theta + 3\sin\theta$ and $\cos^2\theta = 1 - \sin^2\theta$, equation becomes $-0.6275x^3 + 0.9412x^2 - 0.4314x + 0.0588 = 0$ where $x = \sin\theta$. Using `roots` give $x = 0.75, 0.5$ and 0.25. Hence $\theta = 0.8481, 0.5236$ and 0.2527 rad.

CHAPTER 4

4.1. First derivative is 0.2391, the second derivative is −2.8256. The function `diffgen` gives accurate answers using either $h = 0.1$ or 0.01. The function changes slowly over this range of values.

4.2. When $x = 1$, the computed and exact derivative is −5.0488; when $x = 2$ the computed derivative is −176.6375 (exact = −176.6450), and when $x = 3$ the computed derivative is −194.4680 (exact = −218.6079).

4.3. Using the new formula for Problem 4.1, the first derivative estimate is 0.2267 and 0.2390 for $h = 0.1$ and 0.01, respectively. The second derivative is −2.8249 and −2.8256 for $h = 0.1$ and 0.01, respectively. In Problem 4.2 for $x = 1$, 2, and 3 the first derivative estimates are −5.0489, −175.5798, and −150.1775, respectively. Note that these are less accurate than using `diffgen`.

4.4. The approximate derivatives are −1367.2, −979.4472, −1287.7, and −194.4680. If h is decreased to 0.0001, then the values are the same as the exact derivatives to the given number of decimal places.

4.5. The exact partial derivatives with respect to x and y are 593.652 and 445.2395, respectively. The corresponding approximate values are 593.7071 and 445.2933.

4.6. The integral method estimates 6.3470 primes in the range 1 to 10, 9.633 primes in the range 1 to 17 and 15.1851 primes in the range 1 to 30. The actual numbers are 7, 10, and 15.

4.7. Exact values are 1.5708, 0.5890, and 0.2443 for $r = 0$, 1, and 2, respectively. Approximations provided by integral are 1.5338, 0.5820, and 0.2700.

4.8. The exact values are -0.0811 for $a = 1$ and 0.3052 for $a = 2$. Using `simp1` with 512 points gives agreement to 12 decimal places.

4.9. The exact answer is -0.915965591 and the answer given by `fgauss` is -0.9136. Function `simp1` cannot be used because of the singularity at $x = 0$.

4.10. The exact answer is 0.915965591; `fgauss` gives 0.9159655938. Note that the integrals of 4.9 and 4.10 have the same value apart from the sign.

4.11. (i) Using (4.32) with 10 points gives 3.97746326050642, 16-point Gauss gives 3.8145. (ii) Using (4.33) gives 1.77549968921218, 16-point Gauss gives 1.7758.

4.13. Function `filon` gives
2.00000000000098, -0.13333333344440, and $-2.000199980281494 \times 10^{-4}$

4.14. Using Romberg's method with nine divisions gives $-2.000222004003794 \times 10^{-4}$. Using Simpson's rule with 1024 intervals gives $-1.999899106566088 \times 10^{-4}$.

4.18. The solution for (i) is 48.96321182552904 and for (ii) 9726.56492. These compare well with the exact solution which can be computed from the formula $4\pi^{(n+1)}/(n+1)^2$ where n is the power of x and y.

4.19. (i) To fix limits, substitute $y = \left(\sqrt{x/3} - 1\right) z + 1$. Answer: -1.71962748468952. (ii) To fix limits, substitute $y = (2 - x)z$. Answer: 0.22222388780205.

4.20. The answers are (i) 9725.75264, (ii) 0.22222222200993.

4.21. Values of the integral are given in the following table:

z	Exact	16-point Gauss
0.5	0.493107418	0.49310741784618
1.0	0.946083070	0.94608306999140
2.0	1.605412977	1.60541297617644

4.22. Use `gauss2v` and define the following function:

```
z = @(x,y) 1./(1-x.*y);
```

4.23. The following will provide the solution of this problem:

```
% Probability of engine failure
p = [ ];
a = 3.5; b = 8200;
i = 1;
for T = 200:100:4000
P(i) = quad(@(x) a*b^a./((x+b).^(a+1)),0.001,T);
i=i+1;
end
figure(1)
```

```
plot(200:100:4000,P)
xlabel('Time in hours'), ylabel('Probability of failure')
title('plot of probability of failure against time')
grid
```

4.24. The following will provide the solution of this problem, the value of the integral is -0.15415 correct to 5 places.

```
p = 3; q = 4; r = 2;
f = @(x) (x.^p-x.^q).*x.^r ./log(x);
val = quad(f,0,1);
fprintf('\n value of integral = %6.5f\n',val)
check = log((p+r+1)/(q+r+1))
fprintf('\n value of integral = %6.5f\n',check)
```

4.25. The three integrals are approximately equal to 0.91597.
4.26. The integral equals zero to five decimal places.
4.27. A fairly low accuracy result is obtained.
4.28. Accuracy to two decimal places is obtained.
4.29. There is good agreement between values. The script is:

```
f = @(x) -log(x).^3 .* exp(-x)
val = quadgk(f,0,Inf);
fprintf('\n value of integral = %6.5f\n',val)
gam = 0.57722;
S3 = gam^3+0.5*gam*pi^2+2*zeta(3)
fprintf('\n Approximate sum of series = %6.5f\n',S3)
```

4.30. The best result is given by `dblquad` is 2.01131. The script is:

```
R = dblquad(@(x,y) (1-cos(50*x).*cos(100*y))./(2-cos(x)-cos(y)),...,
0.0001,pi,0.0001,pi);R=R/pi^2;
fprintf('\nValue of integral using dblquad = %6.5f\n',R)
R1 = simp2v(@(x,y) (1-cos(50*x).*cos(100*y))./(2-cos(x)-cos(y)),...,
    0.00001,pi,0.00001,pi,64);R1=R1/pi^2;
gamma = -psi(1);
R = (gamma+3*log(2)/2+log(50^2+100^2)/2)/pi;
fprintf('\nValue of integral using simp2v = %6.5f\n',R1)
fprintf('\n Approximate value check = %6.5f\n',R)
```

4.31. Value of integral is 0.46306.
4.32.
4.33.
4.34.
4.35. The solution to a high level of accuracy is $I = 0.538079506912768$. Using 32 panels in the Filon integration gives $I = 0.5380795$.

4.36. You should show that the Lagrendre polynomials are orthogonal so that

$$I_{m,n} = \int_{-1}^{1} P_m(x) P_n(x)\, dx = 2/(2n+1),$$

if $m = n$, else zero.

CHAPTER 5

5.1. When $t = 10$, the exact value is 30.326533.
`feuler`: 29.9368, 30.2885, 30.3227 with $h = 1, 0.1$, and 0.01, respectively.
`eulertp`: 30.3281, 30.3266 with $h = 1$ and 0.1, respectively.
`rkgen`: 30.3265 with $h = 1$.

5.2. Classical method gives 108.9077, Butcher method gives 109.1924, Merson method gives 109.1924.
Exact answer is $2\exp(x^2) = 109.1963$.

5.3. Adams–Bashforth–Moulton method gives 4.1042, Hamming's method gives 4.1043. Exact answer is 4.1042499.

5.4. Using `ode23` gives 0.0254; using `ode45` gives 4.

5.5. Solution of Problem 5.1 with $h = 1$ is 30.3265. Solution of Problem 5.2 with $h = 0.2$ is 108.8906. Solution of Problem 5.2 with $h = 0.02$ is 109.1963.

5.6. (i) 7998.6, exact $= 8000$. (ii) 109.1963.

5.9. Method is stable for $h = 0.1$ and 0.2, and unstable for $h = 0.4$.

5.11. Define the right-hand sides using the following function:

```
function v = p511(t,x)
v = ones(2,1);
v(1) = x(1)*(1-0.001*x(1)-1.8*x(2));
v(2) = x(2)*(.3-.5*x(2)/x(1));
```

5.12. Define the right-hand sides using the following function:

```
function v = p512(t,x)
v = ones(2,1);
v(1) = -20*x(1); v(2) = x(1);
```

5.13. Define the right-hand sides using the following function:

```
function v = p513(t,x)
v = ones(2,1);
v(1) = -30*x(2);
v(2) = -.01*x(1)*x(2);
```

5.14. Define the right-hand sides using the following function:

```
function v = p514(t,x)
global c
k = 4; m = 1; F = 1;
v = ones(2,1);
v(1) = (F-c*x(1)-k*x(2))/m;
v(2) = x(1);
```

The script to solve this equation is

```
global c
i = 0;
for c = [0,2,1]
    i = i+1;
    c
    [t,x] = ode45('q514',[0 10],[0 0]');
    figure(i)
    plot(t,x(:,2))
end
```

5.16. The script to solve this problem is:

```
function prhs = planetrhs(t,x)
% global x0
% NB global is used if initial values x0
% are used to calculate impact probabilities
% rather than x the variable values
for i=1:3
    for j=1:3
        A(i,j)=x(i).*x(j)./(x(i)+x(j))/1000;
    end
end
prhs = zeros(3,1);
prhs(1) = -x(1).*(A(1,2).*x(2)+A(1,3).*x(3));
prhs(2) = 0.5*A(1,1)*x(1).*x(1)-x(2).*(A(2,2).*x(2)+A(2,3).*x(3));
prhs(3) = 0.5*A(1,2)*x(1).*x(2);
```

Together with

```
% Solution of planetary growth
% the coagulation equation three size model
% Let x(1), x(2) and x(3) represent the
% number of planetesimals of the three sizes
global x0
% Initially
x0 = [200,25,1];
tspan = [0,2];
```

```
[t,x] = ode45('planetrhs', tspan,x0);
fprintf('\n number of smallest planets= %3.0f',x(end,1))
fprintf('\n number of intermediate planets=%3.0f',x(end,2))
fprintf('\nlargest planets=%3.0f\n',x(end,3))
figure(1)
plot(t,x)
xlabel('time'), ylabel('planet numbers')
grid
```

5.17. The script to solve this problem is:

```
%Solution of Daiy world problem
span = 10;
[x,t] = ode45('daisyf',span,[0.2, 0.3]);
plot(x,t)
xlabel('Time'), ylabel('black and white daisy areas')
title('daisy world')
grid
```

and the function:

```
function daisyrhs = daisyf(t,x)
daisyrhs = zeros(2,1);
gamma = 0.3;
Tb = 295; Tw = 285;
betab = 1-0.003265*(295.5-Tb);
betaw = 1-0.003265*(295.5-Tw);
barbit = 1-x(1)-x(2);
daisyrhs(1) = x(1).*(barbit.*betab-gamma);
daisyrhs(2) = x(2).*(barbit.*betaw-gamma);
```

CHAPTER 6

6.1. (i) hyperbolic, (ii) parabolic, (iii) $f(x, y) > 0$, hyperbolic, $f(x, y) < 0$ elliptic.
6.2. Initial slope $= -1.6714$. Shooting and FD methods give good results.
6.3. This is an example of a stiff equation. (i) The actual slope when $x = 0$ is 1.0158×10^{-24}. Because we cannot determine this slope accurately, the shooting method gives a very inaccurate solution. (ii) In this case the shooting method provides a good result because the initial slope is -120. In both cases the FD method requires a large number of divisions to give an accurate result.
6.5. Finite difference method gives $\lambda_1 = 2.4623$. Exact $\lambda_1 = (\pi/L)^2 = 2.4674$.
6.6. At $t = 0.5$ the variation of z is almost linear between the boundaries at 0 and 10.
6.7. The exact and FD approximations are very similar.
6.8. The exact and FD approximations are similar with a maximum error of 0.0479.
6.9. With 9 divisions in x, $\lambda = 5.8870, 14.0418, 19.6215, 27.8876, 29.8780$. Even with 36 divisions in x the eigenvalues have not converged.
6.10. [0.7703 1.0813 1.5548 1.583 1.1943 1.5548 1.583 1.194 1.0813 0.7703].

CHAPTER 7

7.1. Using the Aitken function, $E(2°) = 1.5703$, $E(13°) = 1.5507$, $E(27°) = 1.4864$. These are accurate to the places given.

7.2. The root is 27.8235.

7.3. (i) $p(x) = 0.9814x^2 + 0.1529$ and $p(x) = -1.2083x^4 + 2.1897x^2 + 0.0137$. The fourth degree polynomial gives a good fit.

7.4. Interpolation gives 0.9284 (linear), 0.9463 (spline), 0.9380 (cubic polynomial). The MATLAB function `aitken` gives 0.9455. This is the exact value to four decimal places.

7.5. $p(x) = -0.3238x^5 + 3.2x^4 - 6.9905x^3 - 12.8x^2 + 31.1429x$. Note that the polynomial oscillates between data points. The spline does not exhibit this characteristic, suggesting that it better represents any underlying function from which the data might have been taken.

7.6. (i) $f(x) = 3.1276 + 1.9811e^x + e^{2x}$. (ii) $f(x) = 685.1 - 2072.2/(1 + x) + 1443.8/(1 + x)^2$. (iii) $f(x) = 47.3747x^3 - 128.3479x^2 + 103.4153x - 5.2803$.
Plotting these functions shows that the best fit is given by (i). The polynomial fit is a reasonable one.

7.7. The plot should display an airfoil section.

7.8. Product of primes less than P is given by $0.3679 + 1.0182 \log_e P$ approximately.

7.9. $a_0 = 1$, $a_1 = -0.5740$, $a_2 = 0.9456$, $a_3 = -0.6865$, $a_4 = 0.4115$, $a_5 = -0.0966$.

7.10. Exact: -78.3323. Interpolation gives -78.3340 (cubic) or -77.9876 (linear).

7.11. The minimum values of E are approximately -14.95 and -6.45 at points 40 and 170. The maximum values of E are 3.68 and 16.47 at points 110 and 252.

7.12. Estimated production cost in year 6 is \$31.80 using cubic extrapolation and \$20.88 using quadratic extrapolation. Using the revised data the estimated costs are \$24.30 and \$21.57, respectively. These widely varying results, some of which are barely credible, show the dangers of trying to estimate future costs from insufficient data.

7.13. $x = 2.4679$.

7.14. $I = 1.5713$, $\alpha = 8.71406$.

7.15. $f_n = \frac{n}{6}\left(n^2 + 3n + 2\right)$.

7.18. The values are: 22.70, 22.42, 22.42.

CHAPTER 8

8.1. The data is sampled from $y = \sin(2\pi f_1 t) + 2\cos(2\pi f_2 t)$ where $f_1 = 1.25$ Hz and $f_2 = 3.4375$ Hz. At 1.25 Hz, DFT $= -15.9999\imath$ and 3.4375 Hz, DFT $= 32.0001$. The negative complex coefficient is related to the positive size of the coefficient of the sine function and the positive real component is related to the cosine function. To relate the size of the DFT components to the frequency components in the data we divide the DFT by the number of samples (32) and multiply by 2.

8.2. Algebraically,

$$32 \sin^5(30t) = 20 \sin(30t) - 10 \sin(90t) + 2 \sin(150t)$$

and

$$32\sin^6(30t) = 10 - 15\cos(60t) + 6\cos(120t) - \cos(180t)$$

To verify these results from the DFT it is necessary to divide it by n and multiply by 2. The real components are the values of the cosine coefficients. The imaginary components in the DFT are the negative of the values of the sine coefficients. Note also that the coefficient at zero frequency is 20, not 10. This is a consequence of the definition of the DFT; see Section 8.2.

8.3. Components in spectrum at 30 Hz and 112 Hz. The reason for the large component at 112 Hz is that the component in the data at 400 Hz is above the Nyquist frequency and is folded back to give a spurious component, i.e., 400 Hz is 144 Hz above the Nyquist frequency of 256 Hz; 112 Hz is 144 Hz below it.

8.5. With 32 points the frequency increment is 16 Hz and the significant components are at 96 Hz and 112 Hz (the largest amplitude). With 512 points the frequency increment is reduced to 1 Hz and the significant components are at 106, 107, and 108 Hz with the largest amplitude at 107 Hz. With 1024 points the frequency increment is reduced to 0.5 Hz and the component with the largest amplitude is at 107.5 Hz. The original data had a frequency component of 107.5 Hz.

8.7. In each case the sequency is 2 z/s but in Case (a) the predominant sequency is a SAL function; in Case (b) it is a CAL function.

8.8. For both functions the major components at sequencies of 40 z/s, 100 z/s, and 140 z/s equivalent to 20 Hz, 50 Hz, and 70 Hz, but the spectra show many other components, even though there is no random noise in these cases. This is to be expected when using the Walsh transform to analyze data comprising harmonic functions. It is less easily explained in the case where the function is composed of square waves.

8.9. Case (a) components are at level −1 and 3. Case (b) components start at level 3, then increase to 4 and finally decrease to 3. Case (c) is much more difficult to interpret because sampling is not over an integer number of periods. From the wavelet map the most significant component is level 3, but components at level 4 and even 5 appear to be present. This is misleading.

8.10. The script for this problem is

```
load sunspot.dat
year = sunspot(:,1);
sunact = sunspot(:,2);
figure(1)
plot(year,sunact)
xlabel('Year'), ylabel('Sunspots')
title('Sunspot activity by year')
Y = fft(sunact);
N = length(Y);
Power = abs(Y(1:N/2)).^2;
freq = (1:N/2)/(N/2)*0.5;
figure(2)
plot(freq,Power)
xlabel('freq'), ylabel('Power')
```

CHAPTER 9

9.1. Objective is 21.6667. Solution is $x_1 = 3.6667$, $x_3 = 0.3333$, other variables are zero.

9.2. Objective is 21.6667. Solution is $x_1 = 3.3333$, $x_2 = 1.6667$, $x_4 = 0.3333$, other variables are zero. Thus this problem and the previous one have objective functions of equal magnitude.

9.3. Objective is 100. Solution is $x_1 = 10$, $x_3 = 20$, $x_5 = 22$, other variables are zero.

9.4. This is a difficult function for the conjugate gradient method and this is why the accuracy of the line-search for the built-in MATLAB function `fminsearch` was changed to produce more accurate results. Solution is [1.0007 1.0014] with gradient [0.33860.5226] $\times 10^{-3}$.

9.5. Exact and computed solutions are both [−2.9035 −2.9035 1 1 1].

9.6. Solution is $[-0.4600\ 0.5400\ 0.3200\ 0.8200]^\top$. Norm$(\mathbf{b} - \mathbf{Ax}) = 1.3131 \times 10^{-14}$.

9.11. The solution obtained values are 0.1605 and 0.4931.

9.12. The script to solve this question is

```
clf
[x,y] = meshgrid(-4:0.1:4,-4:0.1:4);
p = x.^2+y.^2;
z = (1-x).^2.*exp(-p)- p.*exp(-p) - exp(-(x+1).^2 - y.^2);
figure(1)
surf(x,y,z)
xlabel('x-axis'), ylabel('y-axis'), zlabel('z-axis')
title('mexhat plot')
figure(2)
contour(x,y,z,20)
xlabel('x-axis'), ylabel('y-axis')
title('contour plot')
optp = ginput(3);
x = optp(:,1); y = optp(:,2);
p = x.^2+y.^2;
z = (1-x).^2.*exp(-p)- p.*exp(-p) - exp(-(x+1).^2 - y.^2)
fprintf('maximum value= %6.2f\n',max(z))
fprintf('minimum value= %6.2f\n',min(z))
x
y
P=x(1).^2+x(2).^2;
fopt=@ (x)(1-x(1)).^2 .*exp(-(x(1).^2+x(2).^2))...
    - (x(1).^2+x(2).^2).*exp(-(x(1).^2+x(2).^2))...
    - exp(-(x(1)+1).^2 - x(2).^2) ;
[x,fval] = fminsearch(fopt,[-4;4])
fprintf('\nNon global solution= %8.6f\n',fval)
```

9.14. The minimum is achieved at 63.8157.

9.15. The minimum is achieved at 63.8160, a very similar result to Problem 9.14.

CHAPTER 10

10.1. Use `>>collect((x-1/a-1/b)*(x-1/b-1/c)*(x-1/c-1/a))`.

10.2. Use `>>y = x^4+4*x^3-17*x^2+27*x-19; z = x^2+12*x-13;`
`>>horner(collect(z*y))`

10.3. Use `>>expand(tan(4*x))` `>>expand(cos(x+y))`
`>>expand(cos(3*x))` `>>expand(cos(6*x))`

10.4. Use `expand(cos(x+y+z))`.

10.5. Use `>>taylor(asin(x),8)` `>>taylor(acos(x),8)`
`>>taylor(atan(x),8)`

10.6. Use `taylor(log(cos(x)),13)`.

10.7. Use `>>[solution, how] = simple(symsum((r+3)/(r*(r+1)*(r+2))*(1/3)^r,1,n))`.

10.8. Use `symsum(k^10,1,100)`.

10.9. Use `symsum(k^(-4),1,inf)`.

10.10. Use `a = [1 a a^2;1 b b^2;1 c c^2]; factor(inv(a))`.

10.11. Set `a = [a1 a2 a3 a4;1 0 0 0;0 1 0 0;0 0 1 0]` and use `ev = a-lam*eye(4)` and `det(ev)`.

10.12. Set `trans = [cos(a1) sin(a1);-sin(a1) cos(a1)];` and use
`>>[solution,how] = simple(trans^2)`
`>>[solution,how] = simple(trans^4)`

10.13. Set `r = solve('x^3+3*h*x+g=0')` and use
`[solution,s] = subexpr(r,'s')`.

10.14. Use `solve('x^3-9*x+28 = 0')`.

10.15. Use `p = solve('z^6 = 4*sqrt(2)+i*4*sqrt(2)');`
`res = double(p)`.

10.16. Use `f5 = log((1-x)*(1+x^3)/(1+x^2)); p = diff(f5);`
`factor(p)`. Then use `pretty(ans)` to help interpret this result.

10.17. Use
```
>>f = log(x^2+y^2);
>>d2x = diff(f,x,2)
>>d2y = diff(f,y,2)
>>factor(d2x+d2y)
>>f1 = exp(-2*y)*cos(2*x);
>>r = diff(f1,'x',2)+diff(f1,'y',2)
```

10.18. Use
```
>>z = x^3*sin(y);
>>dyx = diff(diff(z,'y'),'x')
>>dxy = diff(diff(z,'x'),'y')
>>dxy = diff(diff(z,'x',4),'y',6)
>>dxy = diff(diff(z,'y',6),'x',4)
```

10.19. (i) Use
```
>>p = int(1/((a+f*x)*(c+g*x)));
>>[solution,how] = simple(p)
>>[solution,how] = simple(diff(solution))
```
(ii) Use
```
>>solution = int((1-x^2)/(1+x^2))
>>p = diff(solution); factor(p)
```

10.20. Use `int(1/(1+cos(x)+sin(x)))` and `int(1/(a^4+x^4))`.

10.21. Use `int(x^3/(exp(x)-1),0,inf)`.
10.22. Use `int(1/(1+x^6),0,inf)` and `int(1/(1+x^10),0,inf)`.
10.23. Use

```
>>taylor(exp(-x*x),7)
>>p = int(ans,0,1); vpa(p,10)
>>taylor(exp(-x*x),15)
>>p = int(ans,0,1); vpa(p,10)
```

10.24. Use `int(sin(x^2)/x,0,inf)`.
10.25. Use

```
>>taylor(log(1+cos(x)),5)
>>int(ans,0,1)
```

10.26. Use

```
>>dint = 1/(1-x*y)
>>int(int(dint,x,0,1),y,0,1)
```

10.27. Use

```
>>[solution,s] = subexpr(dsolve('D2y+(b*p+a*q)*Dy+a*b*(p*q-1)*
y = c*A ', 'y(0)=0', 'Dy(0)=0','t'),'s')
```

Using the `subs` function

```
>>subs(solution,{p,q,a,b,c,A},{1,2,2,1,1,20})
```

we obtain the solution for the given values as

```
ans =
10-5*s(2)/s(1)^(1/2)*exp(-1/2*s(3)*t)+5*s(3)/s(1)^(1/2)*exp(-1/2*s(2)*t)
```

In addition, since we also require the values of `s(1)`, `s(2)`, `s(3)`, we again use `subs` as follows:

```
>>s = subs(s,{p,q,a,b,c},{1,2,2,1,1})

s =
[          17]
[ 5+17^(1/2)]
[ 5-17^(1/2)]
```

10.28. Using

```
>>sol = dsolve('2*Dx+4*Dy = cos(t),4*Dx-3*Dy = sin(t)','t')
```

gives the solution in the form

```
sol =
    x: [1x1 sym]
    y: [1x1 sym]
```

To see the specific elements of the solution use

```
>>sol.x

ans =
C1+3/22*sin(t)-2/11*cos(t)
```

and

```
>>sol.y

ans =
C2+2/11*sin(t)+1/11*cos(t)
```

10.29. Use `dsolve('(1-x^2)*D2y-2*x*Dy+2*y = 0','x')`.

10.30. (i) Use

```
>>laplace(cos(2*t))
```

and then

```
>>p = solve('s^2*Y+2*s+2*Y = s/(s^2+4)','Y');
>>ilaplace(p)
```

(ii) Use

```
>>laplace(t)
```

and then

```
>>p = solve('s*Y-2*Y = 1/s^2','Y');
>>ilaplace(p)
```

(iii) Use

```
>>laplace(exp(-2*t))
```

and then

```
>>p = solve('s\ 2*Y+3*s-3*(s*Y+3)+Y = 1/(s+2)','Y');]
>>ilaplace(p)
```

(iv) The Laplace transform of zero is zero. Thus take the Laplace transform of the equation and then use

```
>>p = solve('(s*Y-V)+Y/c=0','Y');
>>ilaplace(p)
```

10.31. (i) The Z-transform of zero is zero. Thus take the Z-transform of the equation and then use

```
>>p = solve('Y=-2*(Y/z+4)','Y');
>>iztrans(p)
```

(ii) Use

```
>>ztrans(n)
```

and then use

```
>>p = solve('Y+(Y/z+10) = z/(z-1)^2','Y');
>>iztrans(p)
```

(iii) Use

```
>>ztrans(3*heaviside(n))
```

and then use

```
>>p = solve('Y-2*(Y/z+1)=3*z/(z-1)','Y');
>>iztrans(p)
```

(iv) Use

```
>>ztrans(3*4^n)
```

and then use

```
>>p = solve('Y-3*(Y/z-3)+2*(Y/z^2+5-3/z) = 3*z/(z-4)','Y');
>>iztrans(p)
```

Bibliography

Abramowitz, M., Stegun, I.A., 1965. Handbook of Mathematical Functions, 9th Ed. Dover, New York.

Adby, P.R., Dempster, M.A.H., 1974. Introduction to Optimisation Methods. Chapman and Hall, London.

Addison, P.S., 2005. Wavelet transforms and the ECG: a review. Physiological Measurement 26, R155–R199.

Anderson, D.R., Sweeney, D.J., Williams, T.A., 1993. Statistics for Business and Economics. West Publishing Co., Minneapolis.

Armstrong, R., Kulesza, B.L.J., 1981. An approximate solution to the equation $x = \exp(-x/c)$. Bulletin of the Institute of Mathematics and Its Applications 17 (2–3), 56.

Bailey, D.H., 1988. The computation of π to 29,360,000 decimal digits using Borweins' quadratically convergent algorithm. Mathematics of Computation 30, 283–296.

Barnes, E.R., 1986. Affine transform method. Mathematical Programming 36, 174–182.

Beauchamp, K.G., 1975. Walsh Functions and Their Applications. Academic Press, London.

Beltrami, E.J., 1987. Mathematics for Dynamic Modelling. Academic Press, Boston.

Bracewell, R.N., 1978. The Fourier Transform and Its Applications. McGraw-Hill, New York.

Brent, R.P., 1971. An algorithm with guaranteed convergence for finding the zero of a function. Computer Journal 14, 422–425.

Brigham, E.O., 1974. The Fast Fourier Transform. Prentice Hall, Englewood Cliffs, N.J.

Brin, S., Page, L., 1998. Anatomy of a large scale hypertextual web search engine. Computer Networks and ISDN Systems 33, 107–117.

Butcher, J.C., 1964. On Runge Kutta processes of high order. Journal of the Australian Mathematical Society 4, 179–194.

Chudnovsky, D.V., Chudnovsky, G.V., 1989. The computation of classical constants. Proceedings of the National Academy of Sciences of the United States of America (ISSN 0027-8424) 86 (21), 8178–8182. https://doi.org/10.1073/pnas.86.21.8178. JSTOR 34831, PMC 298242, PMID 16594075.

Coley, D.A., 1999. An Introduction to Genetic Algorithms for Scientists and Engineers. World Scientific, Singapore.

Cooley, P.M., Tukey, J.W., 1965. An algorithm for the machine calculation of complex Fourier series. Mathematics of Computation 19, 297–301.

Dantzig, G.B., 1963. Linear Programming and Extensions. Princeton University Press, Princeton, N.J.

Daubechies, I., 1988. Orthonormal bases of compactly supported wavelets. Communications on Pure and Applied Mathematics 41 (7), 909–996.

Dekker, T.J., 1969. Finding a zero by means of successive linear interpolation. In: Dejon, B., Henrici, P. (Eds.), Constructive Aspects of the Fundamental Theorem of Algebra. Wiley-Interscience, New York.

Dowell, M., Jarrett, P., 1971. A modified regula falsi method for computing the root of an equation. BIT 11, 168–174.

Draper, N.R., Smith, H., 1998. Applied Regression Analysis, 3rd Ed. Wiley, New York.

Feldman, M., 2011. Hilbert transform in vibration analysis. Mechanical Systems and Signal Processing 25, 735–802.

Fiacco, A.V., McCormick, G., 1968. Non-Linear Programming, Sequential Unconstrained Minimization Techniques. Wiley, New York.

Fiacco, A.V., McCormick, G., 1990. Non-Linear Programming: Sequential Unconstrained Minimization Techniques. SIAM Classics in Mathematics. SIAM, Philadelphia (reissue).

Fletcher, R., Reeves, C.M., 1964. Function minimisation by conjugate gradients. Computer Journal 7, 149–154.

Fox, L., Mayers, D.F., 1968. Computing Methods for Scientists and Engineers. Oxford University Press, Oxford, UK.

Froberg, C.-E., 1969. Introduction to Numerical Analysis, 2nd Ed. Addison-Wesley, Reading, Mass.

Gabor, D., 1946. Theory of communication. Journal of the Institution of Electrical Engineers 93, 429–441.

Gander, W., Gautschi, W., 2000. Adaptive quadrature – revisited. BIT 40, 84–101.

Gear, C.W., 1971. Numerical Initial Value Problems in Ordinary Differential Equations. Prentice Hall, Englewood Cliffs, N.J.

Gilbert, J.R., Moler, C.B., Schreiber, R., 1992. Sparse matrices in MATLAB: design and implementation. SIAM Journal of Matrix Analysis and Application 13 (1), 333–356.

Gill, S., 1951. Process for the step by step integration of differential equations in an automatic digital computing machine. Proceedings of the Cambridge Philosophical Society 47, 96–108.

Goldberg, D.E., 1989. Genetic Algorithms in Search, Optimization and Machine Learning. Addison-Wesley, Reading, Mass.

Golub, G.H., Van Loan, C.F., 1989. Matrix Computations, 2nd Ed. John Hopkins University Press, Baltimore.

Gragg, W.B., 1965. On extrapolation algorithms for ordinary initial value problems. SIAM Journal of Numerical Analysis 2, 384–403.

Guyan, R.J., 1965. Reduction of stiffness and mass matrices. AIAA Journal 3 (2), 380.

Hamming, R.W., 1959. Stable predictor–corrector methods for ordinary differential equations. Journal of the ACM 6, 37–47.

Higham, D.J., Higham, N.J., 2017. MATLAB Guide, 3rd Ed. SIAM, Philadelphia.

Hopfield, J.J., Tank, D.W., 1985. Neural computation of decisions in optimisation problems. Biological Cybernetics 52 (3), 141–152.

Hopfield, J.J., Tank, D.W., 1986. Computing with neural circuits: a model. Science 233, 625–633.

Huang, N.E., Shen, Z., Long, S.R., Wu, M.C., Shih, H.H., Zheng, Q., Yen, N.-C., Tung, C.C., Liu, H.H., 1998. The empirical mode decomposition and the Hilbert spectrum for nonlinear and nonstationary time series analysis. Proceedings of the Royal Society of London Series A 454, 903–995. https://doi.org/10.1098/rspa.1998.0193.

Ingber, L., 1993. Simulated annealing: practice versus theory. Mathematical and Computer Modelling 18, 29–57.

Iqbal, M., 1999. Numerical solution of Nagumo's equation. Journal of Applied Mathematics and Decision Sciences 3, 189–193.

Jeffrey, A., 1979. Mathematics for Engineers and Scientists. Nelson, Sunbury-on-Thames, UK.

Jolley, L.B.W., 1961. Summation of Series, 2nd Revised Edition. Dover Publications, New York (reprinted in 2005).

Kalman, R.E., 1960. A new approach to linear filtering and prediction problems. Transactions of the ASME, Journal of Basic Engineering 82 (Series D), 35–45.

Kalman, R.E., Bucy, R.S., 1961. New results in linear filtering and prediction theory. Transactions of the ASME, Journal of Basic Engineering 83 (Series D), 95–108.

Karmarkar, N.K., 1984. A New Polynomial Time Algorithm for Linear Programming. AT&T Bell Laboratories, Murray Hill, N.J.

Karmarkar, N.K., Ramakrishnan, K.G., 1991. Computational results of an interior point algorithm for large scale linear programming. Mathematical Programming 52 (3), 555–586.

Kirkpatrick, S., Gellat, C.D., Vecchi, M.P., 1983. Optimisation by simulated annealing. Science 220, 206–212.

Kronrod, A.S., 1965. Nodes and Weights of Quadrature Formulas: Sixteen Place Tables. Consultants' Bureau.

Kumar, P., Foufoula-Georgiou, E., 1997. Wavelet analysis for geophysical applications. Reviews of Geophysics 35 (4), 385–412.

Lambert, J.D., 1973. Computational Methods in Ordinary Differential Equations. John Wiley & Sons, London.

Lasdon, L., Plummer, J., Warren, A., 1996. Nonlinear programming. In: Avriel, M., Golany, B. (Eds.), Mathematical Programming for Industrial Engineers. Marcel Dekker, New York, pp. 385–485. Chapter 6.

Leondes, C.T., 1970. Theory and applications of Kalman filtering. Full Text: http://www.dtic.mil/dtic/tr/fulltext/u2/704306.pdf.

Lindfield, G.R., Penny, J.E.T., 1989. Microcomputers in Numerical Analysis. Ellis Horwood, Chichester, UK.

Lorenz, E.N., 1993. The Essence of Chaos. University of Washington Press.

Mahoney, J., 2014. The maximum velocity of a falling body. Mathematics Today 50 (2), 96–97.

Marple Jr., S.L., 1999. Computing the discrete-time analytic signal via the FFT. IEEE Transactions on Signal Processing 47 (9), 2600–2603.

MATLAB User's Guide, 1989. The MathWorks, Inc.

Merson, R.H., 1957. An operational method for the study of integration processes. In: Proceedings of a Conference on Data Processing and Automatic Computing Machines. Weapons Research Establishment, Salisbury, South Australia.

Moller, M.F., 1993. A scaled conjugate gradient algorithm for fast supervised learning. Neural Networks 6 (4), 525–533.

Newmark, N.M., 1959. A method of computation for structural dynamics. Journal of Engineering Mechanics, ASCE 85 (EM3), 67–94.

Olver, F.W.J., Lozier, D.W., Boisvert, R.F., Clark, C.W., 2010. NIST Handbook of Mathematical Functions. National Institute of Standards and Cambridge University Press, New York. See also NIST Digital Library of Mathematical Functions. http://dlmf.nist.gov/.

Pearce, P., Shearer, T., 2016. Maths in medicine. Mathematics Today 52 (3), 135–139.

Percy, D.F., 2011. Prior elicitation: a compromise between idealism and pragmatism. Mathematics Today 47 (3), 142–147.

Press, W.H., Flannery, B.P., Teukolsky, S.A., Vetterling, W.T., 1990. Numerical Recipes: The Art of Scientific Computing in Pascal. Cambridge University Press, Cambridge, UK.

Ralston, A., 1962. Runge Kutta methods with minimum error bounds. Mathematics of Computation 16, 431–437.

Ralston, A., Rabinowitz, P., 1978. A First Course in Numerical Analysis. McGraw-Hill, New York.

Ramirez, R.W., 1985. The FFT, Fundamentals and Concepts. Prentice Hall, Englewood Cliffs, N.J.

Salvadori, M.G., Baron, M.L., 1961. Numerical Methods in Engineering. Prentice Hall, London.

Short, L., 1992. Simple iteration behaving chaotically. Bulletin of the Institute of Mathematics and Its Applications 28 (6–8), 118–119.

Simmons, G.F., 1972. Differential Equations with Applications and Historical Notes. McGraw-Hill, New York.

Stakhov, A., Rozin, B., 2005. The golden shofar. Chaos, Solitons and Fractals 26, 677–684.

Stakhov, A., Rozin, B., 2007. The golden hyperbolic models of the universe. Chaos, Solitons and Fractals 34, 159–171.

Storn, R., Price, K., 1997. Differential evolution – a simple and efficient heuristic for global optimization over continuous spaces. Journal of Global Optimization 11 (4), 341–359.

Styblinski, M.A., Tang, T.-S., 1990. Experiments in non-convex optimisation: stochastic approximation with function smoothing and simulated annealing. Neural Networks 3 (4), 467–483.

Sultan, A., 1993. Linear Programming – An Introduction with Applications. Academic Press, San Diego.

Swift, A., 1977. Course Notes. Mathematics Dept., Massey University, New Zealand.

Taylor, K., 2018. Moments of inertia in rotating frames. Mathematics Today 54 (1), 30–32.

Thompson, I., 2010. From Simpson to Kronrod: an elementary approach to quadrature formulae. Mathematics Today 46 (6), 308–313.

Unser, M., Aldroubi, A., 1996. A review of wavelets in biomedical applications. Proceedings of the IEEE 84 (4).

Walpole, R.E., Myers, R.H., 1993. Probability and Statistics for Engineers and Scientists. Macmillan, New York.

Weisstein, E.W., 2017. Gray code. From MathWorld–A Wolfram Web Resource. http://mathworld.wolfram.com/GrayCode.html. (Accessed 14 December 2017).

Wiener, N., 1949. Extrapolation, Interpolation, and Smoothing of Stationary Time Series. Wiley, New York. ISBN 0-262-73005-7.

Yan, R., Gao, R.X., Chen, X., 2014. Wavelets for fault diagnosis of rotary machines: a review with applications. Signal Processing 96 (A), 1–15.

Index

N